D1423474

Ultrasonic Guided Waves in Solid Media

Joseph L. Rose
The Pennsylvania State University

CAMBRIDGE
UNIVERSITY PRESS

CAMBRIDGE
UNIVERSITY PRESS

32 Avenue of the Americas, New York NY 10013-2473, USA

Cambridge University Press is part of the University of Cambridge.

It furthers the university's mission by disseminating knowledge in the pursuit of education, learning, and research at the highest international levels of excellence.

www.cambridge.org
Information on this title: www.cambridge.org/9781107048959

© Joseph L. Rose 2014

This publication is in copyright. Subject to statutory exception and to the provisions of relevant collective licensing agreements, no reproduction of any part may take place without the written permission of Cambridge University Press.

First published 2014

Printed in the United States of America

A catalog record for this publication is available from the British Library.

Library of Congress Cataloging in Publication data
Rose, Joseph L.
Ultrasonic guided waves in solid media / Joseph L. Rose, The Pennsylvania State University.
  pages   cm
Includes bibliographical references and index.
ISBN 978-1-107-04895-9 (hardback)
1. Wave mechanics.   2. Ultrasonic testing.   3. Attenuation (Physics)   I. Title.
QC174.2.R665   2014
534'.22–dc23   2013040589

ISBN 978-1-107-04895-9 Hardback

Cambridge University Press has no responsibility for the persistence or accuracy of URLs for external or third-party Internet Web sites referred to in this publication and does not guarantee that any content on such Web sites is, or will remain, accurate or appropriate.

UNIVERSITY
OF
GLASGOW
LIBRARY

Contents

Plate section follows page 266

Nomenclature

Latin script

$1, 2$	Indicates material 1 or 2
$\mathbf{A}, \mathbf{B}$	Matrices in a first-order eigensystem (Ch 9)
$A, A_1, A_2, \ldots$	Amplitude constants
A_+^{mn}	Mode weighting functions for finite size loading of a hollow cylinder
$A0, A1, A2, \ldots$	Antisymmetric plate modes
$A(\theta_i)$	Discrete weighting function for element i (Ch 16)
a	Acceleration (Ch 2); Coefficient vector for incident wave amplitude (Ch 5); Dimensionless length variable (Ch 17)
B	Amplitude constant; Ratio of acoustic impedances, $B = \dfrac{W_2}{W_1}$ (Ch 4)
$[\mathbf{C}]$	Global damping matrix (Ch 8)
C	Amplitude constant; Elastic constant (material stiffness) matrix; Viscous damping coefficient (Ch 2)
C_{iklm}, C_{nm}	Single entry in the elastic constant matrix (Ch 3)
c, c_0	Velocity of wave propagation
c_E	Velocity of energy transport
c_f	Fluid bulk wave velocity
c_g	Group velocity
c_L	Bulk longitudinal wave velocity
c_p	Phase velocity
c_{plate}	Plate mode wave velocity
c_R	Rayleigh wave velocity
c_T, c_S	Bulk transverse (shear) wave velocity
$\mathbf{D}(p,\omega)$	Coefficient matrix (Ch 11)
D	Amplitude constant; Cross-sectional area of a hollow cylinder (Ch 10)
$D_B^{(m)}$	Coefficient matrix relating to the inner boundary of layer m (Ch 11)
$D_T^{(m)}$	Coefficient matrix relating to the outer boundary of layer m (Ch 11)

$\{\mathbf{d}\}^{(e)}$	Nodal displacement vector (Ch 8)
$\{\dot{\boldsymbol{d}}\}$	Velocity vector (Ch 8)
$\{\ddot{\boldsymbol{d}}\}$	Acceleration vector (Ch 8)
d	Plate or layer thickness; Piston diameter (App A)
dP_i	Change in phase of wave component i (Ch 2)
ds	Arc length (Ch 2)
$\boldsymbol{E}$	Green-Lagrange strain (Ch 20)
E	Young's modulus; Mode excitability function (Ch 19)
$\hat{E}$	Energy density (Ch 2)
e	Mathematical constant, $e \approx 2.71828$
$\mathbf{F}$	External force (load) vector (Ch 8, 9); Deformation gradient (Ch 20)
F	Tension force (Ch 2); Excitation spectrum of transducer (Ch 19)
f	Frequency; Body force; Unknown coefficients that are part of the potentials equation (Ch 10)
$f^{(1,1)}$	Nonlinear forcing function associated with nonlinear terms from the primary wave field (Ch 20)
$f^{\text{2D comb}}$	Two-dimensional comb transducer loading geometry (Ch 19)
f^{ann}	Annular array transducer loading geometry (Ch 19)
f^{comb}	Comb transducer loading geometry (Ch 19)
$f_n^{\ surf}$	Nonlinear surface force (Ch 20)
$f_n^{\ vol}$	Nonlinear volume force (Ch 20)
fd	Frequency-thickness product
G	Amplitude factor relative to transducer loading amplitude (Ch 19)
$\mathbf{G}(\omega)$	Fourier transform of $G(\theta)$ (Ch 16)
$G(\theta)$	Total angular profile of the phased array on a hollow cylinder (Ch 16)
$\boldsymbol{H}$	Displacement gradient (Ch 20)
H	Layer thickness; Hankel function (Ch 19)
$\mathbf{H}(\omega)$	Fourier transform of $H(\theta)$ (Ch 16)
$\vec{H}$	Equivoluminal vector potential (Ch 10)
$H(\theta)$	Angular profile at a certain distance in the cylinder for element 0 in the phased array (Ch 16)
h	Half plate thickness; Layer thickness (Ch 12, 15); Unknown coefficients that are part of the potentials equation (Ch 10)
$\mathbf{I}$	Unit matrix (Ch 9)
I	Incident waveform amplitude (Ch 4); Wave intensity (Ch 4)
I, i, J, j	Index values
i	Imaginary number, $i = \sqrt{-1}$; Mode number
J	Bessel function of the first kind
$[\mathbf{K}]$	Global stiffness matrix (Ch 8)
$\mathbf{K}_1, \mathbf{K}_2, \mathbf{K}_3$	Stiffness matrices (Ch 9)
K	Elastic spring constant (Ch 2)
k	Wave number; Circular wave number (Ch 11); Index for transmitter-receiver pair (Ch 21)
$\bar{k}$	Complex wave number (Ch 2); Wave number vector (Ch 3)

k, l, m, n	Index values
k_{Im}	Imaginary component of wave number
k_r, k_{Re}	Real component of wave number
$\mathbf{L}_x, \mathbf{L}_y, \mathbf{L}_z$	Matrices in the strain-displacement equation (Ch 9)
L	Half length of loading in the axial direction on a hollow cylinder (Ch 10, 16); Length of plate (Ch 18); Element transverse length (Ch 19)
$L(m,n)$	Longitudinal mode of order (m,n) in a hollow cylinder (Ch 10)
l_z	Ratio of z-direction and x-direction wavenumbers, $l_z = \dfrac{k_z}{k_x}$ (Ch 6)
$[\mathbf{M}], \mathbf{M}$	Global mass matrix (Ch 8, 9)
M	Coefficient matrix for wave amplitudes (Ch 5); Mode number (Ch 6)
M, m	Mass (Ch 2); Circumferential order of a wave mode (Ch 10, 16)
$\mathbf{N}$	Normal vector of the loading surface; Shape function matrix (Ch 9)
N	Number array elements; Shape functions (Ch 8, 9); Number of nodes through plate thickness in SAFE analysis (Ch 18); Number of data points (Ch 21)
$\boldsymbol{n}$	Outward normal to a surface (Ch 20)
n	Mode number; Unit normal (Ch 6); Mode group index (Ch 16); Element number in an array (Ch 21)
n_k, n_l	Direction cosines of the normal to the wave front (e.g., $k_k = k n_k$)
$\{\mathbf{P}\}$	Body force in volume V (Ch 8)
$\mathbf{P}, P$	Acoustic Poynting vector, also called power flow or flux
p	Variable comparing the phase and bulk longitudinal wavenumbers, $p = \sqrt{\dfrac{\omega^2}{c_L^2} - k^2}$ (Ch 6); Angular wave number, $p = kR$ (Ch 11); Loading distribution (Ch 19); Pressure field amplitude (App A)
$p_1(\theta)$	Circumferential loading distribution function (Ch 10, 16)
$p_2(z)$	Axial loading distribution function (Ch 10, 16)
$\mathbf{Q}$	Nodal displacement vector (Ch 9)
q	Variable comparing the phase and bulk torsional (shear) wavenumbers, $q = \sqrt{\dfrac{\omega^2}{c_T^2} - k^2}$ (Ch 6); Body force or external loading per unit length (Ch 2)
R	Oblique incidence reflection factor (Ch 6); Array radius (Ch 21)
$R_{n\alpha}^M(r)$	Distribution of the particle displacement produced by mode (M, n) in the α direction (Ch 16)
r	Radial coordinate direction (cylindrical coordinate system); Variable comparing the phase and bulk longitudinal wavenumbers, $r = \sqrt{k^2 - k_L^2}$ (Ch 6)
r_m	Inner radius of the mth layer of a hollow cylinder (Ch 11)

S	Symmetric Lamb mode term in oblique incidence reflection factor equation; Surface over which calculation is to take place (Ch 8)
$S0, S1, S2, \ldots$	Symmetric plate modes
s	Variable comparing the bulk longitudinal and torsional (shear) wavenumbers, $s = \sqrt{k_L^2 - k_T^2}$; Array element spacing (pitch) (Ch 18, 19); Data set (Ch 21)
^{T}	Superscript indicating transpose (Ch 8)
$\mathbf{T}$	Unitary transformation matrix (Ch 9); Particle stress tensor (Ch 16)
$\mathbf{T}^{(e)}$	Nodal external tractions (Ch 9)
$\tilde{T}$	Second Piola-Kirchhoff stress (Ch 20)
T	Stress transmission coefficient (Ch 4)
T_{o}	First Piola-Kirchhoff stress (Ch 20)
$\hat{T}$	Stress field (Ch 10)
$T(m,n)$	Torsional mode of order (m,n) in a hollow cylinder (Ch 10)
T_m	Modal stress (Ch 20)
t	Time
$t, \overline{t}$	Surface traction (Ch 6, 19)
t_i	Physical time delay applied to element i (Ch 16)
$\overline{U}, \overline{u}, \boldsymbol{u}$	Displacement vector
$U_{\alpha\beta}$ ($\alpha = x, y, z$; $\beta = 1, 2, 3$)	Nodal displacement of the node β in the α direction (Ch 9)
u	Displacement
$\dot{u}$	Velocity, the first derivative of displacement with respect to time
$\ddot{u}$	Acceleration, the second derivative of displacement with respect to time
u'	First derivative of displacement with respect to coordinate direction x (Ch 2)
$u_{,xx}$	Second derivative of displacement with respect to coordinate direction x (Ch 2)
$u_{\mathrm{R}}(\theta_i, t)$	Signal received from the ith transducer segment (Ch 16)
$u_{\mathrm{s}}(\theta, z)$	Synthesized pipe image (Ch 16)
$\mathbf{V}$	Particle velocity vector (Ch 5)
V	Volume of an element (Ch 9)
$\vec{v}$	Particle velocity field (Ch 10)
$\tilde{v}_\nu$	Normalized surface velocity of guided wave mode ν (Ch 19)
v	Displacement along coordinate direction 2
W	Acoustic impedance, $W = \rho c$ (Ch 4)
W_m	Bessel function of mth order (Ch 10)
w	Displacement along coordinate direction 3; Array element width (Ch 18, 19)
x	Cartesian coordinate direction 1; Distance; An unknown
Y	Fluid influence term in oblique incidence reflection factor equation (Ch 6); Bessel function
y	Cartesian coordinate direction 2

Z_m	Bessel function of mth order (Ch 10)
z	Cartesian coordinate direction 3; Axial coordinate direction (cylindrical coordinate system)

Greek script

α — Wedge angle (Ch 1); Imaginary component of wavenumber (Ch 2, 6); Reflected wave angle (Ch 5); Half length of loading in the angular direction on a hollow cylinder (Ch 10); Variable comparing the phase and bulk longitudinal wavenumbers,

$$\alpha = \sqrt{\frac{\omega^2}{c_L^2} - k^2} \quad \text{(Ch 10)}$$

; Ratio of wavenumber in the x_3 direction to the wavenumber in x_1 direction (Ch 12, 15); Attenuation factor (Ch 17); Complex amplitude coefficient (Ch 18)

α_g	Angular group velocity (Ch 11)
$\alpha_g(\phi)$	Angular dependence of the guided wave amplitude (Ch 21)
α_i	Component of the eigenvector from the solution to the Christoffel equation
α_n	Complex wavenumbers which are poles to the integrands in Equation (17.24b) (Ch 17)
α_p	Angular phase velocity (Ch 11)

β — Refracted wave angle (Ch 5); Stiffness matrix coefficient for Rayleigh damping (Ch 8); Variable comparing the phase and bulk torsional (shear) wavenumbers, $\beta = \sqrt{\dfrac{\omega^2}{c_T^2} - k^2}$ (Ch 10);

Transformation matrix (Ch 15); Nonlinearity parameter (Ch 20); Scaling parameter for RAPID (Ch 21)

β_n	Complex wavenumbers which are poles to the integrands in Equation (17.24b) (Ch 17)
Γ	Surface of an element (Ch 9)
Γ_{im}	Christoffel acoustic tensor
γ	Reflection coefficient (Ch 21)
Δ	Length of a 1-D element (Ch 8); Area of a 2-D element (Ch 8); Dilatation, $\Delta = \nabla \cdot u$ (Ch 11)
δ	Signal magnification coefficient (Ch 21)
δ_{im}	Kronecker delta, which is 0 for $i \neq m$ and 1 for $i = m$
$\delta\mathbf{u}$	Virtual displacement (Ch 9)
$\delta\varepsilon$	Virtual strain (Ch 9)
$\varepsilon, \{\varepsilon\}$	Strain; Strain vector (Ch 8)
ζ	Dimensionless length variable (Ch 17)
η	Coefficient of viscosity (Ch 17); Surface domain for applied traction (Ch 19)
$\Theta_\xi^M(M\theta)$	Angular distribution function ($\cos(n\theta)$ or $\sin(n\theta)$) of the particle displacement produced by mode (n, M) in the α direction (Ch 16)

θ	Wave angle; Angular coordinate direction (cylindrical coordinate system) (Ch 10)
θ_{cr}	Critical angle
Λ	Wavelength (Ch 12); Coefficient matrix (Ch 18)
λ	Lamé constant; Wavelength; Eigenvalue (Ch 18)
λ_{im}	Christoffel acoustic tensor
λ_R	Rayleigh surface wave wavelength
μ	Shear modulus; Guided wave mode number (Ch 18)
ν	Poisson's ratio
ξ, ξ	Finite element shape function variable (Ch 9); Dimensionless length variable(Ch 17)
Π_a	Rectangle function (Ch 19)
ρ	Density; Mass density per length (Ch 2); Transducer traction density (Ch 19); Correlation coefficient (Ch 21)
ρ_f	Fluid density (Ch 6)
$\{\sigma\}, \bar{\sigma}$	Stress vector (Ch 8); Transverse stress field (Ch 18)
σ	Stress; Cauchy stress (Ch 20); Standard deviation (Ch 21)
τ	Surface traction (Ch 14); Time delay (Ch 19)
$\{\Phi\}$	Surface traction (Ch 8)
Φ	Dilatational scalar potential for Helmholtz decomposition, associated with longitudinal waves; Skew angle (Ch 15)
$\Phi_g(\phi)$	Angular dependence of phase variations (Ch 21)
ϕ	Skew angle (Ch 3); Phase angle for sinusoidal wave (Ch 2)
ϕ_0	Beam steering angle (Ch 21)
ϕ_i	Phase of function $A(\theta_i)$ (Ch 16)
ϕ_n	Phase delay applied to element n (Ch 19)
ψ	Eigenvectors from the SAFEM eigenvalue problem (Ch 9)
ψ_n	Angular locations of array elements (Ch 21)
$\psi, \bar{\psi}$	Vector potential for Helmholtz decomposition, associated with torsional (shear) waves
$\boldsymbol{\omega}$	Rotation vector, $\omega = \dfrac{1}{2}\nabla \cdot u$ (Ch 11)
ω	Circular (angular) frequency

Other symbols and notations

$*$	Complex conjugate
$<>$	Time average of variable inside bracket
$\otimes$	Convolution operator (Ch 16)
$\otimes^{-1}$	Deconvolution operator (Ch 16)

Preface

This book builds on my 1999 book, *Ultrasonic Waves in Solid Media*. Like its predecessor, this book is intended to bring people up to speed with the latest developments in the field, especially new work in ultrasonic guided waves. It is designed for students and for researchers and managers familiar with the field in order to serve as a baseline for further work already under way. I hope to journey with you to provide more breakthroughs in the understanding and application of ultrasonic guided waves. The goal is to improve the health of individuals, industries, and national infrastructures through improved methods of Non-destructive Evaluation (NDE). The purpose of this book is to expand on many of the topics that were introduced in my first book. Several chapters are almost the same, but there are many new fundamental topic chapters with a total emphasis in this book being directed toward the basic principles of ultrasonic guided waves. The field of ultrasonic guided waves itself is treated as a new and separate field compared to ultrasonics and other inspection disciplines as indicated in some of the efforts put forward in inspection certification by the American Society for Non-destructive Testing (ASNT) and also in code requirements in such groups as the American Society for Mechanical Engineers (ASME) and the Department of Transportation (DOT).

The book begins with an overview and background materials in Chapters 1 through 7 and then continues on to more advanced topics in Chapters 8 through 21.

I have had the good fortune to witness the growth of ultrasonic guided waves in Non-destructive Testing (NDT) and Structural Health Monitoring (SHM) since 1985. I have been deeply interested in safety and improved diagnostics utilizing wave propagation concepts. Wave phenomena can be used to evaluate material properties nondestructively as well as to locate and measure defects in critical structures. This work has led to devices that have become valuable quality control tools and/or in-service inspection procedures for structures such as critical aircraft, pipeline, bridge, and nuclear power components whose integrity is vital to public safety.

My first exposure (1970 to 1985) to ultrasonic NDE – beyond basic pulse-echo and through-transmission testing – focused on signal processing and pattern recognition. New tomographic ultrasonic imaging procedures were developed that employed special features to assist in defect classification; these procedures supplemented or replaced the standard more localized ultrasonic test methods. In the late 1970s, ultrasonic research was extended to medical applications. I explored linear phased array transducer systems used in real-time medical imaging. Of special interest to me at the time was tissue classification, in which we worked on differentiating malignant from benign tissue growth.

Around 1985, a newer version of ultrasonics in waveguides was conceived for faster and more sensitive ultrasonic examination. Some pioneering work on oblique incidence of the more localized ultrasonic method onto a bonded structure was carried out that could easily place longitudinal and shear energy into the bondline. The process was tedious and difficult to carry out. It was found that ultrasonic guided waves, however, could easily impinge both longitudinal and shear energy into the structure. Hence, guided wave activity was further developed for such adhesively bonded structures. Further research also revealed that guided waves – waves that travel along a surface or along a rod, tube, or platelike structure – could not only produce the same kind of two-dimensional particle velocity as that in oblique incidence but could also be much more efficient than the traditional technique of point-by-point examination. These guided wave research and application efforts continue today.

Guided wave concepts have been applied to examine the tubing in power plants and pipelines in chemical processing facilities and, importantly, to ensure the safety of large petroleum and gas pipelines. Because of their unique capabilities, guided wave techniques can be used to find tiny defects – over large distances, under adverse conditions, in structures with insulation and coatings, and in harsh environments.

Engineers, technicians, and students involved in ultrasonic NDE will appreciate the usefulness of this textbook. Even though the mathematics is sometimes detailed and sophisticated, the treatment can also be read by managers without detailed understanding of the concepts. They may find this book useful as it is designed to be read from a "black box" point of view so they can develop an understanding of what engineers, technicians, and students are talking about.

Overall, the material presented here in wave mechanics – and, in particular, guided wave mechanics – establishes a framework for the creative data collection and signal processing needed to solve many problems using ultrasonic NDE and SHM. I therefore hope that this book will be used as a reference in ultrasonic NDE by individuals at any level and as a textbook for seniors and graduate students. It is also hoped that this book will expand and promote the use of guided wave technology on both national and international levels.

Acknowledgments

Thanks are given to many individuals for their work efforts, discussions, and contributions in wave mechanics over the past twenty years. A special tribute is made to Dr. Aleksander Pilarski, who passed away on January 6, 1994. "Olek" worked with me as a visiting professor at Drexel University and at The Pennsylvania State University from 1986 to 1988 and from 1992 to 1994. His energetic and enthusiastic style, as well as his technological contributions, had a strong influence on many of us. He was a dear friend whose memory will remain forever.

Thanks are given to all of my PhD students and many MS students for their work efforts and valuable discussions. In particular, special thanks for assistance in the preparation of this text are given to the following very talented individuals, with a brief description of their backgrounds.

Dr. Michael Avioli has worked with me for more than twenty-five years providing signal processing and pattern recognition support in guided wave analysis. He made special contributions in transform methods.

Cody Borigo is currently an engineer at FBS, Inc., and is conducting his PhD thesis research with me at The Pennsylvania State University. His research experience includes guided wave NDE in composites, guided wave tomography, ultrasonic vibrations, phased annular array transducers, and ultrasonic ice sensing and deicing for helicopters and fixed-wing aircraft.

Dr. Jason Philtron received his PhD in Acoustics with me from The Pennsylvania State University in 2013. He is currently a postdoc in my ultrasonic research group. Dr. Philtron's research has focused on ultrasonic guided wave bond evaluation in thick structures, the use of phased arrays for optimal guided wave mode and frequency selection, guided wave tomography, and ultrasonic ice sensing.

Huidong Gao was born in Nantong, China, in 1978. He received his BS and MS degrees from Nanjing University, China, and his PhD degree with me from The Pennsylvania State University in 2007. Dr. Gao is now a principal research engineer at Innerspec Technologies, Inc. His primary research interest is advanced ultrasonic NDT techniques including guided waves, electromagnetic acoustic transducers (EMATs), and high-power UT applications. Dr. Gao is the 2011 Young NDT Professional Award recipient and the author of *Ultrasonic Testing*, a two-volume series book for NDT personnel training published by the ASNT.

Cliff Lissenden is a professor of engineering science and mechanics at The Pennsylvania State University. He came to The Pennsylvania State University in 1995 with expertise in mechanical behavior of materials. Dr. Lissenden now specializes in the use of ultrasonic guided waves for SHM and NDE. His current research investigates monitoring adhesively bonded or mechanically fastened joints in platelike structures and the generation of wave modes at higher harmonics to characterize precursors to macroscale damage.

Yang Liu is currently a research assistant on nonlinear methods in the Guided Wave NDE Lab, The Pennsylvania State University with Dr. Lissenden and myself.

Vamshi Chillara is a PhD candidate in the Engineering Science and Mechanics Department at The Pennsylvania State University.

Dr. Jing Mu, a scientist at FBS, Inc., obtained her PhD degree with me from The Pennsylvania State University in August 2008. Her research experience includes guided wave mechanics analysis and Finite Element Method (FEM) simulations. Dr. Mu specializes in ultrasonic guided wave inspection techniques of pipe structures including active phased array focusing, synthetic focusing, and advanced signal processing for pipe imaging.

Dr. Jason K. Van Velsor received his PhD in engineering science and mechanics with me from The Pennsylvania State University in 2009. He is currently an employee of Structural Integrity Associates. Dr. Van Velsor is an authority in the field application of guided wave technology for the long-range inspection of piping and holds multiple domestic and international certifications in this area. His practical experience includes the application of guided wave methods in nuclear and fossil power generation, oil and gas (on-shore and off-shore), gas transmission, water and wastewater, and pulp and paper industries.

Dr. Fei Yan is a scientist at FBS, Inc. He obtained his PhD degree with me in engineering mechanics from The Pennsylvania State University in 2008. Dr. Yan's research focuses on ultrasonic guided wave NDE and SHM applications including a variety of structures and composite materials. In particular, he has been involved in the development of guided wave phased arrays for isotropic and anisotropic composite plate structures, phased comb and annular array transducers, guided wave tomography SHM systems, and an ultrasonic vibration method.

Dr. Li Zhang is a scientist for FBS, Inc., and has focused on theoretical calculations and numerical simulations of guided wave behavior in various structures, phased array focusing and synthetic focusing in pipelines, and numerical simulations of ultrasonic sensor characteristics. She also obtained a PhD with me at The Pennsylvania State University.

Thanks also to The Pennsylvania State University and to all who have funded my research over the years. Finally, of course I thank my wife Carole and my entire family for their patience, love, and support in all of my activities.

1 Introduction

1.1 Background

The field of ultrasonic guided waves has created much interest this past decade. The number of publications, research activities, and actual product quality control and in-service field inspection applications has increased significantly. Investigators worldwide are considering the possibilities of using ultrasonic guided waves in nondestructive testing (NDT) and structural health monitoring (SHM), and in many other engineering fields. Tremendous opportunities exist because of the hundreds of guided wave modes and frequencies that are available for certain waveguides. Researchers have made tremendous advancements in utilizing mode and frequency selection to solve many problems, for example, in applications for testing pipe, rail, plate, ship hull, aircraft, gas entrapment detection in pipelines, and even ice detection and deicing of rotorcraft and fixed-wing aircraft structures. These have become possible by examining special wave structures that are available via certain modes and frequencies that are capable of effectively carrying out these special work efforts.

Ultrasonic guided waves in solid media have become a critically important subject in NDT and SHM. New faster, more sensitive, and more economical ways of looking at materials and structures have become possible when compared to the previously used normal beam ultrasonic or other inspection techniques. For example, the process of inspecting an insulated pipe required removing all the insulation and using a single probe to check with a normal beam along the length of the pipe with thousands of waveforms. Now, one can use a guided wave probe at a single location, leave the insulation intact, and perhaps inspect the entire pipe by examining just a few waveforms. The knowledge presented in this book will lead to creative ideas that can be used in new inspection developments and procedures.

The tremendous advances made in ultrasonic guided wave technologies in the past three decades are possible because of the tremendous computational power that has evolved over the past two decades and our improved ability to interpret and understand those mathematical guided wave computational results. Many of the problems solved today couldn't have been tackled ten or twenty years ago because the computations would have taken weeks, if they were possible to complete at all.

The finite element methods available today are absolutely amazing. Scientists can study so many problems impossible to solve decades ago. Special structural symmetries and specific loading functions are not necessary. Any configuration can be evaluated.

Wave propagation studies are not limited to NDT and SHM, of course. Many major areas of study in elastic wave analysis are under way, including:

(1) transient response problems, including dynamic impact loading;
(2) stress waves as a tool for studying mechanical properties, such as the modulus of elasticity and other anisotropic constants and constitutive equations (the formulas relating stress with strain and/or strain rate can be computed from the values obtained in various, specially designed, wave propagation experiments);
(3) industrial and medical ultrasonics and acoustic-emission NDT analysis;
(4) other creative applications, for example, in gas entrapment determination in a pipeline, ice detection, deicing of various structures, and viscosity measurements of certain liquids; and
(5) ultrasonic vibration studies that combine traditional low-frequency vibration analysis tools in structural analysis with high-frequency ultrasonic analysis.

Typical problems in wave propagations as waves reflect and continue propagating from boundary to boundary in a long time solution, compared to the short time transient solution, lead to an ultrasonic vibration problem.

Note that ultrasonic bulk wave propagation refers to waves that encounter no boundaries, like waves traveling in infinite media. On the other hand, guided waves require boundaries for propagation as in plates, rods, or tubes, for example. Elastic-wave propagation theory, for example, handles both transient response and the steady-state character of vibration problems.

Historically, the study of wave propagation has interested investigators (engineers and scientists) in the area of mechanics. Early work was carried out by such famous individuals as Stokes, Poisson, Rayleigh, Navier, Hopkinson, Pochhammer, Lamb, Love, Davies, Mindlin, Viktorov, Graff, Miklowitz, Auld, and Achenbach. K. F. Graff presents an interesting history in *Wave Motion in Elastic Solids*. I have included a number of other useful references on history and the basics of wave propagation at the end of this chapter. A detailed literature survey is not presented in the text. With today's tremendously sophisticated information-gathering technology, surveys are easy to perform. Key references enhancing the basic material presented in this text are given throughout the book.

Investigators all over the world now face the challenges of technology transfer and product development in the ultrasonic guided wave field. The basic theory presented in this text prepares us for a theoretically driven approach to sensor, system, and software design. The feedback from field experience and encounters, though, has led to the development of many new problem statements and considerations to meet these challenges effectively. The work presented in this textbook represents a starting point. Hundreds of papers and other work being done today are tremendously useful in meeting our current challenges. The breakthroughs in guided wave application will continue. A paradigm shift from bulk wave ultrasonics in NDT to SHM is triggering this growth in the creative utilization of ultrasonic guided waves. Guided waves will play a critical role in sensor development in the coming decades to improve safety and economics of inspection via self-diagnostics in SHM.

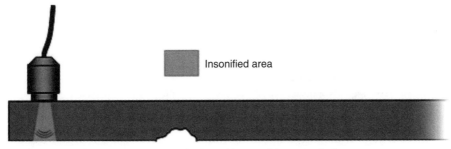

(a) Traditional ultrasonic bulk wave evaluation with normal-beam excitation

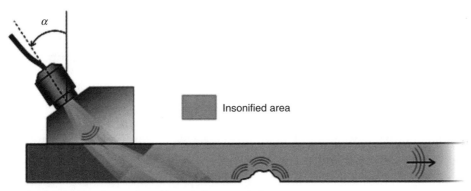

(b) Guided wave inspection with angle-beam excitation

(c) Guided wave inspection with comb excitation

Figure 1.1. Comparison of bulk wave with two guided wave inspection methods.

1.2 A Comparison of Bulk versus Guided Waves

A brief comparison of bulk wave and guided wave ultrasonic inspection is illustrated in Figure 1.1. Note the coverage volume of a structure is huge compared to a local region for a bulk wave. The guided waves cover the total thickness of the structure over a fairly long length compared to a localized area covered in ultrasonic bulk wave studies just below the transducer. Hence, in bulk wave inspection, the transducer must be moved along the surface to collect data, whereas with guided waves the structure can be inspected from a single probe position.

The two guided wave methods shown cover a large area of the structure. Note that the angle beam method could be used in bulk wave evaluation with waves reflecting back and forth inside the structure. Whether a bulk wave or guided wave is generated depends on the frequency used. Lower frequency with larger wavelengths λ would be used in guided wave generation. Wavelength λ would generally be greater than the structural thickness if guided waves are generated.

Table 1.1. Ultrasonic bulk versus guided wave propagation considerations

	BULK	GUIDED
Phase Velocities	Constant	Function of frequency
Group Velocities	Same as phase velocities	Generally not equal to phase velocity
Pulse Shape	Nondispersive	Generally dispersive

In the case of a comb transducer excitation, element spacing is wavelength λ, associated with the frequency used to generate the guided waves. Multiple elements are pulsed whose multiple oscillations lead to the generation of guided waves.

You can easily visualize for guided waves an outcome that is strongly dependent on frequency and impinging wave angles of propagation inside the structure and the resulting complex wave interference phenomenon that occurs in a guided wave situation. The strongly superimposed results are actually points that end up on the wave mechanics solution of the phase velocity dispersion curve for the structure that will be introduced later in this book. Elsewhere there is strong cancellation on destructive interference.

The principal advantages of using ultrasonic guided waves analysis techniques can be summarized as follows.

- Inspection over long distances, as in the length of a pipe, from a single probe position is possible. There's no need to scan the entire object under consideration; all of the data can be acquired from the single probe position.
- Often, ultrasonic guided wave analysis techniques provide greater sensitivity, and thus a better picture of the health of the material, than data obtained in standard localized normal beam ultrasonic inspection or other NDT techniques, even when using lower frequency ultrasonic guided wave inspection techniques.
- The ultrasonic guided wave analysis techniques allow the inspection of hidden structures, structures under water, coated structures, structures running under soil, and structures encapsulated in insulation and concrete. The single probe position inspection using wave structure change and wave propagation controlled mode sensitivity over long distances makes these techniques ideal.
- Guided wave propagation and inspection are cost-effective because the inspection is simple and rapid. In the example described earlier, there would be no need to remove insulation or coating over the length of a pipe or device except at the location of the transducer tool.

A general comparison of bulk and guided waves can be seen in Table 1.1. Key elements of the differences between isotropic and anisotropic media are listed in Table 1.2. *Isotropic* refers to materials with properties independent of direction and *anisotropic* refers to materials with properties dependent on direction like composite materials. Methods of determining characteristic equations for anisotropic waveguides can be found in the literature. See also Rose (1999).

Note that all metals are not isotropic. For example, columnar dendritic centrifugally cast stainless steel is anisotropic. This must be considered in any wave propagation studies.

Table 1.2. Ultrasonic wave considerations for isotropic versus anisotropic media

	ISOTROPIC	ANISOTROPIC
Wave Velocities	Not function of launch direction	Function of launch direction
Skew Angles	No	Yes

Table 1.3. A comparison of the currently used ultrasonic bulk wave technique and the proposed ultrasonic guided wave procedure for plate and pipe inspection

Bulk Wave	Guided Wave
Tedious and time consuming	Fast
Point-by-point scan (accurate rectangular grid scan)	Global in nature (approximate line scan)
Unreliable (can miss points)	Reliable (volumetric coverage)
High-level training required for inspection	Minimal training
Fixed distance from reflector required	Any reasonable distance from reflector acceptable
Reflector must be accessible and seen	Reflector can be hidden

A further practical comparison of the use of bulk and guided waves is presented in Table 1.3, in particular for plate and pipe inspections.

1.3 What Is an Ultrasonic Guided Wave?

Let us go beyond bulk waves traveling in infinite media, *infinite media* meaning that boundaries have no influence on wave propagation, to an explanation of ultrasonic guided waves that require boundaries for propagation. The waves interact with boundaries in a very special way so that boundary conditions can be satisfied. The boundaries could even be the surface of a very thick structure where the structure is considered as a half-space or a semi-infinite media. In this case, Rayleigh surface waves can propagate over the surface of a thick steel plate, for example, or over any thick structure where the frequency is such that the wavelength is very small compared to the thickness of the structure. The Rayleigh surface wave velocity in metals can be estimated as a function of Poisson's ratio, which for steel, as an example, is around 2,900 meters per second. Guided waves can also propagate in many different kinds of waveguides including thin plates, rods, tubes, and multilayered structures. In this case, the ultrasonic waves reflect back and forth inside the waveguide, leading to interference phenomena. Imagine pumping ultrasonic energy into a plate with an initial starting angle and a specific frequency. As the waves reflect back and forth, mode conversion occurs; whereby each time an interface is encountered both longitudinal and shear waves are reflected and/or refracted as in the case of multilayered media. For the particular angle and frequency chosen, the interference phenomena could be totally constructive, destructive, or intermediate in nature. There will be certainly hundreds of solutions of constructive interference points leading to a whole set of incident angles and frequencies that could represent solutions to the guided wave problem. To solve a guided wave problem, we could consider a governing wave equation in solid media, Navier's equations subjected to specific boundary conditions, for example, in a plate with stress-free boundary conditions. Utilizing the theory of elasticity in wave mechanics along with Navier's equation, subsequent strain-displacement equations, and a constitutive equation such as the generalized Hooke's law, with assumed harmonic solutions in satisfying the boundary conditions, one could come up with all of the constructive interference points leading to the dispersion curves for the structure. These constructive interference points can be plotted to produce a wave velocity dispersion curve of phase velocity versus frequency. The relationship between incident angle and phase velocity is simply expressed by Snell's law, so incident angle or phase velocity could be plotted against frequency. As a consequence, each natural waveguide, plate, tube, and so forth has its own unique phase velocity dispersion curve.

An interesting turn of events now takes place. Virtually hundreds of solutions to an inspection problem are available from the phase velocity dispersion curves. How do we pick the best solution? Often, the solution is built into a specific test instrument for a particular application. Every point on a dispersion curve has a different particle velocity vibration characteristic across the thickness of the structure. As an example, maximum in-plane vibration could occur on the surface of a structure at a particular phase velocity and frequency value. If this point were selected as a solution and the structure were placed under water, the water would have almost no influence on the wave propagation characteristics as energy leakage into the fluid would not take place. Let's consider one additional example to get a conceptual understanding of the potential of guided wave inspection. Suppose we wanted to examine a weak interface in a multilayer structure; we would have to search the phase velocity dispersion curve space to seek out a special wave structure across a thickness of a multilayer structure in such a way that we would obtain, for example, a maximum shear stress at the interface under consideration. This phase velocity and frequency value would then have excellent sensitivity to the weak interface situation at the particular layer being designated. Each guided wave problem could be approached in a similar fashion in searching for a particular variable with appropriate sensitivity in a certain problem. Upon selection of a particular point in the phase velocity dispersion curve space, it becomes possible to design an ultrasonic transducer that excites that particular point. Precise excitation is often difficult, however, because of the existence of a phase velocity spectrum and a frequency spectrum. These are concerns, though, for another day. The sensor design could be an angle beam transducer with an excitation line at a constant phase velocity value on the dispersion curve. As frequency is swept across the frequency axis, the specific modes and frequencies will be generated. An alternative probe design could be a comb transducer or an inter-digital design where the excitation line goes from the origin of the phase velocity dispersion curve at an angle of wavelength as the excitation line crosses many modes in the phase velocity dispersion curve space. Again, as the frequency is swept along the frequency axis, the modes crossed by the excitation lines will be generated. The comb spacing will be wavelength, which is the actual slope of the excitation line in the phase velocity dispersion curve space. Note that it becomes possible to move freely over the entire phase velocity dispersion curve space by changing angle in the horizontal excitation line approach or by changing element spacing in the sloped line of slope wavelength from the origin in the phase velocity dispersion curve space.

Table 1.4. Natural waveguides

- Plates (aircraft skin)
- Rods (cylindrical, square, rail, etc.)
- Hollow cylinder (pipes, tubing)
- Multilayer structures
- An interface
- Layer or multiple layers on a half-space

Let's now consider the long time solution to a wave propagation problem simply to add to our understanding of a wave propagation problem versus a vibrations problem. In the bulk wave case, because waves are traveling in infinite space, there is no vibration aspect of the problem to be considered because there are no wave reflection and transmission factors.

When you think about it, many structures are really natural waveguides provided the wavelengths are large enough with respect to some of the key dimensions in the waveguide. If the wavelengths are very small, then bulk wave propagation can be considered. Development of ultrasonic guided wave technology moved slowly until recently because of a lack of understanding and insufficient computational power. One very interesting major difference of many associated with guided waves is that many different wave velocity values can be obtained as a function of frequency, whereas for most practical bulk wave propagation purposes the wave velocity is independent of frequency. In fact, tables of wave velocities are available from most manufacturers of ultrasonic equipment that are applicable to bulk wave propagation in materials, showing just a single wave velocity value for longitudinal waves and one additional value for shear waves. See Table 1.4.

1.4 The Difference between Structural Health Monitoring (SHM) and Nondestructive Testing (NDT)

It seems worthwhile at this point to introduce the strategies of SHM and NDT. NDT is difficult as you carry equipment to a site and are asked to find defects in often very complex structures. For SHM, on the other hand, a baseline is available that can often handle very complex structures. See Table 1.5 for a summary.

1.5 Text Preview

A brief outline and discussion of the material included in this text is presented next. We begin with a discussion of dispersion principles in Chapter 2. Note that in guided wave propagation, basic dispersion concepts are encountered whenever wave velocity becomes a function of frequency or angle of propagation. The phase and group velocities change significantly as a result of the studying of the boundaries of the waveguide, which leads to many possible modes of wave propagation. Criteria must be established for selecting a particular mode and frequency for solving a particular problem. The basic formulas from physics and basic wave mechanics are outlined in Chapter 2.

Chapter 3 outlines wave propagation principles in unbounded isotropic and anisotropic media. Even though this is a subject in bulk wave propagation at this

Table 1.5. The difference between SHM and NDT

NDT	SHM
• Offline evaluation	• Online evaluation
• Time-based maintenance	• Condition-based maintenance
• Find existing damage	• Determine fitness for service and remaining useful time
• More cost and labor	• Less cost and labor
• Baseline not available	• Baseline required
	• Environmental data compensation methods required

point, the concepts will be extended to guided wave analysis in later chapters. The classic Christoffel equations are reviewed in detail to show the steps involved for studying wave propagation in anisotropic media. The wave velocity is no longer independent of angle, as it is in an isotropic material and, in fact, often changes quite drastically with angle. As a result, the interference phenomena as the wave propagates in a waveguide change drastically, affecting the group velocity of the waves in different directions, as well as producing skew angle effects that occur as the wave propagates through the material. Detailed mathematical treatment and sample problems are discussed.

Another subject directed toward bulk wave propagation that becomes critical in guided wave analysis is presented in Chapter 4, with emphasis on reflection and refraction factor analysis as waves encounter an interface. The initial emphasis is on isotropic media, followed by Snell's law and mode conversion. A variety of different models and boundary conditions are used to tackle the various wave propagation problems in different structures and in anisotropic materials, a topic that will be discussed in later chapters while still utilizing some of the concepts presented here.

Chapter 5 treats the more general problem of reflection and refraction analysis for oblique incidence including the study of slowness profiles and critical angle analysis. The energy partitioning into the different modes is treated here. Again, this is a topic presented from a bulk wave ultrasonic wave propagation point of view, the concepts of which are extended to guided wave analysis in later chapters and also considered in current research activity.

Chapter 6 covers the classic problem of wave propagation in a plate, where the Rayleigh–Lamb wave propagation problem is covered in detail. Some of the most significant aspects of guided wave analysis are covered in this chapter, which illustrates the development of the dispersion curves associated with phase velocity and group velocity, along with wave structure computation to show how the choice of mode and frequency changes the problem being investigated quite significantly in having different sensitivity, resolution, and penetration power for certain defects in different structures.

Chapter 7 covers various aspects of surface and subsurface waves in detail. These waves treat a wave traveling in a half-space. Surface waves of course have been used for years, and have often been covered in the more traditional books on ultrasonics. They are covered here as a guided wave problem because of the boundary involved and the similar treatment of guided waves in general.

In Chapter 8, an introduction to and pertinent details of finite element analysis are presented to help us move forward with wave propagation studies in guided waves. The finite element analysis tool is a significantly powerful one that allows us to do many interesting things in guided wave analysis. The computational efficiency available today makes this a unique and extremely useful tool for advancing the state of the art in ultrasonic guided wave analysis. Quite often, when combined with analytical tools to get us started in what we call a hybrid analytical FEM approach to the problem, the analytical work allows us to come up with the phase and group velocity dispersion curves and wave structures from which mode and frequency selection can take place, which leads to an actuator design and eventually a problem and systems solutions. The finite element analysis can take over from the analytical studies because the actuator design serves as the boundary conditions used in the finite element problem. We can then evaluate our choice of mode and frequency to solve a particular problem by looking at the wave propagation in the structure and the potential response from certain defects. All sorts of anomalies encountered in field application can be modeled with FEM assisting greatly in a final system design.

In Chapter 9, a fairly new concept is presented associated with a semi-analytical finite element (SAFE) method that allows us to calculate the wave structures and dispersion curves for a particular structure. It also provides an alternative to calculating dispersion curves for almost any waveguide in going beyond the global matrix technique presented in Chapter 6 for the traditional problem of waves in a plate. The SAFE technique is a very powerful computational process that can assist us greatly in studying and understanding unusually shaped waveguides like a rail or a multilayered anisotropic structure.

Chapter 10 describes the subject of waves in hollow cylinders. The emphasis here is on tubes and pipelines. This probably treats one of today's most popular practical applications in using guided waves in pipeline inspection. A hollow cylinder or tubular structure is a superb waveguide as the energy wraps around on itself and hence the propagation distances can be very large. The basic theoretical concepts presented in this chapter are classic in allowing us to study all of the different axisymmetric longitudinal and torsional modes along with the flexural modes for each that can propagate in a hollow cylinder.

Chapter 11 deals with circumferential guided waves, an important subject dealing with waves over a curved surface. The dispersion curves and wave structures are calculated with a description of a sample problem in advising us how to come up with mode and frequency choice for solving a particular problem, of optimizing coating detection on a pipe.

Chapter 12 covers guided waves in layered systems, which include multilayer structures along with interface waves and a layer on a half-space problem. Classic problems like Stoneley wave and Love wave propagation are discussed in this chapter. The computational methods are presented along with a description of the practical aspects of wave propagation in these layered systems.

Chapter 13 examines source influence on guided wave excitation in detail. This very important subject illustrates what happens when a finite source is used to load a waveguide compared to the theoretically popular analytical approach considering a plane infinite wave excitation. In this case, beyond the frequency spectrum that is considered for a pulse traveling in a structure, there is also a phase velocity spectrum, often with side lobes that can occur. The computational procedures associated with excitation and the phase velocity spectrums are covered. The work here allows us to efficiently get onto specific points on a dispersion curve for best possible sensitivity and penetration power in a particular waveguide. Note that, quite often, in guided wave analysis the signals appear noisy, but the noise is really associated with coherent guided wave propagation because of multiple mode wave propagation, and is not random in nature. This chapter outlines the ability to get onto a specific portion of a dispersion curve.

Chapter 14 tackles the subject of horizontal shear waves. This considers shear activation, but only in a platelike structure. It turns out this is one of the closed-form solution possibilities for calculating phase and group velocity in a waveguide, along with the wave structures and cutoff frequencies. Horizontal shear waves have not received much attention in past years because of experimental wave generation difficulties, but with so many new generation possibilities now realized by way of special shear-type transducers, including, for example, magnetostrictive or electromagnetic acoustic transducers (EMAT), the waves are becoming more popular and have very special applications in ice detection, in deicing, and in structures where water loading or accumulation is a problem in NDT and SHM.

Chapter 15 considers guided waves in anisotropic media. Dispersion curves become a function of direction. This is where the Christoffel equations allow us to look at wave-skewing influences in anisotropic media in certain anisotropic waveguides. Single-layer isotropic and multilayer isotropic structures are treated.

Chapter 16 discusses guided wave phased array focusing in piping. With the onslaught of phased array technology, where electronics are used for scanning and beam steering, tremendous interest is being generated on this subject. So beyond the bulk phased array analysis that relies primarily on simple line of sight computation from the source to the focal point in question, a technique is presented to allow focusing to occur in a pipe. The problem here is more complex than in a simple infinite media or even in a plate. In a plate, of course, you have to deal with the specific modes that you would like to use to produce focusing and appropriate sensitivity, for example, as is the case with piping. But in this case, in piping, a convolution concept is introduced to look at the summation of all the waves that turn around on themselves that cause the superposition and constructive interference phenomena to occur at the focal point. The details are presented on how to calculate the time delays associated with the particular elements around the circumference of the pipe, along with sample results on controlling the focused beam as it travels in the pipe. An understanding of flexural modes is critical to accomplish this focusing.

Chapter 17 consists of a discussion of guided waves in viscoelastic media. The overall viscoelastic approach will be introduced for bulk waves and for waveguides. The emphasis will be placed here on looking at a viscoelastic composite material along with a viscoelastic coating on a structure.

Chapter 18 presents a fairly new subject associated with ultrasonic vibration. The ultrasonic vibration approach goes beyond traditional vibrations studies utilizing vibrations under 20 kHz, but many of the concepts associated with the resonance and modal vibration character of the structure are similar. It turns out, though, that with ultrasonic vibrations the mode and modal pattern depend strongly on the loading function that is taken from transient ultrasonic guided wave analysis, beyond which, after multiple reflections occur, an ultrasonic vibration problem is introduced.

This can be used in NDT and SHM in quality control for new products, as well as in service evaluation of certain structures.

Chapter 19 introduces the subject of guided wave sensor arrays. Because of phasing, sensor arrays have become more crucial in the development of ultrasonic guided wave systems because of the ability to seek out certain modes and frequencies with special wave propagation characteristics. Discussed in this chapter are linear arrays and annular arrays. Quite often, you could use either a linear or an annular array to get onto a dispersion curve, depending on the particular method of analysis you seek to carry out. As an example, if you're using an angle beam transducer, the activation line in the dispersion curve space would be horizontal, and simply sweeping frequency at a particular angle entry that corresponds to a specific phase velocity curve would allow you to move through the phase velocity dispersion curve space. Incident angles are also changed in the search for a special point in the phase velocity dispersion curve space. On the other hand, there is much more versatility if you could use an array structure consisting of many elements, either a linear or an annular array. In this case, the activation line in the phase velocity dispersion curve space is a line drawn from the origin of the phase velocity versus frequency space at a specific angle across the set of curves. The angle is associated with the spacing of two elements in the array. To move that angled line over the phase velocity dispersion curve space, all you have to do is to insert various time delays and it treats the transducer effectively as having a different spacing between the elements. Computations to do this are presented. In this case, however, a certain guided wave travels in one direction and a different guided wave in the other direction. It is possible to make a choice in the phase velocity dispersion curve space to have waves traveling in one direction and seriously suppressed in the opposite direction. These are useful in system design to solve specific problems associated with inspection and SHM of defects in certain waveguides.

Chapter 20 puts forward a fairly new promising method using nonlinear principles of wave propagation. It turns out that for certain kinds of structures that do not have discontinuities similar to corrosion and crack-like reflectors, that if there is overall degradation of a structure because of fatigue, overall corrosion, graphitization, or hydrogen embrittlement, for example, or other large zoned area influences, that nonlinear methods might be used to solve the problem. In this case, in working with a waveguide it becomes critical to select a specific phase velocity that has modes separated in an integer multiple sense, so you might be able to produce these harmonics at higher, say second, third, or fourth harmonics, and the theoretical aspects show that the magnitudes of those harmonics would grow with propagation distance, which is very encouraging as far as detection potential is concerned. Longitudinal and shear waves are discussed. Going beyond very successful experiments in bulk wave analysis, the state of the art is now moving forward in evaluating material variations with nonlinear methods in various waveguides.

Chapter 21 deals with popular imaging methods associated with guided wave analyses that are entering the marketplace today. One of the topics is guided wave tomography, whereby placement of a series of sensors on a test object in some random or preferred set of positions allows us to image the regions around the sensors by having each sensor send energy to every other sensor followed by appropriate analysis; it can then produce an image of thickness variation or corrosion zones with the computational process introduced in tomography. It turns out, of course, that specific modes and frequencies should be used from each transducer to optimize the sensitivity and penetration power for a particular defect type in certain waveguides. This chapter introduces some of this technology. Another subject is guided wave phased arrays in plates – another phased array approach, but with the importance of selecting certain modes and frequencies that can produce a radar-type scan from the middle of a plate and then detect defects that could lie anywhere in the structure. Other imaging ideas are also covered in this chapter.

A series of appendices is also presented in this book. Appendix A is on ultrasonic NDT principles, analysis, and display technology. This is the basic background with a comprehensive, though brief, coverage of bulk wave ultrasonic testing. Many other handbooks are available on this subject, but this is a section required at this point as background reading for understanding some of the guided wave material presented in this book. Appendix B presents basic formulas and concepts in the theory of elasticity. This topic, of course, is critical in the understanding of finite element analysis, and also in guided wave analytical development, by way of understanding the constitutive equations and other aspects of elasticity that are crucial in understanding wave propagation in isotropic and anisotropic materials. Scholars have written excellent textbooks on this subject, but presented here are the basic elements for quick reference for understanding some of the material presented in this book. Appendix C presents a description of physically based signal processing methods for guided waves that deals with topics that are highly popular today, including a Short Time Fourier transform (STFFT) in generating spectrograms, a two-dimensional (2-D) FFT, and a wavelet transform (WT). Doing these computations presents specific features from the described wave feature sources that allow us to implement pattern recognition studies for identifying and classifying certain defects in a structure. They provide a physical basis for understanding guided wave propagation. The images produced can themselves become portions of either a phase velocity or group velocity dispersion curve, also useful by examining imaging shifts for certain variations in the waveguide that might occur as a result of material changes or defect presence. Appendix D outlines a brief description of our hybrid analytical FEM approach to guided wave problem solving and system design that illustrates the process and flow chart of how one could, after coming up with a specific problem statement, approach the problem from a computational point of view initially, onto a final system testing viewpoint. Emphasis is on mode and frequency selection tips for solving a variety of different problems.

1.6 Concluding Remarks

Great breakthroughs on the use of ultrasonic guided waves in NDT and SHM are under way. Advances are possible because of increased understanding and significant advances in computational power. Very few investigators were involved from 1985 to 2000 as indicated in literature survey work, but since 2000 the work and interest has exploded. A brief outline of recent successes and ongoing challenges associated with practical application of ultrasonic guided waves is presented next.

Table 1.6. Successes – Guided waves in general

Increased computational efficiency developments and understanding basic principles
Phased array and focusing developments in plates and pipes
Demonstration of optimal mode and frequency selections for penetration power, fluid loading
 influences, and other defect detection sensitivity requirements

Table 1.7. Successes – Composite materials

Understanding guided wave behavior in anisotropic media (slowness profiles and skew angle
 influence)
Development of ultrasonic guided wave tomographic imaging methods
Comb sensor designs for optimal mode and frequency selection (linear comb and annular arrays)

Table 1.8. Successes – Aircraft applications

Demonstration of feasibility studies in composites and lap splice, tear strap, skin to core
delamination, corrosion detection, and other applications.

Table 1.9. Successes – Pipe inspection

Understanding and utilization of both axisymmetric and non-axisymmetric modes
Achieving excellent penetration power with special sensors, focusing, and mode and frequency
 choices
Handling fluid loading with torsional modes
Defect sizing accomplishments to less than 5 percent cross-sectional area
Reduced false alarm calls in inspection due to focusing for confirmation
Circumferential location and length of defect estimations with focusing
Testing of pipe under insulation, coatings, and/or soil

Ultrasonic guided waves for aircraft and composite material inspections have
come a long way during the past decade or so. Many successes have come about, but
many challenges remain, some of which are outlined later in this chapter. The same
is true for pipeline inspection.

Successes in guided waves, in general, over the past decade, are listed in Table 1.6,
in composite materials in Table 1.7, in aircraft applications in Table 1.8, and in pipeline
inspection in Table 1.9. The topics speak for themselves.

Finally, although many promising methods are evolving into promising inspection
tools, numerous challenges remain. Many of the challenges focus on transferring
technology work tasks to a realistic practical environment. Practical challenges are
listed in Table 1.10 for guided waves in general, in Table 1.11 for composite materials,
in Table 1.12 for aircraft applications, and in Table 1.13 for pipeline inspection. The
topics are self-explanatory and can be used to define future work efforts.

Table 1.10. Practical challenges – Guided waves in general

Modeling accuracy is critically dependent on accurate input parameters often difficult to obtain – (especially for anisotropic and viscoelastic properties, interface conditions, and defect characteristics)

Signal interpretations are often difficult (because of multimode propagation and mode conversion, along with special test structure geometric features)

Sensor robustness to environmental situations like temperature, humidity to high stress, mechanical vibrations, shock, and radiation

Adhesive bonding challenges for mounting sensors and sustainability in an SHM environment

Merger of guided wave developments with energy harvesting and wireless technology

Penetration power requirements

Table 1.11. Practical challenges – Composite materials

Dealing with complex anisotropy and wave velocity and skew angle as a function of direction

Viscoelastic influences

Penetration power due to anisotropy, viscoelasticity, and inhomogeneity

Differentiating critical composite damage such as delamination defects from structural variability during fabrication (including minor fiber misalignments, ply-drops, inaccurate fiber volume fraction, and so on)

Guided wave inspection of composites with unknown material properties

Dealing with bonded structures

Table 1.12. Practical challenges – Aircraft applications

Robustness of guided wave sensors under in-flight conditions

Influences of aircraft paint and embedded metallic mesh in composite airframes for lightning protection

Table 1.13. Challenges – Pipe inspection

Tees, elbows, bends, and number of elbows and inspection beyond elbows

Quantification in defect location, characterization, and sizing, especially depth determination

Inspection reliability and false alarms (because of multimode propagation, mode conversions, and many pipe features like welds, branches, etc.)

Reducers, expanders, unknown layout drawings, cased pipes, and sleeves

1.7 REFERENCES

Achenbach, J. D. (1976). Generalized continuum theories for directionally reinforced solids, *Arch. Mech.* 28(3): 257–78.

(1984). *Wave Propagation in Elastic Solids*. New York: North-Holland.

(1992). Mathematical modeling for quantitative ultrasonics, *Nondestr. Test. Eval.* 8/9: 363–77.

Achenbach, J. D., and Epstein, H. I. (1967). Dynamic interaction of a layer of half space, *J. Eng. Mech. Division* 5: 27–42.

Achenbach, J. D., Gautesen, A. K., and McMaken, H. (1982). *Ray Methods for Waves in Elastic Solids*. Boston, MA: Pitman.

Achenbach, J. D., and Keshava, S. P. (1967). Free waves in a plate supported by a semi-infinite continuum, *J. Appl. Mech.* 34: 397–404.

Auld, B. A. (1990). *Acoustic Fields and Waves in Solids*. 2nd ed., vols. 1 and 2. Malabar, FL: Krieger.

Auld, B. A., and Kino, G. S. (1971). Normal mode theory for acoustic waves and their application to the interdigital transducer, *IEEE Trans.* ED-18: 898–908.

Auld, B. A., and Tau, M. (1978). Symmetrical Lamb wave scattering at a symmetrical pair of thin slots, in *1977 IEEE Ultrasonic Sympos. Proc.* vol. 61.

Beranek, L. L. (1990). *Acoustics*. New York: Acoustical Society of America, American Institute of Physics.

Davies, B. (1985). *Integral Transforms and Their Applications*. 2nd ed. New York: Springer-Verlag.

Eringen, A. C., and Suhubi, E. S. (1975). *Linear Theory* (Elastodynamics, vol. 2). New York: Academic Press.

Ewing, W. M., Jardetsky, W. S., and Press, F. (1957). *Elastic Waves in Layered Media*. New York: McGraw-Hill.

Federov, F. I. (1968). *Theory of Elastic Waves in Crystals*. New York: Plenum.

Graff, K. F. (1991). *Wave Motion in Elastic Solids*. New York: Dover.

Kino, C. S. (1987). *Acoustic Waves: Devices, Imaging and Digital Signal Processing*. Englewood Cliffs, NJ: Prentice-Hall.

Kinsler, L. E., Frey, A. R., Coppens, A. B., and Sanders, J. V. (1982). *Fundamentals of Acoustics*. New York: Wiley.

Kolsky, H. (1963). *Stress Waves in Solids*. New York: Dover.

Love, A. E. H. (1926). *Some Problems of Geodynamics*. Cambridge University Press.

(1944a). *Mathematical Theory of Elasticity*. 4th ed. New York: Dover.

(1944b). *A Treatise on the Mathematical Theory of Elasticity*. New York: Dover.

Miklowitz, J. (1978). *The Theory of Elastic Waves and Waveguides*. New York: North-Holland.

Mindlin, R. D. (1955). *An Introduction to the Mathematical Theory of Vibrations of Elastic Plates*. Fort Monmouth, NJ: U.S. Army Signal Corps Engineers Laboratories.

Musgrave, M. J. P. (1970). *Crystal Acoustics*. San Francisco, CA: Holden-Day.

Pollard, H. F. (1977). *Sound Waves in Solids*. London: Pion Ltd.

Rayleigh, J. W. S. (1945). *The Theory of Sound*. New York: Dover.

Redwood, M. (1960). *Mechanical Waveguides*. New York: Pergamon.

Rose, J. L. (1999). *Ultrasonic Waves in Solid Media*. Cambridge University Press.

(2002). A baseline and vision of ultrasonic guided wave inspection potential, *Journal of Pressure Vessel Technology* 124: 273–82.

Stokes, G. G. (1876). Smith's prize examination, Cambridge. [Reprinted 1905 in *Mathematics and Physics Papers* vol. 5, p. 362, Cambridge University Press.]

Viktorov, I. A. (1967). *Rayleigh and Lamb Waves – Physical Theory Applications*. New York: Plenum.

2 Dispersion Principles

2.1 Introduction

Before studying stress wave propagation in such waveguide structures as solid rods, bars, plates, hollow cylinders, or multiple layers, it is useful and interesting to review some applicable concepts taken from studies of dispersive wave propagation where wave velocity is a function of frequency. Wave propagation characteristics in waveguides are functions of frequency.

Let's first, however, consider wave propagation in a taut string where some basic dispersive concepts can be studied. Models of a taut string, a string on an elastic base, a string on a viscoelastic foundation will be discussed.

Even though wave dispersion can be considered for anisotropic media (where wave velocity is a function of direction), the emphasis in this chapter is on dispersion due to structural geometry. Some basic terms are introduced, including wave velocity, wavenumber, wavelength, material and geometrical dispersion, phase velocity, group velocity, attenuation, cutoff frequency, frequency spectrum, and energy transmission, all of which will be useful in further studies. Graphical interpretations and analysis of phase and group velocity are also covered in this chapter. Additional details can be found in other texts including Graff (1991).

2.2 Waves in a Taut String

2.2.1 Governing Wave Equation

In order to derive a governing wave equation for wave propagation in a string, consider a differential string element as shown in Figure 2.1. Let's initially consider infinite length; later, we'll consider specific boundary conditions. Displacement due to tension F will also be assumed as negligible. Recall that

$$u_{,xx} = \frac{\partial^2 u}{\partial x^2}, \quad \ddot{u} = \frac{\partial^2 u}{\partial t^2}, \quad \text{and} \quad u' = \frac{\partial u}{\partial x}.$$

The equation of motion in the u direction, following Newton's second law ($\Sigma F = ma$), is as follows:

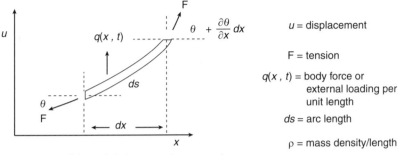

Figure 2.1. Differential element of a taut string.

$$-\mathrm{F}\sin\theta+\mathrm{F}\sin\left(\theta+\frac{\partial\theta}{\partial x}dx\right)+q\ ds=\rho\ ds\ \ddot{u}. \tag{2.1}$$

We assume small deflections ($ds \approx dx$, $\sin\theta \approx \theta$, and $\theta \cong \partial u/\partial x$) and so obtain

$$-\mathrm{F}\theta+\mathrm{F}\left(\theta+\frac{\partial\theta}{\partial x}dx\right)+q\ dx=\rho\ dx\ \ddot{u};$$

as a result,

$$\mathrm{F}\frac{\partial^2 u}{dx^2}+q=\rho\ddot{u}. \tag{2.2}$$

This is a second-order hyperbolic partial differential equation that is homogeneous if $q = 0$. Without external forcing we have

$$u_{,xx}=\frac{1}{c_0^2}\ddot{u}, \quad \text{where } c_0=\sqrt{\mathrm{F}/\rho}. \tag{2.3}$$

This is a one-dimensional, homogeneous, simple wave equation. Here c_0, which arises as a natural consequence of the mathematical solution, denotes the velocity of wave propagation.

2.2.2 Solution by Separation of Variables

Let $u(x, t) = X(x)T(t)$. Substituting this into (2.3) yields

$$\frac{\partial^2 u}{\partial x^2}=X''T, \quad \frac{\partial^2 u}{\partial t^2}=XT'', \quad \text{and } TX''=\frac{1}{c_0^2}\ XT'',$$

where $X = X(x)$ and $T = T(t)$. This simplifies as

$$\frac{X''}{X}=\frac{T''}{c_0^2 T}=\text{const}=-k^2. \tag{2.4}$$

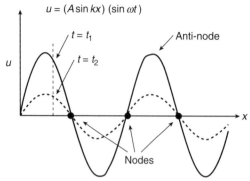

Figure 2.2. Standing or stationary wave example.

Note that the constant in (2.4) must equal $-k^2$ in order to guarantee a solution. Consequently,

$$X'' + k^2 X = 0 \quad \text{and} \quad T'' + k^2 c_0^2 T = 0.$$

These are two simple equations of harmonic motion, about which a great deal is known. For example,

$$X(x) = A_1 \sin kx + A_2 \cos kx, \quad T(t) = A_3 \sin kc_0 t + A_4 \cos kc_0 t.$$

In general, choosing $+k^2$ would produce sinh and cosh (hyperbolic) solutions, which are often used for guided waves.

These results may be interpreted as follows. We first present the relationships between wavenumber k, phase velocity c_p, circular frequency ω, and wavelength λ:

$$k = \frac{2\pi}{\lambda} = \frac{\omega}{c_p}, \quad \text{where } \omega = 2\pi f \text{ and } c_p = f\lambda. \tag{2.5}$$

Since $\omega = kc_0$, for the special case of $c_p = c_0$ we may write the the general solution $u(x, t) = X(x)T(t)$ as

$$u(x, t) = (A_1 \sin kx + A_2 \cos kx)(A_3 \sin \omega t + A_4 \cos \omega t).$$

Regrouping and multiplying, we obtain

$$u(x, t) = A_1 A_4 \sin kx \cos \omega t + A_2 A_3 \cos kx \sin \omega t$$
$$+ A_2 A_4 \cos kx \cos \omega t + A_1 A_3 \sin kx \sin \omega t. \tag{2.6}$$

Consider any of the four terms in (2.6) and examine the illustration of Figure 2.2. As shown in the figure, there is no shifting of the waveform and hence standing or stationary waves result.

By using such trigonometric identities as $\sin(\alpha + \beta) = \sin\alpha \cos\beta + \cos\alpha \sin\beta$, Equation (2.6) becomes

$$u(x, t) = B_1 \sin(kx + \omega t) + B_2 \sin(kx - \omega t)$$
$$+ B_3 \cos(kx + \omega t) - B_4 \cos(kx - \omega t). \tag{2.7}$$

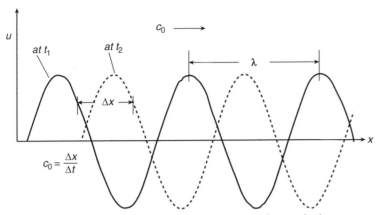

Figure 2.3. A disturbance propagation at constant phase velocity.

Consider a typical version of the four terms in (2.7): $u(x, t) = A \cos(kx - \omega t)$. Because $\omega = kc_0$,

$$u(x, t) = A \cos k(x - c_0 t). \tag{2.8}$$

This is a typical term showing wave propagation in the positive direction of x for a particular wavelength λ. Since $k = 2\pi/\lambda$, we could then plot $u(x)$ for various t values. The phase angle is $\phi = kx - \omega t$ or $k(x - c_0 t)$, the argument of the cosine function. Note that, as time increases, x also must increase by $c_0 * \Delta t$ in order to maintain a constant ϕ. Since $x = c_0 * \Delta t$, we have:

(a) $f(x - c_0 t)$ is a right-traveling wave, since x must be positive;
(b) $f(x + c_0 t)$ is a left-traveling wave, since x must be negative.

In order to introduce c_p, the phase velocity, we now consider waves traveling with constant phase. The physical explanation is as follows. For (2.8), note that t increases as x increases for constant phase ϕ, and that the propagation velocity of constant phase is c_0. See Figure 2.3, where $\Delta t = t_2 - t_1$.

The exponential representation of the solution equivalent to (2.6) is

$$u = C_1 e^{i(kx+\omega t)} + C_2 e^{-i(kx-\omega t)},$$

which may be simplified as $u = (C_1 e^{ikx} + C_2 e^{-ikx}) e^{i\omega t}$. This can be shown using Euler's formula $e^{i\theta} = \cos\theta + i \sin\theta$.

2.2.3 D'Alembert's Solution

We will now discuss an alternative approach to Equation (2.3). The D'Alembert solution to the simple wave equation, which was considered as early as 1750, is

$$u(x, t) = f(x - ct) + g(x + ct). \tag{2.9}$$

Equation (2.9) satisfies (2.3) for any arbitrary function f and g, as long as the initial and boundary conditions can eventually be satisfied. Alternatively, let $\xi = x - ct$ and $\eta = x + ct$. Equation (2.3) then becomes $\partial^2 u/\partial\xi\,\partial\eta = 0$ and hence has the same solution. These waves propagate without distortion, and f and g are arbitrary functions; see Figure 2.4.

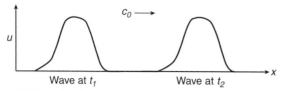

(a) Example of wave traveling without distortion

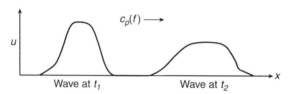

(b) Example of wave envelope traveling with distortion

Figure 2.4. Wave motion possibilities.

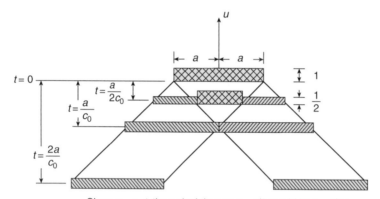

Since $x = c_0\, t$, the arrival time at $x = a/2$ would be $t = a/2c_0$

Figure 2.5. Propagation of an initial condition displacement in a string.

2.2.4 Initial Value Considerations

Consider now the following initial value problem:

$$u(x,0)=U(x), \quad \dot{u}(x,0)=V(x).$$

We can substitute these expressions into D'Alembert's solution, (2.9), to obtain

$$f(x) + g(x) = U(x) \text{ (since } t = 0);$$
$$-c_0 f'(x) + c_0 g'(x) = V(x).$$

Because $u(x, t) = f(x - c_0 t) + g(x + c_0 t)$, we could proceed to find f and g.

We will now discuss the propagation of an initial condition displacement in a string, where $u(x, 0) = U(x) = +1$ over the region $-a < x < a$ and $U(x) = 0$ for $x > a$. The solution,

$$u(x,\ t) = \frac{1}{2}U(x - c_0 t) + \frac{1}{2}U(x + c_0 t),$$

is shown in Figure 2.5.

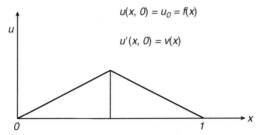

Figure 2.6. An initial value problem.

Figure 2.6 depicts a more general example of an initial value problem in a finite-length string, one that utilizes Fourier series to obtain a solution, and the result of (2.6). Boundary conditions for the string in Figure 2.6 are $u = 0$ for $x = 0$ and $x = l$ for all $t \geq 0$. The string is fastened at $x = 0$ and l. Consider the general solution given $\omega = kc_0$:

$$u(x, t) = (A_1 \sin kx + A_2 \cos kx)(A_3 \sin \omega t + A_4 \cos \omega t).$$

At $t = 0$, we have $u(x, 0) = (A_1 \sin kx + A_2 \cos kx)A_4$. It follows from the boundary conditions that $A_2 = 0$ and $\sin kl = 0$, so the solution is $k = n\pi/l$ for $n = 0, 1, 2, \ldots$ and hence

$$\omega = kc_0 = \frac{n\pi}{l}c_0.$$

From the initial condition $\dot{u}(x, 0) = 0$ we obtain that $A_3 = 0$.

Finally, the solution can be written as

$$u(x, t) = \sum_{n=1}^{\infty} a_n \sin \frac{n\pi x}{l} \cos \frac{n\pi c_0 t}{l}.$$

We can now find a_n from the initial condition for the displacement:

$$\sum_{n=1}^{\infty} a_n \sin \frac{n\pi x}{l} = f(x).$$

From the Fourier series analysis, the coefficients can be obtained as follows:

$$a_n = \frac{2}{l}\int_0^l f(x) \cdot \sin \frac{n\pi x}{l} \, dx \quad (n = 1, 2, \ldots).$$

This completes the solution.

2.3 String on an Elastic Base

Consider now a different problem that might involve wave distortions: a string on an elastic base (see Figure 2.7). The relationship between force q and displacement is

$$q(x, t) = -Ku(x, t). \tag{2.10}$$

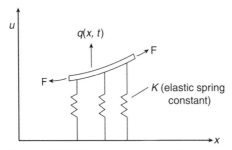

Figure 2.7. Differential element of a string on an elastic base.

Substituting (2.10) into (2.2), we obtain

$$u,_{xx} - \frac{K}{F}u = \frac{1}{c_0^2}\ddot{u}, \quad \text{where } c_0 = \sqrt{\frac{F}{\rho}}. \tag{2.11}$$

Comparing (2.11) with the simple wave Equation (2.3), we see that

(1) there is an extra term;
(2) a solution of the form $f(x - c_0 t)$ may not work; and
(3) wave propagation with no distortion may not be possible.

We may now examine possible conditions for the propagation of harmonic waves. Assume that

$$u(x, t) = A e^{i(kx - \omega t)}, \tag{2.12}$$

Substituting (2.12) into (2.11), we have

$$\frac{\partial u}{\partial x} = A e^{i(kx - \omega t)}(ik),$$

$$\frac{\partial^2 u}{\partial x^2} = A e^{i(kx - \omega t)}(ik)(ik) = -A e^{i(kx - \omega t)}k^2,$$

$$\frac{\partial u}{\partial t} = A e^{i(kx - \omega t)}(-i\omega),$$

$$\frac{\partial^2 u}{\partial t^2} = A e^{i(kx - \omega t)}(-i\omega)(-i\omega) = -A e^{i(kx - \omega t)}\omega^2.$$

Now, substitute the foregoing terms into (2.11):

$$-A e^{i(kx - \omega t)}k^2 - \frac{K}{F}A e^{i(kx - \omega t)} = \frac{1}{c_0^2}(-A e^{i(kx - \omega t)}\omega^2),$$

$$-k^2 e^{i(kx - \omega t)} - \frac{K}{F}e^{i(kx - \omega t)} = \frac{-\omega^2}{c_0^2}e^{i(kx - \omega t)},$$

$$\left(-k^2 - \frac{K}{F} + \frac{\omega^2}{c_0^2}\right)e^{i(kx - \omega t)} = 0. \tag{2.13}$$

This expression is called a *characteristic equation* (sometimes called a *dispersive, frequency,* or *secular* equation). For a general nontrivial solution,

$$\left(-k^2 - \frac{K}{F} + \frac{\omega^2}{c_0^2}\right) = 0.$$

There are many ways to tackle this expression. For example, we may write

$$\omega^2 = c_0^2\left(k^2 + \frac{K}{F}\right) \quad \text{where } \omega = \omega(k) \tag{2.14}$$

or, alternatively,

$$k^2 = \frac{\omega^2}{c_0^2} - \frac{K}{F} \quad \text{where } k = k(\omega); \tag{2.15}$$

note that $c_p \neq c_0$ as in the earlier taut string example. Hence, given $\omega = kc_p$, by Equation (2.14) we have

$$c_p^2 = c_0^2\left(1 + \frac{K}{Fk^2}\right). \tag{2.16}$$

If $k \to \infty$ then

$$\omega = c_0 k\sqrt{1 + \frac{K}{Fk^2}} \to c_0 k. \tag{2.17}$$

The same result is obtained when $K = 0$ in (2.14), and this gives $c_p = c_0$ (see (2.16)). Hence, $c = c(k)$ or $k = k(c)$, and thus

$$k^2 = \frac{K/F}{(c^2/c_0^2) - 1}.$$

This yields three important results.

(1) Wave velocity is a function of frequency. Therefore, from a consideration of Fourier harmonic analysis, pulse distortion must occur.
(2) There are two real roots of (2.15):

$$k = \pm\sqrt{\frac{\omega^2}{c_0^2} - \frac{K}{F}}.$$

If $\omega^2/c_0^2 > K/F$ then propagation is possible to the right or left and so $u = Ae^{-i(\pm kx + \omega t)}$. On the other hand, if $\omega^2/c_0^2 < K/F$ then k is imaginary. We do not deal with this case, which represents a nonpropagating disturbance. If we write $\bar{k}^2 = -k^2$ then $u(x, t) = A\left(e^{\pm \bar{k}x}\right)\left(e^{-\omega t}\right)$ as before, thus producing standing waves.
(3) For the special case of $k = 0$, by using $\omega^2/c_0^2 = K/F$ from (2.15) and recalling that $k = 2\pi/\lambda$, we can see that, as $\lambda \to \infty$ (the long wavelength limit), $u = Ae^{-i\omega_0 t}$. By (2.14) this represents uniform vibration, and the cutoff frequency $\omega_0 = c_0\sqrt{K/F}$.

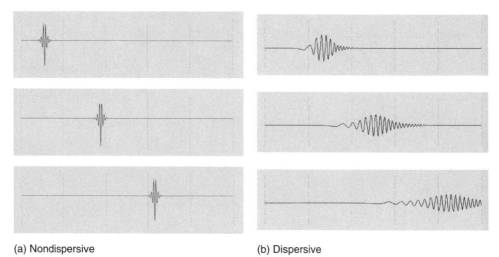

(a) Nondispersive   (b) Dispersive

Figure 2.8. Wave propagation possibilities as the wave travels along a waveguide.

2.4 A Dispersive Wave Propagation Sample Problem

A sample nondispersive and dispersive wave propagation result is illustrated in Figure 2.8. Even though pulse shape changes can be observed, the basic premise of wave propagation in lossless media in a waveguide is that energy is conserved.

The propagation of a dispersive guided wave can be easily simulated using a dispersion curve and any specified frequency spectrum. See Figure 2.9 for example.

The amplitudes, a_i, represent the excitation strength of the mode at the corresponding frequency, f_i. Also corresponding to the excitation, is the phase velocity c_i. Knowing the excitation frequency and the velocity enables the calculation of the wavenumber k_i. See equations below.

$$\omega_i = 2\pi f_i \tag{2.18}$$

$$k_i = \frac{\omega_i}{c_i} \tag{2.19}$$

An animation of dispersive propagation can be generated using equation below.

$$f(x,t) = \sum_{i=1}^{n} a_i \cos\left(k_i x - \omega_i t\right) \tag{2.20}$$

Fixing the point x, a waveform can be generated by running t from 0 to T with T being the width of the waveform. Make a bitmap of this waveform. Change x to $x + \Delta x$ and create the next waveform and make a bitmap of it. Continue this process until the desired length of propagation has been reached.

To view the animation, run the set of bitmaps in rapid succession or create a *avi* file for example. Sample static shots are illustrated in Figure 2.8 at a few specific times.

Because $c_p = \omega_0/k$, it follows that c_p increases if k decreases; hence, c_p approaches infinity as k approaches zero.

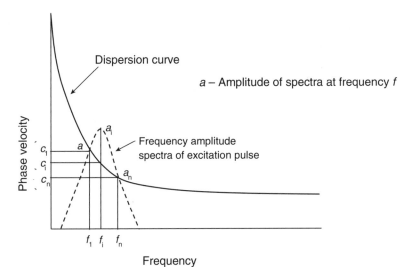

Figure 2.9. Using a dispersion curve and a frequency spectrum to obtain parameters for generating an animation of a dispersive propagating wave.

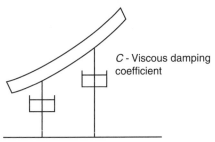

Figure 2.10. String on a viscous foundation.

2.5 String on a Viscous Foundation

Figure 2.10 shows a string on a viscous foundation, an example of a highly dispersive and attenuative system. Let $q(x, t) = -C\dot{u}(x, t)$. If we consider the governing wave equation and a possible solution, we obtain

$$Fu_{,xx} - C\dot{u} = \rho\ddot{u}. \tag{2.21}$$

Assume that

$$u = Ae^{i(kx - \omega t)}, \tag{2.22}$$

Substituting (2.22) into (2.21) yields

$$k^2 = i\frac{C\omega}{F} + \frac{\omega^2}{c_0^2}.$$

Therefore, k is complex and so free propagation of waves is not possible.

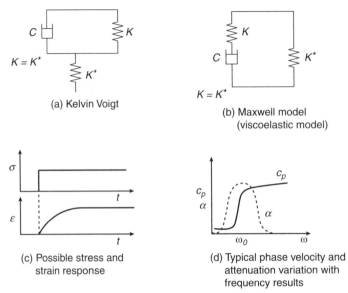

Figure 2.11. Possible viscoelastic models and response functions.

In a more conventional notation, we can consider the following expression for wavenumber with real k_r and imaginary components:

$$\bar{k} = k_r + i\alpha.$$

Substituting this into (2.22) and (2.21), we obtain $u = Ae^{(i(k_r + i\alpha)x - \omega t)}$, which can be simplified as follows:

$$u = \underbrace{Ae^{(-\alpha x)}}_{\substack{\text{new element} \\ \text{resposible for} \\ \text{attenuation}}} * \underbrace{e^{i(k_r x - \omega t)}}_{\substack{\text{old form of} \\ \text{wave motion}}}.$$

2.6 String on a Viscoelastic Foundation

Now imagine a string on a viscoelastic foundation. It is useful to introduce some basic aspects of viscoelastic model analysis in considerations of material dispersion. In dispersive systems, the phase velocity c_p is a function of frequency. Parts (a) and (b) of Figure 2.11 show mechanical analogs of the material system. For the classical Kelvin and Maxwell models, attenuation occurs as a function of frequency; this is difficult to deal with both theoretically and experimentally. Parts (c) and (d) of Figure 2.11 show sample stress and strain response and corresponding sample phase velocity and attenuation curves as a function of frequency. These topics will be discussed more thoroughly in Chapter 17.

2.7 Graphical Representations of a Dispersive System

In order to study some basic aspects of dispersive systems, we now consider some graphical representations of the results for the string on an elastic foundation. Two major representations are used to graph dispersive character: a frequency spectrum and a dispersion curve. We will first consider the frequency spectrum.

Visualize a string on an elastic foundation and recall (2.14):

$$\omega^2 = c_0^2 \left(k^2 + \frac{K}{F} \right).$$

If $k = 0$ then $\omega = \omega_0 = c_0\sqrt{K/F}$ as a cutoff frequency and, as $k \to \infty$, $\omega = c_0 k$ is a straight line (an asymptote). Now consider k_{Re} and k_{Im} and the relationship with ω_0 from (2.15):

$$k^2 = \frac{\omega^2}{c_0^2} - \frac{K}{F}.$$

Since $\omega_0 = c_0\sqrt{K/F}$, it follows that $\omega_0^2 = c_0^2(K/F)$ and $K/F = \omega_0^2/c_0^2$. As a result,

$$k^2 = \frac{\omega^2}{c_0^2} - \frac{\omega_0^2}{c_0^2} \text{ and } k = \sqrt{\frac{\omega^2}{c_0^2} - \frac{\omega_0^2}{c_0^2}}.$$

The term k_{Im} exists when $\omega < \omega_0$, and k_{Re} exists when $\omega > \omega_0$. In a frequency spectrum, we consider frequency versus wavenumber; in a dispersion curve, phase velocity versus wavenumber.

A graphical form of the frequency spectrum is illustrated in Figure 2.12. Phase velocity information can be extracted from the frequency spectrum in order to produce a dispersion curve as follows. Use $\tan\theta = \omega/k = c_p$ to derive phase velocity from the frequency spectrum. Figure 2.13 shows that, as $k \to 0$, $c_p \to \infty$.

Note that, for small k and long λ, the cutoff frequency is ω_0 (here, $c_p \to \infty$), giving rise to uniform vibration. For $k \gg 1$ and $\lambda \ll 1$, the foundation effect is minimized. Note that we can have waves (e.g., Lamb waves) with multimode characteristics.

Later in the text, phase velocity, group velocity, and attenuation dispersion curves will all be presented as a function of frequency while drawing upon many of the concepts presented in this chapter on dispersion principles.

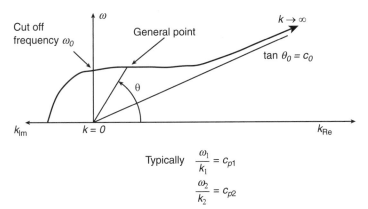

Figure 2.12. A typical frequency spectrum profile for a string on an elastic foundation.

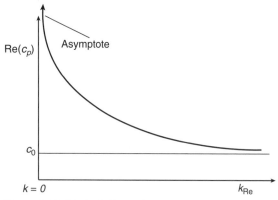

Figure 2.13. Typical dispersion curve of phase velocity versus wavenumber.

2.8 Group Velocity Concepts

Group velocity is associated with the propagation velocity of a group of waves of similar frequency. From Lord Rayleigh: "It has often been remarked that when a group of waves advances into still water, the velocity of the group is less than that of the individual waves of which it is composed; the waves appear to advance through the group, dying away as they approach its interior limit" (1945, vol. I, p. 475). Also note that, in wave mechanics, Heisenberg used the term "velocity of wave packets" (see Serway 1990).

The simplest analytical explanation (Stokes 1876) is to consider two propagating harmonic waves of equal amplitude but of slightly different frequency, ω_1 and ω_2. Then

$$u = A\cos(k_1 x - \omega_1 t) + A\cos(k_2 x - \omega_2 t), \tag{2.23}$$

where $k_1 = \omega_1/c_1$ and $k_2 = \omega_2/c_2$. Using trigonometric identities,

$$A(\cos\alpha + \cos B) = 2A\left[\cos\left(\frac{\alpha - B}{2}\right) * \cos\left(\frac{\alpha + B}{2}\right)\right].$$

We can rewrite (2.23) as

$$u = 2A\cos\left\{\frac{1}{2}(k_2 - k_1)x - \frac{1}{2}(\omega_2 - \omega_1)t\right\} * \cos\left\{\frac{1}{2}(k_2 + k_1)x - \frac{1}{2}\omega_2 + \omega_1)t\right\}, \tag{2.24}$$

noting that the cosine is an even function. We now make the following substitutions:

$$\Delta\omega = \omega_2 - \omega_1, \quad \Delta k = k_2 - k_1;$$

$$\frac{1}{2}(\omega_2 + \omega_1) = \omega_{AV}, \quad \frac{1}{2}(k_2 + k_1) = k_{AV};$$

$$c_{AV} = \frac{\omega_{AV}}{k_{AV}}.$$

Hence,

$$u = 2A\underbrace{\cos\{\tfrac{1}{2}\Delta kx - \tfrac{1}{2}\Delta\omega t\}}_{\text{low-frequency term}} * \underbrace{\cos(kx - \omega t)}_{\text{high-frequency term}} \tag{2.25}$$

(note that the low-frequency term has a propagation velocity). The group velocity is

$$c_g = \frac{\Delta\omega}{\Delta k},$$

which in the limit becomes $c_g = d\omega/dk$. The high-frequency term also has a propagation velocity, $c_p = \omega/k$.

Consider now an alternative approach to a definition of group velocity. At some time increment $t = t_0 + dt$, we may represent changes in phase of any individual component as follows:

$$dP_i = \{k_i(x_0 + dx) - \omega_i(t_0 + dt)\} - \{k_i x_0 - \omega_0 t\} = k_i\, dx - \omega_i\, dt.$$

In order for the wave group to be maintained, the changes in phase for all components should be the same: $dP_j - dP_i = 0$.

With regard to the phase angle $kx - \omega t$, we have

$$\underbrace{(k_j - k_i)}_{dk}\, dx - \underbrace{(\omega_i - \omega_i)}_{dw}\, dt = 0 \text{ and}$$

$$\frac{dx}{dt} = \frac{d\omega}{dk} = c_g \Rightarrow \frac{\Delta\omega}{\Delta k} \qquad (2.26)$$

This represents a classical definition of group velocity.

We may further compare $c_p = \omega/k$ and $c_g = d\omega/dk$ as follows:

$$c_g = \frac{d(kc_p)}{dk} = c_p + k\frac{dc_p}{dk}, \qquad (2.27)$$

where $c_p = c_p(k)$. Three cases ($c_g > 0$, $c_g = 0$, or $c_g < 0$) can be graphed on a frequency spectrum of ω versus k.

Superposition of a group of waves of similar frequency leads to the typical result shown in Figure 2.14. The individual harmonics travel with different phase velocities c_p, but the superimposed packet travels with the group velocity c_g. Realistically, we should therefore consider a superposition of a number of waves, rather than just two as used in the earlier example. Thus,

$$u = \sum_{i=1}^{n} A_i \cos(k_i x - \omega_i t), \qquad (2.28)$$

where k_i and ω_i differ only slightly.

Now we consider phase and group velocity from the perspective of a frequency spectrum. Group velocity is indicated as a local slope to the curve at the point, as illustrated in Figure 2.15. Physical examples of the three cases can now be supplied:

(1) $c_p > c_g$ – *classical* (normal) dispersion, as in a Rayleigh example, that appears to originate behind the group, travel to the front, and disappear;
(2) $c_p = c_g$ – no dispersion;
(3) $c_p < c_g$ – anomalous dispersion, which appears to originate at the front, travel to the rear, and then disappear.

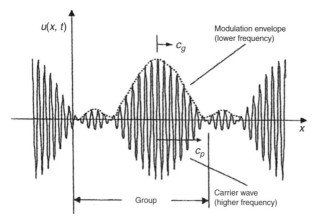

Figure 2.14. Group velocity example.

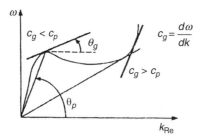

Figure 2.15. Group velocity variation with phase velocity.

Alternative forms of the expressions for cg should also be addressed. Starting with (2.27),

$$c_g = \frac{d(kc_p)}{dk} = c_p + k\frac{dc_p}{dk},$$

we note that $\lambda = 2\pi/k = 2\pi k^{-1}$. Thus

$$c_g = c_p + k\frac{dc_p}{d\lambda}\frac{d\lambda}{dk} = c_p + \frac{2\pi}{\lambda}\frac{dc_p}{d\lambda}(-2\pi k^{-2})$$

$$= c_p + \frac{2\pi}{\lambda}\frac{-2\pi}{(2\pi)^2}\lambda^2\frac{dc_p}{d\lambda} = c_p - \lambda\frac{dc_p}{d\lambda}$$

and so

$$c_g = c_p - \lambda\frac{dc_p}{d\lambda}. \qquad (2.29)$$

Next, we look at group velocity and energy transmission. Consider the simple group

$$u = 2A\cos\left(\frac{1}{2}\Delta kx - \frac{1}{2}\Delta\omega t\right)\cos(kx - \omega t).$$

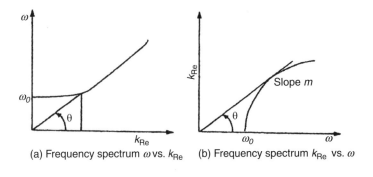

(a) Frequency spectrum ω vs. k_{Re} (b) Frequency spectrum k_{Re} vs. ω

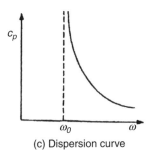

(c) Dispersion curve

Figure 2.16. Construction of dispersion curve from the frequency spectrum.

Define the energy density as

$$\hat{E} = \rho \dot{u}^2$$
$$= 4\rho\omega^2 A^2 \cos^2\left(\frac{1}{2}\Delta kx - \frac{1}{2}\Delta\omega t\right)\sin^2(kx - \omega t) + \cdots$$

Taking the time average of this expression over several periods T (during which the modulation does not change much), from Miklowitz (1978) we find for lossless media

$$\langle \hat{E} \rangle \cong 2\rho\omega^2 A^2 \cos^2\left(\frac{1}{2}\Delta kx - \frac{1}{2}\Delta\omega t\right).$$

This suggests that the time-averaged energy density propagation has the velocity

$$c_E = c_g = \frac{d\omega}{dk}. \tag{2.30}$$

Hence, group velocity is the velocity of energy transportation.

Now imagine the possibility of going from a frequency spectrum to the more popular engineering dispersion curve of $c_p(\omega)$ (see Figure 2.16 for graphical details). In order to plot c_p versus f, it might be easier to change the frequency spectrum from ω versus k to k versus ω, as shown in parts (a) and (b) of Figure 2.16. The curve c is then plotted from (b) on a point-by-point basis, using $c_p = \omega/k$ from this curve. A

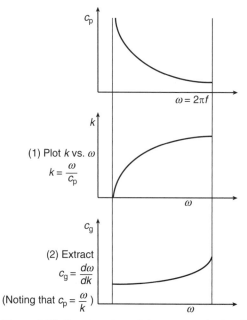

Figure 2.17. Construction of the group velocity dispersion curve from the phase velocity dispersion curve.

graphical procedure for plotting a group velocity dispersion curve from the phase velocity dispersion curve is illustrated in Figure 2.17.

2.9 Exercises

1. (a) What percent tension increase is necessary to double the frequency of a guitar string of length L?
   (b) What change in cross-sectional diameter of the string would also give the same doubled frequency?
2. (a) What is the wave velocity on a string if tension doubles?
   (b) Determine the maximum wave velocity on a string.
3. Derive the wave equation for a string on a viscoelastic foundation.
4. Select a specific frequency spectrum. Derive the dispersion curve from the frequency spectrum. Also, determine the group velocity curve.
5. Solve the wave propagation problem in a string, given boundary conditions $u(0, t) = u(l, t) = 0$ and initial conditions $u(x, 0) = f(x)$ and $(du/dt)(x, 0) = 0$.
6. Derive an expression for group velocity as a function of phase velocity and frequency.
7. For the specific mode shown in Figure 2.18, estimate the numerical value of the group velocity at point A. Also estimate the value of the cutoff frequency for the curve shown.
8. Solve the initial value problem shown in Figure 2.19. The string is fixed at $x = 0$ and $x = 10$; displacement function is shown at $t = 0$.

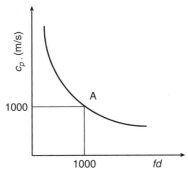

Figure 2.18. Exercise 7.

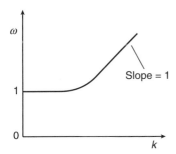

Figure 2.19. Exercise 8.

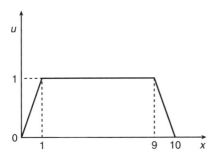

Figure 2.20. Exercise 16.

9. Derive the characteristic equation for a string on a viscoelastic foundation.
10. Given a sample frequency spectrum of $\omega(k)$, plot $c_p(k)$.
11. Given a sample frequency spectrum of $\omega(k)$, plot $c_g(k)$.
12. Given a sample dispersion curve $c_p(\omega)$, plot the group velocity dispersion curve $c_g(\omega)$.
13. What are possible conditions in $k(\omega)$ or $\omega(k)$ curves for a negative group velocity?
14. What are possible conditions in $c_p(f)$ curves for a negative group velocity?
15. Plot a frequency spectrum similar to that shown in Figure 2.12 for sample realistic values of tension, density, and spring constant.

Table 2.1. Given parameters

Frequency (f_i) MHz	Magnitude (a)	Phase Velocity (c) mm/μsec
1.7	4.49	7.55
1.8	5.28	6.10
1.9	5.97	5.36
2.0	6.25	4.91
2.1	5.96	4.59
2.2	5.24	4.37
2.3	4.45	4.20

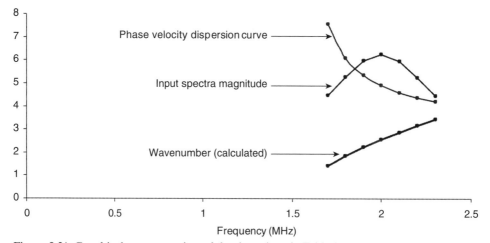

Figure 2.21 Graphical representation of the data given in Table 2.1

16. Given the sample frequency spectrum curve of ω versus k shown in Figure 2.20, graphically estimate the phase and group velocity curves as a function of the wavenumber k. (Concentrate on values when $k = 0$, k large, and then a general trend in between.)

17. Given the information in Table 2.1, confirming the illustration in Figure 2.21. Using this data and appropriate formulas, calculate waveforms for two or three positions between 0 mm and 10 mm. Use a time sequence 0, .1, .2, ... 8 μsec. {81 points}

18. See Figure 2.22 which shows a partial solution to Exercise 17. Consider the appropriate group velocity value over the frequency range 1.7 to 2.3 MHz. Could you estimate this group velocity value from Figure 2.22?

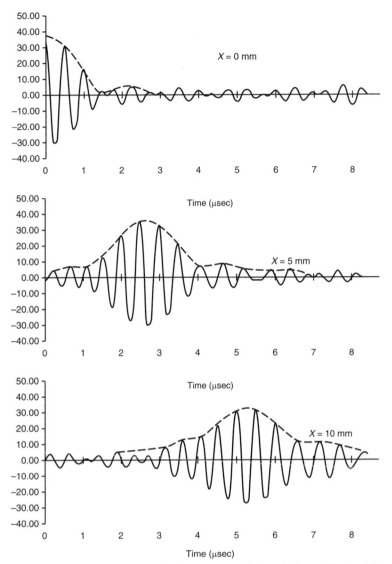

Figure 2.22 Dispersive wave displacements at 0, 5, and 10 mm. Dashed lines added to emphasize wave spreading.

2.10 REFERENCES

Graff, K. F. (1991). *Wave Motion in Elastic Solids*. New York: Dover.

Miklowitz, J. (1978). *The Theory of Elastic Waves and Waveguides*. New York: North-Holland.

Rayleigh, J. W. S. (1945). *The Theory of Sound*. New York: Dover.

Serway, R. A. (1990). *Physics for Scientists and Engineers with Modern Physics*. 3rd ed., chap. 41. Philadelphia: Saunders.

Stokes, G. G. (1876). Smith's prize examination, Cambridge. [Reprinted 1905 in *Mathematics and Physics Papers*, vol. 5, p. 362, Cambridge University Press.]

3 Unbounded Isotropic and Anisotropic Media

3.1 Introduction

Bulk wave propagation refers to wave propagation in infinite media; guided waves are those that require a boundary for their existence, such as surface waves, Lamb waves, and interface waves. This chapter will focus on bulk wave propagation in infinite (or semi-infinite) media. Keep in mind that a thin structure can, for all practical purposes, still be considered a half-space or semi-infinite media if the wavelength of excitation is small with respect to the thickness of the test object.

We shall explore some interesting phenomena of phase velocity variation with angle of propagation into solid media. This leads to a dispersive influence as a result of differences in phase velocity and energy velocity. For isotropic materials, phase velocity is independent of entry angle. For lossless media, the energy velocity is equal to the group velocity. However, because of the wave velocity variations with angle, interference phenomena will lead to a skew angle. Trying to send waves or ultrasonic energy in a specific direction may be more difficult than you think!

3.2 Isotropic Media

3.2.1 Equations of Motion

The development of the equation of motion for an elastic isotropic solid is a classical topic covered in many elasticity textbooks (e.g., Kolsky 1963 and Pollard 1977; see also Appendix B). The Navier governing equations are:

$$(\lambda + \mu)u_{j,ij} + \mu u_{i,jj} + \rho f_i \quad (i, j = 1, 2, 3). \tag{3.1}$$

In scalar Cartesian notation, this represents three equations:

$$(\lambda + \mu)\frac{\partial}{\partial x_1}\left(\frac{\partial u_1}{\partial x_1} + \frac{\partial u_2}{\partial x_2} + \frac{\partial u_3}{\partial x_3}\right) + \mu\nabla^2 u_1 + \rho f_x = \rho\frac{\partial^2 u_1}{\partial t^2},$$

$$(\lambda + \mu)\frac{\partial}{\partial x_2}\left(\frac{\partial u_1}{\partial x_1} + \frac{\partial u_2}{\partial x_2} + \frac{\partial u_3}{\partial x_3}\right) + \mu\nabla^2 u_2 + \rho f_y = \rho\frac{\partial^2 u_2}{\partial t^2}, \tag{3.2}$$

$$(\lambda + \mu)\frac{\partial}{\partial x_3}\left(\frac{\partial u_1}{\partial x_1} + \frac{\partial u_2}{\partial x_2} + \frac{\partial u_3}{\partial x_3}\right) + \mu\nabla^2 u_3 + \rho f_z = \rho\frac{\partial^2 u_3}{\partial t^2},$$

where

$$\nabla^2 = \frac{\partial^2}{\partial x_1^2} + \frac{\partial^2}{\partial x_2^2} + \frac{\partial^2}{\partial x_3^2}. \tag{3.3}$$

We introduce the following notation:

$$\nabla = \bar{i}_1 \frac{\partial}{\partial x_1} + \bar{i}_2 \frac{\partial}{\partial x_2} + \bar{i}_3 \frac{\partial}{\partial x_3};$$
$$\bar{u} = (u_1, u_2, u_3). \tag{3.4}$$

In the absence of body forces, the system of equations (3.2) can be expressed in vector form as

$$(\lambda + \mu)\nabla\nabla \bullet \bar{u} + \mu\nabla^2\bar{u} = \rho\frac{\partial^2\bar{u}}{\partial t^2}. \tag{3.5}$$

The solution for the body force problem is discussed for example in Graff (1991). One can use the vector identity

$$\nabla^2\bar{u} = \nabla\nabla \bullet \bar{u} - \nabla \times \nabla \times \bar{u}, \tag{3.6}$$

where

$$\nabla \times \bar{u} = \begin{vmatrix} \bar{i}_1 & \bar{i}_2 & \bar{i}_3 \\ \dfrac{\partial}{\partial x_1} & \dfrac{\partial}{\partial x_2} & \dfrac{\partial}{\partial x_3} \\ u_1 & u_2 & u_3 \end{vmatrix}.$$

By substituting (3.6) into (3.5), the equation of motion can alternatively be expressed as

$$(\lambda + 2\mu)\nabla\nabla \bullet \bar{u} - \mu\nabla \times \nabla \times \bar{u} = \rho\frac{\partial^2\bar{u}}{\partial t^2}. \tag{3.7}$$

The equation of motion can also be expressed in a more simplified form. First, the vector displacement $\bar{u}$ can be expressed via Helmholtz decomposition as the gradient of a scalar and the curl of the zero divergence vector (see Morse and Feshbach 1953):

$$\bar{u} = \nabla\Phi + \nabla \times \bar{H}, \nabla \bullet \bar{H} = 0, \tag{3.8}$$

where Φ and $\bar{H}$ are scalar and vector potentials, respectively. Then, substituting (3.8) into Navier's equation of motion (3.5), we obtain

$$(\lambda + \mu)\nabla\nabla \bullet (\nabla\Phi + \nabla \times \bar{H}) + \mu\nabla^2(\nabla\Phi + \nabla \times \bar{H})$$
$$= \rho\left(\nabla\frac{\partial^2\Phi}{\partial t^2} + \nabla \times \frac{\partial^2\bar{H}}{\partial t^2}\right). \tag{3.9}$$

Next, using (3.6), Equation (3.9) can be regrouped as

$$\left[(\lambda + 2\mu)\nabla\nabla \bullet (\nabla\Phi) - \rho\nabla\frac{\partial^2\Phi}{\partial t^2} \right] - \mu\nabla\times\nabla\times\nabla\Phi$$
$$+ (\lambda + \mu)\nabla\nabla \bullet \nabla\times\bar{H} + \left[\mu\nabla^2\nabla\times\bar{H} - \nabla\times\frac{\partial^2\bar{H}}{\partial t^2} \right] = 0. \tag{3.10}$$

The following identities can be used:

$$\nabla \bullet \nabla\Phi + \nabla^2\Phi; \quad \nabla\times\nabla\times\nabla\Phi = 0; \quad \nabla \bullet \bar{H} = 0. \tag{3.11}$$

Finally, by using (3.11), from (3.10) we have

$$\nabla\left[(\lambda + 2\mu)\nabla^2\Phi - \rho\frac{\partial^2\Phi}{\partial t^2} \right] + \nabla\times\left[\mu\nabla^2\bar{H} - \rho\frac{\partial^2\bar{H}}{\partial t^2} \right] = 0. \tag{3.12}$$

which is satisfied if both terms vanish. This leads to the equations

$$\nabla^2\Phi = \frac{1}{c_L^2}\frac{\partial^2\Phi}{\partial t^2} \quad \text{and} \tag{3.13}$$

$$\nabla^2\bar{H} = \frac{1}{c_T^2}\frac{\partial^2\bar{H}}{\partial t^2}, \tag{3.14}$$

where

$$c_L^2 = \frac{\lambda + 2\mu}{\rho} \quad \text{and} \quad c_T^2 = \frac{\mu}{\rho}. \tag{3.15}$$

As a result, the equation of motion (3.5) is decomposed as two simplified wave equations, (3.13) and (3.14).

3.2.2 Dilatational and Distortional Waves

Now suppose that the rotational part $\nabla\times\bar{H}$ in (3.8) is zero and that

$$\bar{u} = \nabla\Phi. \tag{3.16}$$

In this case, (3.12) and (3.13) give:

$$\nabla^2\bar{u} = \frac{1}{c_L^2}\frac{\partial^2\bar{u}}{\partial t^2}. \tag{3.17}$$

This means that dilatational disturbance propagates with the velocity c_L. In a similar way, suppose that the displacement in (3.8) has *only* a rotational part:

$$\bar{u} = \nabla\times\bar{H}, \nabla \bullet \bar{H} = 0. \tag{3.18}$$

In this case, from (3.12) and (3.14) it follows that:

$$\nabla^2 \bar{u} = \frac{1}{c_T^2} \frac{\partial^2 \bar{u}}{\partial t^2}. \tag{3.19}$$

Equation (3.19) shows that rotational waves propagate with velocity c_T. Equations (3.17) and (3.19) are independent of each other, which means that the longitudinal and shear (or torsional) waves propagate without interaction in unbounded media. These two types of waves are coupled only on the boundary of the elastic body, an obvious consequence of satisfying the boundary conditions.

3.3 The Christoffel Equation for Anisotropic Media

Imagine waves in an infinite elastic anisotropic solid. We shall consider waves in pure crystals, where homogeneity and pure anisotropy is assumed. There are many different ways to approach this problem, but for now we consider the following.

Using indicial (or tensor) notation and Newton's law, set

$$\frac{\partial \sigma_{ik}}{\partial x_k} = \rho \ddot{u}_i \quad \text{or} \quad \rho \ddot{u}_i = \sigma_{ik,k} \tag{3.20}$$

as a governing wave equation, where σ denotes stress. Using Hooke's law, we have

$$\sigma_{ik} = C_{iklm} \varepsilon_{lm} \tag{3.21}$$

as a constitutive equation. Combining (3.20) and (3.21) yields

$$\rho \ddot{u}_i = C_{iklm} \frac{\partial \varepsilon_{lm}}{\partial x_k} = C_{iklm} \varepsilon_{lm,k}$$

From the definition of strain in the strain displacement equations,

$$\varepsilon_{lm} = \frac{1}{2} \left(\frac{\partial u_l}{\partial x_m} + \frac{\partial u_m}{\partial x_l} \right) = \frac{1}{2} (u_{l,m} + u_{m,t}); \tag{3.22}$$

again, combining (3.22) and (3.21), we find the result:

$$\rho \ddot{u}_i = \frac{1}{2} C_{iklm} \left(\frac{\partial^2 u_l}{\partial x_k \partial x_m} + \frac{\partial^2 u_m}{\partial x_k \partial x_l} \right) \tag{3.23}$$

or

$$\rho \ddot{u}_i = \frac{1}{2} C_{iklm} (u_{l,km} + u_{m,kl}). \tag{3.24}$$

Note that C_{iklm} is symmetrical with respect to l and m, so

$$C_{iklm} = C_{iklm} = C_{kilm}.$$

Therefore, we can interchange l and m.

Let us assume plane harmonic traveling waves to see if a solution of this form is possible. Put

$$u_i = A_i \exp\{i(k_j x_j - \omega t)\}, \tag{3.25}$$

where k_j is the wavevector. (Note that $A_i = A\alpha_i$, where the α_i are direction cosines of particle displacement.) Substituting (3.25) into (3.24) gives an eigenvalue problem. From earlier work in harmonic motion, in considering the amplitude of the second derivative we know that $\rho\ddot{u}_i = \rho\omega^2 u_i$. Note that $k_j x_j$ is a single dot product, which is useful in differentiating to obtain

$$\rho\ddot{u}_i = -\rho\omega^2 u_i = -C_{iklm}k_k k_l u_m.$$

Tensor analysis can be used to establish that $C_{iklm}u_{m.kl}$ is equivalent to $C_{iklm}k_k k_l u_m$, with i the free index and with summing over k, l, and m. For one term we can show the equivalence quite easily as follows. If $u_m = A \exp i(k_1 x_1 + k_2 x_2 + k_3 x_3 - \omega t) = A \exp Q$ then it is easy to see that

$$\frac{\partial u_m}{\partial x_1} = A \exp Q k_1 \quad \text{and} \quad \frac{\partial^2 u_m}{\partial x_1 \partial x_2} = A \exp Q k_1 k_2;$$

hence $u_{m.kl} = k_k k_l u_m$ (note that $u_i = u_m \delta_{im}$). Therefore,

$$(\rho\omega^2\delta_{im} - C_{iklm}k_k k_l)u_m = 0. \tag{3.26}$$

This is the famous Christoffel equation for anisotropic media.

The Christoffel acoustic tensor may be defined as

$$\lambda_{im} = \Gamma_{im} = C_{iklm}n_k n_l, \tag{3.27}$$

where n_k are direction cosines of the normal to the wavefront (since $k_k = kn_k$, $k_l = kn_l$, and $c^2 = \omega^2/k^2$; we will eventually solve for the wave velocity c). Therefore, from $(C_{iklm}k_k k_l - \rho\omega^2\delta_{im})u_m = 0$ we have

$$(\Gamma_{im}k^2 - \rho\omega^2\delta_{im})u_m = 0 \quad \text{or} \quad (\Gamma_{im} - \rho c^2\delta_{im})u_m = 0.$$

This gives us three homogeneous equations, three real roots, and three different velocities, all from the cubic equation in c^2; this leads to an orthogonal classical eigenvalue problem.

For a nontrivial solution, we must set the determinant of the coefficient matrix equal to zero:

$$|\Gamma_{im} - \rho c^2\delta_{im}| = 0. \tag{3.28}$$

Because Γ_{im} depends on crystal symmetry and the orientation of the waves, we have

$$\begin{vmatrix} (\lambda_{11} - \rho c^2) & \lambda_{12} & \lambda_{13} \\ \lambda_{21} & (\lambda_{22} - \rho c^2) & \lambda_{23} \\ \lambda_{31} & \lambda_{32} & (\lambda_{33} - \rho c^2) \end{vmatrix} = 0.$$

where $\lambda_{11}, \lambda_{12}, \lambda_{13}, \ldots$ are obtained from the expression for the acoustic tensor. Recall from (3.27) that $\lambda_{im} = C_{iklm}n_k n_l$.

Expand this acoustic tensor carefully for all elements of the matrix. The indices i and m are free; k and l are summers:

$$k = 1 \qquad k = 2 \qquad k = 3$$

$$
\begin{aligned}
\lambda_{11} = {} & C_{1111}n_1 n_1 + C_{1211}n_2 n_1 + C_{1311}n_3 n_1 & l = 1 \\
& + C_{1121}n_1 n_2 + C_{1221}n_2 n_2 + C_{1321}n_3 n_2 & l = 2 \\
& + C_{1131}n_1 n_3 + C_{1231}n_2 n_3 + C_{1331}n_3 n_3. & l = 3
\end{aligned}
$$

For ease of expansion, specify i and m as $11, 12, 13, 21, 22, 23, 31, 32$, or 33 for the nine components. Consider $l = 1, l = 2$, and $l = 3$ for the first, second, and third rows, with $k = 1, k = 2$, and $k = 3$ for the first, second, and third columns.

We can simplify the results even further by converting C_{ikjl} to C_{nm} as follows: if $i = k$ then $n = i$, and if $j = l$ then $m = j$; if $i \neq k$ then $n = 9 - (i + k)$, and if $j \neq l$ then $m = 9 - (j + l)$. Therefore,

$$
\begin{aligned}
& C_{1111} = C_{11}, C_{1211} = C_{61}, C_{1311} = C_{51}, \\
& C_{1121} = C_{16}, C_{1221} = C_{66}, C_{1321} = C_{56}, \\
& C_{1131} = C_{15}, C_{1231} = C_{65}, C_{1331} = C_{55}.
\end{aligned}
$$

As an example,

$$
\begin{aligned}
\lambda_{11} = {} & C_{11}n_1^2 + C_{16}n_2 n_1 + C_{15}n_3 n_1 \\
& + C_{16}n_1 n_2 + C_{66}n_2^2 + C_{56}n_3 n_2 \\
& + C_{15}n_1 n_3 + C_{65}n_2 n_3 + C_{55}n_3^2.
\end{aligned}
$$

Continuing with these computations would allow us to plot a phase profile. Given constants and directions, various c can be calculated in the wavevector $\bar{k}$ directions. This is a tedious process that calls for an efficient computer program.

Once the phase velocity values are extracted for specific directions, we would like to see if we have a pure mode – that is, if the particle velocity direction is aligned perfectly with the chosen $\bar{k}$ direction for the phase velocity computation. To see if there are pure modes, we must expand and extract all roots of the bi-cubic equation. For a given direction $\bar{k}$ of propagation, there are three waves, $k_l = kn_l$, possibly with mutually perpendicular displacement vectors. It may be possible to find a special direction with one pure longitudinal and two pure transverse waves. Toward this end, we will find the direction cosines α_i of the particle displacements.

Substitute each value of c^2 back into the system of original homogeneous equations to calculate the eigenvectors:

$$
\begin{bmatrix}
(\lambda_{11} - \rho c^2) & \lambda_{12} & \lambda_{13} \\
\lambda_{21} & (\lambda_{22} - \rho c^2) & \lambda_{23} \\
\lambda_{31} & \lambda_{32} & (\lambda_{33} - \rho c^2)
\end{bmatrix}
\begin{Bmatrix}
\alpha_1 \\
\alpha_2 \\
\alpha_3
\end{Bmatrix} = 0.
$$

This yields

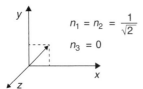

Figure 3.1. Plane waves in the $[1, 1, 0]$ direction.

$$(\lambda_{11} - \rho c^2)\alpha_1 + \lambda_{12}\alpha_2 + \lambda_{13}\alpha_3 = 0,$$
$$\lambda_{21}\alpha_1 + (\lambda_{22} - \rho c^2)\alpha_2 + \lambda_{23}\alpha_3 = 0, \qquad (3.29)$$
$$\lambda_{31}\alpha_1 + \lambda_{32}\alpha_2 + (\lambda_{33} - \rho c^2)\alpha_3 = 0.$$

Solve three times for each value of c^2 (the result can be checked, since A is orthogonal). We can arrange the solution as follows:

$$A = \begin{array}{ccc} c_1^2 & c_2^2 & c_3^2 \\ \begin{bmatrix} \alpha_{11} & \alpha_{12} & \alpha_{13} \\ \alpha_{21} & \alpha_{22} & \alpha_{23} \\ \alpha_{31} & \alpha_{32} & \alpha_{33} \end{bmatrix} \end{array}.$$

Note that determinant $A = +1$ for a right-handed Cartesian coordinate system.

3.3.1 Sample Problem

How can we determine phase velocity in a specific direction for a given level of anisotropy? More particularly, we shall discuss two problems with the aid of C_{ij} matrices as presented in elasticity theory (for waves that are orthotropic, hexagonal, isotropic, etc.; see Appendix B).

(1) Calculate directions for a pure longitudinal wave and two pure transverse waves for a cubic crystal. *Note:* pure longitudinal, $\bar{u} \times \bar{n} = 0$; pure transverse, $\bar{u} \cdot \bar{n} = 0$.
(2) Develop equations for c_1, c_2, and c_3 as a function of the wavevector $\bar{k}$ directions in the (x_1, x_2)-plane. *Note:* $\bar{n}$ represents the direction cosine of the angle between the wavevector and the x_1, x_2, and x_3 axes. Therefore, we compute Γ_{im} for a cubic crystal. When we solve the determinant for specific $\bar{n}$, we must let $\bar{n}$ vary in the plane for small increments. In general, waves are not pure L or pure S and so are often termed quasi-longitudinal or quasi-shear. However, if $\bar{k}$ is an eigenvector of λ_{ik} then the waves are pure.

In order to solve these problems, a specific wavevector must now be considered. Therefore, imagine plane waves in the $[1, 1, 0]$ direction for a cubic crystal (see Figure 3.1). Next, evaluate the terms of the acoustic tensor,

$$\lambda_{im} = C_{iklm}n_k n_l,$$

where i and m are free indices with double summations over k and l from 1 to 3. Using tensor-to-matrix notation yields

$$11 \rightarrow 1, \quad 22 \rightarrow 2, \quad 33 \rightarrow 3, \quad 23, 32 \rightarrow 4, \quad 31, 13 \rightarrow 5, \quad 12, 21 \rightarrow 6.$$

The elastic constant matrix for a cubic crystal may be found in Appendix B:

$$n_1 = n_2 = \frac{1}{\sqrt{2}}$$

$$n_3 = 0$$

Therefore,

$$\begin{bmatrix} C_{11} & C_{12} & C_{12} & 0 & 0 & 0 \\ C_{12} & C_{11} & C_{12} & 0 & 0 & 0 \\ C_{12} & C_{12} & C_{11} & 0 & 0 & 0 \\ 0 & 0 & 0 & C_{44} & 0 & 0 \\ 0 & 0 & 0 & 0 & C_{44} & 0 \\ 0 & 0 & 0 & 0 & 0 & C_{44} \end{bmatrix}.$$

$$\lambda_{11} = C_{11}n_1^2 + C_{66}n_2^2 + C_{55}n_3^2 + 2C_{56}n_2n_3 + 2C_{15}n_1n_3 + 2C_{16}n_1n_2,$$
$$\lambda_{12} = C_{16}n_1^2 + C_{26}n_2^2 + C_{45}n_3^2 + (C_{46} + C_{25})n_2n_3$$
$$+ (C_{14} + C_{56})n_1n_3 + (C_{12} + C_{66})n_1n_2,$$

$\lambda_{21} = \lambda_{12}$, and so forth.

By Christoffel's equations we have $(\lambda_{im} - \delta_{im}\rho c^2)u_m = 0$ and so

$$\begin{vmatrix} \lambda_{11} - \rho c^2 & \lambda_{12} & \lambda_{13} \\ \lambda_{12} & \lambda_{22} - \rho c^2 & \lambda_{23} \\ \lambda_{13} & \lambda_{23} & \lambda_{33} - \rho c^2 \end{vmatrix} = 0.$$

Using $\lambda_{11}, \lambda_{12}, \dots$ and $n_1 = n_2 = 1/\sqrt{2}$ and $n_3 = 0$, we reach:

$$\begin{vmatrix} \frac{1}{2}(C_{11} + C_{44}) - \rho c^2 & \frac{1}{2}(C_{12} + C_{44}) & 0 \\ \frac{1}{2}(C_{12} + C_{44}) & \frac{1}{2}(C_{11} + C_{44}) - \rho c^2 & 0 \\ 0 & 0 & C_{44} - \rho c^2 \end{vmatrix} = 0;$$

therefore,

$$\left[\frac{1}{2}(C_{11} + C_{44}) - \rho c^2\right]\left[\frac{1}{2}(C_{11} + C_{44}) - \rho c^2\right][C_{44} - \rho c^2]$$
$$- \left[\frac{1}{2}(C_{12} + C_{44})\right]\left[\frac{1}{2}(C_{12} + C_{44})\right][C_{44} - \rho c^2] = 0.$$

Through factoring, we solve for three c; the eigenvector solution is then required to know which wave is in which direction or alignment with $\overline{k}$. Since $c_i = \sqrt{C_{44}/\rho}$.

$$\left(\frac{1}{2}C_{11} + \frac{1}{2}C_{44} - \rho c^2\right) \pm \left(\frac{1}{2}(C_{12} + C_{44})\right) = 0.$$

The solutions are:

$$c_1 = \sqrt{\frac{C_{11} + C_{12} + 2C_{44}}{2\rho}}, \quad c_2 = \sqrt{\frac{C_{44}}{\rho}}, \quad c_3 = \sqrt{\frac{C_{11} - C_{12}}{2\rho}}.$$

To confirm wave type and character (pure or quasi), we must solve for the three eigenvectors:

$$\begin{bmatrix} \frac{1}{2}(C_{11} + C_{44}) - \rho c^2 & \frac{1}{2}(C_{12} + C_{44}) & 0 \\ \frac{1}{2}(C_{12} + C_{44}) & \frac{1}{2}(C_{11} + C_{44}) - \rho c^2 & 0 \\ 0 & 0 & C_{44} - \rho c^2 \end{bmatrix} \begin{Bmatrix} u_x \\ u_y \\ u_z \end{Bmatrix} = 0.$$

One wave in closest alignment with $\bar{n}$ is the pure or quasi-longitudinal wave; the other two are shear or quasi-shear. Although the task is tedious, we must solve the equations three times (for u_x, u_y, and u_z) for each c_i. (*Note:* If $\bar{u}_1 \times \bar{n} = 0$ then we have a pure l wave, or use $\bar{u} \times \bar{n} = |u||n| \sin \theta$. If $\bar{u}_2 \bullet \bar{n} = 0$ or $\bar{u}_3 \bullet \bar{n} = 0$ then we have a pure transverse wave.) Consequently, if α_i are direction cosines then we can solve the following.

(1) For c_1: $\alpha_1 = \alpha_2 = 1/\sqrt{2}$ and $\alpha_3 = 0$; therefore, c_1 is a pure l wave. (In the $[1, 1, 0]$ direction, the direction cosine vector is in the same direction as the wavevector.)

(2) For c_2: $\alpha_3 = 1$ and $\alpha_1 = \alpha_2 = 0$. Therefore, when considering the dot product $[0, 0, 1] \bullet [1, 1, 0] = 0$, we see that c_2 is pure shear.

(3) Finally, for c_3: $\alpha_1 = \alpha_2 = 1/\sqrt{2}, \alpha_2 = -1/\sqrt{2}$, and $\alpha_3 = 0$. The dot product is again $[1, -1, 0] \bullet [1, 1, 0] = 0$; this is also a pure shear wave.

These results may be confirmed by checking the eigenvector direction that leads to three mutually perpendicular directions.

In general, we can solve the eigenvector problem by solving three times for each c (and hence the direction cosines α_i for each c):

$$\alpha_1 \lambda_{11} + \alpha_2 \lambda_{12} + \alpha_3 \lambda_{13} = \alpha_1 \rho c^2,$$
$$\alpha_1 \lambda_{21} + \alpha_2 \lambda_{22} + \alpha_3 \lambda_{23} = \alpha_2 \rho c^2,$$
$$\alpha_1 \lambda_{31} + \alpha_2 \lambda_{32} + \alpha_3 \lambda_{33} = \alpha_3 \rho c^2.$$

This general solution was derived by Chistoffel many years ago.

Sample problems can be studied for a variety of different materials. Stiffness coefficients for a few selected materials can be found in Table 3.1.

Table 3.1. *Stiffness coefficients C_{ij} for selected materials*

Material	Stiffness (10^{10} N/m^2)											
	C_{11}	C_{22}	C_{33}	C_{44}	C_{55}	C_{66}	C_{12}	C_{13}	C_{14}	C_{16}	C_{23}	C_{25}
Barium titanate[a][b]	15.0		14.6	4.4			6.6	6.6				
Diamond	102			49.2			25					
Lead titanate–zirconate (PZT-2)[a][b]	13.5		11.3	2.22			6.79	6.81				
Quartz[a]	8.674		10.72	5.794			0.699	1.191	1.791			
Rochelle salt[a]	2.8	4.14	3.94	0.666	0.285	0.96	1.74	1.50			1.97	
Sapphire	49.4		49.6	14.5			15.8	11.4	−2.3			
Cadmium sulfide[a]	9.07		9.38	1.504			5.81	5.10				
Germanium	12.89			6.71			4.83					
Silicon	16.57			7.956			6.39					
Aluminum, crystal	10.80			2.85			6.13					
Aluminum, polycrystal	11.1			2.5								
Gold, crystal	18.6			4.20			15.7					
Titanium, crystal	16.2		18.1	4.67			9.2	6.9				
Titanium, polycrystal	16.59			4.4								
Tungsten, crystal	50.2			15.2			19.9					
Tungsten, polycrystal	58.1			13.4								
Graphite/epoxy (AS4/3501–6)	16.209	1.63			0.774		0.671				0.795	

Sources: Most data from Auld (1990). Data for graphite/epoxy from Mal, Lih, and Bar-Cohen (1994).
[a] Piezoelectric.
[b] Poled ceramic. The stiffness matrix has the same form as for the hexagonal crystal system, with Z along the poling axis.

45

3.4 On Velocity, Wave, and Slowness Surfaces

We will now contemplate several possible phase velocity profiles for anisotropic media, all nonspherical in nature. Several possible velocity surfaces are illustrated in Figure 3.2.

If we are given specific phase velocity values as a function of wavevector direction or a specific angle θ then we can visualize a slowness profile, which is simply a plot of $1/c_p$ versus $\bar{k}$ or θ. Consider the sketch in Figure 3.3, which was generated by solving the Christoffel equation discussed previously. Useful information can then be extracted from the resulting curve.

If we are given the wave surface and the $\bar{k}$ or θ direction, then we can calculate c_p and ϕ. For example, the energy velocity and group velocity vectors are the same for lossless media. This can be seen by examining a normal to the slowness profile – as illustrated in Figure 3.3, which also gives us the skew angle ϕ. If we were now able to plot (on a point-by-point basis) a locus of points as a function of θ, we would be able to produce group velocity and skew angle as a function of θ. If we were instead to plot c_E versus the sum of θ plus ϕ (which we call ψ), then we would obtain the wave surface term illustrated in Figure 3.4. This gives us the group or energy velocity variation with angle (with no mention of launch angle if we didn't know θ and ϕ).

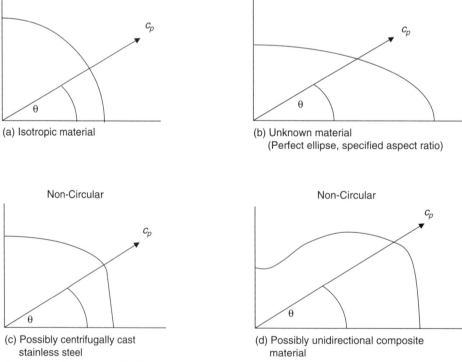

Figure 3.2. Sample longitudinal or shear wave velocity surfaces in anisotropic media. (Polar representation of c_p and θ)

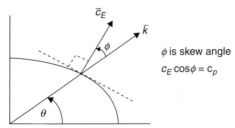

ϕ is skew angle

$c_E \cos\phi = c_p$

Figure 3.3. Sample slowness profile of $1/c_p$ versus θ.

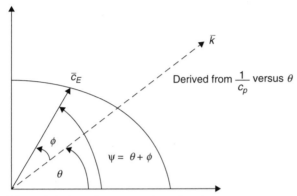

Derived from $\dfrac{1}{c_p}$ versus θ

$\psi = \theta + \phi$

Figure 3.4. Sample wave surface showing c_E versus ψ.

For some physical insight, imagine the actual wave surface propagating from an acoustic source, sending waves in one particular direction that could be sensed with some sensor for a field distribution. As a result of a superposition of all of the phase velocity contributions from different directions, if the source had some finite size then the energy velocity vector would be normal to the slowness profile for the material. See Love (1926), Musgrave (1959), or Pollard (1977) for more details.

We can now explore briefly the fact that the group velocity is normal to the slowness profile. For many different kinds of problems, we'll be extracting information similar to that found to produce a frequency spectrum of $\omega(\bar{k})$. For example, $\omega(k_x)$ is obtained for the one-dimensional string problem; $\omega(k_x, k_y)$ is the result for the anisotropic media problem in one plane. In general, $\omega(k_x, k_y, k_z)$ could be obtained for a three-dimensional problem.

Let's consider $\omega(k_x, k_y)$. This result could be plotted in three dimensions: for a particular ω value, a plane parallel to the (k_x, k_y)-plane could be used to intersect the conical-like $\omega(k_x, k_y)$-surface. For anisotropic media, an intersection plane parallel to the (k_x, k_y)-plane does not produce a circle; see Figure 3.5.

Any ω could be selected, since only a scale factor would be changed and the general shape of the curve would be the same. Note that the line drawn from the origin to a point in question would give c_p in that particular direction. From three-dimensional calculus, the normal to the tangent plane to the $\omega(k_x, k_y)$ function at the point of interest would have a slope equal to $(\partial\omega/\partial k_x, \partial\omega/\partial k_y, 1)$. The intersection of the plane could achieve the slowness curve, $(1/c_p)(k_x, k_y)$; this projection onto the (k_x, k_y)-plane is what we normally extract from the Christoffel equation solution

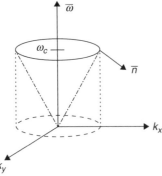

Figure 3.5. Typical intersection of the $\omega(k_x, k_y)$-surface with a plane parallel to the (k_x, k_y)-plane, showing slowness profile and projection onto the (k_x, k_y)-plane.

results. The projection of $\bar{n}$ onto the (k_x, k_y)-plane or dot product is simply $(\partial\omega/\partial k_x, \partial\omega/\partial k_y, 0)$, which we recognize as the group velocity.

It is easy to show that this group velocity vector direction is normal to the slowness profile. Imagine an intersection of a plane parallel to the (k_x, k_y)-plane with the tangent plane to the $\omega(k_x, k_y)$ function at the point in question. This projects as a line onto (k_x, k_y)-space that is tangent to the curve $(1/c_p)(k_x, k_y)$. Consider $\bar{t}$ as the tangent line with slope $(t_x, t_y, 0)$. Then $\bar{t} \bullet \bar{n} = 0$ and

$$t_x\left(\frac{\partial\omega}{\partial k_x}\right) + t_y\left(\frac{\partial\omega}{\partial k_y}\right) + 0 = 0.$$

An alternative argument is presented by Auld (1990) whereby $\bar{k} + \Delta\bar{k}$ is considered in a limit process to show that c_g is normal to the slowness profile.

In the wave propagation diagrams of Figure 3.6, wave packets are sent out in all directions but the wavevector is in the phase velocity direction $\bar{k}$ only. If we view (via Huygens's principle) the acoustic source as a series of point sources, then potential constructive interference path directions are as illustrated in Figure 3.6. Note that an infinite number of point sources is contemplated for a plane wave. Waves across the face of the transducer emanate in all directions. If the velocity surface, or phase velocity profile, is symmetric with respect to the direction $\bar{D}$ (sometimes called the *director*), giving us a sense of the orientation of an anisotropic material, then there would be no skew angle in that particular direction and the phase velocity would be equal to the energy velocity. On the other hand, if the wavevector $\bar{k}$ is *not* in line with the director of the material then the wave interference pattern becomes much more difficult to evaluate. The waves propagating from each point source are nonspherical in nature, but in this case are inclined at an angle Γ. The selected wave path for energy concentration will be (a), (b), (c), or some other path, depending on the actual phase velocity profile, the inclined angle Γ, and the resulting interference patterns.

Consequently, from a piston or point source (for example), three wave surfaces are produced – each with its own phase velocity variation with direction. The faster wave surface is for longitudinal waves, followed by two shear wave surfaces, possibly

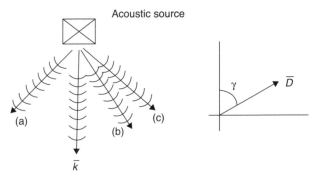

Figure 3.6. Possible directions of wave propagation from an acoustic source in anisotropic media.

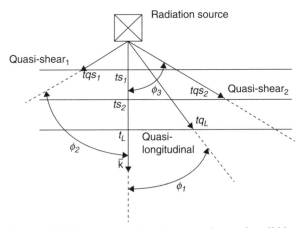

Figure 3.7. Wave propagation into an anisotropic solid in a particular wavevector direction.

with different phase velocity profiles. All of this is possible owing to interference phenomena and to the changes taking place as a result of wave velocity variations with direction, which modify the interference patterns.

For a particular direction $\bar{k}$, consider the sketch in Figure 3.7. The distances illustrated to times ts_1, ts_2, and t_L are proportional to the phase velocities for the directions $\bar{k}$. The three skew angles are also shown. The distances tqs_1, tqs_2, and tq_L are proportional to the group velocities for the three wave types.

A number of interesting presentations can now be made. For example, examining in Figure 3.8 a wave surface profile of c_E versus $\theta + \phi$ shows us how to extract θ and the phase velocity value if needed. Because $\bar{k}$ is normal to the surface at the point in question for a particular $\bar{c}_E$, all information is available. If we wanted to achieve a certain energy velocity value – or, more importantly, a specific energy velocity direction – this technique shows us how to place a transducer at a specified θ or phase velocity direction, as illustrated in the figure.

That is, if we are given $\bar{c}_E$ or the angle of $\bar{c}_E$, we can find out what transducer angle could be used to achieve this. We need only carry out the following steps, which are illustrated in Figure 3.8.

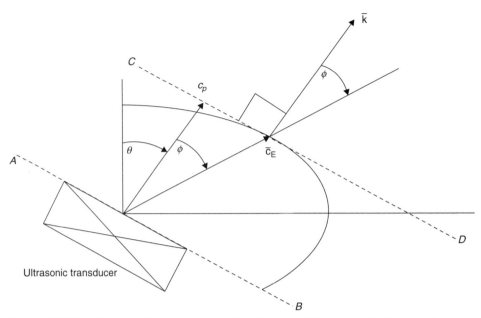

Figure 3.8. Transducer location angle calculation for a specific energy velocity magnitude or direction.

(1) Plot CD tangent to wave surface at $\bar{c}_E$ tip.
(2) Plot AB parallel to CD.
(3) We now know θ and c_p (where $c_p = |\bar{c}_E| \cos \varphi$) and hence the $\bar{k}$ direction from known θ.
(4) The angle φ between $\bar{k}$ and the $\bar{c}_E$ direction is also known and plotted.

3.5 Exercises*

1. Derive the following expressions for the velocities of longitudinal and transverse waves traveling in the $[1, 1, 1]$ direction in a cubic crystal:
$$c_1 = \sqrt{(C_{11} + 2C_{12} + 4C_{44})/3\rho},$$
$$c_2 = c_3 = \sqrt{(C_{11} - C_{12} + C_{44})/3\rho}.$$

2. For Problem 1, define the wave types c_1, c_2, c_3 as pure or quasi-longitudinal and as pure or quasi-shear.

3. Verify the formula
$$c_1 = \sqrt{C_{11}/\rho}, \quad c_2 = \sqrt{C_{44}/\rho}, \quad c_3 = \sqrt{(C_{11} - C_{12})/2\rho}$$

for longitudinal and shear waves traveling in the $[1,0,0]$ direction in a hexagonal crystal.

4. Carefully list all steps necessary to produce a computer program that calculates points on a curve of phase velocity versus angle in the (x, y)-plane for a transversely isotropic material. What assumptions are made in the analysis? How can you define the modes and the pure or quasi nature of the modes? Is the acoustic tensor symmetric?

5. For a cubic tungsten crystal, plot approximately the three slowness profile curves.

---

* To solve problems 15 through 18, it might be necessary to study Section 4.5 in the book.

1 inch = .25 μsec/mm

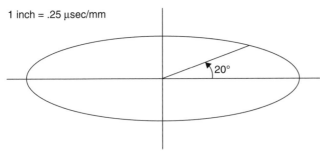

Figure 3.9. Exercise 17.

6. From the cubic tungsten slowness profile, graphically estimate the wave surface of $\bar{c}_E$ versus ψ or $\theta + \varphi$ for the quasi-longitudinal case.
7. Graphically estimate skew angle φ versus θ for the quasi-longitudinal case in cubic tungsten.
8. What angle θ is required to produce a resulting energy velocity at 45° for the quasilongitudinal case in cubic tungsten?
9. Calculate the resulting phase velocity in the $[1,1,0]$ direction for transversely isotropic unidirectional graphite epoxy.
10. For a plane wave in a transversely isotropic crystal, derive an expression for the acoustic tensor and the wave velocities in the $[1,0,0]$ direction.
11. For cubic silicon, develop a velocity surface of c_1, c_2, and c_3 as a function of wavevector direction in the (x_1, x_2)-plane. Use at least three points followed by interpolating estimates.
12. Calculate the phase velocity in a particular direction for transversely isotropic titanium or a barium titanate crystal.
13. Calculate the phase velocity in a particular direction for orthotropic barium sodium niobate. $[1,1,0]$
14. Solve for the elastic constants of a cubic diamond material if the wave velocities were measured for plane waves in the $[1, 1, 0]$ direction as follows: $c_1 = 17{,}890$ m/s; $c_2 = 11{,}930$ m/s; $c_3 = 10{,}330$ m/s.
15. For Plexiglas over tungsten, determine all critical angles. Do likewise for tungsten over Plexiglas.
16. For typical slowness profiles in isotropic media say 1 and 2, measure critical angles for L and S input and also for total reflection (1 to 2 and 2 to 1). Show on a sketch.
17. (a) For the slowness profile shown in anisotropic media (Figure 3.9), calculate the skew angle and energy velocity at the wavevector direction shown.
    (b) Estimate the energy velocity value in the wavevector direction shown.
    (c) How could this result be used to produce a wave surface?
18. Given an ultrasonic transducer emitting plane wave segments on the surface of an anisotropic medium with the slowness curves shown in Figure 3.10, determine the ideal location (of a receiving transducer in through-transmission) to receive maximum energy from quasi-longitudinal wave QL and from quasi-transverse waves QT_1 and QT_2.

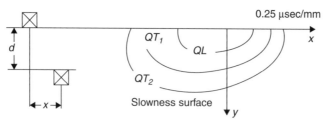

Figure 3.10. Exercise 18.

3.6 REFERENCES

Auld, B. A. (1990). *Acoustic Fields and Waves in Solids*, 2nd ed., vols. 1 and 2. Malabar, FL: Kreiger.

Kolsky, H. (1963). *Stress Waves in Solids*. New York: Dover.

Love, A. E. H. (1926). *Some Problems of Geodynamics*. Cambridge University Press.

Mal, A. K., Xu, P. C., and Bar-Cohen, Y. (1990). *Leaky Lamb waves for the ultrasonic nondestructive evaluation of adhesive bonds*, J. Eng. Mat. Tech. 112: 255-9.

Musgrave, M. J. P. (1959). *The Propagation of Elastic Waves in Crystals and Other Anisotropic Media*. Teddington, Middlesex: National Physical Laboratory.

Pollard, H. F. (1977). *Sound Waves in Solids*. London: Pion Ltd.

4 Reflection and Refraction

4.1 Introduction

Wave reflection and refraction considerations are fundamental to the study of stress wave propagation in solids. This chapter presents basic concepts with an emphasis on physical phenomena. In this chapter we examine normal beam incidence reflection factors as well as computation of refraction angles. Reflection factor concepts are outlined first, followed by angle beam analysis and mode conversion as an ultrasonic wave encounters an interface between two materials. For more details, see Auld (1990), Brekhovskikh (1960), Graff (1991), or Kolsky (1963).

4.2 Normal Beam Incidence Reflection Factor

A plane wave encountering an interface between two materials is divided into two components: some energy at the interface is transmitted and some is reflected. The formula allowing us to compute reflection factor at an interface for normal incidence is presented in Figure 4.1.

This equation can be derived by matching normal stress at the interface as well as by matching displacement or particle velocity. Consider an incident harmonic plane wave σ_I traveling in an x direction to an interface between two media, as shown in Figure 4.1. Stress is reflected σ_R and transmitted σ_T. Since the elastic field is independent of the y direction, all derivatives with respect to y will vanish from the equations of motion. In this very simple case, the governing equation is the simple wave equation, which is applicable for either longitudinal or shear waves:

$$\frac{\partial^2 u_x}{\partial x^2} = \frac{1}{c_L^2} \frac{\partial^2 u_x}{\partial t^2} \tag{4.1}$$

or

$$\frac{\partial^2 u_y}{\partial x^2} = \frac{1}{c_T^2} \frac{\partial^2 u_y}{\partial t^2}, \tag{4.2}$$

where u_x and u_y are displacement vector components along the x- and y-axis, respectively.

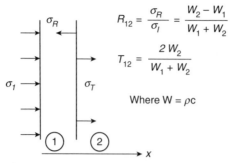

$$R_{12} = \frac{\sigma_R}{\sigma_I} = \frac{W_2 - W_1}{W_1 + W_2}$$

$$T_{12} = \frac{2 W_2}{W_1 + W_2}$$

Where $W = \rho c$

Figure 4.1. Reflection factor.

We will now address the compressional wave. The solution for a compressional harmonic wave in Equation (4.1) is

$$u_x = A_1 e^{i(kx-\omega t)} + A_2 e^{-i(kx+\omega t)}, \tag{4.3}$$

where

$$\text{wavenumber } k = \frac{\omega}{c_L}, \; c_T^2 = \frac{\mu}{\rho}, \; \text{ and } c_L^2 = \frac{\lambda+2\mu}{\rho}. \tag{4.4}$$

The first term in (4.3) describes wave propagation in the positive x direction while the second term describes propagation in the negative direction.

Consider the incident waveform as

$$u_x^I = I e^{i(k_1 x-\omega t)}, \; k_1 = \frac{\omega}{c_L^{(1)}}. \tag{4.5}$$

In this case, the reflected field can be written as

$$u_x^{(R)} = A_R e^{-i(k_1 x+\omega t)}. \tag{4.6}$$

The transmitted field in the second medium is

$$u_x^{(T)} = A_T e^{i(k_2 x-\omega t)}, \; k_2 = \frac{\omega}{c_L^{(2)}}; \tag{4.7}$$

here, A_R and A_T are unknown.

We must satisfy boundary conditions on the interface. The entire elastic field in medium 1 is

$$u^1 = u_x^{(I)} + u_x^{(R)} = I e^{i(k_1 x-\omega t)} + A_R e^{-i(k_1 x+\omega t)}. \tag{4.8}$$

In medium 2,

$$u^2 = u_x^{(T)} = A_T e^{i(k_2 x-\omega t)}. \tag{4.9}$$

Boundary conditions are as follows:

$$u^{(1)}\big|_{x=0} = u^{(2)}\big|_{x=0}, \tag{4.10}$$

$$\sigma_x^{(1)}\big|_{x=0} = \sigma_x^{(2)}\big|_{x=0}, \tag{4.11}$$

$$\sigma_{xy}^{(1)}\big|_{x=0} = \sigma_{xy}^{(2)}\big|_{x=0}. \tag{4.12}$$

For the one-dimensional case, the generalized Hooke's law is

$$\sigma_x = (\lambda + 2\mu)\frac{\partial u_x}{\partial x}, \tag{4.13}$$

$$\sigma_{xy} = 0. \tag{4.14}$$

By substituting (4.8) and (4.9) into (4.10), we obtain

$$I + A_R = A_T. \tag{4.15}$$

From (4.8), (4.9), and (4.13), it follows that

$$\sigma_x^{(1)} = i(\lambda_1 + 2\mu_1) \cdot k_1 [Ie^{i(k_1 x - \omega t)} - A_R e^{-i(k_1 x + \omega t)}], \tag{4.16}$$

$$\sigma_x^{(2)} = i(\lambda_2 + 2\mu_2) \cdot k_2 \cdot A_T e^{i(k_2 x - \omega t)}. \tag{4.17}$$

Substituting (4.16) and (4.17) into (4.11) yields

$$(\lambda_1 + 2\mu_1) \cdot k_1 [I - A_R] = (\lambda_2 + 2\mu_2)k_2 A_T, \tag{4.18}$$

$$\text{wavenumber } k_n = \frac{\omega}{c_L^n} \ (n = 1,\, 2), \tag{4.19}$$

$$\lambda_n + 2\mu_n = \rho_n \cdot [c_L^{(n)}]^2, \tag{4.20}$$

where n is the medium number. Finally, using (4.19) and (4.20), equations (4.15) and (4.18) lead to the following system:

$$\begin{aligned} I + A_R &= A_T \\ \rho_1 c_L^{(1)}(I - A_R) &= \rho_2 c_L^{(2)} \cdot A_T. \end{aligned} \tag{4.21}$$

Solution of this system gives

$$A_R = \frac{\rho_1 c_L^{(1)} - \rho_2 c_L^{(2)}}{\rho_1 c_L^{(1)} + \rho_2 c_L^{(2)}} \cdot I, \tag{4.22}$$

$$A_T = \frac{2\rho_1 c_L^{(1)}}{\rho_1 c_L^{(1)} + \rho_2 c_L^{(2)}} \cdot I. \tag{4.23}$$

The reflected and transmitted wave fields can be obtained from (4.6) and (4.7) by using expressions (4.22) and (4.23).

We may now consider incident, reflected, and transmitted stresses. From (4.13) it follows that

$$\sigma_x^{(1)} = (\lambda_1 + 2\mu_1)\frac{\partial u_x^{(1)}}{\partial x} = ik_1(\lambda_1 + 2\mu_1)e^{i(k_1 x - \omega t)} \cdot I_0$$

(4.24)

Similarly, the reflected stress field is obtained as

$$\sigma_x^{(R)} = -ik_1(\lambda_1 + 2\mu_1)e^{-i(k_1 x + \omega t)} \cdot A_R$$

(4.25)

The transmitted wave field is presented in Equation (4.17). The stress reflection and transmission coefficients are therefore obtained from (4.17), (4.24), and (4.25) as follows:

$$R = \left.\frac{\sigma_x^{(R)}}{\sigma_x^{(1)}}\right|_{x=0} = -\frac{A_R}{I} = -\frac{\rho_1 c_1 - \rho_2 c_2}{\rho_1 c_1 + \rho_2 c_2},$$

(4.26)

$$T = \left.\frac{\sigma_x^{(T)}}{\sigma_x^{(1)}}\right|_{x=0} = \frac{(\lambda_2 + 2\mu_2)k_2}{(\lambda_1 + 2\mu_1)k_1} \cdot \frac{A_T}{I} = +\frac{2\rho_2 c_2}{\rho_1 c_1 + \rho_2 c_2}.$$

(4.27)

Defining acoustic impedance as $W = \rho c_L$, the reflection coefficient can thus be written as

$$R = \frac{W_2 - W_1}{W_1 + W_2};$$

(4.28)

the transmission coefficient is

$$T = \frac{2W_2}{W_1 + W_2}.$$

(4.29)

The formula for energy partition at the interface into transmission and reflection modes can be derived by considering energy as proportional to the square of the pressure magnitude. The energy flow per unit of time through a unit area normal to the direction of propagation is defined as the intensity of the wave (I_I, I_R, or I_T). The intensity is evaluated over one cycle and depends on the amplitude as follows (see Pollard 1977):

$$I_R = \frac{A_R^2}{2\rho_1 c_L^{(1)}}, \quad I_T = \frac{A_T^2}{2\rho_2 c_L^{(2)}}, \quad I_1 = \frac{I^2}{2\rho_1 c_1^{(L)}}.$$

(4.30)

Equations (4.26), (4.27), and (4.30) then yield

$$\frac{I_R}{I_I} = \left(\frac{B-1}{B+1}\right)^2 \text{ and } \frac{I_T}{I_I} = \frac{4B}{(B+1)^2},$$

where $B = W_2/W_1$.

The wave velocity and acoustic impedance values of ultrasound in various materials are listed in Table 4.1; sample reflection factor results for various materials are presented in Table 4.2. The expression for acoustic impedance, W, can be considered as a characteristic property of the material being studied:

Table 4.1. Wave velocity values for selected material samples

Material	Density (g/cm³)	Wave velocity (m/s)	
		Longitudinal	Shear
Air	0.001	330	—
Aluminum	2.7–2.8	6,250–6,350	3,100
Beryllium	1.82	12,800	8,710
Bone	1.738	2.240 ± 8%	—
Brass	8.1	4,430	2,120
Bronze	8.86	3,530	2,230
Cast iron	7.7	4,500	2,400
Chocolate (dark)	1.302	2,584	960
Copper	8.9	4,660	2,260
Cork	0.24	510	—
Glass	2.23–2.51	5,570–5.770	3,430–3,440
Glycerin	1.261	1,920	—
Gold	19.3	3,240	1,200
Ice	1.00	3,980	1,990
Lead	11.4	2,160	700
Lead zirconate titanate	7.65	3,791	—
Magnesium	1.74	5,790	3,100
Nickel	8.3	5,630	2,960
Oil	0.92–0.953	1,380–1,500	—
Plexiglas	1.18	2,670	1,120
Polyethylene (typical)	0.920	2,000	—
Quartz	2.65	5,736	—
Silver	10.5	3,600	1,590
Soft tissue	1.06	1,540	—
Steel	7.8	5,850	3,230
Stainless steel	7.67–8.03	5,660–7,390	2,990–3,120
Tin	7.3	3,320	1,670
Titanium	4.54	6,100	3,120
Tungsten	19.25	5,180	2.870
Water (20°C)	1.00	1,480	—
Wood, oak (with grain)	0.4615	4,640	1,750
Wood, oak (against grain)	0.4615	1,630–2,150	1,460–1,750
Zinc	7.1	4,170	2,410

Table 4.2. Reflection and transmission ratios for selected material interfaces

Reflected surface (1)	σ_R/σ_I (2)	I_R/I_I (3)	σ_T/σ_I (4)	I_T/I_I (5)
Cork–steel	0.995	0.989	1.995	0.011
Steel–cork	−0.995	0.989	0.005	0.011
Aluminum–Plexiglas	−0.692	0.479	0.308	0.521
Plexiglas–aluminum	0.692	0.479	1.692	0.521

Notes: Column (1), material 1 to material 2; column (2), amplitude ratio of reflected to incident wave; column (3), energy ratio of reflected to incident wave; column (4), amplitude ratio of transmitted to incident wave; column (5), energy ratio of transmitted to incident wave.

$$W = \rho c \sim \sqrt{\rho E}.$$

The inclusion of acoustic impedance in this equation gives us some indication of material stiffness. The stiffness is often related to a Young modulus, E, of the material; in general, the stiffer the material, the higher the wave velocity.

4.3 Snell's Law for Angle Beam Analysis

Several things happen when an ultrasonic wave encounters an interface between two materials at some inclined angle. First of all, refraction occurs in much the same way as it occurs for optical light waves. Consider the diagram in Figure 4.2. The refracted angle can be computed from Snell's law: $c_1 \sin \theta_2 = c_2 \sin \theta_1$.

The second thing that occurs at the interface is mode conversion. Energy is distributed into longitudinal and shear waves in the second material. Some energy is also reflected, where the angle of reflection is equal to the angle of incidence. The concept of mode conversion is illustrated in Figure 4.3. As an ultrasonic wave encounters an interface between two materials, the incident energy is portioned into different kinds of ultrasonic energy. Consider a force acting on the interface at an angle θ. This force independently generates a very complicated wave motion in the structure. The complicated wave motion can be considered as a superposition of two specific motions: one associated with a normal force and normal wave propagation, the second associated with a shear force and shear wave propagation. Keep in mind that pulse-type excitation at distances far away from the loading source causes separation of the two waveforms, since the shear wave velocity is less than the normal wave velocity. With continuous-wave excitation, however, the two wave motions are superimposed and fairly complex throughout the entire structure. A complete analysis of the mode conversion problem would consider two waves on each side of the interface. In this case, both normal and shear waves would be produced in both materials, and the superposition process would generate an interface wave that could travel along the interface of the two materials.

Snell's law and the qualitative aspects of mode conversion can be understood by reviewing aspects of the mathematical derivation of Snell's law. Consider a finite portion of an inclined wavefront in the plane of the paper as segment AB,

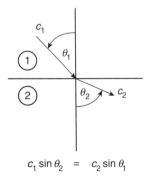

$$c_1 \sin \theta_2 \ = \ c_2 \sin \theta_1$$

Figure 4.2. Snell's law for angle beam analysis.

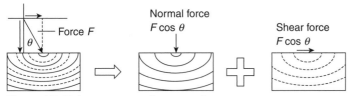

Figure 4.3. Point-source mode conversion concept.

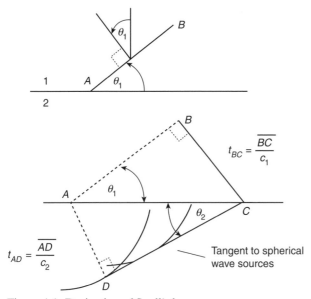

Figure 4.4. Derivation of Snell's law.

traveling toward the interface between materials 1 and 2 as shown in Figure 4.4. Using Huygens's principle of spherical wave propagation, imagine a spherical wave generating from point A and a subsequent generator of spherical waves moving in a direction from A to C as the wave segment AB moves toward the interface between bodies 1 and 2. Construction of the initial wavefront AB and the wave from CD in material 2 at some later time (established when the point B travels to C just at the interface) allows us to derive Snell's law by trigonometry. Since the time traveled from B to C in material 1 must be the same as the time it takes for a wave to move from A to D in material 2, we can easily set these times equal to derive Snell's law:

$$\sin \theta_1 = \frac{\overline{BC}}{\overline{AC}} \quad \text{and} \quad \sin \theta_2 = \frac{\overline{AD}}{\overline{AC}}$$

and so $\overline{AC} = \overline{AC}$; therefore,

$$c_1 \sin \theta_2 = c_2 \sin \theta_1.$$

Snell's law, derived on wave speed analysis only, can also be used to calculate refracted shear angles in material 2. The two equations in Figure 4.5 can therefore be deduced from Snell's law.

Table 4.3. Refraction angle values for various incident angles of a longitudinal wave

Interface	θ_{2L} value			$\theta_{2L} = 90°$
	for $\theta_l = 15°$	for $\theta_l = 30°$	for $\theta_l = 45°$	$\theta_l = \theta_{cr}$
Steel–Plexiglas	6.8°	13.2°	18.8°	—[a]
Plexiglas–steel	34.5°	—	—	$\theta_{cr} = 27°$

[a] No critical angle.

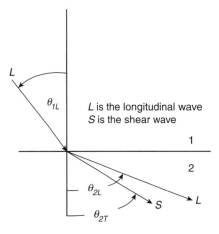

$$c_{1L} \sin \theta_{2L} = c_{2L} \sin \theta_{1L}$$

$$c_{1L} \sin \theta_{2T} = c_{2T} \sin \theta_{1L}$$

Figure 4.5. Snell's law and mode conversion.

Table 4.3 lists sample refraction angle values for ultrasonic waves traveling from one material to another at incidence angles of 15°, 30°, and 45°. Waves traveling from a media with slower wave speed than media 2 are bent away from the normal; this is illustrated in Figure 4.6. Waves traveling from a higher wave speed than media 2 are bent toward the normal.

4.4 Critical Angles and Mode Conversion

We will now examine some critical angle concepts, as illustrated in Figure 4.6, in order to further explore what happens as an ultrasonic wave encounters an interface between two materials.

Consideration of Snell's law indicates that two critical angles exist with respect to the refraction process.

(1) The first critical angle can be defined as

$$\theta_{cr} = \sin^{-1} \frac{c_1}{c_{2L}} \quad \text{when } \theta_2 = 90°.$$

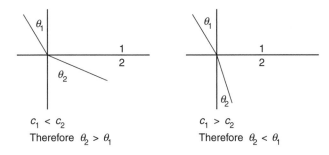

$c_1 < c_2$

Therefore $\theta_2 > \theta_1$

$c_1 > c_2$

Therefore $\theta_2 < \theta_1$

(a) First critical angle can be defined as:

$$\theta_c = \sin^{-1} \frac{c_1}{c_2} \quad \text{when} \quad \theta_2 = 90°$$

(b) Second critical angle occurs when the shear
refracted angle is 90°

Figure 4.6. Critical angle concepts.

In this case, all of the longitudinal energy is either reflected or converted to an interface wave. Only shear waves remain in the second material.

(2) The second critical angle occurs when the shear refracted angle is 90°. In other words, no significant energy is propagated through the second material; all of the energy is either reflected or transformed into interface wave propagation. The second critical angle can be computed by evaluating the inverse sine function of the ratio c_1/c_{2S}.

If one were to compute the critical angle between steel and Plexiglas, one would find that such an angle does not exist ($\sin \theta > 1$). In other words, refracted angles are less than incident angles if the wave velocity in the second material is slower than in the first material. On the other hand, refracted angles are greater than incident angles when wave velocities in the second material are greater than in the first material.

We may now summarize some important topics on mode conversion associated with the subject of angle beam analysis. Figure 4.7 depicts four possibilities of mode conversion. Part (a) shows the possible waveforms produced from a longitudinal incident wave at an angle θ_L. Refracted waves in a second material may be longitudinal or shear; reflected waves also may be longitudinal or shear. The two refracted and two reflected angles can be calculated from Snell's law. For longitudinal incident waves, the angle of incidence is obviously equal to the angle of reflection.

Part (b) of Figure 4.7 illustrates the situation for an incident angle wave at the first critical angle. Only shear waves are propagated into the second material, with longitudinal waves going off at an angle of 90° and so producing an interface wave. In this case, of course, both longitudinal and shear waves are reflected. Part (c) illustrates an incident longitudinal wave at the second critical angle. In this case, it is not possible to produce ultrasonic energy in the second material, because both shear and longitudinal waves are reflected. In part (d), the shear wave reflection angle is equal to the shear wave incident angle.

Figure 4.8 illustrates an interesting feature of mode conversion: an interface wave can actually be generated before the first critical angle is attained as a wave

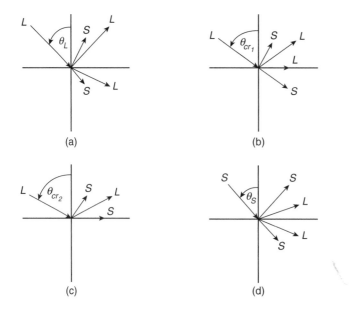

Figure 4.7. Mode conversion concepts: (a) general L input; (b) at θ_{cr_1}; (c) at θ_{cr_2}; (d) general S input.

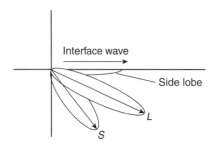

Interface wave before θ_{cr_1} (also before θ_{cr_2})
showing side lobe interaction with interface

Figure 4.8. Mode conversion and possible interface wave generation.

travels from material 1 to material 2. The interface wave is produced as a result of beam spreading and beam angle of divergence. The refracted longitudinal wave in the second material is shown with a principal lobe of ultrasonic energy surrounding it. Notice that, as the refracted longitudinal angle increases, a portion of the principal lobe of ultrasonic energy encounters the interface *before* the center point of the principal lobe reaches the 90° value.

The subject of energy partitioning is receiving attention in some areas of applied mechanics. It is often possible to evaluate what portions of the energy are converted into reflected shear and longitudinal, into refracted shear and longitudinal, and into interface waves from both the shear and longitudinal wave interaction with the interface. (Reflection and refraction factors for oblique incidence are discussed in Chapter 5.)

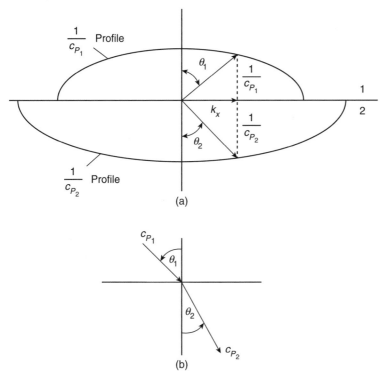

Figure 4.9. Slowness profiles for calculating refraction angle.

4.5 Slowness Profiles for Refraction and Critical Angle Analysis

We will now illustrate how slowness profiles can assist us in oblique incidence studies and subsequent critical angle analysis. If a wave encounters an interface between two anisotropic media then wave reflection and refraction will occur – in addition to any skew that may already be occurring. Let's consider symmetric but different elliptic-type slowness profiles, as illustrated in Figure 4.9. If a wave is incident to material 1 at a specific angle θ (as measured from the normal) then the wavevector component along the interface must be preserved. This is simply a restatement of Snell's law:

$$\frac{1}{c_{p1}} \sin \theta_1 = \frac{1}{c_{p2}} \sin \theta_2,$$

so $c_{p1} \sin \theta_2 = c_{p2} \sin \theta_1$. This allows us to find the refraction angle in the second material.

The same approach can also be used to evaluate critical angles. In Figure 4.9, for example, no critical angle exists; it is impossible to have θ_2 become 90°, despite the increase in θ_1, because the wave velocity in media 2 is less than that in media 1. Figure 4.10, however, shows an example where a critical angle can occur. When the incidence angle in medium 1 reaches θ_{cr}, the refraction angle in medium 2 is 90°; this is the definition of critical angle.

Figure 4.11 shows how slowness curves can be used to evaluate reflection angles as well as refraction and critical angles. For an incident shear wave at θ_S, the reflected shear angle will, of course, still be θ_S. Mode conversion does take place, however, and

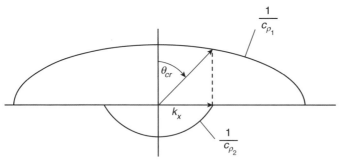

Figure 4.10. Slowness curves for calculating critical angles.

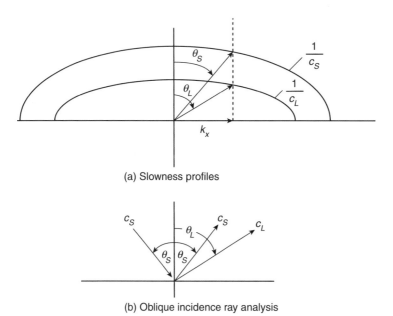

(a) Slowness profiles

(b) Oblique incidence ray analysis

Figure 4.11. Slowness curves for calculating reflection angles for shear input.

some longitudinal waves are reflected also. The reflected longitudinal wave angle will be θ_L, as illustrated in Figure 4.11, with k_x being preserved.

Numerous examples can now be studied; some are left as exercises. For realistic examples, solutions to Christoffel equations for the phase velocity functions could be combined with sample elastic constants to evaluate aspects of skew angle, energy velocity, oblique incidence, critical angles, and so forth. See Table 3.1 for some material constants that can be useful in problem solving.

4.6 Exercises

1. Calculate the reflection factor from steel to air. What does the negative sign mean? Think about boundary conditions on the free surface.
2. Calculate reflection factors (as in Table 4.2) for steel to air, water to Plexiglas, and Plexiglas to water.
3. Calculate refraction angle values for water to Plexiglas and Plexiglas to water at the angles indicated in Table 4.3.

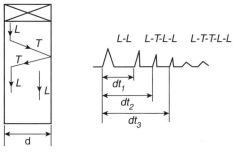

Figure 4.12. Exercise 8.

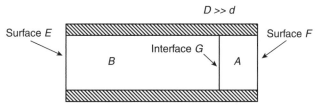

Figure 4.13. Exercise 9.

4. Calculate the reflected longitudinal wave angle for an incident shear wave at 30^c onto a steel–air interface.

5. Develop an equation for oblique incidence reflection angles for longitudinal or shear waves, using either longitudinal or shear waves as input.

6. Calculate the first and second critical angle from Plexiglas to steel. What special situation occurs if wave entry is limited to this range?

7. What would you observe when comparing reflected signals from steel to Plexiglas and from Plexiglas to steel? Explain!

8. Solve for the dt values in the resulting pulse-echo rectified low-pass–filtered RF wave-front pattern shown in Figure 4.12. Assume longitudinal wave input to a rod and mode conversion along a rough outer surface.

9. In Figure 4.13, cladding (of thickness d) of material A is perfectly diffusion bonded onto a material B of length D (shading is for sound absorption). Let $c_B = 6,000$ m/s, $c_A = 2,000$ m/s, $\rho_B = 8,000$ kg/m³, and $\rho_A = 2,000$ kg/m³. If surface E is loaded with a 3-cycle 1-MHz sine wave pulse, consider the pulse-echo wave reflection pattern at E.

   (a) Calculate the pressure amplitude reflection factor at the interface G and also at the back surface F.

   (b) Considering phase variation only (not amplitude), what thickness d produces a maximum response? A minimum response? What thickness value d could be resolved with the 3-cycle 1-MHz sine wave?

   (c) Show sample results for $\lambda \ll d$ on a wave diagram for at least two reflections in D. At what time value does echo identification become difficult? Illustrate and explain.

   (d) What are the longitudinal and shear reflection factor values at the interface G for a smooth interface? (Consider G as an infinitesimally thin fluid film.)

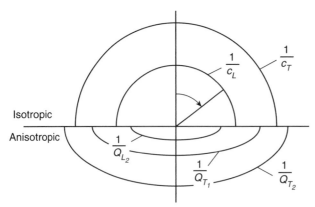

Figure 4.14. Exercise 14.

10. What are the reflection and transmission factors for normal beam shear wave impingement onto an interface between two media?
11. Suppose we were to calculate wave velocity values in the (x_1, x_2)-plane for some orthotropic media. For at least one particular direction – say, $[1, 1, 0]$ – in this plane, detail the steps of the computational process. How would you determine if the wave velocity vector had pure or quasi-longitudinal or shear components? Also, solve for one element – say, Γ_{23} – of the acoustic tensor.
12. Plot reflection and refraction angle for a specific angle of incidence of an L or S wave onto a steel–Plexiglas–steel structure. Calculate θ_1 (the first critical angle) and θ_2 (the second critical angle). Also give a graphical solution using slowness profiles.
13. Compute critical angles for waves impinging onto a silicon–steel interface that is perfectly diffusion bonded and onto a steel–silicon interface (use data from interpolation estimates if necessary). Consider L and S inputs and all possible critical angles. (Consider the definition of critical angle as a phase velocity going to $90°$.)
14. For the slowness profiles illustrated in Figure 4.14, use a graphical technique to solve for all reflected and refracted angles for the incident angle shown. Also, how would you define and calculate the first, second, and third critical angles?

4.7 REFERENCES

Auld, B. A. (1990). *Acoustic Fields and Waves in Solids*, 2nd ed., vols. 1 and 2. Malabar, FL: Kreiger.
Brekhovskikh, L. M. (1960). *Waves in Layered Media*. New York: Academic Press.
Graff, K. F. (1991). *Wave Motion in Elastic Solids*. New York: Dover.
Kolsky, H. (1963). *Stress Waves in Solids*. New York: Dover.
Pollard, H. F. (1977). *Sound Waves in Solids*. London: Pion Ltd.

5 Oblique Incidence

5.1 Background

One of the most important topics associated with the subject of stress wave propagation in solid materials is the wave reflection and refraction at an interface between two different media. (For more details, see Auld 1990; Graff 1991; Pilarski, Rose, and Balasubramaniam 1990; or Rose 1999.) The subject is important to the study of ultrasonic guided waves since oblique incidence via appropriate angles of incidence and frequency selection can be used to generate guided waves in a variety of different waveguides. Introductory topics and concepts are therefore presented in this chapter. If incident angles are selected properly, long enough wavelengths are used, and the material being inspected has a phase velocity larger than the dilatational velocity in the wedge material, then guided waves can be generated in the test material.

A general introduction to oblique incidence in ultrasonic wave analysis will be presented. The reflection (refraction) factor, or coefficient, is defined as the ratio of the amplitude of the reflected (refracted) wave to the amplitude of the incident wave. The factor depends on the angle of incidence, wave velocities, and possibly frequency, depending on the interface condition. In this chapter, we introduce a boundary condition approach for calculating these factors. We use this approach for the interface between two semi-infinite medium spaces: solid–solid, solid–liquid, and liquid–solid. If the reader would like to calculate reflection and refraction factors for a thin interface solid (and liquid) layer between two different media, it is recommended to follow guidelines established by Jiao and Rose (1991) and from a "spring" model (Pilarski and Rose 1998a,b; Pilarski et al. 1990). These cases are also discussed by Rose (1999).

From physics and wave mechanics (Timoshenko and Goodier 1987; Graff 1991), a new wave may be generated from an incident wave at an interface, depending on the wave velocities in the two media and the angle of incidence. For example, a longitudinal wave in Plexiglas impinges the interface between Plexiglas and steel. The reflected and refracted waves could consist of both longitudinal and shear waves. At a certain angle of incidence, the refracted longitudinal wave disappears (i.e., the angle of refraction is 90°). This is the first critical angle. When the refracted shear wave disappears, the angle of incidence is the second critical angle.

With contributions from Jason Philtron.

What are the magnitudes of these waves in the different directions? Oblique incidence reflection and refraction factor calculations must be carried out to determine the mode conversion ratios. Several oblique incidence reflection and refraction sample problems are presented in this chapter. In Chapter 4 we saw how – from Snell's law and a study of the slowness profiles between two media – we can determine all reflection and refraction angles, all critical angles, total reflection possibilities, and so forth. The problem now is to address the energy partitioning into the different modes (i.e., mode conversion).

It is possible to generate both longitudinal and shear waves in solids. However, in nonviscous liquid (e.g., water), only longitudinal waves can exist. An isotropic material has properties that are independent of direction in the material. Our emphasis is primarily on homogeneous isotropic materials; the approach would be similar for anisotropic materials (although the constitutive equations would change, and the algebra would become more cumbersome). For more information on this topic, see Auld (1990), Brekhovshikh (1960), Henneke (1972), Jiao and Rose (1991), Pilarski and Rose (1988a,b), and Rokhlin and Marom (1986).

5.2 Reflection and Refraction Factors

5.2.1 Solid–Solid Boundary Conditions

In this section, the reflection and refraction factors are derived for a perfect solid–solid interface between two semi-infinite media. A plane wave of amplitude A_N is incident with angle θ onto the interface between two isotropic half spaces, as shown in Figure 5.1. The interface is at $y = 0$. There are reflected waves of amplitudes A_L and A_T and angles α_L and α_T for longitudinal and transverse waves, respectively. In addition, the refracted waves have amplitudes B_L and B_T and angles β_L and β_T. The reflected and refracted angles are calculated according to Snell's Law,

$$\frac{\sin \theta}{c_N} = \frac{\sin \alpha_T}{c_{1T}} = \frac{\sin \alpha_L}{c_{1L}} = \frac{\sin \beta_T}{c_{2T}} = \frac{\sin \beta_L}{c_{2L}}, \tag{5.1}$$

where $c_{1T}, c_{1L}, c_{2T},$ and c_{2L} are the transverse and longitudinal wave speeds in materials 1 and 2. c_N is the wave speed of the incident wave, either c_{1T} or c_{1L}, depending on the incident wave type. We will start by considering transverse wave incidence.

The wave velocities of the five waves will be used to calculate reflection and transmission factors. For transverse wave incidence, the particle velocities of the different waves are

$$\vec{V}_N = \frac{-\hat{x} \times k_N}{|k_N|} A_N e^{-ik_N \vec{r}}, \tag{5.2}$$

$$\vec{V}_{1T} = \frac{\hat{x} \times k_{1T}}{|k_{1T}|} A_T e^{-ik_{1T} \vec{r}}, \tag{5.3}$$

$$\vec{V}_{1L} = \frac{k_{1L}}{|k_{1L}|} A_L e^{-ik_{1L} \vec{r}}, \tag{5.4}$$

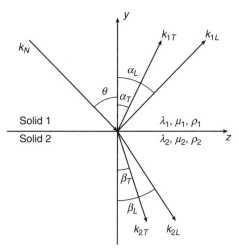

Figure 5.1. Reflection and refraction at an interface between two solid media.

$$\vec{V}_{2T} = \frac{-\hat{x} \times k_{2T}}{|k_{2T}|} B_T e^{-ik_{2T}\vec{r}}, \text{ and} \tag{5.5}$$

$$\vec{V}_{2L} = \frac{k_{2L}}{|k_{2L}|} B_L e^{-ik_{2L}\vec{r}}. \tag{5.6}$$

Note that for longitudinal wave incidence, Equation 5.2 will change. From here on, we will write $|k|$ simply as k.

The incident wave is traveling downward $(-\hat{y})$ and to the right $(\hat{z})$ in the y-z plane. For analysis, we separate the wavevector, k_N, into its components traveling in the y- and z-directions as

$$k_N = -\hat{y}k_N \cos\theta + \hat{z}k_N \sin\theta. \tag{5.7}$$

Substitution into the above equations, for example, leads to the following

$$\vec{V}_N = (\hat{y}\sin\theta + \hat{z}\cos\theta)A_N e^{-i(-yk_T\cos\theta + zk_T\sin\theta)}. \tag{5.8}$$

The boundary conditions for a perfect interface between two solids are continuity of normal and transverse particle velocity and continuity of normal and transverse stress at the interface. These boundary conditions can be expressed as

$$(V_y)_N + (V_y)_{1T} + (V_y)_{1L} = (V_y)_{2T} + (V_y)_{2L}, \tag{5.9}$$

$$(V_z)_N + (V_z)_{1T} + (V_z)_{1L} = (V_z)_{2T} + (V_z)_{2L}, \tag{5.10}$$

$$(\sigma_{yy})_N + (\sigma_{yy})_{1T} + (\sigma_{yy})_{1L} = (\sigma_{yy})_{2T} + (\sigma_{yy})_{2L}, \text{ and} \tag{5.11}$$

$$(\sigma_{yz})_N + (\sigma_{yz})_{1T} + (\sigma_{yz})_{1L} = (\sigma_{yz})_{2T} + (\sigma_{yz})_{2L}. \tag{5.12}$$

Our notation is such that, for example, $(V_y)_N$ represents the particle velocity in the y-direction (V_y) for the incident (N) wave.

Before proceeding with the derivation, we must recall the elastic constant-velocity relationships for isotropic materials,

$$C_{11} = \rho c_L^2 = \lambda + 2\mu, \quad C_{44} = \rho c_T^2 = \mu, \quad \text{and} \quad C_{12} = \lambda; \tag{5.13}$$

the stress-strain relationships (from Hooke's law) for two-dimensional plane strain in isotropic media,

$$\sigma_{yy} = \lambda \left(\varepsilon_{yy} + \varepsilon_{zz} \right) + 2\mu\varepsilon_{yy}, \quad \text{and}$$
$$\sigma_{yz} = 2\mu\varepsilon_{yz}; \tag{5.14}$$

the strain-particle displacement relationships,

$$\varepsilon_{yy} = \frac{\partial u_y}{\partial y},$$
$$\varepsilon_{zz} = \frac{\partial u_z}{\partial z}, \quad \text{and} \tag{5.15}$$
$$\varepsilon_{yz} = \frac{1}{2}\left(\frac{\partial u_y}{\partial z} + \frac{\partial u_z}{\partial y} \right);$$

and the particle velocity-displacement relationship (for harmonic motion),

$$\vec{V} = \frac{\partial \vec{u}}{\partial t} = i\omega\vec{u}. \tag{5.16}$$

From Equations 5.14 to 5.16, we obtain the stress-particle velocity relationships,

$$\sigma_{yy} = \frac{\lambda + 2\mu}{iw}\frac{\partial V_y}{\partial y} + \frac{\lambda}{iw}\frac{\partial V_z}{\partial z}, \quad \text{and}$$
$$\sigma_{yz} = \frac{\mu}{iw}\left(\frac{\partial V_y}{\partial z} + \frac{\partial V_z}{\partial y} \right). \tag{5.17}$$

Additionally, in the derivation that follows we will use the identity

$$\lambda + 2\mu \cos^2 \theta_L = (\lambda + 2\mu) \cos 2\theta_T. \tag{5.18}$$

The next step in solving the problem is to apply the boundary conditions. For the first boundary condition, continuity of normal particle velocity across the interface, we obtain the equation

$$A_N(-\sin \alpha_T) + A_L(-\cos \alpha_L) + A_T(\sin \alpha_T) = B_L(\cos \beta_L) + B_T(-\sin \beta_T) \tag{5.19}$$

for transverse wave incidence, where the time-dependent term, $e^{-i\omega t}$, has been dropped.

Equation 5.19 can be rewritten as

$$A_L(M_{11}) + A_T(M_{12}) + B_L(M_{13}) + B_T(M_{14}) = A_N(a_1), \tag{5.20}$$

where $M_{11}, M_{12}, M_{13}, M_{14}$, and a_1 are entries in the matrix formulation of the problem $Mx = a$. At this point it is useful to define the reflected and refracted wave amplitudes normalized by our incoming wave amplitude as:

$$\text{RL} = \frac{A_L}{A_N}, \quad \text{RT} = \frac{A_T}{A_N} \quad \text{SL} = \frac{B_L}{A_N}, \quad \text{and} \quad \text{ST} = \frac{B_T}{A_N}. \tag{5.21}$$

These four variables comprise our vector of unknowns, $x = [\text{RL}, \text{RT}, \text{SL}, \text{ST}]^{\text{T}}$.

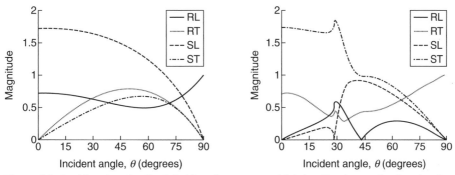

Figure 5.2. Incident longitudinal (left) and transverse (right) reflection and refraction factor results for oblique incidence onto an aluminum-Plexiglas interface.

We apply each of the four boundary conditions in a similar manner to obtain our full system of equations which can be written in matrix form $Mx = a$ where

$$M = \begin{bmatrix} -\cos\alpha_L & \sin\alpha_T & -\cos\beta_L & \sin\beta_T \\ -\sin\alpha_L & -\cos\alpha_T & \sin\beta_L & \cos\beta_T \\ -k_{1L}(\lambda_1+2\mu_1)\cos2\alpha_T & k_{1T}\mu_1\sin2\alpha_T & k_{2L}(\lambda_2+2\mu_2)\cos2\beta_T & -k_{2T}\mu_2\sin2\beta_T \\ -k_{1L}\mu_1\sin2\alpha_L & -k_{1T}\mu_1\cos2\alpha_T & -k_{2L}\mu_2\sin2\beta_L & -k_{2T}\mu_2\cos2\beta_T \end{bmatrix}$$

$$(5.22)$$

M is the same for both transverse and longitudinal incident waves because the reflected and refracted waves are similar. However, the vector a changes because it represents the incident wave. For incident transverse and longitudinal waves we have

$$a_T = \begin{bmatrix} \sin\alpha_T \\ \cos\alpha_T \\ -k_{1T}\mu_1\sin2\alpha_T \\ -k_{1T}\mu_1\cos2\alpha_T \end{bmatrix} \quad \text{and} \quad a_L = \begin{bmatrix} -\cos\alpha_L \\ \sin\alpha_L \\ k_{1L}(\lambda_1+2\mu_1)\cos2\alpha_T \\ -k_{1L}\mu_1\sin2\alpha_L \end{bmatrix} \quad (5.23)$$

respectively. The equation $Mx = a$ can now be solved for the reflection and refraction factors, x, for a particular incident wave type and material properties at any incident angle.

A sample result is shown in Figure 5.2, for both longitudinal and transverse waves incident onto an aluminum-Plexiglas interface. Note that the curves are smooth for an incident longitudinal wave, but show an abrupt change in slop at the critical angle (29.2°) for an incident transverse wave.

5.2.2 Solid–Liquid Boundary Conditions

For oblique plane wave incidence on a solid–liquid boundary (Figure 5.3) the equations for the reflection and refraction factors can be derived in a similar way to Subsection 2.1. There are two main differences between the solid–liquid and the solid–solid problem. First, liquids cannot support shear waves. Second, the boundary condition of

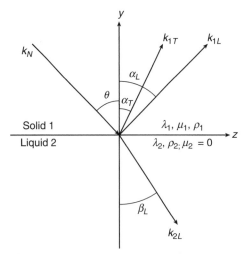

Figure 5.3. Reflection and refraction at an interface between a solid and liquid media.

continuity of transverse particle velocity is no longer true at the interface. However, with several changes, our derivation from the solid–solid case can still be used.

Since there is no shear wave in the liquid, ST = 0. Therefore we can remove column 4 of M and row 4 of x and a. Additionally, since there is no continuity of transverse displacement across the interface, (our second boundary condition,) we remove row 2 of M. Noting that $\mu_2 = 0$ for our liquid, and realizing that Snell's Law gives the trivial result $\beta_T = 0$, our expressions simplify. We obtain

$$M = \begin{bmatrix} -\cos\alpha_L & \sin\alpha_T & -\cos\beta_L \\ -k_{1L}(\lambda_1+2\mu_1)\cos2\alpha_T & k_{1T}\mu_1\sin2\alpha_T & k_{2L}\lambda_2 \\ -k_{1L}\mu_1\sin2\alpha_L & -k_{1T}\mu_1\cos2\alpha_T & 0 \end{bmatrix}, \qquad (5.24)$$

which is still true for both longitudinal and transverse wave incidence. Additionally, we now have

$$a_L = \begin{bmatrix} -\cos\alpha_L \\ k_{1L}(\lambda_1+2\mu_1)\cos2\alpha_T \\ -k_{1L}\mu_1\sin2\alpha_L \end{bmatrix} \quad \text{and} \quad a_T = \begin{bmatrix} \sin\alpha_T \\ -k_{1T}\mu_1\sin2\alpha_T \\ -k_{1T}\mu_1\cos2\alpha_T \end{bmatrix} \qquad (5.25)$$

for longitudinal and transverse wave incidence.

A sample result is shown in Figure 5.4, for both longitudinal and transverse waves incident onto an aluminum-water interface. Note that the reflection and refraction factor results are again smooth for an incident longitudinal wave, and show an abrupt change in slope at the critical angle (29.2°) for an incident transverse wave.

5.2.3 Liquid–Solid Boundary Conditions

Oblique plane wave incidence on a liquid–solid boundary is shown in Figure 5.5. The equations for the reflection and refraction factors can again be derived from the results of the solid–solid case with several changes. Since there is no shear wave

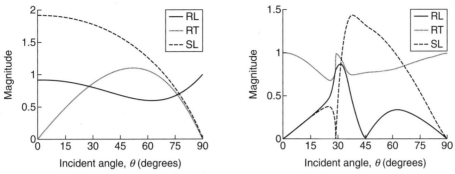

Figure 5.4. Incident longitudinal (left) and transverse (right) reflection and refraction factor results for oblique incidence onto an aluminium-water interface.

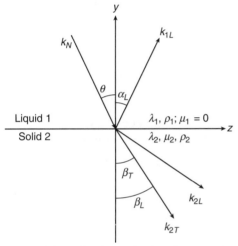

Figure 5.5. Reflection and refraction at an interface between a liquid and solid media.

in the liquid, RT = 0. Therefore we can remove column 2 of M and row 2 of x and a. Additionally, since there is (again) no continuity of transverse displacement, we remove row 2 of M. Noting that $\mu_1 = 0$ for our liquid, and realizing that Snell's Law gives the trivial result $\alpha_T = 0$, our expressions simplify.

We obtain

$$M = \begin{bmatrix} -\cos\alpha_L & -\cos\beta_L & \sin\beta_T \\ -k_{1L}\lambda_1 & k_{1L}(\lambda_2 + 2\mu_2)\cos 2\beta_T & -k_{2T}\mu_2\sin 2\beta_T \\ 0 & -k_{2L}\mu_2\sin 2\beta_L & -k_{2T}\mu_2\cos 2\beta_T \end{bmatrix}, \qquad (5.26)$$

and

$$a_L = \begin{bmatrix} -\cos\alpha_L \\ k_{1L}\lambda_1 \\ 0 \end{bmatrix}. \qquad (5.27)$$

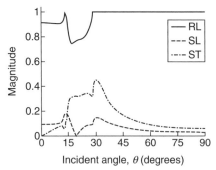

Figure 5.6. Incident longitudinal reflection and refraction factor results for oblique incidence onto a water-aluminium interface.

A sample result is shown in Figure 5.6, for longitudinal wave incidence onto a water-aluminum interface. Note that the reflection and refraction factor results show abrupt changes in slope at two critical angles, $13.5°$ and $28.5°$.

5.3 Moving Forward

A general introduction to oblique incidence in ultrasonic wave analysis has been presented. We have described analytically the wave phenomena that occur at solid–solid, solid–liquid, and liquid–solid boundaries. Analysis of ultrasonic wave reflection and refraction factors further an intuitive understanding of ultrasonic guided wave excitation, resonance, and propagation. The oblique incidence concepts presented in this chapter form an underlying foundation from which we move forward in later chapters as specific geometries are analyzed.

5.4 Exercises

(1) Write out the general boundary condition equations for reflection and refraction at a solid–liquid interface for both shear and longitudinal input.
(2) Write boundary conditions for oblique incidence onto a liquid–solid interface. For a low or high incident angle, what wave types (i.e., shear and/or longitudinal) would you expect to be reflected and refracted?
(3) Derive the reflection and refraction factor equation for a liquid–solid interface.
(4) What changes in the reflection factor derivations are necessary if we are to calculate oblique incidence reflection and refraction factors between two anisotropic media?
(5) Explain, intuitively, the reflection and transmission factors for $\theta = 90°$.
(6) Verify the values of reflection and transmission of normal incident L and T waves shown in Figure 5.2 using equations from Chapter 4. Why does the magnitude of these values decrease as θ increases from $0°$?
(7) In Figure 5.2, why can SL or ST have a magnitude greater than 1?
(8) Prove the relationship given in Equation 5.18. (Hint: you will need to use trigonometric identities, Snell's law, and expressions relating the Lamè constants and sound speeds.)

(9) Use computer programs to generate the reflection and refraction curves shown in Figures 5.2 and 5.6.

(10) Using the multireflection approach across a thin liquid layer, derive an expression for the overall reflection and/or refraction factor from the layer for oblique longitudinal and/or shear input. (Use a geometric progression series to develop the result.)

(11) Calculate the reflection and transmission factors for a normal incident longitudinal wave onto a liquid layer between two solids. (Use a geometric progression series to develop the result.)

5.5 REFERENCES

Auld, B. A. (1990). *Acoustic Fields and Waves in Solids,* 2nd ed., vols. 1 and 2. Malabar, FL: Kreiger.

Brekhovshikh, L. M. (1960). *Waves in Layered Media.* New York: Academic Press.

Graff, K. F. (1991). *Wave Motion in Elastic Solids.* New York: Dover.

Henneke, E. G. II (1972). Reflection-refraction of a stress wave at a plane boundary between anisotropic media, *J. Acoust. Soc. Am.* 51:210.

Jiao, D., and Rose, J. L. (1991). An ultrasonic interface layer model for bond evaluation, *J. Adhesion Sci. Tech.* 5(8): 631–46.

Pilarski, A., Rose, J. L., and Balasubramaniam, K. (1990). The angular and frequency characteristics of reflectivity from a solid layer embedded between two solids with imperfect boundary conditions, *J. Acoust. Soc. Am.* 87(2): 532–42.

Pilarski, A., and Rose, J. L. (1998a). A transverse-wave ultrasonic oblique-incidence technique for interfacial weakness detection in adhesive bonds, *J. Appl. Phys.* 63(2): 300–7.

(1998b). Ultrasonic oblique-incidence for improved sensitivity interface weakness determination, *NDT Int.* 21(4): 241–6.

Rokhlin, S. I., and Marom, D. (1986). Study of adhesive bonds using low-frequency obliquely incident ultrasonic waves, *J. Acoust. Soc. Am.* 80: 245–58.

Rose, J. L. (1999). *Ultrasonic Waves in Solid Media,* Cambridge University Press.

Timoshenko, S. P., and Goodier, J. N. (1987). *Theory of Elasticity,* 3rd. ed. New York: McGraw-Hill.

6 Waves in Plates

6.1 Introduction

This chapter presents the governing equations of elastodynamics for waves in plates, along with a series of sample problems and practical discussions. The method of displacement potentials is used to obtain a solution for the case of propagation in a free plate (see e.g., Achenbach 1984 for more detail). Also, we give a brief outline of the method of partial waves (see Auld 1990).

The classical problem of Lamb wave propagation is associated with wave motion in a traction-free homogeneous and isotropic plate. The procedures we use to develop the governing equations and dispersion curve results of phase velocity versus frequency are similar to those used in a countless number of guided wave problems that incorporate bars, tubes, multiple layers, and anisotropic media. In this chapter we shall therefore detail the basic concepts of guided wave analysis. Interpretation procedures and mathematical analysis of phase and group velocity dispersion curves and wave structure can then be extended to a variety of different guided wave problems. An alternative technique of developing dispersion curves is presented Chapter 9.

We will now briefly re-visit the fundamental differences between guided waves and bulk waves. Bulk waves travel in the bulk of the material – hence, away from the boundaries. However, often there is interaction with boundaries by way of reflection and refraction, and mode conversion occurs between longitudinal and shear waves. Although bulk and guided waves are fundamentally different, they are actually governed by the same set of partial differential wave equations. Mathematically, the principal difference is that, for bulk waves, there are no boundary conditions that need to be satisfied by the proposed solution. In contrast, the solution to a guided wave problem must satisfy the governing equations as well as some physical boundary conditions.

It is the introduction of boundary conditions that makes the guided wave problem difficult to solve analytically; in many cases, analytic solutions cannot even be found. Another interesting feature of guided wave propagation is that, unlike the finite number of modes (primarily longitudinal shear, shear perhaps being horizontal or vertical) that might be present in a bulk wave problem, there are generally an infinite number of modes associated with a given guided wave problem. That is, a finite body can support an infinite number of different guided wave modes.

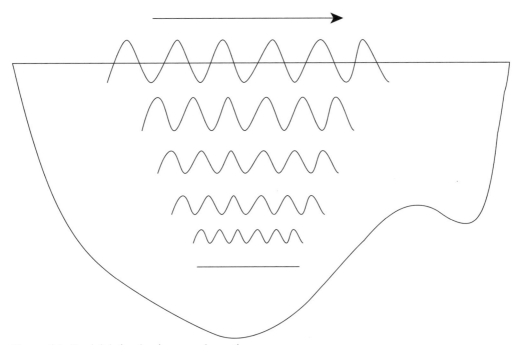

Figure 6.1. Rayleigh (surface) wave schematic.

Figure 6.2. Lamb wave schematic.

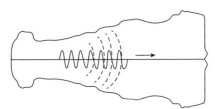

Figure 6.3. Stoneley wave schematic.

Some examples of guided wave problems that have been solved – and whose solution has inherited the name of the investigator – are Rayleigh, Lamb, and Stoneley waves. *Rayleigh waves* are free waves on the surface of a semi-infinite solid. Traction forces must vanish on the boundary, and the waves must decay with depth (see Figure 6.1). *Lamb waves* are waves of plane strain that occur in a free plate, and the traction force must vanish on the upper and lower surface of the plate. Different mode structures occur from point to point as wave entry angle and frequency are varied (see Figure 6.2). *Stoneley waves* are free waves that occur at an interface between two media (see Figure 6.3). Continuity of traction and displacement is required at the interface, and a radiation condition must be satisfied.

We recall the following from the theory of elasticity (here we are using Cartesian tensor notation):

$$\sigma_{ij,j} + \rho f_i = \rho \ddot{u}_i, \quad \text{3 equations of motion } (i = 1, 2, 3); \tag{6.1}$$

$$\varepsilon_{ij} = \frac{1}{2}\left(u_{i,j} + u_{j,i}\right), \quad \text{6 independent strain displacement equations;} \tag{6.2}$$

$$\sigma_{ij} = \lambda \varepsilon_{kk}\delta_{ij} + 2\mu\varepsilon_{ij}, \text{6 independent constitutive equations (isotropic materials).} \tag{6.3}$$

The first two equations are valid for any continuous medium; the specific type of medium concerned is introduced via (6.3). If we eliminate the stress and strain factors from these equations, then we have

$$\mu u_{i,jj} + (\lambda + \mu)u_{j,ji} + \rho f_i = \rho \ddot{u}_i. \tag{6.4}$$

The equations of motion (6.4), which contain only the particle displacements, are the governing partial differential equations for displacement. If the domain in which a solution is sought is infinite, then these equations are sufficient. If the domain is finite, then boundary conditions are needed for a well-posed problem. The boundary conditions take the form of prescribed tractions and/or displacements on the boundaries of the domain of interest. The general forms of such boundary conditions are as follows:

$$u(x, t) = u_0(x, t) \quad \text{on surface displacements;} \tag{6.5}$$

$$t_i = \sigma_{ji}n_j. \quad \text{on surface tractions;} \tag{6.6}$$

$$u(x, t) = u_0(x, t) \quad \text{on } S_1$$
$$\text{and } t_i = \sigma_{ji}n_j. \quad \text{on } S_2 \quad \text{as a mixed boundary condition.} \tag{6.7}$$

6.2 The Free Plate Problem

The geometry of the free plate problem is illustrated in Figure 6.4. This problem is governed by the equations of motion (6.4), with boundary conditions of type (6.6). The surfaces at the coordinates $x_3 = d/2 = h$ and $x_3 = -d/2 = -h$ are considered traction free. Ultrasonic excitation occurs at some point in the plate; as ultrasonic energy from the excitation region encounters the upper and lower bounding surfaces of the plate, mode conversions occur (L wave to T wave, and vice versa). After some travel in the plate, superpositions cause the formation of "wave packets," or what are commonly called *guided wave modes* in the plate. Based on entry angle and frequency used, we can predict how many different modes can be produced in the plate.

The exact solution of this problem has been obtained through the use of several different approaches. The most popular methods of solution are the displacement

Figure 6.4. Geometry of the free plate problem.

potentials and the partial wave techniques (see Achenbach 1984 and Auld 1990, respectively).

6.2.1 Solution by the Method of Potentials

If the displacement vector (field) is decomposed according to Helmholtz decomposition and the result substituted into (6.4), as demonstrated previously, we obtain two uncoupled wave equations. For plane strain, these are

$$\frac{\partial^2 \phi}{\partial x_1^2} + \frac{\partial^2 \phi}{\partial x_3^2} = \frac{1}{c_L^2} \frac{\partial^2 \phi}{\partial t^2}, \quad \text{governing longitudinal waves;} \tag{6.8}$$

$$\frac{\partial^2 \psi}{\partial x_1^2} + \frac{\partial^2 \psi}{\partial x_3^2} = \frac{1}{c_T^2} \frac{\partial^2 \psi}{\partial t^2}, \quad \text{governing shear waves.} \tag{6.9}$$

The case of plane strain is not the most general for the problem at hand, but the analysis is greatly simplified in this case. Achenbach (1984) shows that taking the general state of strain as a starting point results in the same set of solutions presented here plus some additional modes (infinite in number), known as horizontal shear modes, that can exist independently of the other wave modes.

As a result of our assumption of plane strain, the displacements and stresses can be written in terms of the potentials as

$$u_1 = u = \frac{\partial \phi}{\partial x_1} + \frac{\partial \psi}{\partial x_3}. \tag{6.10a}$$

$$u_2 = v = 0, \tag{6.10b}$$

$$u_3 = w = \frac{\partial \phi}{\partial x_3} + \frac{\partial \psi}{\partial x_1} : \tag{6.10c}$$

$$\sigma_{31} = \mu \left(\frac{\partial u_3}{\partial x_1} + \frac{\partial u_1}{\partial x_3} \right) = \mu \left(\frac{2\partial^2 \phi}{\partial x_1 \partial x_3} - \frac{\partial^2 \psi}{\partial x_1^2} + \frac{\partial^2 \psi}{\partial x_3^2} \right), \tag{6.11a}$$

$$\sigma_{33} = \lambda \left(\frac{\partial u_1}{\partial x_1} + \frac{\partial u_3}{\partial x_3} \right) + 2\mu \frac{\partial u_3}{\partial x_3}$$
$$= \lambda \left(\frac{\partial^2 \phi}{\partial x_1^2} + \frac{\partial^2 \phi}{\partial x_3^2} \right) + 2\mu \left(\frac{\partial^2 \phi}{\partial x_3^2} - \frac{\partial^2 \psi}{\partial x_1 \partial x_3} \right), \tag{6.11b}$$

where λ and μ are Lamé constants.

We begin the analysis by assuming infinite plane harmonic wave solutions to (6.8) and (6.9) in the form

$$\phi = \Phi(x_3) \exp[i(kx_1 - \omega t)]. \tag{6.12}$$

$$\Psi = \Psi(x_3) \exp[i(kx_1 - \omega t)]. \tag{6.13}$$

Note that these solutions represent traveling waves in the x_1 direction and standing waves in the x_3 direction. This is evident from the fact that, although there is a complex exponential term (hence sines and cosines) containing the time variable for the x_1 dependencies, there is only an unknown "static" function of x_3 for the x_3

dependence. This phenomenon is referred to in many texts as *transverse resonance* and is exploited in many ways to arrive at a solution. Again, these solutions represent waves that travel along the direction of the plate and that have fixed (as yet, unknown) distributions in the transverse directions.

Substitution of these assumed ϕ and Ψ solutions into (6.8) and (6.9) yields equations governing the unknown functions ϕ and Ψ The solutions to these equations are

$$\phi = \Phi(x_3) = A_1 \sin(px_3) + A_2 \cos(px_3), \tag{6.14}$$

$$\Psi = \Psi(x_3) = B_1 \sin(qx_3) + B_2 \cos(qx_3), \tag{6.15}$$

where

$$p^2 = \frac{\omega^2}{c_L^2} - k^2 \quad \text{and} \quad q^2 = \frac{\omega^2}{c_T^2} - k^2. \tag{6.16}$$

With these results, the displacements and stresses can be obtained directly from (6.10) and (6.11). Omitting the term $\exp[i(kx_1 - \omega t)]$ to simplify all expressions, the results are as follows:

$$u_1 = \left[ik\Phi + \frac{d\Psi}{dx_3} \right]. \tag{6.17}$$

$$u_3 = \left[\frac{d\Phi}{dx_3} + ik\Psi \right]; \tag{6.18}$$

$$\sigma_{31} = \mu\left(2ik\frac{d\Phi}{dx_3} + k^2\Psi + \frac{d^2\Psi}{dx_3^2} \right) \tag{6.19}$$

$$\sigma_{33} = \left[\lambda\left(-k^2\Phi + \frac{d^2\Phi}{dx_3^2} \right) + 2\mu\left(\frac{d^2\Phi}{dx_3^2} - ik\frac{d\Psi}{dx_3} \right) \right] \tag{6.20}$$

Now, since the field variables involve sines (resp. cosines) with argument x_3, which are odd (resp. even) functions about $x_3 = 0$, we split the solution into two sets of modes: symmetric and antisymmetric modes. Specifically, for displacement in the x_1 direction, the motion will be symmetric (with respect to the midplane of the plate) if u_1 contains cosines but will be antisymmetric if u_1 contains sines. The reverse is true for displacements in the x_3 direction. Thus, we split the modes of wave propagation in the plate into two systems:

Symmetric modes

$$\Phi = A_2 \cos(px_3),$$
$$\Psi = B_1 \sin(qx_3),$$
$$u = u_1 - ikA_2 \cos(px_3) + qB_1 \cos(qx_3), \tag{6.21}$$
$$w = u_3 = -pA_2 \sin(px_3) - ikB_1 \sin(qx_3).$$
$$\sigma_{31} = \mu[-2ikpA_2 \sin(px_3) + (k^2 - q^2)B_1 \sin(qx_3)].$$
$$\sigma_{33} = -\lambda(k^2 + p^2)A_2 \cos(px_3) - 2\mu[p^2A_2 \cos(px_3) + ikqB_1 \cos(qx_3)];$$

Antisymmetric modes

$$\Phi = A_1 \sin(px_3),$$
$$\Psi = B_2 \cos(qx_3),$$
$$u = u_1 = ikA_1 \sin(px_3) - qB_2 \sin(qx_3), \tag{6.22}$$
$$w = u_3 = pA_1 \cos(px_3) - ikB_2 \cos(qx_3).$$
$$\sigma_{31} = \mu[2ikpA_1 \cos(px_3) \div (k^2 - q^2)B_2 \cos(qx_3)].$$
$$\sigma_{33} = -\lambda(k^2 + p^2)A_1 \sin(px_3) - 2\mu[p^2 A_1 \sin(px_3) - ikqB_2 \sin(qx_3)].$$

For the symmetric modes, note that the wave structure across the thickness of the plate is symmetric for u and antisymmetric for w. On the other hand, for the antisymmetric modes, the wave structure across the thickness is symmetric for w and hence antisymmetric for u.

It should be noted that this separation of waves into symmetric and antisymmetric modes is an exception rather than a rule. In hollow cylinders, the lack of structure symmetry does not allow this separation. Plate wave modes do exist in an anisotropic plate, but the separation into symmetric and antisymmetric modes is not possible unless the wave propagates along a symmetry axis of the plate. (See Solie and Auld 1973 for an excellent discussion of plate waves in anisotropic plates.)

The constants A_1, A_2, B_1, B_2, as well as the dispersion equations, are still unknown. They can be determined by applying the traction-free boundary condition, which reduces to

$$\sigma_{31} = \sigma_{33} \equiv 0 \quad \text{at} \quad x_3 = \pm d/2 = \pm h \quad \text{(for convenience)} \tag{6.23}$$

in the case of plane strain. The resulting displacement, stress, and strain fields depend upon the type of mode (i.e., symmetric or antisymmetric). However, applying the boundary conditions will give a homogeneous system of two equations for the appropriate two constants A_2, B_1 (for the symmetric case) and A_1, B_2 (antisymmetric case). For homogeneous equations we require that the determinant of the coefficient matrix vanish in order to ensure solutions other than the trivial one. From (6.23) we thus have

$$\frac{(k^2 - q^2)\sin(qh)}{2ikp(\sin(ph))} = \frac{-2\mu ikq(\cos(qh))}{(\lambda k^2 + \lambda p^2 + 2\mu p^2)\cos(ph)}. \tag{6.24}$$

After some manipulation, this may be rewritten as

$$\frac{\tan(qh)}{\tan(ph)} = \frac{4k^2 qp\mu}{(\lambda k^2 + \lambda p^2 + 2\mu p^2)(k^2 - q^2)}. \tag{6.25}$$

The denominator on the right-hand side of (6.25) can be further simplified by using wave velocities and the definitions of p and q from (6.16). From the definition of c_L, we obtain

$$\lambda = c_L^2 \rho - 2\mu. \tag{6.26}$$

Then

$$\lambda k^2 + \lambda p^2 + 2\mu p^2 = \lambda(k^2 + p^2) + 2\mu p^2$$
$$= (c_L^2 \rho - 2\mu)(k^2 + p^2) + 2\mu p^2 \tag{6.27}$$
$$= \rho c_L^2 (k^2 + p^2) + 2\mu k^2. \tag{6.28}$$

Using (6.16) and $c_T^2 = \mu/\rho$ yields

$$\lambda k^2 + \lambda p^2 + 2\mu p^2 = \rho \omega^2 - 2\rho c_T^2 k^2, \tag{6.29}$$

which therefore implies

$$\rho c_T^2 \left[\left(\frac{\omega}{c_T} \right)^2 - 2k^2 \right] = \rho c_T^2 (q^2 - k^2) = \mu(q^2 - k^2). \tag{6.30}$$

Now, substituting (6.30) into an initial form of the dispersion equation (6.25), we obtain a transcendental equation

$$\frac{\tan(qh)}{\tan(ph)} = -\frac{4k^2 pq}{(q^2 - k^2)^2} \quad \text{for symmetric modes.} \tag{6.31}$$

Proceeding along analogous lines, we can show that

$$\frac{\tan(qh)}{\tan(ph)} = -\frac{(q^2 - k^2)^2}{4k^2 pq} \quad \text{for antisymmetric modes.} \tag{6.32}$$

Recall that p and q are as defined in (6.16).

For a given ω and derived k, the displacements can be calculated using the expressions for u and w in (6.21) and (6.22). More explicit expressions are given in Auld (1990).

These equations are known as the Rayleigh–Lamb frequency relations, and they were first derived at the end of the nineteenth century. These equations can be used to determine the velocity (or velocities) at which a wave of a particular frequency (*fh* or *fd* product) will propagate within the plate. Equations of this nature are known as *dispersion relations*. Although the equations look simple, they can be solved only by numerical methods.

6.2.2 The Partial Wave Technique

Although the method just presented is quite simple and elegant, its usefulness is restricted to isotropic plates: only then will the governing equations (6.4) be in such a simple form. For the problem of plate waves in anisotropic plates, the only suitable technique is the partial wave (or transverse resonance) technique. As pointed out in Solie and Auld (1973), the partial wave technique has two major advantages over the method of displacement potentials: (1) it leads more directly to wave solutions, and (2) it provides more insight into the physical nature of the waves.

Keep in mind that the formulation of the free plate problem has in no way changed: we are merely trying another solution method. In the partial wave technique, we try to construct solutions to the problem defined by (6.4) and (6.6) from simple

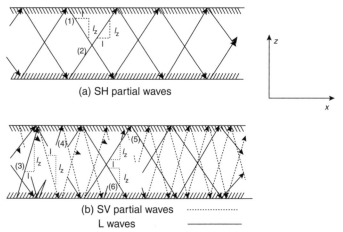

(a) SH partial waves

(b) SV partial waves
L waves

Figure 6.5. Types of partial waves used in the isotropic problem.

exponential-type waves that reflect back and forth between the boundaries of the plate (see Figure 6.5).

We begin by assuming that each of the waves depicted in Figure 6.5 can be expressed as

$$u_j = a_j \exp\left[ik(x + l_z z)\right], \tag{6.33}$$

where $j = x, y, z$ and $l_z = k_z/k_x$ (we are now using $x, y,$ and z, instead of the x_i, to denote position). Also, we are solving the more general problem, that is, with no assumption of plane strain. Note that the x component of each assumed partial wave is the same, which is exactly the statement of Snell's law. Substituting these solutions in (6.33) into a form of Christoffel equations,

$$(k_{il}c_{IJ}k_{Jj} - \rho\omega^2\delta_{ij})u_j = 0, \tag{6.34}$$

where $i, j = x, y, z, I, J = xx, yy, zz, yz, xz, xy$, and

$$k_{il} = \begin{pmatrix} k_x & 0 & 0 & 0 & k_z & k_y \\ 0 & k_y & 0 & k_z & 0 & k_x \\ 0 & 0 & k_z & k_y & k_x & 0 \end{pmatrix}$$

(which is equivalent to using Equation (6.4) after a plane wave solution is assumed) yields a linear homogeneous system of three equations in the three polarization components for each partial wave. The coefficients are functions of the material properties of the plate and also of the (unknown) phase velocity of the plate wave mode. Requiring the determinant to vanish for nontrivial solutions yields a sixth-order equation for l_z that defines the propagation direction of the six partial waves (see Figure 6.5).

Now we know the direction of propagation for each of the partial waves. Hence we can take a linear combination of them in the form

$$u_j = \sum_{n=1}^{6} C_n \alpha_j^{(n)} \exp[ik(x + l_z^{(n)} z)] \quad (j = x, y, z) \tag{6.35}$$

and so try to determine the coefficients C_n ($n = 1, 2, ..., 6$), or wave amplitudes, and thereby satisfy the traction-free boundary conditions (Equation (6.6) evaluated at z

$= \pm h$). Note that the traction-free condition must be satisfied along the entire upper and lower surfaces, so that the partial waves must reflect in such a manner that they reconstruct themselves after returning to the top of the plate. That is, they must be standing waves in the transverse direction (hence the term "transverse resonance"). This explains why, when using the method of displacement potentials, we assumed that the potentials (6.12) and (6.13) had not only a static dependence on x_2 but were also allowed to propagate in the x_1 direction.

The last step in our problem is to substitute this assumed linear combination of partial waves into the boundary condition equations. This gives a system of six (remember that we no longer assume plane strain, so to (6.23) we must add $\sigma_{xz} = 0$) homogeneous linear equations in which the coefficients C_n are now functions of the density, the elastic constants of the plate, and the product hk. Requiring the determinant of this "boundary condition matrix" to vanish (and thus yield nontrivial solutions for the wave amplitudes) gives us the dispersion relations that we seek.

Just as in the solution found by the method of potentials, an infinite number of modes are defined by our dispersion relations. In this case, however, we pick up the additional modes that were lost in that previous method by our assumption of plane strain. (We did not actually need to make that assumption in the method of potentials, but did so for simplicity.) The dispersion relations for these extra modes, known as shear horizontal (SH) modes, are

$$(M\pi)^2 = (\omega h/c_T)^2 - (kh)^2. \tag{6.36}$$

These modes are plotted in Figure 6.6. As can be seen, they are simple hyperbolas in a $(kh, \omega h)$-plane. Complete details of a horizontal shear wave solution in a plate are presented in Chapter 14.

The dispersion relations governing the symmetric and antisymmetric "in-plane" modes (known as Lamb waves) are of course the same as the ones resulting from the method of potentials, equations (6.31) and (6.32). Figure 6.7 shows a sample plot of the dispersion curves for these modes. As mentioned previously, the dispersion equations for the Lamb wave modes are simple in appearance but can be solved only by numerical methods.

6.3 Numerical Solution of the Rayleigh–Lamb Frequency Equations

Recall that the Rayleigh–Lamb frequency equations can be written as

$$\frac{\tan(qh)}{\tan(ph)} = -\frac{4k^2 pq}{(q^2 - k^2)^2} \quad \text{for symmetric modes.} \tag{6.31}$$

$$\frac{\tan(qh)}{\tan(ph)} = -\frac{(q^2 - k^2)^2}{4k^2 pq} \quad \text{for antisymmetric modes.} \tag{6.32}$$

Here p and q are given by

$$p^2 = \left(\frac{\omega}{c_L}\right)^2 - k^2 \quad \text{and} \quad q^2 = \left(\frac{\omega}{c_T}\right)^2 - k^2.$$

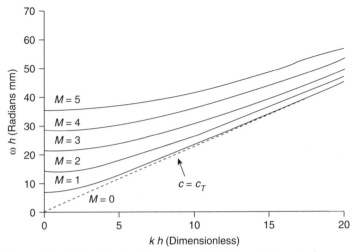

Figure 6.6. SH modes in a free copper plate ($c_T = 2.26$ mm/μs).

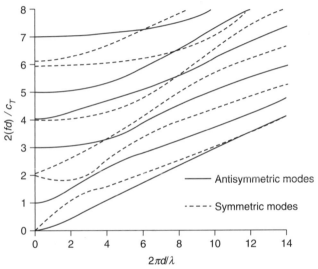

Figure 6.7. Dispersion curves for a free isotropic aluminum plate ($c_L = 6.35$ mm/μs, $c_T = 3.13$ mm/μs, $c_L/c_T = 2.0288$).

The wavenumber k is numerically equal to ω/c_p, where c_p is the phase velocity of the Lamb wave mode and ω is the circular frequency. The phase velocity is related to the wavelength by the simple relation $c_p = (\omega/2\pi)\lambda$.

Equations (6.31) and (6.32) can be considered as relating the frequency ω to the wavenumber k of the Lamb wave modes, resulting in the frequency spectrum, or as relating the phase velocity c_p to the frequency ω, resulting in the dispersion curves. It is known that, for any given frequency, there are an infinite number of wavenumbers that will satisfy (6.31) and (6.32). A finite number of these wavenumbers will be real or purely imaginary, while infinitely many will be complex.

It is often useful to consider various regions of the Rayleigh–Lamb equations for k compared with ω/c_L or ω/c_T (see Graff 1991). Let region 1 be $k > \omega/c_T$; region 2, $\omega/c_T > k > \omega/c_L$; and region 3, $k < \omega/c_L$. In region 1, where $c_p < c_T$, we therefore have

$$\frac{\tanh(q'h)}{\tanh(p'h)} = \left\{ \frac{4p'q'k^2}{(k^2-q'^2)^2} \right\}^{\pm 1} ;$$

from (6.32), $p = ip'$, $q = iq'$, $p'^2 = -p^2$, and $q'^2 = -q^2$ (the exponent +1 is for symmetric and −1 for antisymmetric modes). In region 2, where $c_T < c_p < c_L$, we have

$$\frac{\tan(q'h)}{\tanh(p'h)} = \pm \left\{ \frac{4p'qk^2}{(k^2-q^2)^2} \right\}^{\pm 1} ;$$

in region 3, where $c_p > c_L$, equations (6.31) and (6.32) are unaltered.

A key element to understanding the physical character of dispersion curves is associated with the time harmonic factor $\exp[i(kx - \omega t)]$. Equations (6.31) and (6.32) are functions of k and ω. Noting that k could possibly assume complex values, the physical meaning of k can be discussed as follows.

Let $k = k_r + ik_{im}$. Then the time-harmonic factor becomes:

$$\exp[i(k_r x - \omega t)] \exp[-k_{im}x]. \tag{6.37}$$

There are three possible signed values for k_{im}, and each has a physical interpretation:

$k_{im} < 0$,   the waves grow exponentially with distance;
$k_{im} = 0$,   the waves propagate with no damping;
$k_{im} > 0$,   the waves decay exponentially with distance.

The decaying waves are called "evanescent," which means they will disappear. Their amplitude decreases exponentially with distance from their source or a scattering center. (When studying scattering, they become important in the sense of forming a "near field.") The exponentially growing waves have not been physically observed. It can be concluded that, for the simple unloaded plate problem, only real values for k are necessary or supply information about propagating waves. An example of wave decay with distance can be seen in the water-loaded plate problem. Out-of-plane displacement at the surface of the plate produces normal loading on the fluid and hence leakage into the fluid. In this case, k_{im} would be greater than zero.

When plotting the dispersion curves, we are now interested only in the real solutions of these equations, which represent the (undamped) propagating modes of the structure. It is therefore useful to rewrite (6.31) and (6.32) so that they take on only real values for real or pure imaginary wavenumbers k. This is achieved by the following set of equations:

$$\frac{\tan(qh)}{q} + \frac{4k^2p\tan(ph)}{(q^2-k^2)^2} = 0 \quad \text{for symmetric modes,} \tag{6.38}$$

$$q\tan(qh) + \frac{(q^2-k^2)^2\tan(ph)}{4k^2p} = 0 \quad \text{for antisymmetric modes.} \tag{6.39}$$

The numerical solution of these equations is now relatively simple. The steps in the routine may be listed as follows:

(1) Choose a frequency–thickness product $(\omega h)_0$.
(2) Make an initial estimate of the phase velocity $(c_p)_0$.

(3) Evaluate the signs of each of the left-hand sides of (6.38) or (6.39) (assuming they do not equal zero).

(4) Choose another phase velocity $(c_p)_1 > (c_p)_0$ and re-evaluate the signs of (6.38) or (6.39).

(5) Repeat steps (3) and (4) until the sign changes. Because the functions involved are continuous, a change in sign must be accompanied by a crossing through zero. Therefore, a root m exists in the interval where a sign change occurs. Assume that this happens between phase velocities $(c_p)_n$ and $(c_p)_{n+1}$.

(6) Use some sort of iterative root-finding algorithm (e.g., Newton–Raphson, bisection, ...) to locate precisely the phase velocity in the interval $(c_p)_n < c_p < (c_p)n+1$ where the LHS of the required equation is close enough to zero.

(7) After finding the root, continue searching at this ωh for other roots according to steps (2) through (6).

(8) Choose another ωh product and repeat steps (2) through (7).

This procedure is performed for as many ωh as required.

Here is a helpful tip for performing the calculations: For large ωh, instead of searching for a phase velocity lying in the range $c_R < c_p < \infty$, search for $1/c_p$ in the range $1/c_R < 1/c_p < 0$ (where c_R is the Rayleigh wave velocity of the material). This technique will also be useful for smaller ωh, where c_R should be replaced with the phase velocity of the lowest A0 mode. Note that alternate schemes for the root extraction process are available, including procedures for both real and imaginary roots (if both exist).

As a result of this root extraction, sample dispersion curves for an aluminum plate are presented in Figure 6.8, which depicts modes A0, S0, A1, S1, A2, S2, ... The plate velocity c_{plate}, Rayleigh surface wave velocity c_R, shear (transverse) velocity c_T, and mode cutoff regions are also shown. The classical long wavelength plate velocity; c_{plate}, is equal to $E^{1/2}[\rho(1 - v^2)]^{-1/2}$. Note the limiting values of c_p as fd values become large. The A0 and S0 modes converge to c_R, and the other modes converge to c_T. More will be said about cutoff frequencies in Section 6.7.4.

6.4 Group Velocity

The group velocity c_g can be found from the phase velocity c_p by use of the formula

$$c_g = \frac{d\omega}{dk}.$$

Substituting $k = \omega/c_p$ into this equation yields

$$c_g = d\omega \left[d\left(\frac{\omega}{c_p}\right) \right]^{-1}$$

$$= d\omega \left[\frac{d\omega}{c_p} - \omega \frac{dc_p}{c_p^2} \right]^{-1}$$

$$= c_p^2 \left[c_p - \omega \frac{dc_p}{d\omega} \right]^{-1}.$$

Using $\omega = 2\pi f$, the third equality can be written as

$$c_g = c_p^2 \left[c_p - (fd)\frac{dc_p}{d(fd)} \right]^{-1}, \tag{6.40}$$

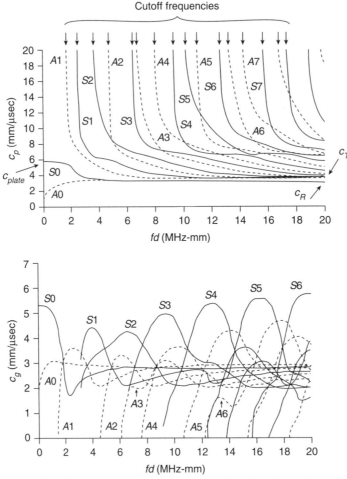

Figure 6.8. Dispersion curves for a traction-free aluminum plate.

where *fd* denotes frequency times thickness. Note that, when the derivative of c_p with respect to *fd* becomes zero, $c_g = c_p$. Note also that, as the derivative of c_p with respect to *fd* approaches infinity (i.e., at cutoff), c_g approaches zero.

6.5 Wave Structure Analysis

It is interesting to study wave structure variation as one increases the *fd* product along a particular mode. The symmetric mode cannot be thought of as simply an in-plane vibration mode. As one moves along the mode, the ratio of the in-plane to out-of-plane displacement changes. Of particular note are the changes on the outside surface of a structure. Similarly, the antisymmetric modes cannot be thought of as a mode with only out-of-plane displacement values.

Figures 6.9 to 6.14 depict solutions at a variety of *fd* values to modes S0, S1, S2, A0, A1, and A2 for an aluminum plate. The values for c_T and c_L are 3.1 mm/μs and 6.3 mm/μs, respectively. Sample points were chosen along the symmetric mode of an

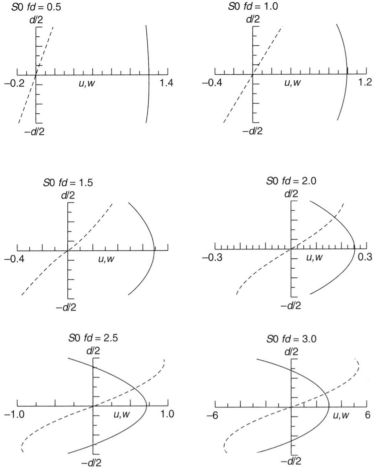

Figure 6.9. Wave structure for various points on the S0 mode of an aluminum plate, showing the in-plane (*u*, solid line) and out-of-plane (*w*, dashed line) displacement profiles across the thickness of the plate.

aluminum plate to show the variation in wave structure that occurs as one moves along the curve (see Figure 6.8 for the basic dispersion curve).

For example, the in-plane displacement is almost constant across the thickness of the plate at low *fd* values, but it becomes heavily concentrated at the center of the plate as the *fd* values increase (see Figure 6.9). In contrast, the out-of-plane displacement component *w*, which is initially close to zero on the outside surface for small *fd* values, becomes dominant on the outside surface as the *fd* values increase to 2 or 2.5. For in-plane distributions along the S1 mode, shown in Figure 6.10, the in-plane displacement on the outside surface is a maximum value while the out-of-plane value is zero. As *fd* increases to 6.0, the gradual changes show out-of-plane dominance with the in-plane component becoming zero. For the S2 mode, however, the in-plane displacement on the outer surface is close to zero for low *fd* values, as shown in Figure 6.11, but becomes dominant as *fd* changes from 5.0 to 7.0. On the other hand, the out-of-plane component goes from dominant to zero as *fd* changes from 5.0 to 7.0.

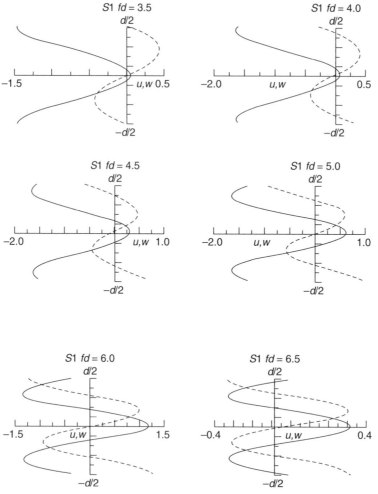

Figure 6.10. Wave structure for various points on the S1 mode of an aluminum plate.

Similar studies are graphed for the antisymmetric modes in Figures 6.12, 6.13, and 6.14. Again, some interesting observations can be made. For example, dominant out-of-plane displacement on the surface for the A1 mode occurs at an *fd* value of 4.5; at this point, the in-plane displacement value is zero.

Use of wave structure can lead to increased wave penetration power along a structure – for example, by avoiding energy leakage from water loading or insulation. Improved sensitivity to certain defects can be obtained as a result of controlling in-plane or out-of-plane impingement at a certain location across the thickness of the structure. Work by Ditri, Rose, and Chen (1991) demonstrates the use of mode selection and wave structure to detect small defects on the surface of a structure. This is accomplished by getting higher energy concentrations on the outside surface. Work is currently in progress to examine other parameter distributions across the thickness of a structure for improved penetration power and/or defect detection sensitivity. Parameters including stresses, strain energy, and power – in addition to the displacement components *u* and *w* – could prove useful.

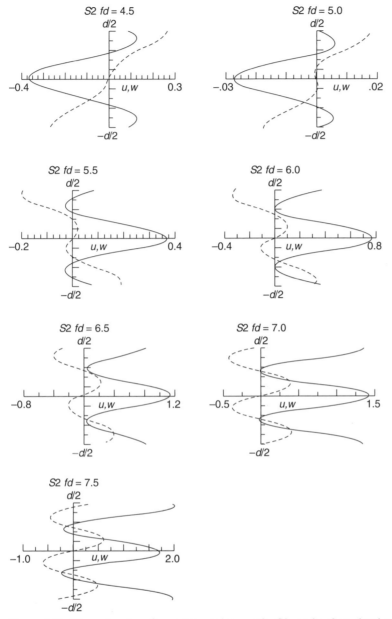

Figure 6.11. Wave structure for various points on the S2 mode of an aluminum plate.

6.6 Compressional and Flexural Waves

Figure 6.15 shows a (highly exaggerated) schematic "snapshot" of the displacement vector field distribution on the surface of a plate and its effect on the shape of the plate. Symmetric mode waves are often termed "compressional" and antisymmetric waves are known as "flexural." The particle displacements shown are the vector sums of the in-plane (u) and out-of-plane (w) particle displacement components.

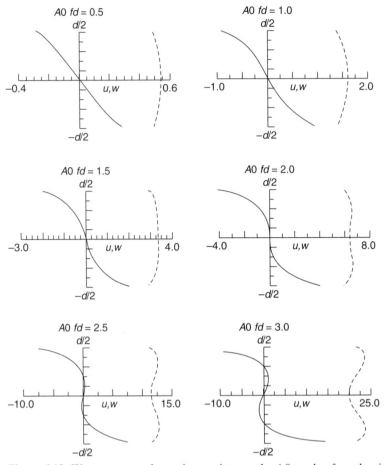

Figure 6.12. Wave structure for various points on the A0 mode of an aluminum plate.

6.7 Miscellaneous Topics

A great challenge exists with respect to the interpretation of results obtained in the Lamb wave propagation problem. Even though one might very carefully formulate the theoretical approach to the problem and develop a detailed numerical solution (for the phase velocity, group velocity, and wave structure values of displacements and stress), the ability to use these curves in a practical sense – for example, in ultrasonic nondestructive testing (NDT) – requires a great deal of study and attention. A topic of great concern is associated with mode selection, isolation, and control in order to study specific wave propagation characteristics in a plate. The subject of generation and reception also becomes critical with respect to the use of the specific modes and frequencies chosen for a particular problem. This section presents miscellaneous topics associated with interpretation and utilization of some of the results associated with Lamb wave propagation. The first topic we discuss concerns the practical reasons for locating Lamb wave propagation characteristics with a dominant longitudinal displacement. We also discuss the physical interpretation of the zeros and poles for a fluid-coupled elastic layer. Finally, we present a problem

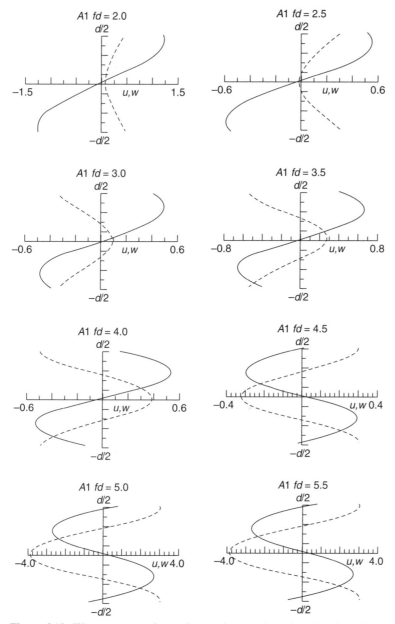

Figure 6.13. Wave structure for various points on the A1 mode of an aluminum plate.

of interest in the immersion testing of a plate involving nonspecular reflection and transmission characteristics for layered media.

6.7.1 Lamb Waves with Dominant Longitudinal Displacements

Of great interest in the utilization of dispersion curves are certain points where phase velocity and frequency have particular or unusual wave structural characteristics. One might think of points where displacement or stress is concentrated on the

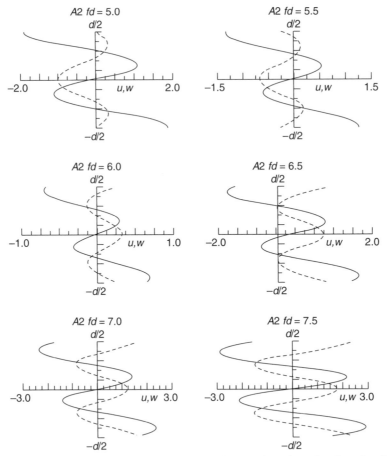

Figure 6.14. Wave structure for various points on the A2 mode of an aluminum plate.

outside surface or perhaps at the center section of the plate. On the other hand, one might examine the in-plane and out-of-plane displacement distributions in such a manner as to achieve dominant in-plane (or out-of-plane) characteristics on the outside surface of the plate.

Imagine, for example, a plate immersed in water. One can appreciate that energy loss would be minimal to the fluid if a dominant in-plane or longitudinal displacement were available on the outside surface of the plate. In fact, from a theoretical standpoint, the loss would be zero because we are actually trying to load the fluid with a shear force. Because shear waves cannot propagate into the fluid, the energy of propagation in the plate would be strongest at these points since the energy is retained by the plate. If we were to locate points with a dominant out-of-plane displacement, it would be easy to see that such normal pressure loading onto the fluid would propagate into the fluid as the Lamb wave propagates along the plate. This is what we call the propagation of a *leaky Lamb wave* in the structure.

In immersion NDT, we might wish to make use of this leaky Lamb wave to look at distortions in the reflection pattern as the wave travels along the plate and encounters defects or inhomogeneities in the structure. On the other hand, if we were using a pulse-echo or through-transmission technique – with the transducer actually located on the plate – then we would want strong penetration power and hence minimal leakage into the fluid; this would call for the elimination or reduction of the leaky Lamb wave component.

Using elasticity concepts, one can embark upon a detailed study of the wave structural characteristics of the in-plane and out-of-plane displacement while moving along a particular mode. The continuous variations and shifting of characteristics from the in-plane and out-of-plane distribution as we move along the modes are interesting and can be useful for ultrasonic testing. Sample results were presented in Figures 6.9 to 6.14.

This section will focus on finding specific points where there is a dominant longitudinal displacement and very minimal out-of-plane component on the surface of the plate. The following discussion is taken from a paper by Pilarski, Ditri, and Rose (1993). Remarks are presented on symmetric Lamb waves with a dominant longitudinal displacement.

We prove that the normal component of the particle displacement vector vanishes on the free surfaces of an isotropic, homogeneous plate for nonzero-order symmetric Lamb waves; this conclusion was stated without proof by Viktorov (1967, p. 121). We also give an expression for the frequency–thickness products where this vanishing occurs, which enables practical (i.e., experimental) selection of such modes. Other interesting features of such modes, such as the independence of their group velocity on the mode's order or its nondispersivity, are also addressed. The features discussed here are of great practical importance in the NDT of plates loaded by liquids and in the ultrasonic tensometry of a thin plate (see Pilarski et al. 1992).

For Lamb waves in an isotropic, homogeneous solid layer immersed in a liquid, Viktorov (1967, p. 121) asserts the following: For symmetric modes, when the phase velocity reaches the value of the velocity of bulk longitudinal waves, the vertical (normal to the plate surface) component of the displacement vector vanishes on the free surfaces. Using Viktorov's notation, this can be written as

$$\lim_{r \to 0} w_s(z) = 0 \text{ for } z = \pm d/2 = \pm h \quad (\text{for convenience}), \tag{6.41}$$

where w_s is the amplitude of the normal component of the displacement vector for a symmetric mode and d is the thickness of the plate. The quantity in this case is r, defined by

$$r^2 = k^2 - k_L^2, \tag{6.42}$$

where k and k_L are the wavenumbers related to the phase velocity of the plate mode $c = c_p$ and to the velocity of bulk longitudinal waves c_L, respectively. The condition $r \to 0$ implies that:

$$\lim_{r \to 0} c_p = c_L. \tag{6.43}$$

We would like to (a) prove the aforementioned conclusion about the vanishing of the vertical displacement component and (b) give an expression for the frequency-thickness products where this occurs.

For NDT of plates loaded by liquids, appropriate mode selection can substantially reduce leakage of energy into the liquid owing to the vanishing normal surface

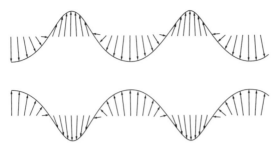

Compressional waves in a plate (symmetric mode)

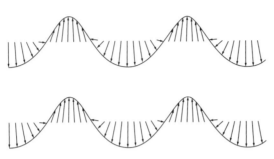

Flexural waves in a plate (antisymmetric mode)

Figure 6.15. Compressional and flexural wave particle displacement schematic.

displacement. For ultrasonic tensometry of a thin plate, symmetric modes that are higher than fundamental order are useful with inhomogeneously stressed plates owing to the dominance of longitudinal particle vibration, for which the values of an acoustoelastic constant (for the longitudinal applied or residual stresses) are the largest, and because the plates' field distributions vary with frequency (see Pilarski et al. 1992).

We will start the analysis from the characteristic equation for symmetric modes found in Viktorov (1967, eq. II.4):

$$\Omega_s = (k^2 + s^2) \cosh(rh) \sinh(sh) - 4k^2 rs \sinh(rh) \cosh(sh) = 0. \qquad (6.44)$$

Here, $s^2 = k^2 - k_T^2$ where k_T is the wavenumber related to the bulk transverse wave velocity c_T. Consequently, for r approaching zero, we arrive at the equality

$$\sinh(sh) = 0, \qquad (6.45)$$

which is satisfied by

$$(sh)_n = in\pi \quad \text{for } n = 0, 1, 2, \dots. \qquad (6.46)$$

Finally, using that

$$s \to k_L^2 - k_T^2 \quad \text{as } r \to 0,$$

we obtain the following expression for the frequency–thickness product at which the nth symmetric mode has a phase velocity equal to c_L:

$$(fd)_n = \frac{nc_T}{\sqrt{1 - (c_T / c_L)^2}} \quad \text{for } n = 1, 2, \dots. \qquad (6.47)$$

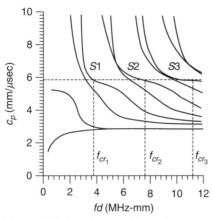

Figure 6.16. Lamb wave dispersion curves for steel, showing dominant in-plane displacement points.

Note that $n = 0$ is excluded: for zero frequency, no mode exists that has a phase velocity equal to the bulk longitudinal wave velocity. Equation (6.47) has great practical significance in that it enables us to select a mode by choosing the appropriate frequency for a given plate thickness and (e.g., in the wedge technique of Lamb wave generation) by adjusting the angle of incidence of the ultrasonic wave onto the plate surface. The first such modes and their frequencies are marked on the phase velocity dispersion curve diagram (Figure 6.16) for a 1-mm-thick steel plate.

To prove the validity of expression (6.41), the first of the two equations given by Viktorov (1967, eq. II.2) will be used to determine the relationship between the two constants A_s and D_s occurring in the expressions for the scalar potentials ϕ_s and Ψ_s:

$$D_s = \frac{-(k^2 + s^2)\cosh(rh)}{2iks\cosh(sh)} A_s. \qquad (6.48)$$

Then, given the well-known relation

$$w_s = \frac{\partial \phi_s}{\partial z} - \frac{\partial \psi_s}{\partial x}, \qquad (6.49)$$

we substitute (6.48) into (6.49) and use the following equations:

$$\begin{aligned} \varphi_s &= A_s \cosh(rz)e^{i(kx-\omega t)}, \\ \Psi_s &= D_s \sinh(sz)e^{i(kx-\omega t)} \end{aligned} \qquad (6.50)$$

(cf. eq. II.1 in Viktorov 1967).

The displacement component w_s may be determined as

$$w_s = \left(r\sinh(rz) - \frac{(k^2 + s^2)\cosh(rh)\sinh(sz)}{2s\cosh(sh)} \right) e^{i(kx-\omega t)} A_s. \qquad (6.51)$$

In the limit, for r approaching zero, (6.49) can be written as

$$\lim_{r \to 0} w_s = \frac{-h(k^2 - n^2\pi^2/h^2)\sin(n\pi z/h)}{2n\pi\cos(n\pi)} e^{i(kx-\omega t)} A_s \qquad (6.52)$$

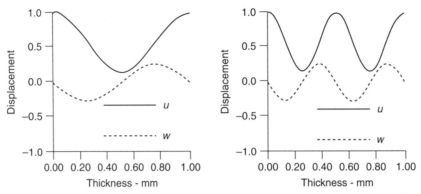

Figure 6.17. Cross-sectional normalized distribution of normal (w) and longitudinal (u) displacements for: (a) mode S1 at $(fd)_1 = 3.798$ MHz-mm and (b) mode S2 at $(fd)_2 = 7.596$ MHz-mm; both modes are shown in Figure 6.16.

(bearing in mind (6.46) and well-known relations between trigonometric and hyperbolic functions of complex variables). It is readily seen that (6.52) can be equal to zero for any n only if z is equal to $\pm h$ or zero – that is, only on the free surfaces or in the midplane. Figure 6.16 shows a sample dispersion curve for discussion purposes.

Figure 6.17 displays calculated results for both the normal w_s and the tangential (longitudinal) u_s displacement distributions across the 1-mm-thick steel plate for the first and second symmetric modes at their respective critical frequency–thickness products. The displacement amplitudes have been normalized by the largest value of the longitudinal displacement. For both modes we can see that the normal displacement of the free surfaces vanishes and that the longitudinal displacement dominates across the entire thickness. Figure 6.18 plots the normalized time-averaged power flux distributions per unit waveguide width for the S1 and S2 modes of Figure 6.16. For these modes, the higher the order, the more the energy is concentrated near the free surfaces. This means that the higher the order of the mode, the smaller the distance between free surfaces and the first minimum displacement value.

Calculations of group velocities for frequencies near the critical frequencies of modes S1, S2, and S3 have revealed that, although for these modes the phase velocity dispersive curves look almost flat, the maxima of the relevant group velocities are shifted slightly to other frequencies. This is shown in Figure 6.19 for the first two nonzero-order symmetric modes. The maximum shift of frequency is less than $\pm 7\%$. Such a small shift from the critical frequency f_{cr} to the frequency of maximum group velocity $f_{g\,max}$ leads to a small nonzero value of the normal displacement on the surfaces, while the distribution of both displacement components remain almost unchanged (see Figure 6.20).

It is also worth noting that, for all nonzero-order symmetric modes, the values of group velocities for any of the fd products given by (6.47) are the same. This conclusion may be proved as follows. The group velocity can be found from the implicit form of the dispersion relation, Equation (6.44), as

$$c_g = -\frac{\partial \Omega_s / \partial k}{\partial \Omega_s / \partial \omega}. \qquad (6.53)$$

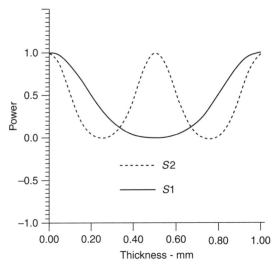

Figure 6.18. Normalized time-averaged power flux distribution for the first two nonzero-order symmetric modes of Figure 6.16: S1 at $(fd)_1 = 3.798$ MHz-mm and S2 at $(fd)_2 = 7.596$ MHz-mm.

In the $r \to 0$ (i.e., $c_p \to c_L$) limit and for any of the fd products given by (6.47), we arrive at the following relation:

$$\lim_{\substack{r \to 0 \\ fd \to (fd)_n}} c_g = c_L \left(\frac{(c_T / c_L)^2 + 8P}{1 + 8P} \right),\tag{6.54}$$

where

$$P = \frac{1 - (c_T / c_L)^2}{[2 - (c_L / c_T)^2]^2}.\tag{6.55}$$

We can see that (6.54) and (6.55) are independent of n, as stated. Note also that these two equations are written in nondimensional form, showing that the group velocity normalized by c_L is a function only of the Poisson ratio v:

$$\frac{c_g}{c_L} = \frac{(1-2v)(v^2 - 4v + 2)}{2[v^2(1-v) + (1-2v)^2]}.\tag{6.56}$$

When we examine the wave structure of various modes (Figure 6.18), it is obvious that there is a dramatic variation between different modes and also between different fd values of the same mode. Of particular interest on the symmetric modes are the points where $c_p = c_L$ (see Figure 6.21). Note that, in Figure 6.22, the group velocity value at these special points (where c_p equals c_L) is the same. Two of these points are S1 ($fd = 3.56$) and S2 ($fd = 7.12$). As expected from previous analysis, there is no out-of-plane displacement at the plate surfaces (see Figure 6.23).

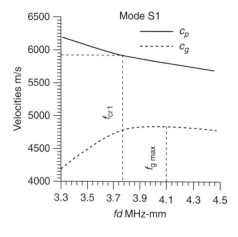

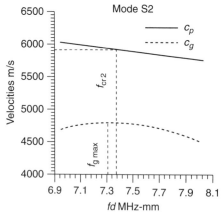

Figure 6.19. Phase and group velocities for the first two nonzero-order symmetric modes marked in Figure 6.16.

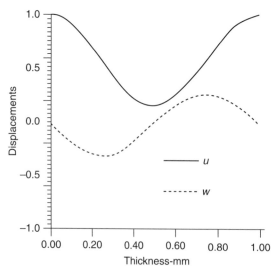

Figure 6.20. Normalized cross-sectional displacement distributions for mode S1 at f_{gmax} as shown in Figure 6.19.

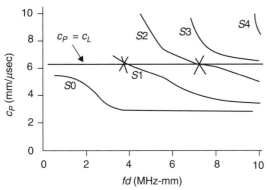

Figure 6.21. Dispersion curve for the symmetric modes of an aluminum layer, showing the intersections of the modes with a phase velocity equal to the longitudinal wave velocity.

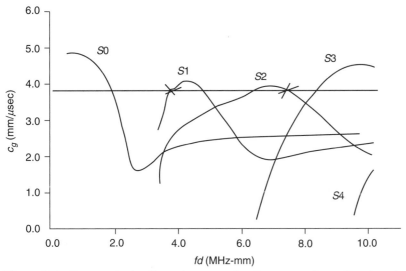

Figure 6.22. Group velocity dispersion curve for the symmetric modes of an aluminum layer, depicting the group velocity of the intersection points shown in Figure 6.21.

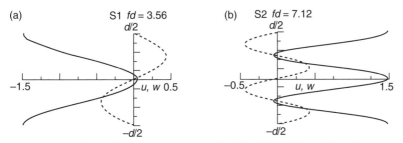

Figure 6.23. Wave structure for the S1 and S2 modes where $c_p = c_L$.

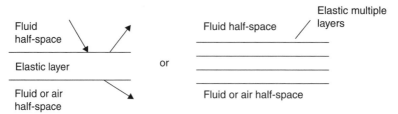

Figure 6.24. Boundary value problem for liquid-loaded layer.

6.7.2 Zeros and Poles for a Fluid-Coupled Elastic Layer

In this section we address the influence of a fluid load on an elastic layer. This problem of oblique incidence reflection factor and a corresponding physical interpretation of zeros and poles was studied by Chimenti and Rokhlin (1990). If one were to tackle the boundary value problems illustrated in Figure 6.24, using the appropriate boundary conditions at each interface, one would come up with an expression for reflection factor that would take on a special form (this expression is given by Chimenti and Rokhlin). Our emphasis will be placed on examining very small densities of the fluid with respect to the elastic layer, which represents the practical problem from an ultrasonic NDT point of view. The form of the oblique incidence reflection factor R in (6.57) can be studied in more detail by examining the formula's poles and zeros. In its general form, R can be written as

$$R = \frac{AS - Y^2}{(S + iY)(A - iY)},\qquad(6.57)$$

where A and S are respectively antisymmetric and symmetric Lamb mode terms (see Chimenti and Rokhlin 1990):

$$A = \frac{(q^2 - 1)^2}{q}\tan\left(kp\frac{d}{2}\right) + 4p\tan\left(kq\frac{d}{2}\right),$$
$$S = \frac{(q^2 - 1)^2}{q}\cot\left(kp\frac{d}{2}\right) + 4p\cot\left(kq\frac{d}{2}\right).$$

Here $p^2 = (c/c_L)^2 - 1$, $q^2 = (c/c_T)^2 - 1$, and $k = \omega/c$; the terms c and d denote phase velocity and thickness of the plate, respectively.

The influence of the fluid is determined by Y:

$$Y = \frac{\rho f}{\rho}\left(\frac{c}{c_T}\right)^4\frac{p}{qm},\quad\text{where } m = \left(\frac{c}{c_f}\right)^2 - 1$$

The index f is associated with the fluid parameters; hence ρ_f = fluid density and c_f = fluid bulk velocity (ρ = plate density).

It can be shown that, for small densities, the real part of the poles may be used to produce the dispersion curves for this fluid-loaded structure, as well as expressions associated with the leaky Lamb waves. The interpretations of these results are quite interesting. For example, if the reflection factor $R = 0$ then (at least from a physical point of view) we could measure these points to produce a dispersion curve of the structure. Keep in mind, though, that the energy may go into the layer as a guided wave but could also be entirely transmitted. Or, in the most general case, a little of each is obtained – with some transmission factor and some energy going into guided waves. For small densities of the fluid with respect to the plate, the poles are close to the zero values. Note that a reflection factor of some amplitude divided by zero would equal infinity, which is not possible; on the other hand, a reflection factor of zero divided by zero is merely undefined.

In studying this material, we can recall the Cremer hypothesis (which is actually Snell's law) and so derive a relationship for oblique incidence and for energy going along the plate. Note that the poles equal to zero correspond to the leaky waves, not to the reflection factor from the plate. If the phase velocity of the waves in the plate is greater than in the fluid, then energy will leak into the fluid. As a result, the wavevector k is complex (with a positive imaginary part) and so the solution will attenuate with distance along the plate. The basic explanation is captured by

$$u = Ae^{i(k_r x - wt)}e^{-\alpha x}, \tag{6.58}$$

where $k = k_r + i\alpha$ and $e^{-\alpha x}$ is the decay term.

For reflection factor values of zero, the wave behaves as if the plate were completely transparent to the incident wave. But for the Lamb wave angle corresponding to the real part of the pole, we can have leaky Lamb waves on both sides of the plate. The pole equaling zero does not correspond to full reflection factor or full transmission factor; in this case, the reflection factor would have intermediate values or even a zero component.

More on this and related subjects will be presented later in Section 12.5.

6.7.3 Mode Cutoff Frequency

Mode cutoff values occur at specific fd values for modes higher than S0 and A0. At these points, the phase velocity approaches infinity as the group velocity approaches zero. Hence, these frequency values occur whenever standing longitudinal or shear waves are present across the thickness of the plate. See Graff (1991) for an excellent discussion on this topic.

Mode cutoff values can be calculated by examining a limiting condition of $k \rightarrow 0$. In this case, the Rayleigh–Lamb frequency equation of the form (6.24) becomes (for the symmetric case)

$$\sin qh \cos ph = 0. \tag{6.59}$$

The solutions for the fd values at cutoff frequency can therefore be calculated. For this symmetric case, let $qh = n\pi$ for $n = 0, 1, 2, \dots$. We then have

$$qh = \frac{\omega}{c_T}\frac{d}{2} = \frac{2\pi}{c_T}\frac{fd}{2} = n\pi \tag{6.60}$$

or $fd = nc_T$; that is,

$$fd = \{c_T, 2c_T, 3c_T, \dots\}. \tag{6.61}$$

Alternatively, let $ph = n(\pi/2)$ for $n = 0, 1, 2, \ldots$. Then

$$ph = \frac{\omega}{c_L}\frac{d}{2} = \frac{2\pi}{c_L}\frac{fd}{2} = \frac{n\pi}{2} \tag{6.62}$$

or $fd = nc_L/2$; that is,

$$fd = \left\{ \frac{c_L}{2}, \frac{3c_L}{2}, \frac{5c_L}{2}, \ldots \right\}. \tag{6.63}$$

For the antisymmetric case, similar algebraic manipulation will yield

$$fd = \left\{ \frac{c_T}{2}, \frac{3c_T}{2}, \frac{5c_T}{2}, \ldots \right\} \text{ and} \tag{6.64}$$

$$fd = \{c_L, 2c_L, 3c_L, \ldots\}. \tag{6.65}$$

These values of the cutoff frequency can all be found on the dispersion curves (see e.g., Figure 6.8).

Many interesting aspects of mode cutoff frequency could be utilized for corrosion detection and thickness measurement in a variety of different structures (sec Rose et al. 1997a); the same can be said for frequency shifting and group velocity measurement. Creative applications of mode-related features to ultrasonic nondestructive evaluation (NDE) have certainly proved useful.

6.8 Exercises

1. From Helmholtz decomposition, show that

$$\bar{\mu} = \nabla\Phi\left(X_1, X_2, X_3, t\right) + \nabla \times \bar{\psi}\left(X_1, X_2, X_3, t\right)$$

   leads to equations (6.8) and (6.9) from (6.4).
2. Outline the development of equations (6.14), (6.15), and (6.16).
3. Show the development of equations (6.24) and (6.25) from (6.23).
4. Show how satisfying the boundary condition in (6.23) can lead to the Rayleigh–Lamb frequency equation (6.32) for antisymmetric modes.
5. Derive expressions for the plate wave velocity and the long wavelength limit compared with the thickness of the plate. Show that $c_{\text{plate}} = \sqrt{E / \rho(1 - v^2)}$. What is the high-frequency wave velocity limit? Show both on a dispersion curve.
6. Make a detailed sketch of symmetric and antisymmetric cross-sectional Lamb wave propagation possibilities in a plate for both u and w displacement components. Explain!
7. List steps for proving that the in-plane vibration is dominant (and that the out-of-plane component of particle velocity is zero) at the intersection points of the longitudinal wave velocity with higher-order symmetrical modes.
8. Why is it possible to generate dominant in-plane particle displacement on the surface of a plate with an oblique incidence wedge over a fluid film?
9. How could you distinguish ice from water as a contaminant on a test surface?
10. Calculate fd values of points of zero out-of-plane displacement on the dispersion curve for aluminum.

11. Assuming that the reflection coefficient can be written as in (6.57), how would you solve for zero reflection? How could you use this expression to produce a dispersion curve?

12. In a reflection factor problem, what does $R = 0$ imply?

13. When could the zeros of the reflection factor formula be used to produce a dispersion curve?

14. What does $T = 0$ imply? (T denotes transmission coefficient.)

15. If the roots of a characteristic equation are complex (say, $k_r + iB$) then what is physically implied with respect to wave propagation in a fluid-loaded structure and in the fluid itself?

16. How could nonspecular reflection occur from a planar object or multilayer planar structure? What instrumentation parameters affect the nonspecular reflection pattern?

17. How could you concentrate ultrasonic energy close to the surface in a Lamb wave plate experiment?

18. Compute the angle of incidence required in a Plexiglas wedge to produce a Lamb wave in a plate with phase velocity equal to 5 mm/μs.

19. How would you select a nondispersive mode from a phase velocity or group velocity dispersion diagram?

20. Make a sketch of the vibration pattern of a plate undergoing antisymmetrical (flexural) vibration motion. Show typical in-plane and out-of-plane particle displacement vectors.

21. Make a sketch of the vibration pattern of a plate undergoing symmetrical (compressional) vibration motion. Show typical in-plane and out-of-plane particle displacement vectors.

22. What might happen as a particular guided wave mode impinges onto the free edge of the plate? Why? Use a sketch to illustrate a possible scenario.

23. What are the specific boundary conditions in the surface wave problem? The plate wave problem? The two-layer plate problem?

24. Dispersion curves for a plate converge to a particular phase velocity value for large fd values. Explain.

25. Show that, for a complex wavevector, there is attenuation of the resulting displacement field. Why does this occur?

26. Plot typical wave structure (in-plane and out-of-plane distribution) for symmetric and antisymmetric modes.

27. What mode and frequency would you select if you were to use guided waves on the wing of an aircraft to detect and discriminate between ice and water? Explain.

28. Derive expressions for the Rayleigh–Lamb frequency equations for three regions: $c_p < cT; cT < cp < cL; cp > c_L$.

29. Illustrate graphically a root extraction procedure to produce a dispersion curve.

30. List experimental methods of generating Lamb waves in a plate.

31. What angle could be used to generate a specific mode and frequency value in an aluminum plate? Give an example.

32. What are the group velocity values at the zero out-of-plane displacement points on the symmetric modes of a dispersion curve for an aluminum plate?

33. Calculate all cutoff frequency values for an aluminum plate and show the results on the dispersion curve in Figure 6.8.

34. Consider a guided wave experiment in an aluminum plate at a fixed angle and phase velocity value. Could thickness be measured by either a tone-burst frequency sweep or a Fourier transform of a shock-excited broad–frequency bandwidth transducer? Explain your answer.

35. How could cutoff frequency be used to estimate remaining wall thickness in a corrosion detection experiment?

36. In general, how does one compute the mode cutoff values?

6.9 REFERENCES

Achenbach, J. D. (1984). *Wave Propagation in Elastic Solids*. New York: North-Holland.

Auld, B. A. (1990). *Acoustic Fields and Waves in Solids*, 2nd ed., vols. 1 and 2. Malabar, FL: Kreiger.

Chimenti, D. E., and Rokhlin, S. I. (1990). Relationship between leaky Lamb modes and reflection coefficient zeroes for a fluid-coupled elastic layer, *J. Acoust. Soc. Am.* 88(3): 1603–11.

Ditri, J., Rose, J. L., and Chen, G. (1991). Mode selection guidelines for defect detection optimization using Lamb waves, in Proceedings of the 18th Annual Review of Progress in Quantitative NDE, vol. 11, pp. 2109–15. New York: Plenum.

Graff, K. F. (1991). *Wave Motion in Elastic Solids*. New York: Dover.

Pilarski, A., Ditri, J. J., and Rose, J. L. (1993). Remarks of symmetric Lamb waves with dominant longitudinal displacements, *J. Acoust. Soc. Am.* 93(4) (part 1): 2228–30.

Pilarski, A., Szelazek, J., Deputat, J., Ditri, J., and Rose, J. L. (1992). *High-frequency Lamb modes for ultrasonic tensometry*, in C. Hallai and P. Kulcsar (Eds.) Non-Destructive Testing 92 (Proceedings of 13th World Conference on NDT), pp. 1044–8. Amsterdam: Elsevier.

Rose, J. L., Jiao, D., Pelts, S. P., Barshinger, J. N., and Quarry, M. J. (1997). Hidden corrosion detection with guided waves, Paper no. 292, NACE International Corrosion 97 (March 10–14), New Orleans.

Solie, L. P., and Auld, B. A. (1973). Elastic waves in free anisotropic plates, *J. Acoust. Soc. Am.* 54:1.

Viktorov, I. A. (1967). *Rayleigh and Lamb Waves – Physical Theory and Applications*. New York: Plenum.

7 Surface and Subsurface Waves

7.1 Background

The existence of surface waves was predicted theoretically over a century ago. Elastic waves propagating along the surface of a half-space were first predicted by Lord Rayleigh in 1885 in his paper *"On waves propagating along the plane surface of an elastic bar,"* submitted to the Proceedings of the London Mathematical Society. It is telling that this paper was submitted to a mathematical society and not a physical society, as such surface waves were primarily a mathematical concept, although Lord Rayleigh did suspect that they would be relevant to seismology. By the middle of the twentieth century, however, surface waves began to enter into mainstream technological applications. These waves, often referred to as Rayleigh waves or surface acoustic waves (SAW), are now being employed in a number of areas of science and technology, including ultrasonic NDE and SHM, seismology, and electronic circuitry. There is much literature on this subject, including for example Chadwick and Smith (1977), Farnell (1970), Pollard (1977), and Viktorov (1967). Experimental evidence was first obtained in observing wave propagation over the surface of the earth (as a result of earthquakes) and subsequent mode conversion at the earth's surface. Observations were made regarding the unusual behavior of energy decay with increased depth and the ability of waves to travel along curved surfaces.

This chapter examines surface waves on an isotropic, homogeneous, linear elastic semi-space. We take a rather classical approach to the problem, one that is based on potential functions and boundary conditions for a free surface. Assumptions of isotropy, homogeneity, and linear elastic response will also be made. For more detail, see Auld (1990), Basatskaya and Ermolov (1980), Couchman and Bell (1978), Heelan (1953), Kolsky (1963), Nikiforov and Kharitonov (1981), Pilarski and Rose (1989), Uberall (1973), and Viktorov (1967).

7.2 Surface Waves

A general derivation of the equation for surface wave propagation will be treated next. As Lord Rayleigh did, we will assume that the material we are concerned with is a homogeneous, isotropic, linear elastic half-space with a free surface. Elastic waves in such media obey Navier's governing equation:

With contributions from Cody Borigo.

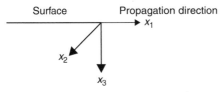

Figure 7.1. Coordinate system used $(x_1, x_2, x_3$ or $x, y, z)$.

$$(\lambda + \mu)\nabla\nabla \cdot \bar{u} + \mu\nabla^2\bar{u} = \rho\frac{\partial^2\bar{u}}{\partial t^2}, \tag{7.1}$$

in which $\bar{u}$ is the displacement.

As a result of Helmholtz decomposition of the displacement vector, which is divided into two components, we have

$$\bar{u} = \nabla\varphi + \nabla \times \bar{\psi}, \quad \text{in which } \nabla \cdot \bar{\psi} = 0. \tag{7.2}$$

Taking the option to use grad, curl, or rot notation, we find that

$$\bar{u} = \text{grad } \phi + \text{rot } \bar{\psi}.$$

Substituting the Helmholtz decomposition into Navier's equation yields

$$(\lambda + \mu)\nabla\nabla \cdot (\nabla\varphi + \nabla \times \bar{\psi}) + \mu\nabla^2(\nabla\varphi + \nabla \times \bar{\psi}) = \rho\left(\nabla\frac{\partial^2\varphi}{\partial t^2} + \nabla \times \frac{\partial^2\bar{\psi}}{\partial t^2}\right). \tag{7.3}$$

By separating like terms and utilizing several vector identities, this can be simplified to

$$\nabla\left[(\lambda + 2\mu)\nabla^2\varphi - \rho\frac{\partial^2\varphi}{\partial t^2}\right] + \nabla \times \left[\mu\nabla^2\bar{\psi} - \rho\frac{\partial^2\bar{\psi}}{\partial t^2}\right] = 0. \tag{7.4}$$

This form of Navier's governing wave equation can be separated into two simple wave equations, one for dilatational waves and one for rotational or shear waves (see e.g., Kolsky 1963; Pollard 1977):

$$\begin{aligned} \nabla^2\phi - \frac{1}{c_L^2}\ddot{\phi} &= 0, \\ \nabla^2\bar{\psi} - \frac{1}{c_T^2}\ddot{\bar{\psi}} &= 0, \end{aligned} \tag{7.5}$$

where c_L is longitudinal and c_T shear velocity.

For the plane two-dimensional case, assuming a plane strain formulation, the only displacement that occurs is in the sagittal plane formed by the x_1 and x_3 unit vectors.

$$\bar{u} = (u_1, 0, u_3)$$

(see the coordinate system in Figure 7.1). Therefore, the solution can be obtained from the two functions ϕ and ψ:

$$\phi = \phi(x_1, x_3, t); \tag{7.6}$$

the values of ψ_1 and ψ_3 are zero or constant because $u_2 = 0$, given

$$\bar{u} = \nabla \times \bar{\psi} \quad \text{and} \quad \nabla \cdot \bar{\psi} = 0.$$

Hence $\psi = \psi_2 = \psi(x_1, x_3, t)$. Note that ψ_2 motion is in the (x_1, x_3)-plane only. In Lamé form (7.2) can therefore be written as

$$\begin{aligned}
u &= u_1 = \varphi_{,1} - \psi_3 = \varphi_{,x} - \psi_{,z} \\
w &= u_3 = \varphi_{,3} + \psi_1 = \varphi_{,z} + \psi_{,x}
\end{aligned} \tag{7.6}$$

We are seeking the harmonic solution of (7.5), which represents harmonic waves traveling in the direction x_1. We have

$$\phi = D_1(Z)e^{i(kx-\omega t)}, \tag{7.7}$$

$$\psi = D_2(Z)e^{i(kx-\omega t)}, \tag{7.8}$$

Substituting trial solutions (7.7) and (7.8) into the dilatational wave equation of (7.5) yields

$$\nabla^2 \varphi - \frac{1}{c_L^2}\ddot{\varphi} = \left[-k^2 D_1(z) + D_1(z)_{,zz} + \frac{\omega^2}{c_L^2}D_1(z) \right] e^{i(kx-\omega t)} = 0. \tag{7.9}$$

This differential equation can be simplified to

$$D_1(z)\xi^2 = D_1(z)_{,zz} \tag{7.10}$$

where

$$\xi^2 = k^2\left(1 - \frac{c^2}{c_L^2}\right).$$

This allows us to guess the function $D_1(z)$ to be

$$D_1(z) = A_1 e^{-\xi z} + A_2 e^{\xi z} = A_1 e^{-k\sqrt{1-\frac{c^2}{c_L^2}}\,z} + A_2 e^{k\sqrt{1-\frac{c^2}{c_L^2}}\,z}. \tag{7.11}$$

The impractical part of the solution, which doesn't attenuate, is discarded. A similar method is used to solve for ψ. Thus we obtain

$$\phi = A_1 e^{-kqz}e^{ik(x-ct)}, \tag{7.12}$$

$$\psi = B_1 e^{-ksz}e^{ik(x-ct)}, \tag{7.13}$$

where

$$q = \sqrt{1 - \left(\frac{c}{c_L}\right)^2}, \quad s = \sqrt{1 - \left(\frac{c}{c_T}\right)^2}, \quad c = \frac{\omega}{k},$$

and A_1, B_1 are arbitrary constants.

From equations (7.7), (7.12), and (7.13), it then follows that

$$u = \frac{\partial\varphi}{\partial x} - \frac{\partial\psi}{\partial z} = k(iA_1 e^{-kqz} + sB_1 e^{-ksz})e^{ik(x-ct)}, \tag{7.14}$$

$$w = \frac{\partial \varphi}{\partial z} + \frac{\partial \psi}{\partial x} = k(-qA_1 e^{-kqz} + iB_1 e^{-ksz})e^{ik(x-ct)}. \tag{7.15}$$

To obtain the stresses in the material due to Rayleigh surface waves, we must use the constitutive equation for homogeneous isotropic materials

$$\sigma_{ij} = \lambda \delta_{ij}\varepsilon_{kk} + 2\mu\varepsilon_{ij}. \tag{7.16}$$

To use Equation (7.16) to calculate the stress components, we must first solve for the strain components:

$$\varepsilon_{xx} = \frac{\partial u}{\partial x} = k^2(-A_1 e^{-kqz} + isB_1 e^{-ksz})e^{ik(x-ct)},$$

$$\varepsilon_{yy} = \frac{\partial v}{\partial y} = 0,$$

$$\varepsilon_{zz} = \frac{\partial w}{\partial z} = k^2(q^2 A_1 e^{-kqz} - isB_1 e^{-ksz})e^{ik(x-ct)},$$

$$\varepsilon_{xy} = \frac{1}{2}\left(\frac{\partial u}{\partial y} + \frac{\partial v}{\partial x}\right) = 0, \tag{7.17}$$

$$\varepsilon_{yz} = \frac{1}{2}\left(\frac{\partial u}{\partial y} + \frac{\partial v}{\partial x}\right) = 0,$$

$$\varepsilon_{xz} = \frac{1}{2}\left(\frac{\partial u}{\partial z} + \frac{\partial w}{\partial x}\right) = \frac{1}{2}k^2(2iqA_1 e^{-kqz} + rB_1 e^{-ksz})e^{ik(x-ct)},$$

where $r = 2 - c^2 / c_T^2$.

We can solve the constitutive Equation (7.16) for stresses in terms of the non-zero strains in (7.17) to obtain:

$$\sigma_{xx} = \lambda(\varepsilon_{xx} + \varepsilon_{zz}) + 2\mu\varepsilon_{xx} = \mu k^2[(s^2 - 2q^2 - 1)A_1 e^{-kqz} + 2isB_1 e^{-ksz}]e^{ik(x-ct)}, \tag{7.18}$$

$$\sigma_{yy} = 0,$$

$$\sigma_{zz} = \lambda(\varepsilon_{xx} + \varepsilon_{zz}) + 2\mu\varepsilon_{zz} = \mu k^2[rA_1 e^{-kqz} - 2isB_1 e^{-ksz}]e^{ik(x-ct)}, \tag{7.19}$$

$$\sigma_{xy} = 0,$$

$$\sigma_{yz} = 0,$$

$$\sigma_{xz} = 2\mu\varepsilon_{xz} = -\mu k^2[2iqA_1 e^{-kqz} + rB_1 e^{-ksz}]e^{ik(x-ct)}, \tag{7.20}$$

We must now satisfy the boundary conditions

$$\sigma_{zz} = \sigma_{xz} = 0 \quad \text{for } z = 0. \tag{7.21}$$

From (7.18)–(7.21) it follows that

$$\sigma_{zz}|_{z=0} = rA_1 - 2isB_1 = 0,$$
$$\sigma_{xz}|_{z=0} = 2iqA_1 + rB_1 = 0. \tag{7.22}$$

Equations (7.22) lead to the equation for the unknown phase velocity c:

$$r^2 - 4sq = 0. \tag{7.23}$$

From (7.22) we also obtain

$$A_1 = \frac{2is}{r}B_1 = \frac{ir}{2q}B_1. \tag{7.24}$$

By substituting (7.24), into (7.14) and (7.15), we find the forms of u and w that satisfy the boundary conditions:

$$u = A[re^{-kqz} - 2sqe^{-ksz}]e^{ik(x-ct)}, \tag{7.25}$$
$$w = iAq[re^{-kqz} - 2e^{-ksz}]e^{ik(x-ct)}, \tag{7.26}$$

where $A = -kB_1/2q$.

Equation (7.23) is the characteristic equation for this surface wave problem. In order to solve this equation, we will introduce new variables:

$$\eta = \frac{k_T}{k} = \frac{c}{c_T}, \quad \zeta = \frac{k_L}{k_T} = \frac{c_T}{c_L} \quad \text{(since } k = \omega/c\text{)}, \tag{7.27}$$

where we have used that $r = 2 - \eta^2$, $q = \sqrt{1 - \eta^2\zeta^2}$ and $s = \sqrt{1 - \eta^2}$. Algebraic manipulation then yields

$$\eta^6 - 8\eta^4 + 8\eta^2(3 - 2\zeta^2) + 16(\zeta^2 - 1) = 0, \tag{7.28}$$

which is the Rayleigh wave velocity equation.

Note now that

$$\zeta = \frac{c_T}{c_L} = \sqrt{\frac{1 - 2v}{2(1 - v)}}; \tag{7.29}$$

since $\eta = c/c_T$, it also equals c_R/c_T with c equal to c_R, which in turn is equal to the Rayleigh *surface* wave velocity. As a result, we have three roots of (7.28) for η as a function of Poisson's ratio v only.

There is an approximate solution from Viktorov (1967) $\eta = (0.87 + 1.12v)/(1 + v)$. Note that, in the cubic equation, there are two complex conjugate roots and one real root for $v > 0.263$; for $v \leq 0.263$, there are three real roots but only one realistic one. We note further that η does not depend on frequency. Therefore, Rayleigh surface waves are nondispersive.

We now travel to the area of displacement analysis. Because the components of the displacement vector are real, from (7.25), and (7.26) we obtain

$$\tilde{u} = \frac{u}{A} = (re^{-kqz} - 2sqe^{-ksz})\cos k(x - ct), \tag{7.30}$$

$$\tilde{w} = \frac{w}{A} = q(-re^{-kqz} + 2e^{-ksz})\sin k(x - ct), \tag{7.31}$$

where $\tilde{u}$ and $\tilde{w}$ are normalized displacements with respect to the unknown constant A.

Consider now some aspects of the solution. From (7.30) and (7.31) it follows that, for any value of z, the vector with coordinates $(\tilde{u}, \tilde{w})$ moves along an ellipse (see Figure 7.2). We have

$$\frac{(\tilde{u})^2}{(re^{-kqz} - 2sqe^{-ksz})^2} + \frac{(\tilde{w})^2}{q^2(re^{-kqz} - 2e^{-ksz})^2} = 1, \tag{7.32}$$

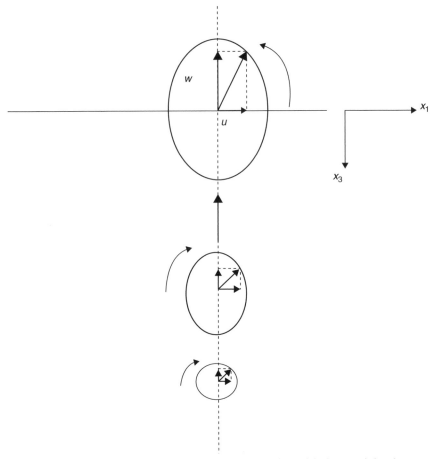

Figure 7.2. Displacement vector elliptical particle motion with time and depth.

which is the equation of an ellipse. Note that the original expressions (7.14) and (7.15) show two waves, one longitudinal and one transverse, propagating along a boundary that attenuates with depth. The superposition, however, leads to the surface waves and special qualities that we now observe.

The equation in u and w on the surface is an ellipse for all time values. This means that the vector sum of u and w gives the ellipse, as shown in Figure 7.2. It can be demonstrated that elliptical particle motion occurs as a function of depth, with the motion actually reversing after a particular depth. Analysis of the displacement equation can be used to plot these graphs. Further analysis can be used to plot displacement profiles with depth that are similar to those shown in Figure 7.3.

Figure 7.4 is a detailed presentation of particle displacements in an aluminum plate. Each displacement vector shown is the vector sum of the two normal components, $\bar{u}$ and $\bar{w}$. The figure can be considered as a snapshot of the resulting particle displacement vectors at different time periods. This gives us an idea of the displacement vector rotation along the surface and elsewhere throughout a thick aluminum plate. Note the decay with depth and the change in rotation of the particle displacement vector in space and time. Note also that, along any horizontal line

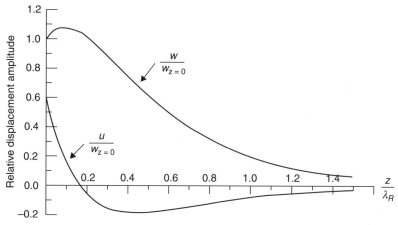

Figure 7.3. Sample displacement values of a Rayleigh surface wave with depth for an aluminum half-space (λ_R is the Rayleigh surface wavelength); curves are normalized with respect to w at $z = 0$.

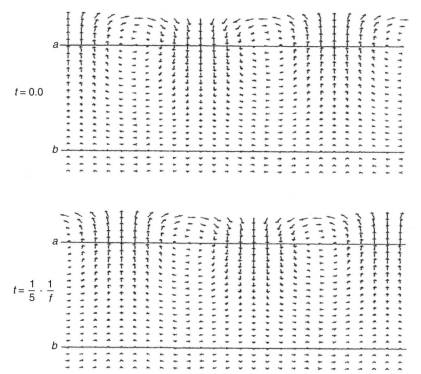

Figure 7.4. Surface wave particle displacement vector in an aluminum half-space for different time periods (line *a* represents the depth at which reversal in particle rotation occurs; line *b* represents one wavelength in depth.

above *a*, the particle displacement vector rotation is clockwise as you move from left to right; below *a*, the particle displacement vector rotation is counterclockwise. On the other hand, if we were to fix a position in space to observe the particle displacement vector rotation, it would be counterclockwise above *a* and clockwise below *a*. The vertical displacement component *w* is also generally larger than the

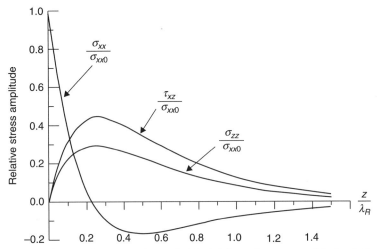

Figure 7.5. Sample stress values of a Rayleigh surface wave with depth for an aluminum half-space (λ_R is the Rayleigh surface wavelength; curves are normalized with respect to σ_{xx} at $z = 0$.

horizontal component u, hence leading to an elliptical particle displacement rotation with a vertical major axis.

It is interesting to plot out the stress components with depth, again on a normalized basis. This can be done using stress equations (7.18)–(7.20); a typical result is presented in Figure 7.5.

7.3 Generation and Reception of Surface Waves

Surface waves can be generated and received through a variety of different techniques, a few of which are discussed here. A normal beam or shear wave transducer can simply be placed on a surface, as in Figure 7.6. A wedge technique might be used as in Figure 7.7, with c_{1w} the wave velocity in the wedge material and c_R the surface wave velocity in medium 2. Snell's law states that $c_1 \sin \theta_2 = c_2 \sin \theta_1$; applied to this case it yields $c_{1w} \sin 90° = c_R \sin \theta_w$, so

$$\sin \theta_w = \frac{c_{1w}}{c_R} = \frac{\lambda_w}{\lambda_R}.$$

Note that c_{1w} must be less than c_R in order for this technique to work. This approach is the most efficient, since surface waves propagate in one direction only (only one angle is required for all frequencies).

A third possibility is to use a periodic array or "comb" transducer; see Figure 7.8. In this case, twice the gap spacing should equal the Rayleigh wavelength. A direct piezoelectric coupling could also be used to produce surface waves, where $2a = \lambda_R$. Yet another technique for generating surface waves is to utilize a mediator, as illustrated in Figure 7.9. A normal beam longitudinal wave transducer should be placed on a wedge at an angle – the third critical angle – that allows a surface wave

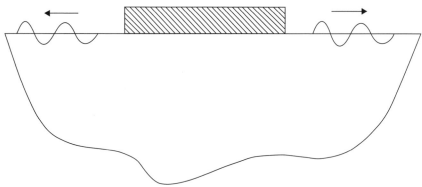

Figure 7.6. Normal beam transducer excitation for producing surface waves.

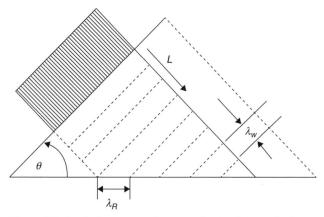

Figure 7.7. Angle beam transducer excitation for producing surface waves.

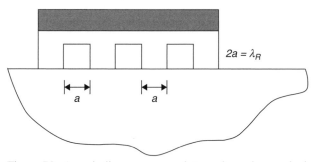

Figure 7.8. A periodic array or comb transducer for producing surface waves.

to be generated on the surface of the mediator. This wave then travels along the mediator to impinge onto the test specimen. A sharp-tipped mediator and proper coupling allows this to happen even for test specimens with low values of surface wave velocity.

Consider now the damping of surface waves. Many interesting studies of this phenomenon can be found in the literature. In Viktorov (1967), for example, if the wavenumbers are treated as complex then the following results are obtained.

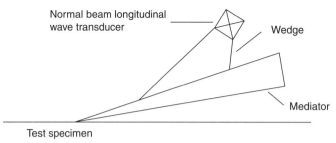

Figure 7.9. Mediator technique of generating surface waves in a test specimen.

$$\overline{k}_L = k'_L + ik''_L, \overline{k}_T = k'_T + ik''_T, \quad \text{and} \quad \overline{k}_R = k'_R + ik''_R.$$

It can be shown that the damping of Rayleigh waves is similar to that of cylindrical waves, as follows: $A_R \sim 1/\sqrt{k_R r}$ is compared to bulk waves, where $A_{L,T} \sim 1/k_{L,T}r$ (L for longitudinal and T for transverse). Rayleigh waves are less attenuated than bulk waves and are thus, for example, responsible for greater damage from earthquakes. Propagation of Rayleigh waves can be studied on curved surfaces. As an example, consider the wave equation in a system using cylindrical coordinates.

A noteworthy analysis of surface waves on anisotropic media may be found in Rose, Pilarski, and Huang (1990). These conditions give rise to variations in surface wave velocity with angle and subsequent differences in phase and in group or energy velocity, as well as to the presence of a skew angle. An inverse problem to evaluate certain composite material properties as a function of skew angle could be carried out.

7.4 Subsurface Longitudinal Waves

Subsurface elastic waves, mostly geoacoustical, are reported under many names: head waves, lateral waves, creeping longitudinal waves, or fast surface waves. Here, the term *subsurface* waves or subsurface longitudinal (SSL) waves is used to describe the field of longitudinal waves excited in a solid half-space by an angle beam transducer with an angle of incidence close to the first critical angle.

Subsurface longitudinal waves have been extensively investigated, not only theoretically but also experimentally to detect defects in subsurface layers of an isotropic material.

It has been established that – at or near the first critical angle, for longitudinal waves incident onto an interface (liquid–solid or solid–solid) from the medium with a smaller velocity of longitudinal waves – there coexist two waves: SSL and head waves. This is shown in part (a) of Figure 7.10. The two wave types cooperatively fulfill the boundary conditions on the free surface of the solid, where all stresses are supposed to equal zero. Any disturbance on the free surface moves with a velocity equal to the velocity of longitudinal waves in the solid. The amplitude of this displacement decreases as distance increases, according to the $1/r^n$ law, where n ranges from 1.5 to 2.0. This means that the SSL waves close to the free surface are strongly attenuated compared with the bulk waves, since the former are proportional

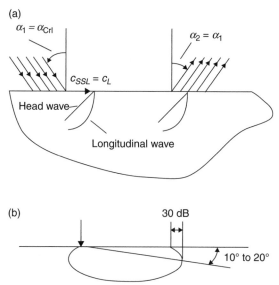

Figure 7.10. Subsurface longitudinal waves at first critical angle: (a) coexistence of head and longitudinal waves; (b) pressure field pattern.

to $1/r^n$ with n ranging from 0.5 to 1.0. The SSL waves can be detected at some other spot on the same surface, but the receiving transducer must be inclined at an angle equal to the first critical angle.

One characteristic of subsurface waves is the distribution of the amplitude of acoustic pressure in the plane of incidence; this is shown in part (b) of Figure 7.10. The shape of the pressure field distribution reveals that the maximum sensitivity of the ray occurs at an angle of 10 to 20 degrees from the free surface. Hence, the name chosen for these waves is appropriate, since they can be utilized for the detection of subsurface defects – and especially since the SSL waves show a relatively small sensitivity to surface roughness.

The definition given here of SSL waves is related to the first critical angle, which (for isotropic media) is given by Snell's law as $\alpha_{cr} = \sin^{-1}(c_1/c_L)$, where c_1 and c_L are longitudinal wave velocities for the upper (shoe of angle beam probe) and lower media, respectively. For an anisotropic medium the critical angle must be redefined, since the phase and group velocity vectors are generally of different orientations.

7.5 Exercises

1. Show that the equation $r^2 - 4sq = 0$ becomes
$$\eta^6 - 8\eta^4 + 8(3 - 2\zeta^2)\eta^2 - 16(1 - \zeta^2) = 0:$$
let $\eta = c/c_T$ and $\zeta = c_T/c_L$.
2. Solve the Rayleigh surface wave velocity equation as a function of Poisson's ratio v. Plot graphs. Compare with Viktorov's approximate solution, $\eta = (0.87 + 1.12v)/(1 + v)$.
3. Where does the particle velocity projection onto an ellipse reverse direction?
4. Study the damping of Rayleigh waves, comparing such media as steel, aluminum, and epoxy. Use Viktorov (1967) or other references.

5. For a wave incident on a Plexiglas specimen, assume that the longitudinal velocity is 2.71 mm/µs and the shear wave velocity is 1.38 mm/µs. Calculate the theoretical Rayleigh surface wave velocity and compare with the measured values.

6. Discuss or explain why the Rayleigh equation is a function of only Poisson's ratio for isotropic material.

7. For $t = 0$, plot the particle displacement on a 12×12 grid as a function of x and z in order to demonstrate concepts of particle motion.

8. Design a comb-type transducer to generate a surface wave in a structure.

9. For a Poisson's ratio of $v = 0.25$, find the surface wave velocity. Which root is correct, and why?

10. Show the development of Equation (7.4) from (7.1).

11. How can one be assured of displacement being zero at a depth z of infinity?

12. An SSL wave can create secondary waves, head waves, that can coexist with SSLs. Make a sketch showing head wave formation and the proper angle of propagation into a test material.

13. How do you select the correct root from the cubic equation in a Rayleigh surface wave problem for isotropic material?

14. What is a typical particle motion profile for a surface wave as we consider particles further away from the surface in question?

15. What spacing is required in a comb transducer to produce a surface wave of specified frequency?

16. Show that attenuation for surface waves is less than that of bulk waves.

17. Compare the attenuation coefficients of bulk longitudinal waves with subsurface longitudinal waves.

18. Using an angle beam transducer to generate a surface wave in a material, how would you achieve the best result with maximum energy into the surface wave? What conditions would be necessary to generate the surface wave with the angle beam transducer?

19. Show on a sketch the SSL wave, longitudinal head wave, SST wave, transverse head wave, and surface wave due to a point source loading on a half-space.

20. Make a plot of surface wave particle displacement for a given time t.

21. How would you proceed to calculate surface wave velocities for excitation on an anisotropic material?

22. Using surface waves, discuss a procedure to identify certain anisotropic material constants as a function of measured skew angles.

7.6 REFERENCES

Auld, B. A. (1990). *Acoustic Fields and Waves in Solids*, 2nd ed., vols. 1 and 2. Malabar, FL: Kreiger.

Basatskaya, L. V., and Ermolov, L. N. (1980). Theoretical study of ultrasonic longitudinal subsurface waves in solid media, *Defektoskopiya* 7: 58–65.

Chadwick, P., and Smith, G. D. (1977). Foundations of the theory of surface waves in anisotropic elastic materials, *Adv. Appl. Mech.* 17: 303–77.

Couchman, J. C., and Bell, J. R. (1978). Prediction, detection and characterization of a fast surface wave produced near the first critical angle, *Ultrasonics* 16: 272–4.

Farnell, G. W. (1970). Properties of elastic surface waves, in W. P. Mason and R. N. Thurston (Eds.), *Physical Acoustics*, vol. 6, pp. 109–66. New York: Academic Press.

Heelan, P. A. (1953). On the theory of head waves, *Geophys.* 18: 871–6.

Kolsky, H. (1963). *Stress Waves in Solids*. New York: Dover.

Nikiforov, L. A., and Kharitonov, A. V. (1981). Parameters of longitudinal subsurface waves excited by angle-beam transducers, *Defektoskopiya* 6: 80–5.

Pilarski, A., and Rose, J. L. (1989). Utility of subsurface longitudinal waves in composite material characterization, *Ultrasonics* 27: 226–33.

Pollard, H. F. (1977). *Sound Waves in Solids*. London: Pion Ltd.

Rayleigh, J. W. S. (1945). *The Theory of Sound*. New York: Dover.

Rose, J. L., Pilarski, A., and Huang, Y. (1990). Surface wave utility in composite material characterization, in *Research in Nondestructive Evaluation*, vol. 1, pp. 247–65. New York: Springer-Verlag.

Uberall, H. (1973). Surface waves in acoustics, in P. Mason and R.N. Thurston (Eds.), *Physical Acoustics*, vol. 10, pp. 1–60. New York: Academic Press.

Viktorov, I. A. (1967). *Rayleigh and Lamb Waves – Physical Theory and Applications*. New York: Plenum.

8 Finite Element Method for Guided Wave Mechanics

8.1 Introduction

The finite element method (FEM), also known as finite element analysis (FEA), is a widely used numerical solution approach for solving problems of wave propagation and vibration. For some complex differential or integral equations without analytical solutions, FEM can provide accurate and computationally efficient solutions that may be improved by refining the elements used in the study. FEM may be combined with analytical analysis or other numerical methods to achieve optimum solutions.

Cook and colleagues (1992) concluded that a successful FEM simulation usually involves the items listed in Table 8.1. Although commercial FEM software has been widely used, the only step commercial software can accomplish is the finite element analysis step. To achieve an accurate FEM simulation, the analyst needs to appropriately understand the physical problem, conduct a basic preliminary study, and select a correct mathematical model. The analyst should also check the calculation results based on either theoretical research or experimental data.

8.2 Overview of the Finite Element Method

8.2.1 Using the Finite Element Method to Solve a Problem

Simple geometry, loading, and boundary conditions are required to achieve an exact analytical solution for a field problem. For a complicated problem that cannot be solved analytically while considering if the entire structure can be separated into small, discrete elements with simple loading and boundary conditions, one may simplify this problem as a field problem in each simple element.

FEM represents a numerical calculation approach based on the principle of virtual work:

With significant contribution from Li Zhang.

Table 8.1. Flow chart of the basic FEM procedure

1. Problem classification: physical phenomena analysis.
2. Mathematical modeling: determining mathematical models, including geometries, governing equations, and appropriate solving approaches.
3. Discretization: dividing a mathematical model into a mesh of finite elements.
4. Preliminary analysis: having some analytical results, experience, or experimental results for comparison.
5. Finite element analysis:
   a) Preprocessing: inputting data of geometry, material properties, boundary conditions, etc.
   b) Numerical calculation: deciding interpolation functions, obtaining a matrix to describe the behavior of each element, assembling these matrices into a global matrix equation, and solving this equation to determine the results.
   c) Postprocessing: listing or graphically displaying the solutions.
6. Check the results: making sure that the FEM simulation has been carried out correctly and then comparing the FEM results to the preliminary analysis.

$$\int \{\delta\varepsilon\}^{T} \{\sigma\} dV = \int \{\delta\mathbf{u}\}^{T} \{\mathbf{P}\} dV + \int \{\delta\mathbf{u}\}^{T} \{\Phi\} dS \qquad (8.1)$$

$\{\varepsilon\}$: strain tensor;
$\{\sigma\}$: stress tensor;
$\{\mathbf{u}\}$: displacement vector;
$\{\mathbf{P}\}$: body force in volume V;
$\{\Phi\}$: surface traction on surface S.

Here δ represents a functional derivative in the calculus of variations, which is similar to a one-dimensional derivative in usual calculus. In the calculus of variations, $\delta\mathbf{u}$ is called the first variation of displacement $\mathbf{u}$.

To numerically solve the problem, FEM separates a structure into discrete elements with nodes. Figures 8.1 and 8.2 show sample discrete elements in 1-D and 2-D structures.

After a structure is meshed into a number of finite elements, all of the loading functions are applied to the nodes of each element instead of on the surface or in the volume. In addition, FEM only calculates field variables on the nodes. Therefore, an interpolation function is needed to calculate values of all of the variables inside the elements. This interpolation function is associated with what we call a shape function.

For a one-dimensional linear element as shown in Figure 8.1, the shape function of a node i can be defined as:

$$N_i(x) = a_i + b_i x \qquad (8.2)$$

$$\text{where: } N_i(x_j) = \begin{cases} 1, & (i = j) \\ 0, & (i \neq j) \end{cases} \qquad (8.3)$$

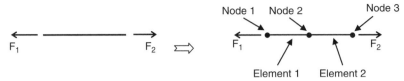

Figure 8.1. Sample one-dimensional problem: the structure is disintegrated to two bar elements.

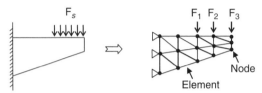

Figure 8.2. Sample two-dimensional problem: the structure is disintegrated to sixteen triangular elements.

If the two nodes of the element are located at x_1 and x_2, then:

$$\begin{aligned}
N_1(x_1) &= a_1 + b_1 x_1 = 1 \\
N_1(x_2) &= a_1 + b_1 x_2 = 0 \\
N_2(x_1) &= a_2 + b_2 x_1 = 0 \\
N_2(x_2) &= a_2 + b_2 x_2 = 1
\end{aligned} \tag{8.4}$$

Solving Equation (8.4) gives:

$$a_1 = \frac{x_2}{\Delta}; \, b_1 = -\frac{1}{\Delta}; \, a_2 = -\frac{x_1}{\Delta}; \, b_2 = \frac{1}{\Delta} \tag{8.5}$$

where $\Delta = x_2 - x_1 = l^{(e)}$, which is the length of the element.

Therefore, the shape functions become:

$$\begin{aligned}
N_1(x) &= \frac{x_2 - x}{\Delta} \\
N_2(x) &= \frac{-x_1 + x}{\Delta}
\end{aligned} \tag{8.6}$$

For a 2-D linear triangular element shown in Figure 8.3, the shape function of node i can be defined as:

$$N_i(x, y) = a_i + b_i x + c_i y \tag{8.7}$$

$$\text{where: } N_i(x_j, y_j) = \begin{cases} 1, & (i = j) \\ 0, & (i \neq j) \end{cases} \tag{8.8}$$

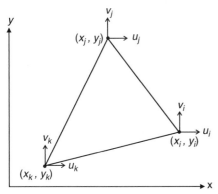

Figure 8.3. Sample two-dimensional linear triangular element. u_i and v_i represent displacements of node i in x and y directions. Node is located at (x_i, y_i).

Substituting Equation (8.7) into Equation (8.8) yields:

$$a_i = \frac{(x_j y_k - x_k y_j)}{2\Delta}$$
$$b_i = \frac{(y_j - y_k)}{2\Delta} \qquad (i = 1,2,3;\ j = 2,3,1;\ (k) = 3,1,2) \qquad (8.9)$$
$$c_i = \frac{(x_k - x_j)}{2\Delta}$$

where Δ is the element area.

Shape functions for the 3-D elements can be derived similarly.

After the shape function is determined for an element, one can obtain the displacement distribution in the element. For the linear triangular element in Figure 8.3, the displacements in the x and y directions are:

$$\mathrm{u} = N_i u_i + N_j u_j + N_k u_k$$
$$\mathrm{v} = N_i v_i + N_j v_j + N_k v_k \qquad (8.10a)$$

or:

$$\{\mathbf{u}\} = \begin{Bmatrix} \mathrm{u} \\ \mathrm{v} \end{Bmatrix} = \begin{bmatrix} N_i & 0 & N_j & 0 & N_k & 0 \\ 0 & N_i & 0 & N_j & 0 & N_k \end{bmatrix} \{\mathbf{d}\}^{(e)} = [\mathbf{N}]\{\mathbf{d}\}^{(e)} \qquad (8.10b)$$

where $\{\mathbf{d}\}^{(e)} = [u_i \quad v_i \quad u_j \quad v_j \quad u_k \quad v_k]^T$ are nodal displacements in the element.

Based on elasticity theory, strains in the element are defined as:

$$\{\varepsilon\} = \begin{Bmatrix} \varepsilon_x \\ \varepsilon_y \\ \gamma_{xy} \end{Bmatrix} = \begin{bmatrix} \dfrac{\partial u}{\partial x} \\[2ex] \dfrac{\partial v}{\partial y} \\[2ex] \dfrac{\partial u}{\partial y} + \dfrac{\partial v}{\partial x} \end{bmatrix} \qquad (8.11)$$

Substituting Equation (8.10b) into Equation (8.11) yields:

$$\{\varepsilon\} = \frac{1}{2\Delta}\begin{bmatrix} b_i & 0 & b_j & 0 & b_k & 0 \\ 0 & c_i & 0 & c_j & 0 & c_k \\ c_i & b_i & c_j & b_j & c_k & b_k \end{bmatrix}\{\mathbf{d}\}^{(e)} = [\mathbf{B}]\{\mathbf{d}\}^{(e)} \tag{8.12}$$

In addition, the constitutive equation in the element can be written as:

$$\{\sigma\} = [\mathbf{D}]\{\varepsilon\} = [\mathbf{D}][\mathbf{B}]\{\mathbf{d}\}^{(e)} \tag{8.13}$$

For a plane strain problem in an isotropic material, $[\mathbf{D}]$ becomes:

$$[\mathbf{D}] = \frac{E}{1-v^2}\begin{bmatrix} 1 & v & 0 \\ v & 1 & 0 \\ 0 & 0 & \frac{1-v}{2} \end{bmatrix} \tag{8.14}$$

where E and v indicate Young's modulus and Poisson's ratio.

For a plane stress problem in an isotropic material, $[\mathbf{D}]$ becomes:

$$[\mathbf{D}] = \frac{E(1-v)}{(1+v)(1-2v)}\begin{bmatrix} 1 & \frac{v}{1-v} & 0 \\ \frac{v}{1-v} & 1 & 0 \\ 0 & 0 & \frac{1-2v}{2(1-v)} \end{bmatrix} \tag{8.15}$$

For an axisymmetric 2-D problem in cylindrical coordinates in an isotropic material, $[\mathbf{D}]$ becomes:

$$[\mathbf{D}] = \frac{E(1-v)}{(1+v)(1-2v)}\begin{bmatrix} 1 & \frac{v}{1-v} & \frac{v}{1-v} & 0 \\ \frac{v}{1-v} & 1 & \frac{v}{1-v} & 0 \\ \frac{v}{1-v} & \frac{v}{1-v} & 1 & 0 \\ 0 & 0 & 0 & \frac{1-2v}{1-v} \end{bmatrix} \tag{8.16}$$

Substituting Equations (8.10b), (8.12), and (8.13) into Equation (8.1) yields:

$$\{\delta\mathbf{d}\}^T\left(\int[\mathbf{B}]^T[\mathbf{D}][\mathbf{B}]dV\{\mathbf{d}\}^{(e)} - \left(\int[\mathbf{N}]^T\{\mathbf{P}\}dV + \int[\mathbf{N}]^T\{\Phi\}dS\right)\right) = 0 \tag{8.17}$$

Let:

$$[\mathbf{k}]^{(e)} = \int[\mathbf{B}]^T[\mathbf{D}][\mathbf{B}]dV \text{ and } \{\mathbf{F}\}^{(e)} = \left(\int[\mathbf{N}]^T\{\mathbf{P}\}dV + \int[\mathbf{N}]^T\{\Phi\}dS\right)$$

then Equation (8.17) becomes:

$$[\mathbf{k}]^{(e)}\{\mathbf{d}\}^{(e)} = \{\mathbf{F}\}^{(e)} \tag{8.18}$$

where $[\mathbf{k}]^{(e)}$ is called the element stiffness matrix and $\{\mathbf{F}\}^{(e)}$ is the equivalent element load vector. Equation (8.18) is the governing equation for a single element.

If there are n nodes in the entire structure, by assembling all of the elements in the entire structure, the governing equation becomes:

$$[\mathbf{K}]_{2n\times2n}\{\mathbf{d}\}_{2n\times1} = \{\mathbf{F}\}_{2n\times1} \tag{8.19}$$

where:

the Nodal displacement vector is as follows:

$$\{\mathbf{d}\} = [u_1 \quad v_1 \quad \cdots \quad u_i \quad v_i \quad \cdots \quad u_j \quad v_j \quad \cdots \quad u_k \quad v_k \quad \cdots \quad u_n \quad v_n]^T \tag{8.20}$$

the Equivalent load vector is as follows:

$$\{\mathbf{F}\} = [f_{x,1} \quad f_{y,1} \quad \cdots \quad f_{x,i} \quad f_{y,i} \quad \cdots \quad f_{x,j} \quad f_{y,j} \quad \cdots \quad f_{x,k} \quad f_{y,k} \quad \cdots \quad f_{x,n} \quad f_{y,n}]^T \tag{8.21}$$

the Global stiffness matrix is as follows:

$$[\mathbf{K}] = \begin{bmatrix} K_{11} & \cdots & K_{1i} & \cdots & K_{1j} & \cdots & K_{1k} & \cdots & K_{1n} \\ \vdots & & \vdots & & \vdots & & \vdots & & \vdots \\ K_{i1} & \cdots & K_{ii} & \cdots & K_{ij} & \cdots & K_{ik} & \cdots & K_{in} \\ \vdots & & \vdots & & \vdots & & \vdots & & \vdots \\ K_{j1} & \cdots & K_{ji} & \cdots & K_{jj} & \cdots & K_{jk} & \cdots & K_{jn} \\ \vdots & & \vdots & & \vdots & & \vdots & & \vdots \\ K_{k1} & \cdots & K_{ki} & \cdots & K_{kj} & \cdots & K_{kk} & \cdots & K_{kn} \\ \vdots & & \vdots & & \vdots & & \vdots & & \vdots \\ K_{n1} & \cdots & K_{ni} & \cdots & K_{nj} & \cdots & K_{nk} & \cdots & K_{nn} \end{bmatrix} \tag{8.22}$$

For plane strain problems, the element in stiffness matrix $[\mathbf{K}]$ is:

$$K_{ls} = \frac{Eh}{4(1-v^2)\Delta}\begin{bmatrix} b_l b_s + \dfrac{1-v}{2}c_l c_s & vb_l c_s + \dfrac{1-v}{2}c_l b_s \\[2ex] vc_l b_s + \dfrac{1-v}{2}b_l c_s & c_l c_s + \dfrac{1-v}{2}b_l b_s \end{bmatrix} \quad (l,s = 1,2,\ldots,n) \tag{8.23}$$

For plane stress problems, one needs to use $E/(1-v^2)$ and $v/(1-v)$ instead of E and v in Equation (8.23).

8.2.2 Quadratic Elements

In FEM, linear elements are simple and easily calculated. However, because the shape functions of these elements are linear, the displacement distributions are also linear. In other words, the deformation of a linear element is restricted. For instance,

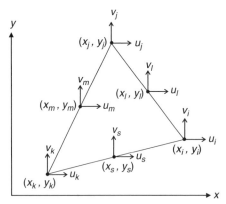

Figure 8.4. Sample two-dimensional quadratic triangular element. u_i and v_i represent displacements of node i in the x and y directions. Node i is located at (x_i, y_i).

a deformed linear triangular element will always be a triangle without any curved edges. In addition, according to Equation (8.12), the strain in a linear element is a constant. Because a real structure usually does not have these restrictions, it requires that FEM uses many more elements to achieve accurate solutions for structures with complex geometries and/or energy distributions.

An efficient approach to reduce the required element number is to employ nonlinear elements. Figure 8.4 illustrates a quadratic triangular six-node element. Besides three vertex nodes, *i, j, k*, three side nodes, *l, m,* and *s*, are added to the element. The shape function of this quadratic triangular element is:

$$N_i(x, y) = a_i + b_i x + c_i y + d_i x^2 + e_i xy + f_i y^2 \qquad (8.24)$$

For structures with simple geometries but complex energy distributions, bilinear elements would be a good choice. The shape function of a bilinear element is:

$$N_i(x, y) = a_i + b_i x + c_i y + d_i xy \qquad (8.25)$$

8.2.3 Dynamic Problem

Wave propagation is a dynamic problem. The governing equation of FEM for a dynamic problem is:

$$[\mathbf{M}]\{\ddot{\mathbf{d}}\} + [\mathbf{C}]\{\dot{\mathbf{d}}\} + [\mathbf{K}]\{\mathbf{d}\} = \{\mathbf{F}\} \qquad (8.26)$$

which can be considered as generalized Newton's second law. Here $[\mathbf{M}]$ and $[\mathbf{C}]$ represent the mass matrix and damping matrix while $\{\dot{\mathbf{d}}\} = d\{\mathbf{d}\}/dt$ is the velocity vector and $\{\ddot{\mathbf{d}}\} = d^2\{\mathbf{d}\}/dt^2$ is the acceleration vector.

If a 2-D element has a thickness *h* and density ρ, the mass matrix becomes:

$$[\mathbf{M}] = \rho h \int\int [\mathbf{N}]^T [\mathbf{N}]\, dxdy \qquad (8.27)$$

For Rayleigh damping, the damping matrix is defined as a linear combination of the mass matrix and the stiffness matrix:

$$[\mathbf{C}] = \alpha[\mathbf{M}] + \beta[\mathbf{K}] \tag{8.28}$$

The coefficients α and β can be determined by experiments. It has been shown that α has very little influence on high-frequency problems. Because ultrasonic waves are of high frequency, the governing equation becomes:

$$[\mathbf{M}]\{\ddot{\mathbf{d}}\} + \beta[\mathbf{K}]\{\dot{\mathbf{d}}\} + [\mathbf{K}]\{\mathbf{d}\} = \{\mathbf{F}\} \tag{8.29}$$

For a harmonic wave with circular frequency ω in an undamped material, Equation (8.29) can be simplified as:

$$\left([\mathbf{K}] - \omega^2[\mathbf{M}]\right)\{\mathbf{d}\} = \{\mathbf{F}\} \tag{8.30}$$

Matrix $[\mathbf{K}] - \omega^2[\mathbf{M}]$ is called the dynamic stiffness matrix. Based on Equation (8.30), the problem can be solved in the frequency domain instead of the time domain.

To solve a dynamic problem in the time domain, numerical calculation methods in the time domain are needed. A direct integration method is the most popular numerical approach for calculating the time domain response. The numerical calculation method is a finite difference method that obtains derivation of time t from Taylor's polynomial. The most widely used direct integration methods include the center difference method, the Newmark method, and the Wilson θ method. We first discuss the classic center difference method here.

In the classic center difference method, if the time increment is Δt, the velocity and acceleration at time step n can be approximated as:

$$\{\dot{\mathbf{d}}\}_n = \frac{d}{dt}\{\mathbf{d}\}_n = \frac{1}{2\Delta t}\left(\{\mathbf{d}\}_{n+1} - \{\mathbf{d}\}_{n-1}\right) \tag{8.31}$$

$$\{\ddot{\mathbf{d}}\}_n = \frac{d}{dt^2}\{\mathbf{d}\}_n = \frac{1}{\Delta t^2}\left(\{\mathbf{d}\}_{n+1} - 2\{\mathbf{d}\}_n + \{\mathbf{d}\}_{n-1}\right) \tag{8.32}$$

If we only use the linear part of Taylor series expansions of $\{\mathbf{d}\}_{n-1}$ and $\{\mathbf{d}\}_{n+1}$ at time step n:

$$\{\mathbf{d}\}_{n-1} = \{\mathbf{d}\}_n - \Delta t\{\dot{\mathbf{d}}\}_n + \cdots \tag{8.33}$$

$$\{\mathbf{d}\}_{n+1} = \{\mathbf{d}\}_n + \Delta t\{\dot{\mathbf{d}}\}_n + \cdots \tag{8.34}$$

substituting Equations (8.31)~(8.34) into Equation (8.29) provides:

$$\begin{aligned}
&\left(\frac{1}{\Delta t^2}[\mathbf{M}] + \frac{1}{2\Delta t}[\mathbf{C}]\right)\{\mathbf{d}\}_{n+1} \\
&= \{\mathbf{F}\} - [\mathbf{C}]\{\mathbf{d}\}_n + \frac{2}{\Delta t^2}[\mathbf{M}]\{\mathbf{d}\}_n - \left(\frac{1}{\Delta t^2}[\mathbf{M}] - \frac{1}{2\Delta t}[\mathbf{C}]\right)\{\mathbf{d}\}_{n-1}
\end{aligned} \tag{8.35}$$

When using the direct integration method, the initial displacement, velocity, and acceleration of the structure must be known to solve the problem.

The classic center difference method is a simple approach, although it is not unconditionally stable. To optimize the stability of the direct integration method, one may employ the Newmark method or the Wilson θ method. Because the Wilson θ method has large "damping" for high-frequency problems, the Newmark method is a better choice for ultrasonic wave simulation.

The Newmark method approximates the velocity and displacement at time step n as:

$$\{\dot{\mathbf{d}}\}_{n+1} = \{\dot{\mathbf{d}}\}_n + \Delta t \left((1-\gamma)\{\ddot{\mathbf{d}}\}_n + \gamma\{\ddot{\mathbf{d}}\}_{n+1} \right) \qquad (0 \leq \gamma \leq 1) \qquad (8.36)$$

$$\{\mathbf{d}\}_{n+1} = \{\mathbf{d}\}_n + \Delta t \{\dot{\mathbf{d}}\}_n + \frac{\Delta t^2}{2} \left((1-2\beta)\{\ddot{\mathbf{d}}\}_n + 2\beta\{\ddot{\mathbf{d}}\}_{n+1} \right) \qquad (0 \leq 2\beta \leq 1) \qquad (8.37)$$

When $0.5 \leq \gamma \leq 1$ and $0.25 \leq \beta \leq 0.5$, the Newmark method is unconditionally stable.

8.2.4 Error Control

FEM is a numerical calculation method that produces results that do not always match the exact solutions, so it is important to reduce the errors in FEM application and to make the FEM solutions converge to the exact solutions. Because most FEM simulations of ultrasonic wave behavior are accomplished by commercial FEM software, in this section, we discuss how to reduce common errors occurring when using FEM software.

Several types of errors must be considered when using FEM software:

1. **Incorrect mathematical model.** If the physical phenomenon of a problem is misunderstood, an incorrect mathematical model may be built for FEM analysis. This type of error includes incorrect FEM model selections, material properties, geometry simplifications, loading conditions, and boundary condition assumptions.

   All successful ultrasonic wave FEM simulations are based on a correct understanding of the theory of wave propagation. For example, one may use a 2-D plane strain FEM model to simulate Lamb wave propagations in an infinite plate. However, 2-D plane strain FEM simulation is not applicable for SH waves in the same plate, because the dominant particle vibration is an in-plane vibration that is normal to the wave propagation direction. Therefore, a FEM model of SH waves in an infinite plate should consider a 3-D model with infinite boundary conditions and cannot be simplified as a 2-D plane strain problem. As we discussed in former sections, the material damping and time domain simulation method depend on the input waves. In addition, other aspects of the mathematic models, including loading conditions, boundary conditions, material properties, simplified geometries of defects, or other complex structures, should all be determined based on the simulated wave modes and frequencies.

2. **Wrong user input.** Besides the modeling errors caused by misunderstanding of physical phenomena, a user of FEM software may also accidently input errors when building a FEM model. These errors can be avoided by carefully checking input data for the FEM models.

3. **Inappropriate element discretization, shape, and type.** As a numerical calculation method, FEM has more restrictions on the degrees of freedom (d.o.f.) than the real structure. Therefore, one needs to employ a sufficient number of elements to make a FEM solution converge to the exact solution. To accurately simulate ultrasonic waves, we usually require at least eight elements per wavelength when using linear elements. For guided wave simulations, because the wave structures may significantly vary in the thickness direction, usually a minimum of four to five nodes in the thickness direction are required to achieve accurate solutions. When using linear elements, one should try to avoid the size of an element that is significantly larger or smaller than the nearby elements, because the stress of each linear element is constant and is usually different from the next element. A sharp change of stress level could cause calculation errors in FEM. Besides the element size difference, the shape of an element is also important. A sharp angle ($\leq 30°$) in an element or an element with one dimension more than ten times larger than another dimension should also be avoided to reduce the possibility of calculation error.

   A large number of element types can be chosen in FEM calculations. Linear elements are good for calculations in simple geometries with gradual variations of the energy distributions because they are simple and can save computational time. For more complicated problems, such as small geometries and energy concentrations, nonlinear elements are more appropriate.

4. **Numerical error.** As we mentioned in the former section, inappropriate element discretization, shape, and type could cause numerical errors. Other aspects that may lead to numerical errors include special geometries, energy concentrations, singularities, and ill-conditioned equations. Although most commercial FEM software can reduce numerical errors, the FEM simulation results should be verified by experiments.

5. **Software bugs.** Bugs are found in all software, so FEM results could be affected by these faults.

8.3 FEM Applications for Guided Wave Analysis

This section demonstrates simulations of guided wave behaviors. In the first example, two-dimensional axisymmetric models were built to simulate surface waves generated by comb transducers in a round plate. The second example simulated axisymmetric guided wave inspection and guided wave phase array focusing in a two-inch steel tube. Both examples were implemented in a commercial FEM software, ABAQUS.

8.3.1 2-D Surface Wave Generation in a Plate

When simulating surface wave generation in a round titanium disk by utilizing annular comb transducers (Figure 8.5), one can use a FEM model with 2-D axisymmetric quadrilateral elements. The comb transducer designs and schematics of the test setup are shown in Figure 8.5.

A commercial FEM software, ABAQUS, was utilized to build the 2-D axisymmetric model and to simulate 1 MHz and 2 MHz surface wave generations. In this case, implicit integration was chosen to solve the dynamic problem. ABAQUS

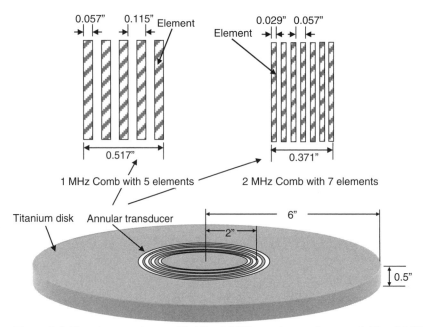

Figure 8.5. Titanium specimen with (1) a 1 MHz comb transducer and (2) a 2 MHz comb transducer located at two inches from the center.

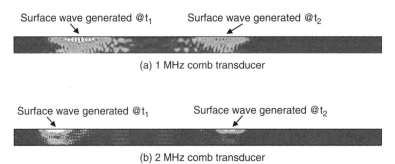

Figure 8.6. Rayleigh surface waves were excited by the 1 MHz and 2 MHz comb transducers on a 1/4″-thick titanium plate. See plates section for color version.

uses the Hilber-Hughes-Taylor method, which is an extension of the Newmark method, for implicit integration. Because ABAQUS lets $\gamma = 1/2$ and $\beta = 1/4$, the implicit integration is unconditionally stable.

The simulation results of wave generations are illustrated in Figure 8.6. Both the transducers efficiently excited surface waves, although a small amount of energy went into Lamb wave modes. Figure 8.7 shows the simulated waveforms of the surface waves.

8.3.2 Guided Wave Defect Detection in a Two-Inch Steel Tube

One important application of guided wave FEM models is to simulate guided wave inspections of defect. In this example, we used FEM simulations to compare and evaluate guided wave axisymmetric and focusing inspections for a saw cut in a two-

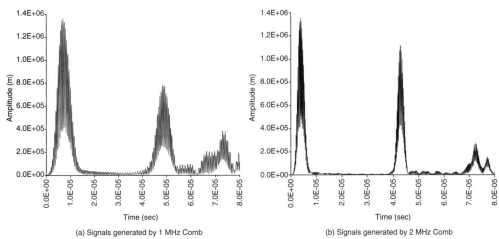

(a) Signals generated by 1 MHz Comb

(b) Signals generated by 2 MHz Comb

Figure 8.7. The FEA simulation results of surface wave signals generated and received by a 1 MHz comb transducer and a 2 MHz comb transducer in a 1/4″-thick titanium plate. The input function is a ten-cycle Hanning signal.

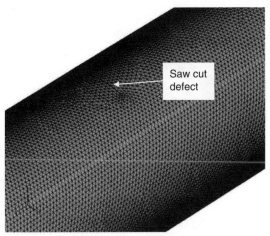

Figure 8.8. Schematic of a 9 percent cross-sectional area (CSA) planar saw cut defect in the finite element model of a two-inch schedule 40 steel tube. Four-node tetrahedral elements were used in ABAQUS to discretize the structure.

inch schedule 40 steel tube. The schematic of the FEM meshing in ABAQUS is shown in Figure 8.8. Three-dimensional four-node linear tetrahedral elements and explicit Newmark integration are employed for the models. The explicit integration usually requires less computer memory usage and computation time than the implicit integration, so it works well for solving large 3-D models. However, the explicit integration in ABAQUS is conditionally stable, so the largest time step highly depends on the wave frequency. It is recommended to have at least five to ten time steps in one wavelength.

In this example, five-element comb transducers are employed to generate 200 kHz T(m, 1) wave groups in a two-inch schedule 40 steel tube. Four transducers, each of which is 90° wide and covers a quadrant of the circumference, are mounted on the end of the tube. Each quadrant has individual excitation signals. If all four

(a) axisymmetric

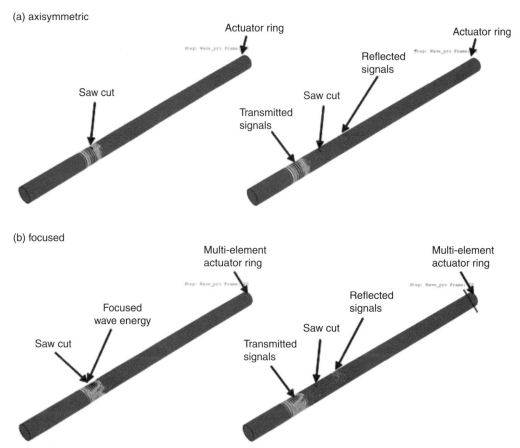

(b) focused

Figure 8.9. FEM simulations of 200 kHz (a) T(0, 1) axisymmetric wave and (b) focused T(m, 1) wave group reflected from the 9 percent CSA saw cut defect in the two-inch schedule 40 tube. Note that stronger reflected signals were obtained when focusing at the defect. See plates section for color version.

quadrants are excited simultaneously with a 200 kHz ten-cycle Hanning signal, the axisymmetric wave T(0, 1) will be generated. If appropriate time delays and amplitude controls are applied to the quadrants, the T(m, 1) wave group will focus at a preselected spot in the tube (see Chapter 10 for details). The simulation results of axisymmetric and focusing inspection are shown in Figures 8.9a and b. The simulations verified that the 200 kHz T(m, 1) waves successfully focused at the defect location and enhanced the reflected signals from the defect. In other words, the FEM results indicate that focusing techniques can improve the sensitivity of guided wave pipeline inspection by increasing energy impinging onto and reflected from defects.

8.4 Summary

Finite element analysis is a powerful computationally efficient tool for studying guided wave mechanics in structures. Although anyone can run a FEM code, it takes serious study to run the code properly with appropriate boundary conditions simulating the desired transducer loading function to achieve desired performance of adequate penetration power, mode excitability, and sensitivity to certain defect and structural situations.

8.5 Exercises

1. When simulating ultrasonic wave propagation by utilizing the finite element method to obtain accurate solutions, if linear elements are used what is the minimum number of elements per wavelength required for accurate solutions?

2. For a bar in pure tension, the displacement is: $u = \frac{FL}{AE}$, where F is external force, L is the length of the bar, A is the cross-sectional area of the bar, and E is the elastic modulus.

    If one finite element with two nodes is used to simulate this problem (as shown in the figure below), the problem can be described as Equation 1 and the stiffness matrix is shown in Equation 2. The $\{F\}$ represents external force vector and $\{U\}$ is displacement vector of the nodes.

    Please derive the stiffness matrix for the following elements in pure tension:

(1)

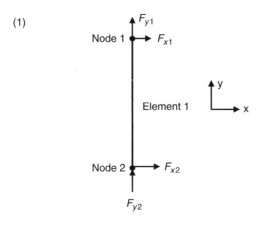

(2)

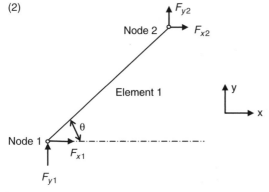

3. If using ABAQUS/Explicit to simulate a 50 kHz guided wave propagation, what is the maximum time step allowed?
4. Can you use a 2-D plane strain FEM model to simulate SH wave propagation in a plate? Why?
5. FEM problems can be very time consuming when using higher frequencies and for long travel distances. Explain! In ABAQUS, which time integration method should be used for these problems?
6. How could a narrower focal region be obtained in a pipe compared to the result shown in Figure 8.9?
7. Explain the results shown in Figure 8.6. How could depth of penetration be increased?

8.6 REFERENCES

Bertolini, A. F. (1998). Review of eigensolution procedures for linear dynamic finite element analysis, *ASME Applied Mechanics Reviews* 51(2): 155–72.

Cook, R. D., and Avrashi, J. (1992). Error estimation and adaptive meshing for vibration problems, *Computers & Structures* 44(3): 619–26.

Cook, R. D. et al. (2002). *Concepts and Applications of Finite Element Analysis*. New York: John Wiley & Sons.

Craig Jr., R. R. (1981). *Structural Dynamics: An Introduction to Computer Methods*. New York: John Wiley & Sons.

Dassault Systèmes Simulia Corp. (2012). *Abaqus Theory Manual*.

Hayashi, T. et al. (2006). Wave structure analysis of guided waves in a bar with an arbitrary cross-section, *Journal of Ultrasonics* 44(1): 17–24.

Hitchings, D. (1992). *A Finite Element Dynamics Primer*. Glasgow, UK: NAFEMS.

Koshiba, M., Karakida, S., and Suruki, M. (1984). Finite-element analysis of Lamb wave scattering in an elastic plate waveguide, *IEEE Trans. Sonics. Ultrason.* SU-31(1): 18–25.

Koshiba, M., Morita, H., and Suruki, M. (1981). Finite-element analysis of discontinuity problem of SH modes in an elastic plate waveguide, *Electron. Lett.* 17(13): 480–3.

Li, X. D., Zeng, L. F., and Wiberg, N.-E. (1993). A simple load error estimator and an adaptive time-stepping procedure for direct integration method in dynamic analysis, *Communications in Numerical Methods in Engineering* 9(4): 273–92.

Lowe, M. J. S., Alleyne, D. N., and Cawley, P. (1998). Defect detection in pipes using guided waves, *Journal of Ultrasonics* 36(1–5): 147–54.

Newmark, N. M. (1959). A method of computation for structural dynamics, *ASCE Journal of the Engineering Mechanics Division* 5(EM3): 67–94.

Rosanoff, R. A., and Ginsburg, T. A. (1965). Matrix error analysis for engineers, *Proceedings of the Conference on Matrix Methods in Structural Mechanics*, Wright-Patterson AFB, Ohio, pp. 887–910.

Strang, G., and Fix, G. J. (1973). *An Analysis of Finite Element Method*. Englewood Cliffs, NJ: Prentice Hall.

Utku, S., and Melosh, R. J. (1987). Estimating the manipulation errors in finite element analysis, part I, *Finite Element in Analysis and Design* 3(4): 285–95.

Williamson Jr., F. (1980). A historical note on the finite element method, *International Journal for Numerical Methods in Engineering* 15(6): 930–4.

Wilson, E. L., Yuan, M.-W., and Dickens, J. M. (1982). Dynamic analysis by direct superposition of Ritz vectors, *Earthquake Engineering and Structural Dynamics* 10(6): 813–21.

Zhang, L., Luo, W., and Rose, J. L. (2005). Ultrasonic guided waves focused beyond welds in pipeline, *Proceedings of QNDE*, August, 25: 877–84.

Zhao, X., and Zhang, L. (2008). Finite element analysis of ultrasonic surface waves for crack inspection on an engine rotor disk, *Journal of Material Evaluation* 66: 67–72.

9 The Semi-Analytical Finite Element Method

9.1 Introduction

The semi-analytical finite element method (SAFEM) has recently become widely adopted for solving wave propagation problems in waveguides. SAFEM was developed as an alternative approach to more traditional methods such as the global matrix method, primarily because of its benefits of solving arbitrary cross-section waveguide problems (see Hayashi, Song, and Rose 2003). In SAFEM, the waveguide is discretized in the cross section, while an analytical solution is adopted in the wave propagation direction. Based on a variational scheme, a system of linear equations can be constructed with the frequency and wavenumber as unknowns. The unknowns can be solved using standard eigenvalue routines. SAFEM can solve problems of wave propagation in waveguides with complex cross sections, for example, multilayered laminates (Shorter 2004) and rails (Gavrić 1995; Hayashi, Song, and Rose 2003), where it is often difficult to obtain analytical solutions. For waveguides that are infinitely long in one dimension, SAFEM is superior to pure FEM in that exact analytical representations are used for one or two dimensions of the waveguide. Therefore, computational cost is reduced. SAFEM is also advantageous compared to analytical matrix methods because it is less prone to missing roots in developing the dispersion curves. Early employment of SAFEM in solving guided wave propagation problems can be found in Nelson and colleagues (1971) and Dong and colleagues (1972). In recent years, SAFEM was applied to the analysis of wave modes across a pipe elbow (Hayashi et al. 2005) and in materials with viscoelastic properties (Shorter 2004; Bartoli et al. 2006). SAFEM was also utilized to model the composite wing skin-to-spar bonded joints in aerospace structures by Matt and colleagues (2005) and to investigate guided wave propagations in rail structures (Damljanović and Weaver 2004; Lee et al. 2006). Applications of the SAFE technique for guided waves in composite plates can also be found in Liu and Achenbach (1994, 1995), Gao (2007), and Yan (2008).

This chapter presents a detailed theoretical derivation of the SAFE formulation for multilayered plate and cylindrical structures. Numerical examples of wave propagation in a multilayered anisotropic plate are also provided. Modal characteristics such as wave structures and attenuation properties are also given

---

With significant contribution from Fei Yan.

and discussed. Because a large number of roots obtained at each frequency depend on the number of nodes used in the SAFE calculation, a mode sorting technique based on guided wave orthogonality is introduced. The orthogonality-based mode sorting method is applicable not only for dispersion curve calculations by SAFE formulations, but also for those obtained from analytical methods.

9.2 SAFE Formulation for Plate Structures

The SAFE method adopts a harmonic exponential term, $e^{i(kx-\omega t)}$, to describe the wave behavior in the wave propagation direction, where x represents the wave propagation direction, k represents the wavenumber, ω is the radial frequency, and t is time. For anisotropic waveguides, it is possible to have different propagation directions for phase velocity and energy velocity. To avoid confusion, the wave propagation direction of phase velocity is hereafter named as the wavevector direction. The finite element discretization of the SAFE method takes place at the cross section of the waveguide that is perpendicular to the wavevector direction. For the problem of plane wave propagation in a plate, a one-dimensional discretization across the plate thickness is sufficient. The coordinate system and the finite element discretization for the SAFE calculation are shown in Figure 9.1. The plate width in the y direction is assumed to be infinite. Three-node line elements are employed here. The shape functions for the three-node line element are:

$$N_1 = \frac{\xi^2 - \xi}{2},$$
$$N_2 = 1 - \xi^2, \qquad\qquad (9.1)$$
$$N_3 = \frac{\xi^2 + \xi}{2},$$

where ξ is the variable in the local coordinate system for the element itself. For a given point in the local coordinate described by ξ, the global coordinate of the point can be calculated from the global coordinates of the three nodes (z_1, z_2, and z_3) and the shape functions in Equation (9.1).

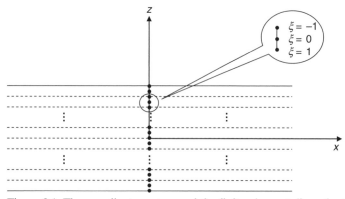

Figure 9.1. The coordinate system and the finite element discretization for the problem of wave propagation in plates. The insert shows the local coordinates of the three nodes of an element.

$$z = \begin{bmatrix} N_1 & N_2 & N_3 \end{bmatrix} \begin{bmatrix} z_1 \\ z_2 \\ z_3 \end{bmatrix}. \tag{9.2}$$

The local coordinates of nodes 1 through 3 are $\xi = -1$, $\xi = 0$, and $\xi = 1$, respectively, as shown in the insert of Figure 9.1.

Combining the time harmonic assumption and the finite element discretization, one can write the particle displacements of any point in an element in terms of the shape functions and the nodal displacements as follows:

$$\mathbf{u}^{(e)} = \begin{bmatrix} u_x^{(e)} \\ u_y^{(e)} \\ u_z^{(e)} \end{bmatrix} = \begin{bmatrix} \sum_{l=1}^{3} N_l(\xi) U_{xl} \\ \sum_{l=1}^{3} N_l(\xi) U_{yl} \\ \sum_{l=1}^{3} N_l(\xi) U_{zl} \end{bmatrix}^{(e)} e^{i(kx - \omega t)} = \mathbf{N}(\xi) \mathbf{Q}^{(e)} e^{i(kx - \omega t)}, \tag{9.3}$$

where:

$$\mathbf{N}(\xi) = \begin{bmatrix} N_1 & 0 & 0 & N_2 & 0 & 0 & N_3 & 0 & 0 \\ 0 & N_1 & 0 & 0 & N_2 & 0 & 0 & N_3 & 0 \\ 0 & 0 & N_1 & 0 & 0 & N_2 & 0 & 0 & N_3 \end{bmatrix}, \tag{9.4}$$

$$\mathbf{Q}^{(e)} = [U_{x1} \ U_{y1} \ U_{z1} \ U_{x2} \ U_{y2} \ U_{z2} \ U_{x3} \ U_{y3} \ U_{z3}]^{\mathrm{T}}, \tag{9.5}$$

and $U_{\alpha\beta}$ ($\alpha = x, y, z$; $\beta = 1, 2, 3$) denotes the nodal displacement of the node β in the α direction. The corresponding strain and stress vectors can then be calculated from the following equations:

$$\varepsilon^{(e)} = \left[\mathbf{L}_x \frac{\partial}{\partial x} + \mathbf{L}_y \frac{\partial}{\partial y} + \mathbf{L}_z \frac{\partial}{\partial z} \right] \mathbf{u}^{(e)}, \tag{9.6}$$

$$\sigma^{(e)} = \mathbf{C}^{(e)} \varepsilon^{(e)}, \tag{9.7}$$

where:

$$\mathbf{L}_x = \begin{bmatrix} 1 & 0 & 0 \\ 0 & 0 & 0 \\ 0 & 0 & 0 \\ 0 & 0 & 0 \\ 0 & 0 & 1 \\ 0 & 1 & 0 \end{bmatrix}, \quad \mathbf{L}_y = \begin{bmatrix} 0 & 0 & 0 \\ 0 & 1 & 0 \\ 0 & 0 & 0 \\ 0 & 0 & 1 \\ 0 & 0 & 0 \\ 1 & 0 & 0 \end{bmatrix}, \quad \mathbf{L}_z = \begin{bmatrix} 0 & 0 & 0 \\ 0 & 0 & 0 \\ 0 & 0 & 1 \\ 0 & 1 & 0 \\ 1 & 0 & 0 \\ 0 & 0 & 0 \end{bmatrix}, \tag{9.8}$$

and $\mathbf{C}^{(e)}$ is the material stiffness matrix of the element in the global coordinate system. For anisotropic materials, stiffness matrix transformations from the principal axes to the global coordinate system are necessary (Auld 1990). Substituting Equation (9.3) into Equation (9.6) results in:

$$\varepsilon^{(e)} = \left(\mathbf{B}_1 + ik\mathbf{B}_2 \right) \mathbf{Q}^{(e)} e^{i(kx - \omega t)}, \tag{9.9}$$

where

$$\mathbf{B}_1 = \mathbf{L}_y \mathbf{N}_{,y} + \mathbf{L}_z \mathbf{N}_{,z}, \quad \mathbf{B}_2 = \mathbf{L}_x \mathbf{N}. \tag{9.10}$$

In Equation (9.10), $\mathbf{N}_{,y}$ and $\mathbf{N}_{,z}$ are the derivatives of the shape function matrix given in Equation (9.4) with respect to the y and z directions, respectively.

A governing equation for the wave motion of each element can be obtained through the virtual work principle (Hayashi, Song, and Rose 2003):

$$\int_\Gamma \delta \mathbf{u}^{(e)T} \mathbf{t}^{(e)} d\Gamma = \int_V \delta \mathbf{u}^{(e)T} \left(\rho^{(e)} \ddot{\mathbf{u}}^{(e)} \right) dV + \int_V \delta \varepsilon^{(e)T} \sigma^{(e)} dV \tag{9.11}$$

where $\delta \mathbf{u}$ and $\delta \varepsilon$ represent the virtual displacement and virtual strain, respectively, $\mathbf{t}^{(e)}$ represents the external traction vector that can also be expressed using shape functions and nodal external tractions $\mathbf{T}^{(e)}$ (Equation (9.12)), $\bullet^T$ denotes a complex conjugate transpose, $\rho^{(e)}$ is density, $\ddot{\bullet}$ is a second derivative with respect to time, Γ is the surface of the element, and V is the volume of the element.

$$\mathbf{t}^{(e)} = \mathbf{N}(\xi) \mathbf{T}^{(e)} e^{i(kx - \omega t)} \tag{9.12}$$

Substituting Equations (9.13), (9.7), (9.9), and (9.12) into Equation (9.11) yields:

$$\int_\Gamma \delta \mathbf{Q}^{(e)T} \mathbf{N}^T \mathbf{N} \mathbf{T}^{(e)} d\Gamma = \int_V \delta \mathbf{Q}^{(e)T} \mathbf{N}^T \left(-k^2 \rho^{(e)} \mathbf{N} \mathbf{Q}^{(e)} \right) dV$$
$$+ \int_V \delta \mathbf{Q}^{(e)T} \left\{ (\mathbf{B}_1 - ik\mathbf{B}_2) \right\}^T \mathbf{C}^{(e)} (\mathbf{B}_1 + ik\mathbf{B}_2) \mathbf{Q}^{(e)} dV \tag{9.13}$$

Equation (9.13) is satisfied for any arbitrarily chosen virtual displacement. Therefore, the virtual displacement term $\delta \mathbf{Q}^{(e)T}$ can be eliminated from the equation:

$$\int_\Gamma \mathbf{N}^T \mathbf{N} \mathbf{T}^{(e)} d\Gamma = \int_V \mathbf{N}^T \left(-k^2 \rho^{(e)} \mathbf{N} \mathbf{Q}^{(e)} \right) dV$$
$$+ \int_V \left\{ (\mathbf{B}_1 - ik\mathbf{B}_2) \right\}^T \mathbf{C}^{(e)} (\mathbf{B}_1 + ik\mathbf{B}_2) \mathbf{Q}^{(e)} dV \tag{9.14}$$

For the line elements used here, the integrands in both sides of Equation (9.14) are functions of the variable ξ only. Equation (9.14) can thus be further simplified to:

$$\mathbf{F}^{(e)} = \left(\mathbf{K}_1^{(e)} + ik \mathbf{K}_2^{(e)} + k^2 \mathbf{K}_3^{(e)} \right) \mathbf{Q}^{(e)} - \omega^2 \mathbf{M}^{(e)} \mathbf{Q}^{(e)}, \tag{9.15}$$

With

$$\mathbf{F}^{(e)} = \int_{-1}^1 \mathbf{N}^T \mathbf{N} \mathbf{T}^{(e)} d\xi, \tag{9.16}$$

$$\mathbf{K}_1^{(e)} = \int_{-1}^1 \mathbf{B}_1^T \mathbf{C}^{(e)} \mathbf{B}_1 d\xi, \tag{9.17}$$

$$\mathbf{K}_2^{(e)} = \int_{-1}^1 \left(\mathbf{B}_1^T \mathbf{C}^{(e)} \mathbf{B}_2 - \mathbf{B}_2^T \mathbf{C}^{(e)} \mathbf{B}_1 \right) d\xi, \tag{9.18}$$

$$\mathbf{K}_3^{(e)} = \int_{-1}^{1} \mathbf{B}_2^{\mathrm{T}} \mathbf{C}^{(e)} \mathbf{B}_2 \, d\xi, \tag{9.19}$$

$$\mathbf{M}^{(e)} = \int_{-1}^{1} \rho^{(e)} \mathbf{N}^{\mathrm{T}} \mathbf{N} \, d\xi. \tag{9.20}$$

Equations (9.16) through (9.20) can be evaluated numerically for each element. Adopting a convertional finite element assembly methodology for all the elements and applying the traction-free boundary conditions on the top and bottom surfaces of the plate, one can form an eigenvalue problem in the global coordinate system:

$$\left(\mathbf{K}_1 + ik\mathbf{K}_2 + k^2\mathbf{K}_3 - \omega^2\mathbf{M}\right)\mathbf{Q} = 0. \tag{9.21}$$

The size of matrices $\mathbf{K}_1, \mathbf{K}_2, \mathbf{K}_3,$ and $\mathbf{M}$ is $3N \times 3N$, where N is the total number of nodes. $\mathbf{Q}$ is a $3N \times 1$ vector representing the particle displacements at the node positions.

It has been shown that the imaginary unit in Equation (9.21) can be eliminated by introducing a $3N \times 3N$ unitary transformation matrix $\mathbf{T}$ (Damljanović and Weaver 2004; Bartoli et al. 2006). All off-diagonal elements of the matrix $\mathbf{T}$ are zero. All diagonal elements of $\mathbf{T}$ are equal to one except for the elements corresponding to the particle displacements in the wavevector direction that are equal to the imaginary unit i. Because $\mathbf{T}^{\mathrm{T}}\mathbf{T} = \mathbf{I}$ ($\mathbf{I}$ is a $3N \times 3N$ unit matrix), by replacing $\mathbf{Q}$ in Equation (9.21) with $\mathbf{T}^{\mathrm{T}}\mathbf{T}\mathbf{Q}$ and then multiplying the equation with $\mathbf{T}$ from the left side followed by applying the properties of the matrices, one can obtain a new eigenvalue equation:

$$\left(\mathbf{K}_1 + k\hat{\mathbf{K}}_2 + k^2\mathbf{K}_3 - \omega^2\mathbf{M}\right)\hat{\mathbf{Q}} = 0, \tag{9.22}$$

with $\hat{\mathbf{K}}_2 = \dfrac{\mathbf{T}^{\mathrm{T}}\mathbf{K}_2\mathbf{T}}{-i}$ and $\hat{\mathbf{Q}} = \mathbf{T}\mathbf{Q}$. Equation (9.22) can be rewritten as a first-order eigensystem:

$$[\mathbf{A} - k\mathbf{B}]\begin{bmatrix} \hat{\mathbf{Q}} \\ k\hat{\mathbf{Q}} \end{bmatrix} = 0, \tag{9.23}$$

where

$$\mathbf{A} = \begin{bmatrix} 0 & \mathbf{K}_1 - \omega^2\mathbf{M} \\ \mathbf{K}_1 - \omega^2\mathbf{M} & \hat{\mathbf{K}}_2 \end{bmatrix}, \quad \mathbf{B} = \begin{bmatrix} \mathbf{K}_1 - \omega^2\mathbf{M} & 0 \\ 0 & -\mathbf{K}_3 \end{bmatrix}. \tag{9.24}$$

For elastic materials, that is, materials with real stiffness matrices, the matrices $\mathbf{A}$ and $\mathbf{B}$ are $6N \times 6N$ real matrices.

In total, $6N$ eigenvalues for wavenumber k can be solved at each frequency ω from Equation (9.23). Among all of the eigenvalues, there are real eigenvalues for propagating guided wave modes and complex eigenvalues including pure imaginary eigenvalues for the evanescent modes. The eigenvalues are also solved in pairs from Equation (9.23), that is, if k_μ is a solution of Equation (9.23), $-k_\mu$ is also a solution. Let k_{real} denote the real solutions from Equation (9.23). Positive wavenumbers among k_{real} are wavenumbers of wave modes propagating in the $+x$ direction, while their negatives are wavenumbers of the corresponding wave modes propagating in the $-x$

direction. For complex solutions $k_{complex}$, if the imaginary parts $\text{Im}\{k_{complex}\} > 0$, the wave modes whose wavenumbers are $k_{complex}$ attenuate in the $+x$ direction. Otherwise, $k_{complex}$ are the wavenumbers of the wave modes that attenuate in the $-x$ direction.

For each eigenvalue, an eigenvector is solved from Equation (9.23) as well. It is explicit that the corresponding wave structure information is contained in the eigenvector. Based on the wave structures, the strain and stress fields can be obtained using Equation (9.9) and Equation (9.7).

9.3 Orthogonality-Based Mode Sorting

Based on Equation (9.23), numerically calculated dispersion relations $k(\omega)$ can be obtained by conducting the eigenvalue extraction for different frequency points in the frequency range of interest. Because dispersion curves for different guided wave modes may cross each other, it is necessary to apply a mode sorting process to sort the eigenvalues extracted from the different frequencies into different wave modes. In this chapter, the orthogonality-based wave mode sorting algorithm, first developed in Loveday and Long (2007), is employed to sort the propagating modes.

An orthogonality relation between different modes can be derived from Equation (9.23):

$$\begin{cases} \psi_m^{\mathrm{T}} B_m \psi_m \neq 0 \\ \psi_n^{\mathrm{T}} B_m \psi_m = 0 \end{cases} \tag{9.25}$$

where

$$\psi = \begin{bmatrix} \hat{\mathbf{Q}} \\ k\hat{\mathbf{Q}} \end{bmatrix}, \tag{9.26}$$

represents the eigenvector solved from Equation (9.23), and the subscripts n and m label different eigenvectors and **B** matrices in Equation (9.23) for different wave modes. Equation (9.25) states that all of the eigenvectors solved at a same frequency are orthogonal to each other.

It can be assumed that the orthogonality relation still approximately holds for eigenvectors solved for two frequencies that are of a small frequency difference $\Delta\omega$:

$$\begin{cases} \psi_m^{\mathrm{T}}(\omega) B_m(\omega) \psi_m(\omega+\Delta\omega) \neq 0 \\ \psi_n^{\mathrm{T}}(\omega) B_m(\omega) \psi_m(\omega+\Delta\omega) \approx 0 \end{cases}. \tag{9.27}$$

Equation (9.27) can be directly applied into a mode sorting algorithm. After extracting eigenvalues and their corresponding eigenvectors from Equation (9.23) for two adjacent frequencies, orthogonality terms $\psi_n^{\mathrm{T}}(\omega) B_m(\omega) \psi_m(\omega+\Delta\omega)$ are calculated between one eigenvector from the low-frequency solutions and every eigenvector from the high-frequency solutions, yielding a vector of orthogonality values. Equation (9.27) guarantees that the maximum value of the vector is achieved when the two eigenvectors used in the orthogonality calculation belong to a same wave mode. Therefore, by seeking the maximum value within the vector of orthogonality values, we can tell which high-frequency eigenvector belongs to the same wave mode as the low-frequency eigenvector. Dispersion curves with continuous lines representing different guided wave modes can be obtained by solving Equation (9.23) for different frequencies and then applying the mode sorting algorithm.

9.4 Group Velocity Dispersion Curves

Group velocities describe the propagating speeds of guided wave packets with components of similar frequency. They are very critical in guided wave applications because the velocities measured by transducers are usually group velocities. Group velocity dispersion curves are quite helpful in identifying guided wave modes in guided wave tests. Group velocity dispersion curves can be calculated from the definition of group velocity:

$$c_g = \frac{d\omega}{dk}, \tag{9.28}$$

where c_g is the group velocity. Substituting $k = \omega/c_p$ into Equation (9.28), one can get Equation (9.29) after some algebraic operations:

$$c_g = c_p^2 \left[c_p - f \frac{dc_p}{df} \right]^{-1}. \tag{9.29}$$

The phase velocity is denoted by c_p in Equation (9.29). Obviously, the group velocity dispersion curves can be calculated from the mode sorted phase velocity dispersion curves based on Equation (9.29). However, the differential operation in Equation (9.29) is prone to numerical errors, especially when the phase velocity dispersion curves are calculated using a large frequency increment step.

An alternative way of calculating group velocity directly for each (k, ω) solution of Equation (9.23) using the SAFE method can be found in Bartoli and colleagues (2006). Taking a differential operation on Equation (9.22) with respect to the wavenumber k, one can get:

$$\frac{\partial}{\partial k} \left(\mathbf{K}_1 + k\hat{\mathbf{K}}_2 + k^2\mathbf{K}_3 - \omega^2\mathbf{M} \right) \hat{\mathbf{Q}}_R = 0 \tag{9.30}$$

$\hat{\mathbf{Q}}_R$ is used in Equation (9.30) to represent the right eigenvector that can be directly solved from Equation (9.23). A corresponding left eigenvector $\hat{\mathbf{Q}}_L$ can be obtained by replacing matrices $\mathbf{A}$ and $\mathbf{B}$ in Equation (9.23) with their transposes and then solving the eigenvalue problem. A left multiplication of $\hat{\mathbf{Q}}_L$ to Equation (9.30) yields:

$$\hat{\mathbf{Q}}_L \left[\frac{\partial}{\partial k} \left(\mathbf{K}_1 + k\hat{\mathbf{K}}_2 + k^2\mathbf{K}_3 \right) - 2\omega\frac{\partial\omega}{\partial k}\mathbf{M} \right] \hat{\mathbf{Q}}_R = 0. \tag{9.31}$$

The group velocity can then be derived from Equation (9.31) as:

$$c_g = \frac{\partial\omega}{\partial k} = \frac{\hat{\mathbf{Q}}_L \left(\hat{\mathbf{K}}_2 + 2k\mathbf{K}_3 \right) \hat{\mathbf{Q}}_R}{2\omega\hat{\mathbf{Q}}_L\mathbf{M}\hat{\mathbf{Q}}_R}. \tag{9.32}$$

All terms in the right-hand side of Equation (9.32) depend only on a single (k, ω) solution. As a result, the group velocity can be calculated for each (k, ω) solution without requiring the knowledge on the phase velocities of the adjacent frequencies as Equation (9.28) does.

9.5 Guided Wave Energy

9.5.1 Poynting Vector

Starting from an electromagnetic-acoustic analogy, Auld derived an acoustic Poynting vector that represents the power flow density vector of acoustic waves in solids (1990). The acoustic Poynting vector is not only useful for investigating guided wave energy, but also plays an important role in guided wave field normalization and guided wave excitability studies, as well as skew angle determination for anisotropic waveguides. The complex Poynting vector can be calculated from the particle velocity and stress fields based on the following equation:

$$\mathbf{P} = -\frac{\tilde{\mathbf{v}} \cdot \sigma_M}{2},$$
(9.33)

where $\tilde{\mathbf{v}}$ represents the conjugate of the particle velocity vector and σ_M represents the stress matrix of format:

$$\sigma_M = \begin{bmatrix} \sigma_{xx} & \sigma_{xy} & \sigma_{xz} \\ \sigma_{yx} & \sigma_{yy} & \sigma_{yz} \\ \sigma_{zx} & \sigma_{zy} & \sigma_{zz} \end{bmatrix}.$$
(9.34)

For plate waveguides, the three components of the Poynting vector provide us with the power flow densities of a guided wave mode in three orthogonal directions. An integral of $\mathbf{P}_x$ (the component in the wavevector direction) across the plate thickness yields the power flow carried by the guided wave mode in the wavevector direction.

9.5.2 Energy Velocity

In elastic waveguides, the group velocity defined by Equation (9.28) also describes the speed of guided wave energy transport. However, for waveguides containing viscoelastic materials, there is no physical meaning of Equation (9.28). Instead, an energy velocity that is different from the group velocity should be used (Bernard et al. 2001). The energy velocity of a guided wave mode in a plate waveguide can be calculated based on the power flow in the wavevector direction and the energy density of the wave field:

$$c_e = \frac{\int_d P_x dz}{\frac{1}{T} \int_T \left(\int_d \left(e_k + e_p \right) dz \right) dt},$$
(9.35)

where P_x represents the power flow density (Poynting value) in the wavevector direction, z represents the thickness direction, d denotes the thickness of the plate, T denotes the time period of the guided wave, and e_k and e_p represent the kinetic energy density and strain energy density of the guided wave mode, respectively.

9.5.3 Skew Effects in Anisotropic Plates

If material anisotropy is not present, all three components of the second row in the stress matrix vanish for a plane wave assumption. In this case, the Poynting vector only has two nonzero components, P_x and P_z, for the wavevector direction and the plate thickness direction, respectively. For the cases of anisotropic plates, however, the P_y component becomes nonzero for some wave modes. The nonzero P_y component introduces wave skew effects to the guided wave modes. The skew angles can be calculated based on the following equation:

$$\varphi_{skew} = \tan^{-1} \left(\frac{\int_d P_y \, dz}{\int_d P_x \, dz} \right). \tag{9.36}$$

9.6 Solution Convergence of the SAFE Method

To verify the SAFE method and to discuss the solution convergence issue, the SAFE technique is applied to a sample dispersion curve computation for a stress-free 2 mm thick aluminum plate. The SAFE-calculated phase velocity dispersion curves are given in Figure 9.2(a) with the results obtained using the partial wave method (PWM) also shown for comparison. A total of seven nodes across the plate thickness is used for the SAFE results. As can be seen, there is a discrepancy between the SAFE results and the PWM results, especially for the high-frequency portion of the higher order wave modes. It is easy to demonstrate that the SAFE solutions converge to the PWM solutions after increasing the number of nodes used in the thickness discretization. While increasing the number of nodes, the convergence is achieved faster at lower frequency for the lower order wave modes.

The wave structure of the S0 mode at 500 kHz and that of the A2 mode at 2.4 MHz are shown in Figure 9.2(b) and (c). It can be observed that for the S0 mode at 500 kHz, the SAFE solution converges to the PWM solution when only seven nodes are used. The wave structure has small variations across the thickness of the plate. By contrast, for the A2 mode at 2.4 MHz, there is a larger discrepancy between the seven-node SAFE solution and the PWM solution; the wave structure varies severely. Obviously, more nodes are needed in the SAFE calculations to accurately represent the wave modes whose wave structures are of larger variations across the plate thickness. The number of elements used in the SAFE calculation should always be chosen to ensure convergence of the results. The convergence is achieved when the dispersion curves do not change with an increasing number of elements.

9.7 Free Guided Waves in an Eight-Layer Quasi-Isotropic Plate

Using an eight-layer, quasi-isotropic, fiber-reinforced composite plate as an example, characteristics of free guided waves including phase velocity dispersion curves, group velocity dispersion curves, wave structures, stress distributions, and skew angle dispersion curves are calculated. The plate is made from Cytec CYCOM 977–3 carbon epoxy prepregs. The stacking sequence of the plate is [0/45/90/−45]$_s$. The engineering properties of the prepreg used in the calculations are given in Table 9.1 (Schoeppner et al. 2001). The density is 1.608 g/cm^3 and the thickness of each layer of the plate is 0.2 mm.

The phase velocity dispersion curves for the 0°, 30°, and −30° wavevector directions are given in Figure 9.3(a), (b), and (c), respectively.

As can be seen, all guided wave modes are effectively sorted by the mode sorting algorithm. Because of material anisotropy, the dispersion curves are different for

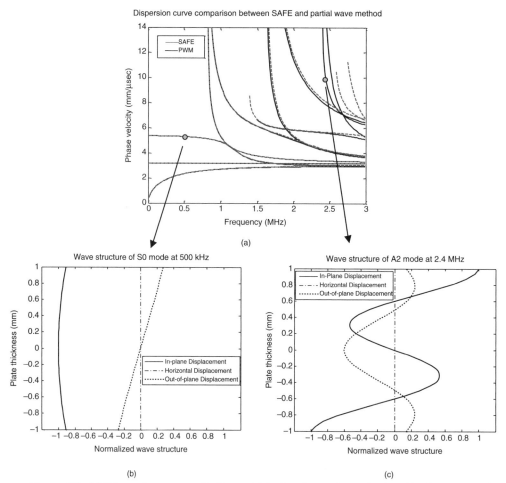

Figure 9.2. (a) Dispersion curves of a 2 mm thick aluminum plate calculated by SAFEM and PWM, (b) wave structure of S0 mode at 500 kHz, (c) wave structure of A2 mode at 2.4 MHz.

all three wavevector directions. Besides, the decoupling effect between Lamb type waves and SH waves in single isotropic plates no longer exists for the composite plate. The corresponding group velocity dispersion curves that are calculated using Equation (9.32) are shown in Figure 9.4.

Example wave structures for guided wave modes propagating in the 0° direction are presented in Figure 9.5.

The wave mode numbering system proposed by Gao (2007) is used in this chapter to name the guided wave modes in the composite plate. The three fundamental modes with no cutoff frequencies are named modes 1 to 3, based on the values of the phase velocities at low frequency. Higher order modes starting from mode 4 are numbered based on their cutoff frequencies for the 0° wavevector direction. The first six modes are labeled with their names in Figure 9.3(a). As demonstrated in Figure 9.5, wave structures vary dramatically from one mode to another even though the frequencies are the same. Because guided wave modes with different wave structures are sensitive to different types of defects, the strong variations of the wave structures among the different guided wave modes and frequencies offer excellent sensitivity potential in guided wave NDE and SHM applications via mode and frequency choices. The wave structures shown in Figure 9.5 have nonzero displacements in all three displacement

Table 9.1. *Material properties of Cytec CYCOM 977–3*
carbon epoxy prepreg

E_1	172 GPa
$E_2 = E_3$	9.8 GPa
G_{23}	3.2 GPa
$G_{12} = G_{13}$	6.1 GPa
v_{23}	0.55
$v_{12} = v_{13}$	0.37

directions, whereas in isotropic plates, Lamb waves only have displacements in the x and z directions while SH waves have y direction displacements only. Nevertheless, analogies between the guided wave modes in composite plates and the wave modes in isotropic plates can still be observed. For instance, the wave structure of mode 1 in the composite plate has a dominant out-of-plane displacement (u_z) that makes mode 1 similar to the Lamb wave mode A0 in isotropic plates. Similarly, mode 2 and mode 3 have analogies with the SH0 mode and the S0 mode, respectively. Because of the analogy based on wave structures, the fundamental wave modes in the composite plate possess somewhat similar characteristics as the A0, S0, and SH0 modes in isotropic plates when considering wave excitabilities, which are discussed elsewhere in the text.

The stress distributions across the plate thickness for the three fundamental modes of the composite plate at 500 kHz are shown in Figure 9.6.

The stress distributions are calculated from the wave structures using Equation (9.7) and Equation (9.9). As has been shown, the calculated stress components σ_{xz}, σ_{yz}, and σ_{zz} vanish at the plate surfaces, which satisfies the traction-free boundary conditions. They also satisfy the continuity interface conditions between different layers of the composite plate. The correctness of the SAFE calculations is thus verified.

The skew angle dispersion curves calculated based on Equation (9.36) are shown in Figure 9.7. The integrations in Equation (9.36) are evaluated numerically. The wave vector direction is zero degrees.

Although the composite plate is of a quasi-isotropic stacking sequence, the skew angles introduced by the material anisotropy actually vary in a large range roughly from –45 degrees to 38 degrees. For guided wave modes with nonzero skew angles, the wave energies propagate in the directions away from the wavevector direction. Hence, a careful investigation of wave skew effects is necessary in guided wave applications when studying anisotropic waveguides.

9.8 SAFE Formulation for Cylindrical Structures

The SAFE formulation for cylindrical structures also starts with the governing equation provided by the virtual work principle. Equation (9.37) holds for a stress-free hollow cylinder. Linear elastic and viscoelastic material behavior is considered here.

$$\int_V \delta\mathbf{u}^T \cdot \rho\ddot{\mathbf{u}}\,dV + \int_V \delta\boldsymbol{\varepsilon}^T \cdot \boldsymbol{\sigma}\,dV = 0, \tag{9.37}$$

where T represents matrix transpose, ρ is density, and $\ddot{\mathbf{u}}$ is the second derivative of displacement $\mathbf{u}$ with respect to time t. $\int_\Gamma d\Gamma$ and $\int_V dV$ are the surface and volume

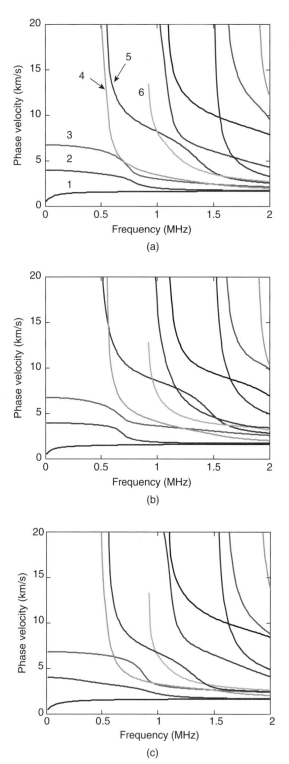

Figure 9.3. Phase velocity dispersion curves of the eight-layer quasi-isotropic plate for the wavevector directions of (a) 0°, (b) 30°, and (c) −30°.

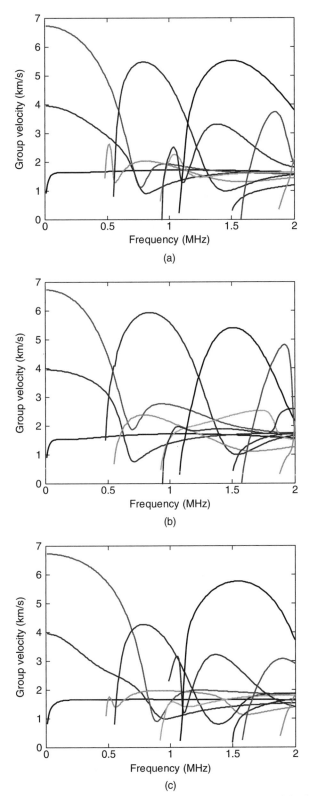

Figure 9.4. Group velocity dispersion curves of the eight-layer quasi-isotropic plate for the wavevector directions of (a) 0°, (b) 30°, and (c) −30°.

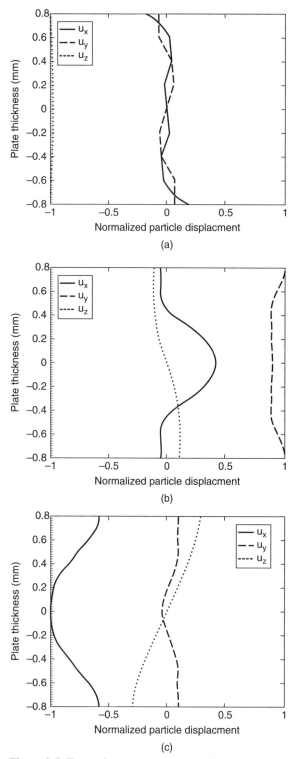

Figure 9.5. Example wave structures of the three fundamental modes at 500 kHz. (a) mode 1, (b) mode 2, (c) mode 3.

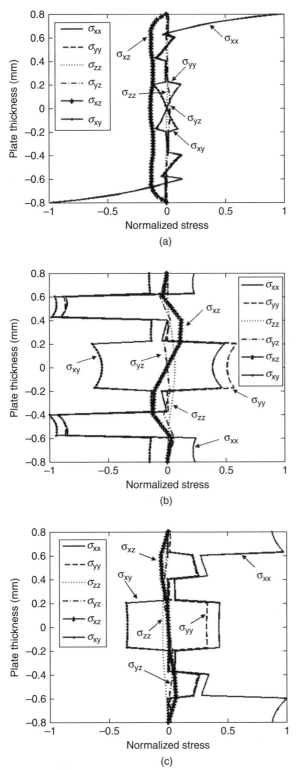

Figure 9.6. Stress distributions of the three fundamental modes at 500 kHz. (a) mode 1, (b) mode 2, (c) mode 3.

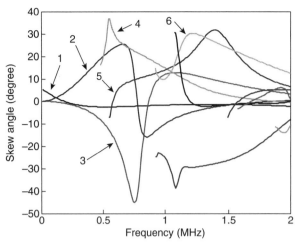

Figure 9.7. Skew angle dispersion curves of the composite plate for the $0°$ wavevector direction. Modes 1–6 are labeled.

integrals of the element, respectively. For wave propagation in cylindrical structures, the wave equation needs to be solved in cylindrical coordinates, $dV = rdrd\theta dz$. The first and second terms on the left-hand side are the corresponding increments of kinetic energy and potential energy.

We adopted an exact analytical solution $e^{in\theta}$ in the circumferential direction. Exact analytical harmonic solutions are therefore used in both the θ and z directions. The finite element approximation reduces to only one dimension, r. This 1-D SAFE formulation not only improves the accuracy in the calculation of flexural modes with higher circumferential orders, but also greatly reduces the computational cost compared to the 2-D SAFE formulation for cylindrical structures. For a harmonic wave propagating in the z direction, the displacement at any point $\mathbf{u}(r, \theta, z, t)$ can be represented by

$$\mathbf{u}(r, \theta, z, t) = \sum_{j=1}^{2} \mathbf{N}(r) \mathbf{U}^j e^{i(kz+n\theta-\omega t)}, \tag{9.38}$$

where $\mathbf{U}^j$ is the nodal displacement vector at the jth element and $\mathbf{N}(r)$ is the shape function in the thickness direction r. For a two-node element, $\mathbf{U}^j$ is a six-element vector and $\mathbf{N}(r)$ is a 3×6 matrix. The shape function matrix is chosen as follows,

$$\mathbf{N} = \begin{bmatrix} N_1 & 0 & 0 & N_2 & 0 & 0 \\ 0 & N_1 & 0 & 0 & N_2 & 0 \\ 0 & 0 & N_1 & 0 & 0 & N_2 \end{bmatrix}, \tag{9.39}$$

using linear shape functions

$$N_1 = \frac{1}{2}(1-\xi), \qquad N_2 = \frac{1}{2}(1+\xi), \tag{9.40}$$

where $-1 \le \xi \le 1$ is the natural coordinate in the r direction. The strain-displacement relations in cylindrical coordinates are:

$$\varepsilon_{rr} = \frac{\partial u_r}{\partial r}$$

$$\varepsilon_{\theta\theta} = \frac{u_r}{r} + \frac{1}{r}\frac{\partial u_\theta}{\partial \theta}$$

$$\varepsilon_{zz} = \frac{\partial u_z}{\partial z}$$

$$\gamma_{\theta z} = \frac{1}{r}\frac{\partial u_z}{\partial \theta} + \frac{\partial u_\theta}{\partial z} \tag{9.41}$$

$$\gamma_{rz} = \frac{\partial u_r}{\partial z} + \frac{\partial u_z}{\partial r}$$

$$\gamma_{r\theta} = \frac{1}{r}\frac{\partial u_r}{\partial \theta} + \frac{\partial u_\theta}{\partial r} - \frac{u_\theta}{r}$$

Substituting Equation (9.38) and Equation (9.40) into the strain-displacement relationships Equation (9.41), the six strain components can be expressed as:

$$\begin{aligned}
\varepsilon &= (\mathbf{L}_1 + \mathbf{L}_2)\mathbf{u} \\
&= (\mathbf{L}_1 + \mathbf{L}_2)\mathbf{N}\mathbf{U}^j e^{i(kz+n\theta-\omega t)} \\
&= (\mathbf{B}_1 + ik\mathbf{B}_2)\mathbf{U}^j e^{i(kz+n\theta-\omega t)}
\end{aligned} \tag{9.42}$$

where

$$\mathbf{L}_1 = \begin{bmatrix}
\dfrac{\partial}{\partial r} & 0 & 0 \\[2mm]
\dfrac{1}{r} & \dfrac{1}{r}\dfrac{\partial}{\partial \theta} & 0 \\[2mm]
0 & 0 & 0 \\[2mm]
0 & 0 & \dfrac{1}{r}\dfrac{\partial}{\partial \theta} \\[2mm]
0 & 0 & \dfrac{\partial}{\partial r} \\[2mm]
\dfrac{1}{r}\dfrac{\partial}{\partial \theta} & \dfrac{\partial}{\partial r} - \dfrac{1}{r} & 0
\end{bmatrix}, \tag{9.43}$$

$$\mathbf{L}_2 = \begin{bmatrix}
0 & 0 & 0 \\[2mm]
0 & 0 & 0 \\[2mm]
0 & 0 & \dfrac{\partial}{\partial z} \\[2mm]
0 & \dfrac{\partial}{\partial z} & 0 \\[2mm]
\dfrac{\partial}{\partial z} & 0 & 0 \\[2mm]
0 & 0 & 0
\end{bmatrix}, \tag{9.44}$$

$$\mathbf{B}_1 = \begin{bmatrix} \dfrac{\partial N_1}{\partial r} & 0 & 0 & \dfrac{\partial N_2}{\partial r} & 0 & 0 \\[2mm] \dfrac{N_1}{r} & i\dfrac{n}{r}N_1 & 0 & \dfrac{N_2}{r} & i\dfrac{n}{r}N_2 & 0 \\[2mm] 0 & 0 & 0 & 0 & 0 & 0 \\[2mm] 0 & 0 & i\dfrac{n}{r}N_1 & 0 & 0 & i\dfrac{n}{r}N_2 \\[2mm] 0 & 0 & \dfrac{\partial N_1}{\partial r} & 0 & 0 & \dfrac{\partial N_2}{\partial r} \\[2mm] i\dfrac{n}{r}N_1 & \dfrac{\partial N_1}{\partial r} - \dfrac{N_1}{r} & 0 & i\dfrac{n}{r}N_2 & \dfrac{\partial N_2}{\partial r} - \dfrac{N_2}{r} & 0 \end{bmatrix}_{6\times6}, \quad (9.45)$$

$$\mathbf{B}_2 = \begin{bmatrix} 0 & 0 & 0 & 0 & 0 & 0 \\ 0 & 0 & 0 & 0 & 0 & 0 \\ 0 & 0 & N_1 & 0 & 0 & N_2 \\ 0 & N_1 & 0 & 0 & N_2 & 0 \\ N_1 & 0 & 0 & N_2 & 0 & 0 \\ 0 & 0 & 0 & 0 & 0 & 0 \end{bmatrix}_{6\times6}. \quad (9.46)$$

Substituting the strain components obtained in Equation (9.42) into the constitutive relation for strain and stress yields the expression for stress components:

$$\sigma = \mathbf{C}\varepsilon = \mathbf{C}(\mathbf{B}_1 + ik\mathbf{B}_2)\mathbf{U}^j e^{i(kz+n\theta-\omega t)}, \quad (9.47)$$

where $\mathbf{C}$ is the stiffness matrix. The values in the matrix $\mathbf{C}$ are real for elastic materials and complex for viscoelastic materials according to the correspondence principle (Christensen 1982).

Substituting the displacements Equation (9.38), strains Equation (9.42), and stresses Equation (9.47) into the governing Equation (9.37), one obtains Equation (9.48) for element j after simplification.

$$(\mathbf{K}_1^j + ik\mathbf{K}_2^j + k^2\mathbf{K}_3^j)\mathbf{U}^j - \omega^2\mathbf{M}^j\mathbf{U}^j = 0, \quad (9.48)$$

Where

$$\mathbf{K}_1^j = \iint_{r\,\theta} \mathbf{B}_1^T \mathbf{C} \mathbf{B}_1 r\,dr\,d\theta,$$

$$\mathbf{K}_2^j = \iint_{r\,\theta} (\mathbf{B}_1^T \mathbf{C} \mathbf{B}_2 - \mathbf{B}_2^T \mathbf{C} \mathbf{B}_1) r\,dr\,d\theta,$$

$$\mathbf{K}_3^j = \iint_{r\,\theta} \mathbf{B}_2^T \mathbf{C} \mathbf{B}_2 r\,dr\,d\theta, \quad (9.49)$$

$$\mathbf{M}^j = \rho \iint_{r\,\theta} \mathbf{N}^T \mathbf{N} r\,dr\,d\theta.$$

Similarly, Equation (9.49) can be assembled into the system of equations given in Equation (9.21) with vector **Q** replaced by **U**. It can also be further reduced to the first-order eigensystem as given in Equation (9.23) and solved using a standard eigenvalue solution routine.

9.9 Summary

This chapter discusses the guided wave solutions to multilayered plate and cylindrical structures using the SAFE method. Both isotropic and multilayer anisotropic composite plate and pipe structures can be modeled by the SAFE technique. Mode sorting algorithms are implemented into the dispersion curve calculations based on the wave mode orthogonality that is derived in the SAFE formulation for both plate and pipe structures. Important characteristics of free guided waves including group velocity, energy velocity, wave structure, stress distribution, Poynting vector, and skew angle are discussed. The wave characteristics form a theoretical foundation for guided wave applications. Example calculations on the guided wave characteristics are performed for an eight-layer, quasi-isotropic, fiber-reinforced composite plate. It is demonstrated that the influence of material anisotropy on guided wave propagation remains strong, even when a quasi-isotropic stacking sequence is used. Complete dispersion curves for a viscoelastic coated pipe are calculated. The solution convergence of the SAFE technique is also discussed. A physically based explanation for the differences in the solution convergence of different wave modes and frequencies is given based on a wave structure analysis.

9.10 Exercises

1. What are the differences between the SAFE method and the conventional finite element method?
2. What are the major benefits of using the SAFE method for ultrasonic guided wave applications?
3. What are the physical meanings of the real and imaginary parts of a wavenumber calculated in a SAFE calculation? How to determine the wavevector directions from the wavenumbers?
4. What are guided wave skew effects? Are there skew effects in composite plates with quasi-isotropic layups?
5. Explain how to use a 1-D SAFE to calculate guided wave dispersion curves for flexural modes in a hollow cylinder.
6. For a calculation with a particular mesh size, why does the SAFEM calculation match an analytical solution at relatively low frequencies, but fail to give an accurate answer at relatively high frequencies?
7. Make a list of several potential waveguide problems with different geometrical cross sections. Indicate whether each problem may be solved using a 1-D or 2-D SAFEM approach.
8. Assume you have a computer code that calculates modes in a plate using the 1-D SAFEM approach. Describe the changes necessary to develop this code to solve 2-D problems with the SAFEM approach.

9. For the eight-layer quasi-isotropic plate mentioned in Section 9.7, why would you expect to have some mode-frequency combinations with a higher skew angle than others?

10. Why does orthogonality-based mode sorting work?

9.11 REFERENCES

Auld, B. A. (1990). *Acoustic Fields and Waves in Solids*. Malabar, FL: Krieger.

Bartoli, I., Marzani, A., Lanza di Scalea, F., and Viola, E. (2006). Modeling wave propagation in damped waveguides of arbitrary cross-section, *J. Sound. Vibr.* 295: 685–707.

Barshinger, J. N., and Rose, J. L. (2004). Guided wave propagation in an elastic hollow cylinder coated with a viscoelastic material, *IEEE Trans. Ultrason. Ferroelectr. Freq. Control* 51: 1547–56.

Bernard, A., Lowe, M. J. S., and Deschamps, M. (2001). Guided waves energy velocity in absorbing and non-absorbing plates, *J. Acoust. Soc. Am.* 110(1): 186–96.

Christensen, R. M. (1982). *Theory of Viscoelasticity: An Introduction*, 2nd ed., New York: Academic Press.

Damljanović, V., and Weaver, R. L. (2004). Forced response of a cylindrical waveguide with simulation of the wavenumber extraction problem, *J. Acoust. Soc. Am.* 115: 1582–91.

Dong, S., and Nelson, R. (1972). On natural vibrations and waves in laminated orthotropic plates, *J. Appl. Mech.* 39: 739–45.

Gao, H. (2007). *Ultrasonic Guided Wave Mechanics for Composite Material Structural Health Monitoring*. PhD thesis, Pennsylvania State University.

Gavrić, L. (1995). Computation of propagative waves in free rail using a finite element technique, *J. Sound. Vibr.* 185(3): 531–43.

Hayashi, T., Kawashima, K., Sun, Z., and Rose, J. L. (2005). Guided wave propagation mechanics across a pipe elbow, Trans. *ASME* 127: 322–7.

Hayashi, T., Song, W.-J., and Rose, J. L. (2003). Guided wave dispersion curves for a bar with an arbitrary cross-section, a rod and rail example, *Ultrasonics* 41: 175–83.

Lee, C. (2006). *Guided Waves in Rail for Transverse Crack Detection*. Department of engineering science and mechanics, PhD thesis, Pennsylvania State University.

Liu, G. R., and Achenbach, J. D. (1994). A strip element method for stress analysis of anisotropic linearly elastic solids, *ASME J. Appl. Mech.* 61: 270–7.

(1995). Strip element method to analyze wave scattering by cracks in anisotropic laminated plates, *ASME J. Appl. Mech.* 62: 607–13.

Loveday, P. W., and Long, C. S. (2007). Time domain simulation of piezoelectric excitation of guided waves in rails using waveguide finite elements, *Proc. SPIE* 6529, 65290V.

Matt, H., Bartoli, I., and Lanza di Scalea, F. (2005). Ultrasonic guided wave monitoring of composite wing skin-to-spar bonded joints in aerospace structures, *Acoust. Soc. Am.* 118(4): 2240–52.

Mu, J., and Rose, J. L. (2008). Guided wave propagation and mode differentiation in hollow cylinders with viscoelastic coatings, *J. Acoust. Soc. Am.* 124(2): 866–74.

Nelson, R. B., Dong, S. B., and Kalra, R. D. (1971). Vibrations and waves in laminated orthotropic circular cylinders, *J. Sound. Vibr.* 18: 429–44.

Schoeppner, G. A., Kim, R., and Donadson, S. L. (2001). *Proceedings of AIAA/ASME/ASCE/AHS/ASC Structures, Structural Dynamics, and Materials Conference and Exhibit*, 42nd, Seattle, WA, Apr. 16–19, 2001, AIAA–2001–1216.

Shorter, P. J. (2004). Wave propagation and damping in linear viscoelastic laminates, *J. Acoust. Soc. Am.* 115(5): 1917–25.

Yan, F., (2008). *Ultrasonic Guided Wave Phased Array for Isotropic and Anisotropic Plates*, PhD thesis, Pennsylvania State University.

10 Guided Waves in Hollow Cylinders

10.1 Introduction

Ultrasonic guided waves are most commonly used in plate, rod, and hollow cylinder (pipeline and tubing) inspections. This subject is receiving much attention recently because of the possibility of inspecting long volumetric lengths of a structure from a single probe position. Components can be inspected if hidden, coated, or under insulation, oil, soil, or concrete. Excellent defect detection sensitivities and long inspection distances have been demonstrated. Guided waves in cylindrical structures may travel in the circumferential or axial direction. Based on boundary conditions, material properties, and geometric properties of the hollow cylinder, the wave behavior can be described by solving the governing wave equations with appropriate boundary conditions. In this chapter, simulations of guided waves propagating in axial directions in cylindrical structures are calculated and evaluated.

10.2 Guided Waves Propagating in an Axial Direction

In this section, a calculation approach is developed for guided wave propagation in the axial direction of a hollow cylinder.

10.2.1 Analytic Calculation Approach

When considering the particle motion direction possibilities in a hollow cylinder, the guided waves propagating in the axial direction may involve longitudinal waves and torsional waves. The longitudinal waves have dominant particle motions in either the r and/or z directions and the torsional waves have dominant particle motions in the θ direction. According to the energy distribution in the circumferential direction, the guided waves contain axisymmetric modes and non-axisymmetric modes (also known as flexural modes). For convenience, a longitudinal mode group will be expressed as L(m, n) and a torsional mode group as T(m, n). Here the integer m denotes the circumferential order of a mode and the integer n represents the group order of a mode. An axisymmetric mode has the circumferential number $m = 0$.

With significant contribution from Li Zhang.

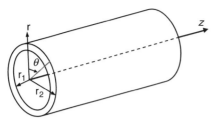

Figure 10.1. Hollow cylinder axes configuration.

If an elastic isotropic hollow cylinder has traction-free boundary conditions on its surfaces (Figure 10.1), the Navier's governing wave equation for guided wave propagation can be written as follows:

$$\mu\nabla^2\vec{U}+(\lambda+\mu)\nabla\nabla\cdot\vec{U}=\rho\left(\partial^2\vec{U}\big/\partial t^2\right)\tag{10.1}$$

where t indicates the time and $\vec{U}$ represents the displacement field, which is a function of the three cylindrical coordinates and time. The density ρ and Lamé constants μ and λ determine the bulk wave velocities in the material. If c_1 indicates the dilatational longitudinal bulk wave velocity and c_2 is the shear bulk wave velocity, then:

$$c_1=\sqrt{\frac{\lambda+2\mu}{\rho}}\tag{10.2}$$

$$c_2=\sqrt{\frac{\mu}{\rho}}\tag{10.3}$$

Because the cylinder considered here is isotropic, Helmholtz decomposition can be utilized to simplify the problem:

$$\vec{U}=\nabla\Phi+\nabla\times\vec{H}\tag{10.4}$$

The dilatational scalar potential Φ and the equivoluminal vector potential $\vec{H}$ are utilized to describe the displacement field.

Boundary conditions are needed to solve the governing equation. There are traction-free boundary conditions on the inner and outer surfaces of the hollow cylinder. To simplify boundary conditions, by assuming that the hollow cylinder is infinitely long, gauge invariance can be defined as:

$$\nabla\cdot\vec{H}=0\tag{10.5}$$

The gauge invariance conditions here are also known as equal volume conditions, which means that the volume of a structure is a constant. Because the hollow cylinder has an infinite length and volume, the gauge invariance is applicable here. By employing gauge invariance conditions, the problem can be solved without considering the boundary conditions on the two ends of the hollow cylinder.

Substituting Equation (10.4) into Equation (10.1) gives:

$$\nabla^2\Phi=\frac{1}{c_1}\frac{\partial^2\Phi}{\partial t^2}\tag{10.6}$$

$$\nabla^2 \vec{H} = \frac{1}{c_2} \frac{\partial^2 \vec{H}}{\partial t^2} \tag{10.7}$$

Based on elasticity theory, the potentials $\vec{H}$ and Φ have the following formats in cylindrical coordinates:

$$\nabla^2 \Phi = \frac{\partial^2 \Phi}{\partial r^2} + \frac{1}{r} \frac{\partial \Phi}{\partial r} + \frac{1}{r^2} \frac{\partial^2 \Phi}{\partial \theta^2} + \frac{\partial^2 \Phi}{\partial z^2} \tag{10.8}$$

$$\vec{H} = H_r \vec{e}_r + H_\theta \vec{e}_\theta + H_z \vec{e}_z \tag{10.9}$$

$$\nabla^2 \vec{H} = \nabla^2 \left(H_r \vec{e}_r + H_\theta \vec{e}_\theta + H_z \vec{e}_z \right)$$

$$= \left(\nabla^2 H_r - \frac{1}{r^2} H_r - 2\frac{1}{r^2} \frac{\partial H_\theta}{\partial \theta} \right) \vec{e}_r + \left(\nabla^2 H_\theta - \frac{1}{r^2} H_\theta + 2\frac{1}{r^2} \frac{\partial H_r}{\partial \theta} \right) \vec{e}_\theta$$

$$+ \nabla^2 H_z \vec{e}_z \tag{10.10}$$

To separate variables, Gazis presented the expressions of the potentials (1959):

$$\Phi = f(r)\Theta(m\theta)e^{i(kz-\omega t)}$$
$$H_r = h_r(r)\Theta_r(m\theta)e^{i(kz-\omega t)} \quad (m=0, 1, 2 \ldots)$$
$$H_\theta = h_\theta(r)\Theta_\theta(m\theta)e^{i(kz-\omega t)} \tag{10.11}$$
$$H_z = h_z(r)\Theta_z(m\theta)e^{i(kz-\omega t)}$$

where the integer m is known as the circumferential order of a wave mode and k is the wavenumber. The $f(r)$ and $h_\xi(r)$ ($\xi = r, \theta, z$) are unknown coefficients.

If considering the continuity conditions for θ and $\theta+2\pi$, the functions $\Theta(m\theta)$ and $\Theta_\xi(m\theta)(\xi = r, \theta, z)$ must only contain $\sin(m\theta)$ and/or $\cos(m\theta)$. Therefore, Gazis makes the following assumptions:

$$\Phi = f(r)\cos m\theta\, e^{i(kz-\omega t)}$$
$$H_r = h_r(r)\sin m\theta e^{i(kz-\omega t)} \quad (m=0, 1, 2 \ldots)$$
$$H_\theta = h_\theta(r)\cos m\theta e^{i(kz-\omega t)} \tag{10.12}$$
$$H_z = h_z(r)\sin m\theta e^{i(kz-\omega t)}$$

Equation (10.12) will make the displacement have forms:

$$U_r = A_r(r)\cos(m\theta)e^{i(kz-\omega t)}$$
$$U_\theta = A_\theta(r)\sin(m\theta)e^{i(kz-\omega t)} \tag{10.13}$$
$$U_z = A_z(r)\cos(m\theta)e^{i(kz-\omega t)}$$

According to Equation (10.13), if m = 0 (axisymmetric mode), one can see that $U_\theta = 0$. Because the dominant displacement of an axisymmetric torsional mode is U_θ, Equation (10.12) is not an appropriate assumption for torsional modes. In other words, Gazis's solutions are only accurate for longitudinal mode calculations, although some researchers argue they hold for both longitudinal and torsional waves.

In 2005, Sun, Zhang, and Rose presented alternative solutions for torsional waves:

$$
\begin{aligned}
\Phi &= f(r)\sin m\theta e^{i(kz-\omega t)} \\
H_r &= h_r(r)\cos m\theta e^{i(kz-\omega t)} \quad (m=0,1,2\,...) \\
H_\theta &= h_\theta(r)\sin m\theta e^{i(kz-\omega t)} \\
H_z &= h_z(r)\cos m\theta e^{i(kz-\omega t)}
\end{aligned}
\tag{10.14}
$$

Let us consider assumptions of complete solutions for both longitudinal and torsional waves:

$$
\begin{aligned}
\Phi &= f(r)e^{im\theta}e^{i(kz-\omega t)} \\
H_r &= h_r(r)e^{im\theta}e^{i(kz-\omega t)} \quad (m=0,1,2\,...) \\
H_\theta &= h_\theta(r)e^{im\theta}e^{i(kz-\omega t)} \\
H_z &= h_z(r)e^{im\theta}e^{i(kz-\omega t)}
\end{aligned}
\tag{10.15}
$$

If considering only the axisymmetric wave modes, Equation (10.15) can be simplified in the form of:

$$
\begin{aligned}
\Phi &= f(r)e^{i(kz-\omega t)} \\
H &= h(r)e^{i(kz-\omega t)}
\end{aligned}
\tag{10.16}
$$

To simplify the solutions, we let:

$$
\begin{aligned}
\alpha^2 &= \omega^2/c_1^2 - k^2,\ \beta^2 = \omega^2/c_2^2 - k^2 \\
\alpha_1 &= |\alpha|, \quad \beta_1 = |\beta|
\end{aligned}
\tag{10.17}
$$

Substituting Equations (10.8) ~ (10.11), (10.15), and (10.17) into Equations (10.6) and (10.7) yields:

$$
r^2 f'' + rf' + \left[\left(\frac{\omega^2}{c_1^2} - k^2\right)r^2 - m^2\right]f = 0
\tag{10.18}
$$

$$
r^2 h_r'' + rh_r' + \left[\left(\frac{\omega^2}{c_2^2} - k^2\right)r^2 - (m^2+1)\right]h_r - 2mih_\theta = 0
\tag{10.19}
$$

$$
r^2 h_\theta'' + rh_\theta' + \left[\left(\frac{\omega^2}{c_2^2} - k^2\right)r^2 - (m^2+1)\right]h_\theta + 2mih_r = 0
\tag{10.20}
$$

$$r^2 h_z'' + r h_z' + \left[\left(\frac{\omega^2}{c_2^2} - k^2 \right) r^2 - m^2 \right] h_z = 0 \tag{10.21}$$

Let:

$$2h_1 = (ih_r - h_\theta), 2h_2 = (ih_r + h_\theta), \quad h_3 = h_z \tag{10.22}$$

then Equations (10.19) through (10.21) become:

$$r^2 h_1'' + r h_1' + \left[\left(\frac{\omega^2}{c_2^2} - k^2 \right) r^2 - (m+1)^2 \right] h_1 = 0 \tag{10.23}$$

$$r^2 h_2'' + r h_2' + \left[\left(\frac{\omega^2}{c_2^2} - k^2 \right) r^2 - (m-1)^2 \right] h_2 = 0 \tag{10.24}$$

$$r^2 h_3'' + r h_3' + \left[\left(\frac{\omega^2}{c_2^2} - k^2 \right) r^2 - m^2 \right] h_3 = 0 \tag{10.25}$$

Solving with Bessel functions gives:

$$
\begin{aligned}
f &= A Z_m(\alpha_1 r) + B W_m(\alpha_1 r) \\
h_3 &= h_z = A_3 Z_m(\beta_1 r) + B_3 W_m(\beta_1 r) \\
h_1 &= \frac{(ih_r - h_\theta)}{2} = A_1 Z_{m+1}(\beta_1 r) + B_1 W_{m+1}(\beta_1 r) \\
h_2 &= \frac{(ih_r + h_\theta)}{2} = A_2 Z_{m-1}(\beta_1 r) + B_2 W_{m-1}(\beta_1 r)
\end{aligned}
\tag{10.26}
$$

where the Z_m and W_m are the mth order Bessel functions, which are determined by the following definitions:

$$Z_m(\alpha_1 r) = \begin{cases} J_m(\alpha_1 r), & \text{if } \dfrac{\omega^2}{c_1^2} - k^2 \geq 0 \\[2mm] I_m(\alpha_1 r), & \text{if } \dfrac{\omega^2}{c_1^2} - k^2 < 0 \end{cases} \tag{10.27}$$

$$W_m(\alpha_1 r) = \begin{cases} Y_m(\alpha_1 r), & \text{if } \dfrac{\omega^2}{c_1^2} - k^2 \geq 0 \\[2mm] K_m(\alpha_1 r), & \text{if } \dfrac{\omega^2}{c_1^2} - k^2 < 0 \end{cases} \tag{10.28}$$

$$Z_m(\beta_1 r) = \begin{cases} J_m(\beta_1 r), & \text{if } \dfrac{\omega^2}{c_2^2} - k^2 \geq 0 \\[2mm] I_m(\beta_1 r), & \text{if } \dfrac{\omega^2}{c_2^2} - k^2 < 0 \end{cases} \tag{10.29}$$

$$W_m(\beta_1 r) = \begin{cases} Y_m(\beta_1 r), & if \ \dfrac{\omega^2}{c_2^2} - k^2 \geq 0 \\[2mm] K_m(\beta_1 r), & if \ \dfrac{\omega^2}{c_2^2} - k^2 < 0 \end{cases} \tag{10.30}$$

Similarly, the gauge invariance conditions can be rewritten as:

$$\frac{i}{r}(-h_1 - h_2 - rh_1' - rh_2' + mh_2 - mh_1 + krh_3) = 0 \tag{10.31}$$

Substituting Equation (10.26) into Equation (10.4) yields the displacement field:

$$\begin{aligned} U_r &= \frac{\partial \Phi}{\partial r} + \frac{1}{r}\frac{\partial H_z}{\partial \theta} - \frac{\partial H_\theta}{\partial z} \\ &= [f' + (im/r)h_3 + ikh_1 - ikh_2]e^{im\theta}e^{i(kz-\omega t)} \end{aligned} \tag{10.32}$$

$$\begin{aligned} U_\theta &= \frac{1}{r}\frac{\partial \Phi}{\partial \theta} + \frac{\partial H_r}{\partial z} - \frac{\partial H_z}{\partial r} \\ &= \left[(im/r)f + kh_1 + kh_2 - h_3'\right]e^{im\theta}e^{i(kz-\omega t)} \end{aligned} \tag{10.33}$$

$$\begin{aligned} U_z &= \frac{\partial \Phi}{\partial z} + \frac{1}{r}\frac{\partial}{\partial r}(rH_\theta) - \frac{1}{r}\frac{\partial H_r}{\partial \theta} \\ &= [ikf + h_2' - h_1' - (m+1)(h_1/r) - (m-1)(h_2/r)]e^{im\theta}e^{i(kz-\omega t)} \end{aligned} \tag{10.34}$$

Based on elasticity theory, one obtains the stress field for longitudinal waves:

$$\begin{aligned} \sigma_{rr} &= \lambda \nabla^2 \Phi + 2\mu\varepsilon_{rr} = \lambda \nabla^2 \Phi + 2\mu\frac{\partial U_r}{\partial r} \\ &= \left\{ -\lambda(\alpha^2 + k^2)f + 2\mu\left[f'' + \frac{m}{r}\left(h_3' - \frac{h_3}{r}\right) + ik(h_1' - h_2')\right]\right\}e^{im\theta}e^{i(kz-\omega t)} \end{aligned} \tag{10.35}$$

$$\begin{aligned} \sigma_{r\theta} &= 2\mu\varepsilon_{r\theta} = 2\mu(1/2)\left[r\frac{\partial}{\partial r}\left(\frac{U_\theta}{r}\right) + \frac{1}{r}\frac{\partial U_r}{\partial \theta}\right] \\ &= \mu\left[\frac{2mi}{r}\left(f' - \frac{f}{r}\right) - k\left(\frac{m-1}{r}h_1 - h_1'\right) + k\left(\frac{m+1}{r}h_2 + h_2'\right)\right. \\ &\left. + \beta^2 h_3\right]e^{im\theta}e^{i(kz-\omega t)} \end{aligned} \tag{10.36}$$

$$\begin{aligned} \sigma_{rz} &= 2\mu\varepsilon_{rz} = 2\mu(1/2)\left[\frac{\partial U_r}{\partial z} + \frac{\partial U_z}{\partial r}\right] \\ &= \mu\left\{ 2ikf' + \frac{m}{r}\left(h_1' + \frac{m+1}{r}h_1\right) + (k^2 - \beta^2)(h_1 - h_2) - \frac{mk}{r}h_3 \right. \\ &\left. - \frac{m}{r}\left(h_2' + \frac{m-1}{r}h_2\right)\right\}e^{im\theta}e^{i(kz-\omega t)} \end{aligned} \tag{10.37}$$

There are six traction-free boundary conditions on the inner ($r = r_1$) and outer ($r = r_2$) surfaces of the hollow cylinder with infinite length:

$$\sigma_{rr} = \sigma_{r\theta} = \sigma_{rz} = 0, \quad at \ r = r_1 \ and \ r = r_2 \tag{10.38}$$

and two gauge invariance condition equations at $r = r_1$ and $r = r_2$.

Therefore, we obtain the following eigenvalue equations:

$$\begin{bmatrix} C_{11} & C_{12} & \cdots & C_{18} \\ C_{21} & C_{22} & \cdots & C_{28} \\ \vdots & \vdots & \ddots & \vdots \\ C_{81} & C_{82} & \cdots & C_{88} \end{bmatrix} \begin{bmatrix} A \\ B \\ \vdots \\ B_3 \end{bmatrix} = \begin{bmatrix} 0 \\ 0 \\ \vdots \\ 0 \end{bmatrix} \tag{10.39}$$

Let: $k_1 = \dfrac{\omega}{c_1}$ and $k_2 = \dfrac{\omega}{c_2}$, then the elements in the matrix $[C]_{8 \times 8}$ have the following forms:

By assuming $r = r_1$, one obtains the elements in the first three rows as:

$$C_{11} = \frac{2\mu}{\lambda} Z_m{}''(\alpha_1 r) - k_1{}^2 Z_m(\alpha_1 r);$$

$$C_{12} = \frac{2\mu}{\lambda} W_m{}''(\alpha_1 r) - k_1{}^2 W_m(\alpha_1 r);$$

$$C_{13} = \frac{2\mu k i}{\lambda} Z_{m+1}{}'(\beta_1 r)$$

$$C_{14} = \frac{2\mu k i}{\lambda} W_{m+1}{}'(\beta_1 r)$$

$$C_{15} = -\frac{2\mu k i}{\lambda} Z_{m-1}{}'(\beta_1 r)$$

$$C_{16} = -\frac{2\mu k i}{\lambda} W_{m-1}{}'(\beta_1 r)$$

$$C_{17} = \frac{2\mu m}{\lambda r} \left(Z_m{}'(\beta_1 r) - \frac{Z_m(\beta_1 r)}{r} \right)$$

$$C_{18} = \frac{2\mu m}{\lambda r} \left(W_m{}'(\beta_1 r) - \frac{W_m(\beta_1 r)}{r} \right)$$

$$C_{21} = \frac{2m i}{r} \left(Z_m{}'(\alpha_1 r) - \frac{Z_m(\alpha_1 r)}{r} \right)$$

$$C_{22} = \frac{2m i}{r} \left(W_m{}'(\alpha_1 r) - \frac{W_m(\alpha_1 r)}{r} \right)$$

$$C_{23} = -k \left(\frac{m-1}{r} Z_{m+1}(\beta_1 r) - Z_{m+1}{}'(\beta_1 r) \right)$$

$$C_{24} = -k \left(\frac{m-1}{r} W_{m+1}(\beta_1 r) - W_{m+1}{}'(\beta_1 r) \right)$$

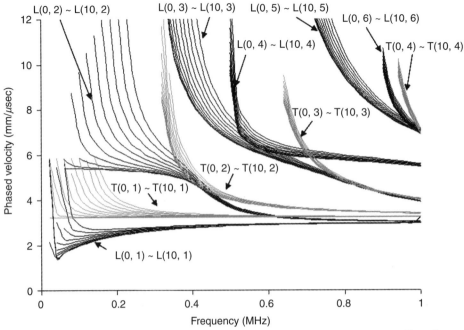

Figure 10.2. Sample dispersion curves of three-inch schedule 40 steel pipe including all of the longitudinal and torsional modes, including axisymmetric modes L(0, n)/T(0, n) (n = 1, 2, 3, ...) and non-axisymmetric modes L(m, n)/T(m, n) (m = 1, 2, 3, ..., n = 1, 2, 3, ...).

$$C_{25} = k\left(\frac{m+1}{r}Z_{m-1}(\beta_1 r) + Z_{m-1}'(\beta_1 r)\right)$$

(10.40)

$$C_{26} = k\left(\frac{m+1}{r}W_{m-1}(\beta_1 r) + W_{m-1}'(\beta_1 r)\right)$$

$$C_{27} = \beta^2 Z_m(\beta_1 r)$$

$$C_{24} = \beta^2 W_m(\beta_1 r)$$

$$C_{31} = 2ikZ_m'(\alpha_1 r)$$

$$C_{32} = 2ikW_m'(\alpha_1 r)$$

$$C_{33} = \left(\frac{m(m+1)}{r^2} + 2k^2 - k_2^2\right)Z_{m+1}(\beta_1 r) + \frac{m}{r}Z_{m+1}'(\beta_1 r)$$

$$C_{34} = \left(\frac{m(m+1)}{r^2} + 2k^2 - k_2^2\right)W_{m+1}(\beta_1 r) + \frac{m}{r}W_{m+1}'(\beta_1 r)$$

$$C_{35} = -\left(\frac{m(m-1)}{r^2} + 2k^2 - k_2^2\right)Z_{m-1}(\beta_1 r) - \frac{m}{r}Z_{m-1}'(\beta_1 r)$$

$$C_{36} = -\left(\frac{m(m-1)}{r^2} + 2k^2 - k_2^2\right)W_{m-1}(\beta_1 r) - \frac{m}{r}W_{m-1}'(\beta_1 r)$$

$$C_{37} = -\frac{km}{r}Z_m(\beta_1 r)$$

$$C_{38} = -\frac{km}{r}W_m(\beta_1 r)$$

By letting $r = r_2$, we have the 4th~6th matrix rows the same as Equation (10.40).

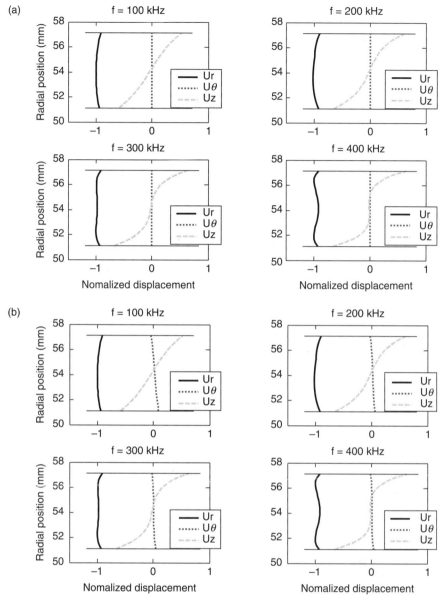

Figure 10.3. Sample wave structures of (a) the L(0, 1) and (b) the L(2, 1) at the frequency = 100 kHz, 200 kHz, 300 kHz, and 400 kHz in a four-inch schedule 40 steel pipe.

In addition, the elements in the last two rows can be obtained from Equation (10.31) as:

$$C_{71} = C_{72} = C_{74} = C_{76} = C_{78} = 0; C_{73} = -\beta_1; C_{77} = k;$$
$$C_{81} = C_{82} = C_{83} = C_{85} = C_{87} = 0; C_{86} = \beta_1; C_{88} = k;$$
$$\text{If } c \geq c_2, C_{75} = \beta_1 \text{ and } C_{84} = -\beta_1; \text{else}, C_{75} = -\beta_1 \text{ and } C_{84} = \beta_1. \tag{10.41}$$

The eigenvalues lead to dispersion curves for the hollow cylinder.

Figure 10.2 illustrates sample dispersion curves and wave structures in a three-inch schedule 40 steel pipe.

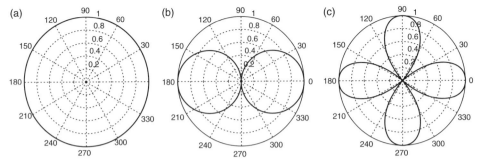

Figure 10.4. Angular profiles of the (a) 0th (known as axisymmetric modes), (b) 1st, and (c) 2nd modes in a group m ($m = 1, 2, 3, ...$). Neither the pipeline properties nor the group number n affects the angular profiles of a single mode.

The real parts of the eigenvectors lead to displacement distributions for every individual wave mode. The energy distributions in the r-direction are wave structures and the energy distributions in the θ-direction are called angular profiles. Figure 10.3 shows that the modes in a sample group have similar wave structures. In Figure 10.3, the L(2, 1) mode has some small θ-direction displacement U_θ, although $U_\theta = 0$ for the axisymmetric mode L(0, 1).

The angular profile of the mth circumferential order wave mode will not change with either the pipe properties or the group numbers. Figure 10.4 illustrates the angular profiles of the wave modes with 0th ~2nd circumferential orders. However, the real propagating angular profiles of a wave group highly depend on the properties of the hollow cylinder and the excitation conditions.

10.2.2 Excitation Conditions and Angular Profiles

Ditri and Rose (1992) have derived the particle velocity field from Equations (10.32) ~ (10.34) as follows:

$$v_r = R_r^{mn}(r)e^{im\theta}e^{i(\omega t - k^{mn}z)}$$

$$v_\theta = R_\theta^{mn}(r)e^{im\theta}e^{i(\omega t - k^{mn}z)} \quad (m = 0, 1, 2,...; n = 1, 2, 3,...) \quad (10.42)$$

$$v_z = R_z^{mn}(r)e^{im\theta}e^{i(\omega t - k^{mn}z)}$$

The orthogonality relation between two modes M(m, n) and M(l, s) (M represents either longitudinal mode L or torsional mode T) can be written as:

$$P_{m\,\ln s} = -\frac{1}{4}\iint_D (\vec{v}_{mn}^* \cdot \hat{T}_{ls} + \vec{v}_{ls}^* \cdot \hat{T}_{mn}) \cdot \vec{e}_z dV = 0, m \neq l \text{ or } k^{mn} \neq k^{ls} \quad (10.43)$$

where D indicates the cross-sectional area of the cylinder and $\vec{v}$ and $\hat{T}$ are the velocity and stress fields of the mode. The asterisk denotes complex conjugation.

For the guided wave generator element shown in Figure 10.5, the loading condition is as follows:

$$\hat{T} \cdot \vec{n} = \begin{cases} -p_1(\theta)p_2(z)\vec{e}_\xi, & |z| \leq L, |\theta| \leq \alpha, \quad r = b \\ 0, & |z| > L, \text{ or } |\theta| > \alpha, \quad \text{or} \quad r \neq b \end{cases} \quad (10.44)$$

where $\vec{e}_\xi (\xi = 1, 2, 3)$ represents the dominant vibration direction.

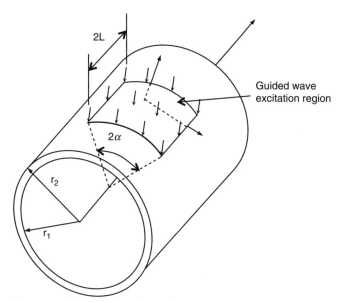

Figure 10.5. An ultrasonic wave generator is loaded on a hollow cylinder with inner radius r_1 and outer radius r_2. The axial length of the loading is 2L and the circumferential length is 2α.

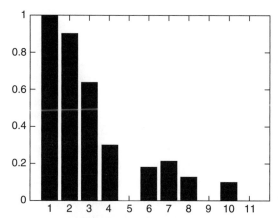

Figure 10.6. Amplitude distributions of 100 kHz L(m, 2) mode group in a sixteen-inch schedule 30 steel pipe with 90° circumferential loading.

Applying the loading conditions gives the stress amplitude of the wave mode M(m, n) in the positive propagation direction:

$$A_+^{mn} = -\frac{R_\xi^{mn}(r)e^{-ik^{mn}z}r_{n+1}}{4P_{mmnn}}\int_{-\alpha}^{-\alpha+2\pi} e^{im\theta}p_1(\theta)d\theta\int_{-\infty}^{\infty}p_2(z)e^{ik^{mn}z}dz,$$

$$z \geq L, \xi = 1, 2, 3$$

(10.45)

Figure 10.6 shows sample amplitude distributions of the 100 kHz L(m, 2) mode group in a sixteen-inch schedule 30 steel pipe with a transmitter, which has a 90° circumferential loading angle and a 10 mm axial length.

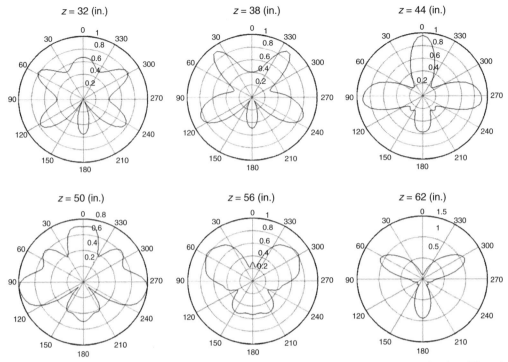

Figure 10.7. Sample angular profile of the T(m, 1) wave mode group propagating in a 2″ steel pipe with 0.125″ wall thickness by using 45° circumferential loading at 545 kHz. The angular profiles are shown in distance z = 32″~62″ at every 6″ step.

In Figure 10.6, one can see that the modes with high circumferential orders have small amplitudes. Therefore, we can ignore the influence of the higher order modes when considering the energy distributions of a wave group. Li and Rose (2001) first presented angular profiles of an entire longitudinal wave group in a hollow cylinder. They obtained the angular profiles by summing up the angular profile of each mode in this group with weighting functions A_+^{mn}, which are the amplitudes calculated in Equation (10.45). Sample angular profiles are shown in Figure 10.7.

Similarly, the wave structures are determined by the wave structures and amplitudes of the individual modes. Sample wave structures of the L(m, 1) group in a four-inch schedule 40 pipe are illustrated in Figure 10.8. By comparing Figures 10.8 and 10.3, one can see that the wave structures of all the modes in the same group are similar.

10.2.3 Source Influence

According to Equation (10.43), at a certain position (r_0, θ_0, z_0), A_+^{mn} in a particular wave group is only determined by the excitation conditions, including frequencies, loading area, and loading functions. If assuming that a time harmonic loading is uniformly distributed within the excitation $2\alpha{*}2L$ region (Figure 10.5) on the pipe surface, we have the following loading distribution functions in the circumferential direction:

$$p_1(\theta) = \begin{cases} p_0, & -\alpha \leq \theta \leq \alpha \\ 0, & -\alpha > \theta, \quad \text{or} \quad \theta > \alpha \end{cases} \tag{10.46}$$

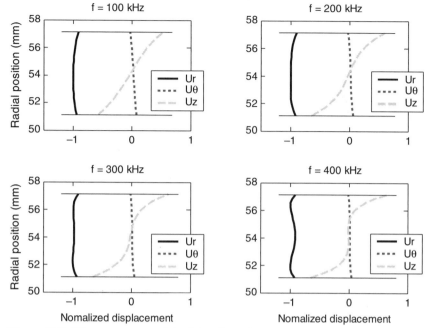

Figure 10.8. Wave structures of the L(m, 1) mode group in a four-inch schedule 40 steel pipe at the frequency = 100 kHz, 200 kHz, 300 kHz, and 400 kHz.

where p_0 is a constant that depends on the loading forces.

The axial loading distribution function $p_2(z)$ of a normal incident transducer is:

$$p_2(z) = \begin{cases} 1, & |z| \leq L \\ 0, & |z| > L \end{cases} \qquad (10.47)$$

At a high excitation frequency, angle beam transducers or comb transducers are widely used to ensure excitations of specific modes. To generate a particular mode M(m, n) (M denotes L or T), the incident wave must have the wavenumber k^{mn}. Because wavenumbers in one group are usually close and the axisymmetric modes always contain large amplitudes, we choose the k^{0n} as the incident wavenumber. Hence, the loading function $p_2(z)$ becomes:

$$p_2(z) = \begin{cases} e^{-ik^{0n}z}, & |z| \leq L \\ 0, & otherwise \end{cases} \qquad (10.48)$$

If normalizing the amplitudes A_+^{mn} in Equation (10.45) by dividing by the amplitude A_+^{0n}, one obtains the following undefined parts:

$$\frac{\int_{-\alpha}^{-\alpha+2\pi} e^{im\theta} p_1(\theta) d\theta}{\int_{-\alpha}^{-\alpha+2\pi} e^0 p_1(\theta) d\theta} \cdot \frac{\int_{-\infty}^{\infty} p_2(z) e^{ik^{mn}z} dz}{\int_{-\infty}^{\infty} p_2(z) e^{ik^{0n}z} dz}, \quad m = 1, 2, 3, \ldots \qquad (10.49)$$

Substituting Equations (10.46) and (10.47)/(10.48) into Equation (10.49) yields:

$$\frac{\int_{-\alpha}^{-\alpha+2\pi} e^{im\theta} p_1(\theta)d\theta}{\int_{-\alpha}^{-\alpha+2\pi} e^{0} p_1(\theta)d\theta} = \frac{\sin(m\alpha)}{m\alpha}, \quad m = 1, 2, 3, \dots \tag{10.50}$$

For normal incidences:

$$\frac{\int_{-\infty}^{\infty} p_2(z)e^{ik^{mn}z}dz}{\int_{-\infty}^{\infty} p_2(z)e^{ik^{0n}z}dz} = \frac{k^{0n}\sin(k^{mn}L)}{k^{mn}\sin(k^{0n}L)}, \quad m, n = 1, 2, 3, \dots \tag{10.51}$$

For certain incident wavenumber:

$$\frac{\int_{-\infty}^{\infty} p_2(z)e^{ik^{mn}z}dz}{\int_{-\infty}^{\infty} p_2(z)e^{ik^{0n}z}dz} = \frac{\sin[(k^{mn} - k^{0n})L]}{(k^{mn} - k^{0n})L}, \quad m, n = 1, 2, 3, \dots \tag{10.52}$$

Based on Equations (10.50) ~ (10.52), we have the following conclusions:

1) A smaller circumferential loading length α will make the higher order modes have larger amplitudes. Figure 10.9 shows sample calculation amplitude distributions of the 100 kHz T(m, 1) mode group generated from an individual segment around the circumference of a sixteen-inch schedule 30 steel pipe. The axial length of the transmitter is 10 mm and the circumferential lengths are from 1° to 360°. There are twenty-one modes for each example circumferential length shown in the figure. More modes must be included in energy distribution calculations in a hollow cylinder as the transducer circumferential length gets smaller.

   A larger number of flexural modes usually make the energy distribution change more sharply in the circumferential direction. Therefore, a large loading length α excites fewer modes and leads to "smooth" angular profiles, as shown in Figure 10.10. When $\alpha = \pi$, only axisymmetric modes will be excited and the angular profiles become circular.

2) If the dispersion curves are "diverging," which means $|k^{mn} - k^{0n}|$ are large, the amplitudes of the higher circumferential order modes become very small. Therefore, the axisymmetric mode M(0, n) is dominant at these regions. When $|k^{mn} - k^{0n}| \to \infty$, only the axisymmetric mode will be generated, even with a partial loading.

3) If $|k^{mn} - k^{0n}|$ are reasonably small, a small axial loading length L will make Equations (10.49) and (10.50) have similar values. In other words, both normally incident transducers and transducers with particular incident wavenumbers have similar excitation efficiencies. However, if $L \to \infty$, they will have different excitation abilities: the transducers with designed incident wavenumbers will only generate axisymmetric modes; whereas the normally incident transducers still have capabilities to generate flexural modes.

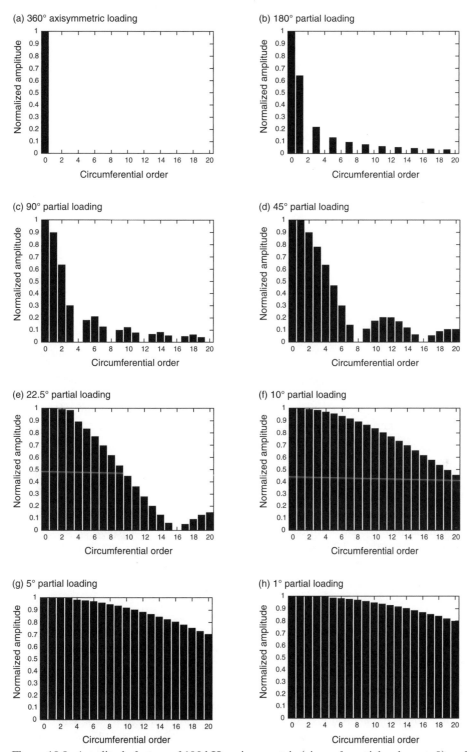

Figure 10.9. Amplitude factors of 100 kHz axisymmetric (circumferential order m = 0) and non-axisymmetric (circumferential order m > 0) torsional wave modes T(m, 1) that excited in a sixteen-inch schedule 30 steel pipe by applying (a) 360° axisymmetric loading, (b) 180° partial loading, (c) 90° partial loading, (d) 45° partial loading, (e) 22.5° partial loading, (f) 10° partial loading, (g) 5° partial loading, and (h) 1° partial loading.

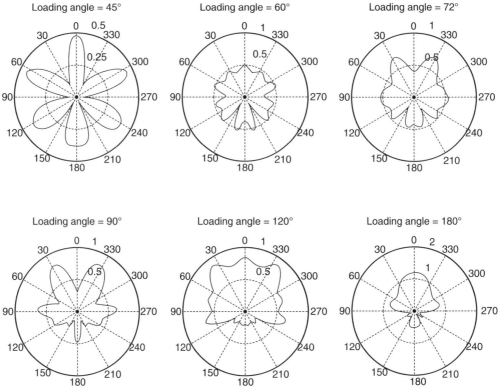

Figure 10.10. Sample angular profiles of the 40 kHz L(m, 1) mode group generated by a partially loaded transducer in a sixteen-inch schedule 30 steel pipe. The circumferential excitation length is from 45° to 180° at z = 240″.

4) Because the wavenumbers usually have much larger values than the axial length of the transducer, the axial length may have an insignificant influence on the angular profile distributions. At a frequency lower than 1 MHz, we can ignore the effect of the axial length variations over [0.01″~12″] in most pipelines.

5) To generate axisymmetric guided waves only, one needs to utilize a transmitter with 360° uniform loading in the circumferential direction. However, many commercial guided wave pipeline inspection tools have gaps between their transducers in the circumferential direction. Therefore, when these transducers are excited synchronously, many wave modes are launched instead of a single axisymmetric wave. A quasi-axisymmetric wave is formed at a reasonably short distance from the transducer array via superposition of the hundreds of wave modes.

Figure 10.11 illustrates a transducer array with eight 22.5° transducers excited at 100 kHz synchronously to excite guided waves in a four-inch schedule 40 steel pipe. The quasi-axisymmetric waves have angular profiles as shown in Figure 10.11 (b)~(f).

Although only quasi-axisymmetric waves will be generated by using transducer array with gaps in the circumferential direction, by decreasing the gap size between transducers, one can reduce the flexural modes and make the quasi-axisymmetric waves become almost axisymmetric waves.

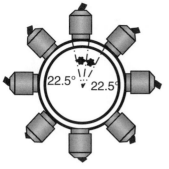

(a) Transducer array

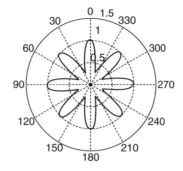

(b) Angular profile at $z = 0$ inch

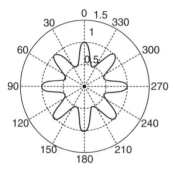

(c) Angular profile at $z = 4$ inch

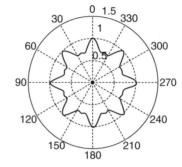

(d) Angular profile at $z = 10$ inch

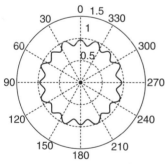

(e) Angular profile at $z = 2$ ft

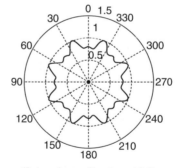

(f) Angular profile at $z = 10$ ft

Figure 10.11. Angular profile simulations of 100 kHz T(m, 1) modes in a four-inch schedule 40 steel pipe at several axial distances from the (a) transducers, including (b) $z = 0$ inch (at the center of the transducer array), (c) $z = 4$ inch, (d) $z = 10$ inch, (e) $z = 2$ ft, and (f) $z = 10$ ft. There are eight transducers excited simultaneously and circumferential length of each transducer is 22.5°. There is a 22.5° space between two transducers.

10.3 Exercises

1. Based on the dominant vibration direction, how many types of wave modes are involved in guided waves propagating in the axial direction of a pipeline? What are the dominant vibration directions of these modes?

2. What are axisymmetric modes and flexural modes? How can you generate axisymmetric guided waves only? How does the circumferential length of a transducer affect amplitudes of the excited guided wave modes in a hollow cylinder?

3. Based on Equation (10.45), is it possible to generate a single flexural mode only? Please explain.

4. What is gauge invariance? Explain why we use gauge invariance conditions to calculate guided wave propagation in pipes.

5. How could Gazis simplify his solutions for guided wave in pipelines as:

$$\Phi = f(r)\cos(m\theta)e^{i(\omega t - kz)}$$

$$H_r = h_r(r)\sin(m\theta)e^{i(\omega t - kz)}$$

$$H_\theta = h_\theta(r)\cos(m\theta)e^{i(\omega t - kz)}$$

$$H_z = h_z(r)\sin(m\theta)e^{i(\omega t - kz)}$$

instead of:

$$\Phi = f(r)e^{im\theta}e^{i(kz - \omega t)}$$

$$H_r = h_r(r)e^{im\theta}e^{i(kz - \omega t)}$$

$$H_\theta = h_\theta(r)e^{im\theta}e^{i(kz - \omega t)}$$

$$H_z = h_z(r)e^{im\theta}e^{i(kz - \omega t)}$$

6. What is the difference between an angular profile of a single mode and an angular profile of a mode group? How can you obtain an angular profile of a mode group based on the single mode angular profiles?

10.4 REFERENCES

Auld, B. A., and Tsao, M. T. (1977). A variational analysis of edge resonance in a semi-infinite plate, *IEEE Trans. Sonics. Ultrason.* SU-24(5): 317–26.

Cooper, R. M., and Naghdi, P. M. (1957). Propagation of nonaxially symmetric waves in elastic cylindrical shells, *J. Acoust. Soc. Am.* 29: 1365–73.

Ditri, J. J., and Rose, J. L. (1992). Excitation of guided elastic wave modes in hollow cylinders by applied surface tractions, *J. Appl. Phys.* 72(7): 2589–97.

Gazis, D. C. (1959a). Three dimensional investigation of the propagation of waves in hollow circular cylinders. I. Analytical foundation, *J. Acoust. Soc. Am.* 31: 568–73.

  (1959b). Three dimensional investigation of the propagation of waves in hollow circular cylinders. II. Numerical results, *J. Acoust. Soc. Am.* 31: 573–8.

Gazis, D. C., and Mindlin, R. D. (1960). Extensional vibrations and waves in a circular disk and a semi-infinite plate, *J. Appl. Mech.* 27: 541–7.

Ghosh, J. (1923). Longitudinal vibrations of a hollow cylinder, *Bulletin of the Calcutta Mathematical Society* 14: 31–40.

Graff, K. F. (1991). *Wave Motion in Elastic Solids.* New York: Dover Publications.

Gridin, D., Craster, R. V., Fong, J., Lowe, M. J. S., and Beard, M. (2003). The high-frequency asymptotic analysis of guided waves in a circular elastic annulus, *Journal of Wave Motion* 38: 67–90.

Herrmann, G., and Mirsky, I. (1956). Three-dimensional and shell-theory analysis of axially symmetric motions of cylinders, *J. Appl. Mech.* 78: 563–8.

Lamb, H. (1917). On waves in an elastic plate, *Proc. Royal Soc. London* A93: 114.

Li, J., and Rose, J. L. (2001). Excitation and propagation of non-axisymmetric guided waves in a hollow cylinder, *J. Acoust. Soc. Am.* 109(2): 457–64.

   (2002). Angular-profile tuning of guided waves in hollow cylinders using a circumferential phased array, *IEEE Trans. Ultrason., Ferroelect., Freq. Contr.* 49(12): 1720–9.

Liu, G., and Qu, J. (1998). Guided circumferential waves in a circular annulus, *J. Appl. Mech* 65: 424–30.

Lowe, M. J. S. (1995). Matrix techniques for modeling ultrasonic waves in multilayered media, *IEEE Trans. Ultrason., Ferroelect., Freq. Contr.* 42(4): 525–41.

Rayleigh, J. (1945). *The Theory of Sound*, Vols. I and II. New York: Dover Publications.

Robinett, R. W. (1999). Periodic orbit theory analysis of the circular disk or annulus billiard: Nonclassical effects and the distribution of energy eigenvalues, *J. Phys.* 67: 67–77.

Rose, J. L. (2002). A baseline and vision of ultrasonic guided wave inspection potential, transactions of the ASME, *Journal of Pressure Vessel Technology* 124: 273–82.

Sun, Z., Zhang, L., and Rose, J. L. (2005). Flexural torsional guided wave mechanics and focusing in pipe, *Journal of Pressure Vessel Technology* 127(4): 471–8.

Viktorov, I. A. (1967). *Rayleigh and Lamb Waves: Physical Theory and Applications.* New York: Plenum Press.

Zemanek, J. J. (1970). An experimental and theoretical investigation of elastic wave propagation in a cylinder, *J. Acoust. Soc. Am.* 52: 265–83.

Zhang, L. (2005). *Guided Wave Focusing Potential in Hollow Cylinders*, PhD thesis, Pennsylvania State University.

Zhang, L., Gavigan, B. J., and Rose, J. L. (2004). Source influence for focusing potential of guided waves in hollow cylinders by using a circumferential phased array, *Proceedings of the Biennial International Pipeline Conference*, 3: 2771–81.

Zhang, L., and Rose, J. L. (2004). Guided wave particle motion in a hollow cylinder, *Proceedings of QNDE*, 24: 1952–7.

Zhao, X., and Rose, J. L. (2004). Guided circumferential shear horizontal waves in an isotropic hollow cylinder, *J. Acoust. Soc. Am.*, 115(5): 1912–16.

Zhuang, W., Shah, A. H., and Datta, S. K. (1997). Axisymmetric guided wave scattering by cracks in welded steel pipes, *Journal of Pressure Vessel Technology* 119(4): 401–6.

11 Circumferential Guided Waves

11.1 Introduction

In this chapter the governing equations for circumferential SH waves and circumferential Lamb type waves are developed. For brevity, circumferential SH and Lamb type waves will be abbreviated as CSH-waves and CLT-waves, respectively, from this point onward. Following the development of the single-layer cases, considerations will be made for n-layer annuli. More details on this practical problem can be found in Appendix D, Section 2.11.

Circumferential guided waves are guided waves that propagate in the circumferential direction of a hollow cylinder. They have many practical applications, including the detection of corrosion in piping from in-pipe or in-line inspection (ILI) vehicles. While researchers often use plate-wave solutions to study circumferentially propagating guided waves, this chapter shows that the two cases are quite different, both physically and mathematically, and that significant differences often exist between the plate-wave and circumferential-wave solutions to the governing wave equations. The case in which the two solutions are similar is also discussed in this chapter.

The amount of work published in the area of circumferential guided waves is relatively terse when compared to the body of work relating to wave propagation in the axial direction of hollow cylinders. The treatment of wave propagation in cylindrical layers is first seen in Viktorov (1967). In his text, Viktorov identifies the major physical differences between wave propagation in cylindrical structures and planar structures and specifically addresses the topics of Rayleigh waves on concave and convex surfaces and Lamb type waves in cylindrical layers. He defines the concept of angular wavenumber, which is a physical phenomenon unique to cylindrically curved waveguides. In his treatment of Lamb type waves in a cylindrical layer, Viktorov forms the characteristic equation for an elastic single layer. Because of the limited computational abilities of the time, Viktorov makes several simplifications to arrive at a first approximation. Specifically, he replaces the Bessel and Neumann functions in the characteristic equation with asymptotic representations in terms of semi-convergent Debye series.

Although several authors revisited the surface wave problem throughout the years following the publication of Viktorov's book, the topic of circumferential

With significant contributions from Jason Van Velsor.

guided wave propagation in a hollow cylindrical structure was not addressed again until Qu, Berthelot, and Li (1996) and Liu and Qu (1998a) presented the first numerical examples of the dispersion curves and wave structures for time-harmonic CLT-waves propagating in a circular annulus. Shortly thereafter, Liu and Qu (1998b) published a solution of transient wave propagation in a circular annulus using mode eigenfunction expansion, also known as normal mode expansion (NME).

Another important class of circumferential wave problem is that of the CSH-wave. Although the solution to this type of problem is fundamentally simpler, it is significant as CSH-waves have desirable properties from a practical perspective, such as excellent mode isolation capabilities, insensitivity to water loading and reduced sensitivity to viscous materials, and pure excitation with electromagnetic acoustic transducers (EMATs). Zhao and Rose (2004) first addressed the problem of time-harmonic CSH-wave propagation in a single-layer annulus.

11.2 Development of the Governing Wave Equations for Circumferential Waves

The development of the characteristic equations for a single-layer annulus is a necessary step toward the development of the multilayer cases. Authors such as Viktorov (1967), Liu and Qu (1998a,b), and Zhao and Rose (2004) have made single-layer considerations for circumferential guided waves. Liu and Qu (1998a) and Zhao and Rose (2004) provide detailed derivations of the characteristic equations and wave structures for the single-layer cases of CLT-wave and CSH-wave propagation, respectively. The development presented here is similar to these two cases but is modified in anticipation of developing the multilayer solutions. Specifically, a generalized boundary value problem will be developed such that the phase term is no longer associated with the boundary of the annulus. All dimensionless quantities must also be removed from the formulation.

The solution approach used here utilizes the method of displacement potentials and, as a result, applies only to isotropic materials. Additionally, all generalized plane-strain assumptions prevail. Figure 11.1 shows the theoretical model of a single-layer annulus used for the development of the CSH and CLT characteristic equations.

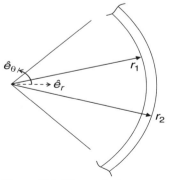

Figure 11.1. Theoretical model used for the development of the governing equations for CSH-wave and CLT-wave propagation in a single-layer annulus.

The development of the characteristic equations begins with the well-known Navier's Equation of Motion, Equation (11.1) (Graff 1991):

$$(\lambda + 2\mu)\nabla\Delta - 2\mu\nabla\times\boldsymbol{\omega} + \rho\mathbf{f} = \rho\ddot{\mathbf{u}}, \tag{11.1}$$

where the dilatation of a material, Δ, is given by:

$$\Delta = \nabla\cdot\mathbf{u}, \tag{11.2}$$

and the rotation vector, $\boldsymbol{\omega}$, is given by:

$$\boldsymbol{\omega} = \frac{1}{2}\nabla\times\mathbf{u} \tag{11.3}$$

In Equation (11.1), ρ is the material density, λ and μ are Lamé constants, $\mathbf{f}$ is the body-force vector, and $\mathbf{u}$ is the displacement vector. Note that the form of Navier's equation shown in Equation (11.1) has the advantage of applicability in any curvilinear coordinate system and the convenience of separated dilatational and rotational components (Graff 1991). From this point forward, it is assumed that no body forces are present.

In the case of circumferential guided wave propagation, the displacement vector is given by Equation (11.4):

$$\mathbf{u} = u_r(r,\theta,t)\hat{\mathbf{e}}_r + u_\theta(r,\theta,t)\hat{\mathbf{e}}_\theta + u_z(r,\theta,t)\hat{\mathbf{e}} \tag{11.4}$$

It is seen that, because generalized plane-strain assumptions have been made, the displacement components are functions of r, θ, and t only. This does not imply that displacement cannot occur in the z-direction but, instead, that any displacement in the z-direction must be uniform throughout the entire z-plane. The specific cases of CSH- and CLT-wave propagation are discussed next.

11.2.1 Circumferential Shear Horizontal Waves in a Single-Layer Annulus

Shear horizontal (SH) waves are waves in which the particle motion is in-plane but orthogonal to the direction of propagation. In the case of CSH-waves, particle displacement would be in the z-direction with no displacement in the r- or θ-directions ($u_r = u_\theta = 0$). Also, from the generalized plane-strain assumption, there must be no variation of any quantity in the z-direction ($\partial/\partial z = 0$). Under these conditions, Equation (11.1) simplifies to:

$$\mu\nabla^2 u_z = \rho\frac{\partial^2 u_z}{\partial t^2} \quad or \quad \nabla^2 u_z = \frac{1}{c_s^2}\frac{\partial^2 u_z}{\partial t^2}. \tag{11.5}$$

Note that in the CSH-wave case, a scalar form of the governing displacement equation of motion has been obtained with no need for the use of Helmholtz decomposition. Two forms of the wave equation are shown in Equation (11.5), from which it can be determined that bulk shear waves propagate with a velocity given by:

$$c_s = \sqrt{\frac{\mu}{\rho}}. \tag{11.6}$$

Assuming time-harmonic motion of the form $e^{-i\omega t}$ and propagation in the θ-direction, one solution of Equation (11.5) is assumed to be:

$$u_z = \psi(r)e^{i(p\theta - \omega t)}, \tag{11.7}$$

where p is the angular wavenumber (Viktorov 1967; Liu and Qu 1998a). It is important to distinguish between the circular wavenumber k and the angular wavenumber p as the terms *circular* and *angular* are often used interchangeably. In this work, the angular wavenumber, p, refers specifically to the circular wavenumber, k, multiplied by some radius, R, at which the linear phase velocity is to be determined. It is a dimensionless quantity. Equation (11.8) summarizes this relationship.

$$p = kR. \tag{11.8}$$

Equation (11.8) shows that there are two different ways to proceed: either assume a radius and solve for the k-roots of the characteristic equation at the assumed radius or find the p-roots and then calculate the k-roots at any radius. For multilayered annuli, finding the roots of the characteristic equation in the ω-p domain provides the most general solution that can subsequently be used to determine the linear phase velocity, c_p, at an arbitrary radius of the multilayered structure by the use of Equations (11.9) and (11.10), first introduced by Viktorov (1967).

$$\alpha_p = \frac{\omega}{p}. \tag{11.9}$$

$$c_p(R) = \alpha_p R. \tag{11.10}$$

Equation (11.10) demonstrates that there is a linear increase in phase velocity, c_p, from the internal to the external surface of the annulus. This increase in phase velocity with radius is necessary to maintain a constant phase front through the thickness of the annulus. This phenomenon is unique to wave propagation along structures that are curved in the direction of wave propagation. Note that this does not include wave propagation in the axial direction of a hollow cylinder as it is typically assumed that the structure is straight in this direction.

Substituting Equation (11.7) into Equation (11.5) yields Equation (11.11):

$$r^2\psi'' + r\psi' + \left[\left(\frac{r\omega}{c_s}\right)^2 - p^2\right]\psi = 0, \tag{11.11}$$

which is a second-order ODE, commonly solved using Bessel or Hankel functions (Hayek 2001). The solution of Equation (11.11) in terms of Bessel functions is of the form:

$$\psi(r) = A_1 J_p(k_s r) + A_2 Y_p(k_s r), \tag{11.12}$$

where:

$$k_s = \omega/c_s \tag{11.13}$$

is the circular wavenumber of a bulk shear wave.

To solve for the unknown coefficients in Equation (11.12), the boundary conditions for the problem of interest must be considered. In the case of a single-layer "free" annulus, traction-free boundary conditions are assumed; that is, stress is required to vanish on the inside and outside surfaces of the annulus. In the case of CSH-waves, shear stress is given by:

$$\sigma_{rz} = \mu \frac{\partial u_z}{\partial r}. \tag{11.14}$$

Using the recurrence relation for Bessel functions (Abramowitz and Stegun 1964):

$$2\xi_v'(x) = \xi_{v-1}(x) - \xi_{v+1}(x), \quad where \quad \xi = J \ or \ Y, \tag{11.15}$$

and with the use of Equation (11.7) and Equation (11.12), Equation (11.14) can be written as:

$$\sigma_{rz} = s(r)e^{ip\theta}, \tag{11.16}$$

where:

$$s(r) = \mu \frac{k_s}{2} \left\{ A_1 \left(J_{p-1}(k_s r) - J_{p+1}(k_s r) \right) + A_2 \left(Y_{p-1}(k_s r) - Y_{p+1}(k_s r) \right) \right\}. \tag{11.17}$$

The $e^{-i\omega t}$ time dependence is inferred but not explicitly written in Equation (11.16) or Equation (11.17), as will be the case for the remainder of this chapter.

In preparation for the case of multiple layers, it is convenient to summarize the relevant stress and displacement eigenfunctions in a single matrix equation, as seen in Equation (11.18):

$$\begin{bmatrix} J_p(k_s r) & Y_p(k_s r) \\ \mu \frac{k_s}{2} \left(J_{p-1}(k_s r) - J_{p+1}(k_s r) \right) & \mu \frac{k_s}{2} \left(Y_{p-1}(k_s r) - Y_{p+1}(k_s r) \right) \end{bmatrix} \begin{Bmatrix} A_1 \\ A_2 \end{Bmatrix} = \begin{Bmatrix} \psi(r) \\ s(r) \end{Bmatrix}. \tag{11.18}$$

In the single-layer case, only the stress-related component of Equation (11.18) is needed, whereas in the multilayer case, both the displacement- and stress-related components will be needed.

In the case of the single-layer annulus, the traction-free boundary conditions require that

$$\sigma_{rz}\big|_{r_1, r_2} = 0, \tag{11.19}$$

resulting in the set of linear homogeneous algebraic equations given by Equation (11.20) and Equation (11.21).

$$\mathbf{D}(p, \omega)\mathbf{A} = \mathbf{0}, \tag{11.20}$$

where

$$\mathbf{D}(p,\omega) = \begin{bmatrix} J_{p-1}(k_s r_1) - J_{p+1}(k_s r_1) & Y_{p-1}(k_s r_1) - Y_{p+1}(k_s r_1) \\ J_{p-1}(k_s r_2) - J_{p+1}(k_s r_2) & Y_{p-1}(k_s r_2) - Y_{p+1}(k_s r_2) \end{bmatrix}, \qquad (11.21)$$

and $\mathbf{A} = (A_1, A_2)^T$. Nontrivial solutions to this set of equations are found by setting the determinant of the coefficient matrix, $\mathbf{D}(p,\omega)$, to zero:

$$\det(\mathbf{D}(p,\omega)) = 0. \qquad (11.22)$$

Equation (11.22) is the characteristic equation for CSH-waves in a single-layer annulus, the eigenvalues of which form the frequency-wavenumber dispersion curves. The angular phase and group velocity dispersion curves can then be calculated using the relations seen in Equation (11.9) and Equation (11.23), respectively.

$$\alpha_g = \frac{\partial \omega}{\partial p}. \qquad (11.23)$$

Once the eigenvalues of the characteristic equation have been determined, it is possible to solve for the displacement and stress distribution throughout the wall thickness of the annulus for any given mode and frequency combination. The displacement field is given by Equation (11.24):

$$u_z = \left(J_p(k_s r) - \Lambda Y_p(k_s r)\right) e^{i(p\theta - \omega t)}, \qquad (11.24)$$

where

$$\Lambda = \frac{J_{p-1}(k_s r_1) - J_{p+1}(k_s r_1)}{Y_{p-1}(k_s r_1) - Y_{p+1}(k_s r_1)}. \qquad (11.25)$$

The nonzero stress components are given by Equation (11.26) and Equation (11.27).

$$\sigma_{rz} = \frac{k_s \mu}{2}\left\{ J_{p-1}(k_s r) - J_{p+1}(k_s r) - \Lambda\left(Y_{p-1}(k_s r) - Y_{p+1}(k_s r)\right) \right\} e^{i(p\theta - \omega t)}. \qquad (11.26)$$

$$\sigma_{\theta z} = \frac{\mu}{r}\frac{\partial u_z}{\partial \theta} = \frac{ip\mu}{r}\left(J_p(k_s r) - \Lambda Y_p(k_s r)\right) e^{i(p\theta - \omega t)}. \qquad (11.27)$$

In Equations (11.24), (11.26), and 11.27 the constant A_1 has been arbitrarily set to unity, hence the solutions are not unique and multiplication by any constant will yield a valid solution.

This concludes the development of the characteristic equation and the displacement and stress relations for the case of CSH-wave propagation in a single-layer annulus. The next section addresses the development of the Lamb type dispersion equation, displacements, and stresses for the single-layer case.

11.2.2 Circumferential Lamb Type Waves in a Single-Layer Annulus

Lamb waves are waves with displacements in two directions: in-plane and out-of-plane. Unlike SH-waves, the in-plane component in the Lamb wave case is along the line of propagation. Because Lamb waves are technically a type of plate wave, this work refers to circumferential Lamb type, or CLT-, waves when referring to propagation in an annulus. This distinction is important because of the physical differences in the propagation characteristics, although, as the ratio of inner diameter to outer diameter approaches unity, the annulus solution approaches that of a plate and the two cases become equivalent (Liu and Qu 1998a).

Using the theorem introduced by Helmholtz, it is possible to dissect the displacement field, u, into a sum of the gradient of a scalar and the curl of a zero-divergence vector with the use of the scalar and vector potentials, Φ and $\mathbf{H}$ (Morse and Feshbach 1953). This is shown in Equation (11.28).

$$\mathbf{u} = \nabla\Phi + \nabla \times \mathbf{H}, \ \nabla \cdot \mathbf{H} = 0, \tag{11.28}$$

where $\nabla \cdot \mathbf{H} = 0$ is a necessary condition to determine a unique solution for the three displacement components from the total of four components of Φ and $\mathbf{H}$. Substituting Equation (11.28) into Navier's equation, Equation (11.1), Equation (11.29) is obtained:

$$\nabla\left((\lambda + 2\mu)\nabla^2\Phi - \rho\ddot{\Phi}\right) + \nabla \times \left(-\mu\nabla \times \nabla \times \mathbf{H} - \rho\ddot{\mathbf{H}}\right) = 0. \tag{11.29}$$

For a nontrivial solution of Equation (11.29), it is required that:

$$(\lambda + 2\mu)\nabla^2\Phi = \rho\ddot{\Phi}, \tag{11.30}$$

and

$$-\mu\nabla \times \nabla \times \mathbf{H} = \rho\ddot{\mathbf{H}}. \tag{11.31}$$

Recalling Equation (11.6) and also noting that longitudinal waves propagate at a velocity given by

$$c_l = \sqrt{\frac{(\lambda + 2\mu)}{\rho}}, \tag{11.32}$$

it is possible to rewrite Equation (11.30) and Equation (11.31) as

$$\nabla^2\Phi = \frac{1}{c_l^2}\frac{\partial^2\Phi}{\partial t^2} \tag{11.33}$$

and

$$\nabla^2\mathbf{H} = \frac{1}{c_s^2}\frac{\partial^2\mathbf{H}}{\partial t^2}, \tag{11.34}$$

respectively. In Equation (11.34),

$$\mathbf{H} = H_r\mathbf{e}_r + H_\theta\mathbf{e}_\theta + H_z\mathbf{e}_z. \tag{11.35}$$

Assuming generalized plane-strain and Lamb type wave propagation,

$$H_r = H_\theta = 0, \tag{11.36}$$

because

$$u_z = \frac{\partial}{\partial z} = 0. \tag{11.37}$$

Applying the conditions shown in Equation (11.36) and Equation (11.37), Equation (11.34) reduces to the scalar equation shown in Equation (11.38).

$$\nabla^2 H_z = \frac{1}{c_s^2}\frac{\partial^2 H_z}{\partial t^2}. \tag{11.38}$$

As was done in the CSH-wave case, solutions to Equation (11.33) and Equation (11.38) are assumed to be of the forms

$$\Phi = f(r)e^{i(p\theta - \omega t)} \tag{11.39}$$

and

$$H_z = h_z(r)e^{i(p\theta - \omega t)}, \tag{11.40}$$

respectively, where f and h_z are functions of r, yet to be determined. Substituting Equation (11.39) into Equation (11.33) and Equation (11.40) into Equation (11.38), two second-order linear ODEs are obtained, as seen in Equation (11.41) and Equation (11.42):

$$f''(r) + \frac{1}{r}f'(r) + \left[k_l^2 - \left(\frac{p}{r}\right)^2\right]f(r) = 0, \tag{11.41}$$

$$h_z''(r) + \frac{1}{r}h_z'(r) + \left[k_s^2 - \left(\frac{p}{r}\right)^2\right]h_z(r) = 0, \tag{11.42}$$

where k_s was defined in Equation (11.13) and k_l is the bulk longitudinal wavenumber,

$$k_l = \omega / c_l. \tag{11.43}$$

The Bessel function solutions to Equation (11.41) and Equation (11.42) are shown in Equation (11.44) and Equation (11.45), respectively (Hayek 2001).

$$f(r) = A_1 J_p(k_l r) + A_2 Y_p(k_l r). \tag{11.44}$$

$$h_z(r) = A_3 J_p(k_s r) + A_4 Y_p(k_s r). \tag{11.45}$$

The term k_s was defined in Equation (11.13); k_l is defined in Equation (11.43) and is the wavenumber of a bulk longitudinal wave.

To solve for the four unknown coefficients in Equation (11.44) and Equation (11.45), it is again necessary to apply traction-free boundary conditions. The normal and shear components of stress are required to vanish at the surfaces of the annulus and the expressions for these components are given by Equation (11.46) and Equation (11.47), respectively (Chou and Pagano 1992; Sadd 2005).

$$\sigma_r = \lambda \left[\frac{\partial u_r}{\partial r} + \frac{1}{r} \left(u_r + \frac{\partial u_\theta}{\partial \theta} \right) \right] + 2\mu \frac{\partial u_r}{\partial r}. \tag{11.46}$$

$$\sigma_{r\theta} = \mu \left(\frac{1}{r} \frac{\partial u_r}{\partial \theta} + \frac{\partial u_\theta}{\partial r} - \frac{u_\theta}{r} \right). \tag{11.47}$$

The particle displacement terms, u_r and u_θ, are given by Equation (11.48) and Equation (11.49), respectively.

$$u_r = \frac{\partial \Phi}{\partial r} + \frac{1}{r} \frac{\partial H_z}{\partial \theta}. \tag{11.48}$$

$$u_\theta = \frac{1}{r} \frac{\partial \Phi}{\partial \theta} - \frac{\partial H_z}{\partial r}. \tag{11.49}$$

By inserting Equations (11.48) and (11.49) into Equations (11.46) and (11.47) and collecting the equations into a single matrix equation, Equation (11.50) is formed.

$$\mathbf{D}(p,\omega)\, \mathbf{A} = \begin{Bmatrix} u_r \\ u_\theta \\ \sigma_r \\ \sigma_{r\theta} \end{Bmatrix}, \tag{11.50}$$

where

$$\mathbf{D}(p,\omega) = \begin{bmatrix} d_{11} & d_{12} & d_{13} & d_{14} \\ d_{21} & d_{22} & d_{23} & d_{24} \\ d_{31} & d_{32} & d_{33} & d_{34} \\ d_{41} & d_{42} & d_{43} & d_{44} \end{bmatrix}, \tag{11.51}$$

and $\mathbf{A} = (A_1, A_2, A_3, A_4)^T$. The individual components of the coefficient matrix, $\mathbf{D}(p,\omega)$, are provided in Equations (11.52) through (11.67).

$$d_{11} = \frac{k_L}{2} J_{p-1}(k_L r) - \frac{k_L}{2} J_{p+1}(k_L r) \tag{11.52}$$

$$d_{12} = \frac{k_L}{2} Y_{p-1}(k_L r) - \frac{k_L}{2} Y_{p+1}(k_L r) \tag{11.53}$$

$$d_{13} = i\frac{p}{r}J_p(k_S r) \tag{11.54}$$

$$d_{14} = i\frac{p}{r}Y_p(k_S r) \tag{11.55}$$

$$d_{21} = i\frac{p}{r}J_p(k_L r) \tag{11.56}$$

$$d_{22} = i\frac{p}{r}Y_p(k_L r) \tag{11.57}$$

$$d_{23} = \frac{k_S}{2}J_{p+1}(k_S r) - \frac{k_S}{2}J_{p-1}(k_S r) \tag{11.58}$$

$$d_{24} = \frac{k_S}{2}Y_{p+1}(k_S r) - \frac{k_S}{2}Y_{p-1}(k_S r) \tag{11.59}$$

$$d_{31} = \frac{\mu}{r^2}\left(\frac{r^2\kappa^2 k_L^2}{4}\left(J_{p-2}(k_L r) - 2J_p(k_L r) + J_{p+2}(k_L r)\right) \right.$$
$$\left. + \frac{rk_L(\kappa^2-2)}{2}\left(J_{p-1}(k_L r) - J_{p+1}(k_L r)\right) - p^2(\kappa^2-2)J_p(k_L r) \right)$$

where $\kappa = c_l / c_s$ $\tag{11.60}$

$$d_{32} = \frac{\mu}{r^2}\left(\frac{r^2\kappa^2 k_L^2}{4}\left(Y_{p-2}(k_L r) - 2Y_p(k_L r) + Y_{p+2}(k_L r)\right) \right.$$
$$\left. + \frac{rk_L(\kappa^2-2)}{2}\left(Y_{p-1}(k_L r) - Y_{p+1}(k_L r)\right) - p^2(\kappa^2-2)Y_p(k_L r) \right), \tag{11.61}$$

$$d_{33} = \frac{\mu}{r^2}\left(irpk_S\left(J_{p-1}(k_S r) - J_{p+1}(k_S r)\right) - i2pJ_p(k_S r) \right) \tag{11.62}$$

$$d_{34} = \frac{\mu}{r^2}\left(irpk_S\left(Y_{p-1}(k_S r) - Y_{p+1}(k_S r)\right) - i2pY_p(k_S r) \right) \tag{11.63}$$

$$d_{41} = \frac{\mu}{r^2}\left(irpk_L\left(J_{p-1}(k_L r) - J_{p+1}(k_L r)\right) - i2pJ_p(k_L r) \right) \tag{11.64}$$

$$d_{42} = \frac{\mu}{r^2}\left(irpk_L\left(Y_{p-1}(k_L r) - Y_{p+1}(k_L r)\right) - i2pY_p(k_L r) \right) \tag{11.65}$$

$$d_{43} = \frac{\mu}{r^2}\left(-\frac{r^2 k_S^2}{4}\left(J_{p-2}(k_S r) - 2J_p(k_S r) + J_{p+2}(k_S r)\right) \right.$$
$$\left. + \frac{rk_S}{2}\left(J_{p-1}(k_S r) - J_{p+1}(k_S r)\right) - p^2 J_p(k_S r) \right) \tag{11.66}$$

$$d_{44} = \frac{\mu}{r^2}\left(-\frac{r^2 k_S^2}{4}\left(Y_{p-2}(k_S r) - 2Y_p(k_S r) + Y_{p+2}(k_S r)\right) \right.$$
$$\left. + \frac{rk_S}{2}\left(Y_{p-1}(k_S r) - Y_{p+1}(k_S r)\right) - p^2 Y_p(k_S r) \right) \tag{11.67}$$

As was the case for CSH-waves, the stress-related components of Equation (11.51) are needed in the single-layer case and, therefore, the bottom two rows are required to vanish at the boundaries of the annulus. The eigenvalues of the resulting characteristic equation form the frequency-wavenumber dispersion curves for CLT-waves traveling in a single-layer annulus. The angular phase and group velocity dispersion curves can then be calculated using the relations seen in Equation (11.9) and Equation (11.23), respectively.

After the eigenvalues of the characteristic equation have been determined, the displacements u_r and u_θ and the stresses σ_r and $\sigma_{r\theta}$ can be determined by arbitrarily setting one of the four amplitude coefficients to unity and determining the relative amplitude of the other three. Once the relative values of the four coefficients, $A_1 - A_4$, are known, Equation (11.50) and Equation (11.51) can be used to solve for the displacements and stresses.

This concludes the development of the single-layer CSH and CLT cases. The next section discusses the extension of the theoretical models to the case of multilayered annuli. Numerical examples are presented for both cases in Section 11.3 and limitations on computational capability are discussed.

11.3 Extension to Multilayer Annuli

The extension of single-layer cases to multilayer cases is accomplished by one of two fundamental matrix methods: the transfer matrix method (Thomson 1950; Haskell 1953) or the global matrix method (GMM) (Knopoff 1964). Although it may take longer to arrive at a solution, the GMM is used here as it is more stable and can handle many categories of problems without modification (Lowe 1995). The underlying strategy of the GMM is to develop the displacement and stress equations for each individual layer and then, by applying the boundary and continuity conditions, to assemble a global matrix representing the entire layered system. The characteristic equation of the layered system is then obtained by setting the determinant of the global matrix to zero.

Consider the multilayer annulus shown in Figure 11.2. The GMM approach is illustrated using the case of CSH-wave propagation as this case is simpler and makes the development of the global matrix more transparent. The same approach is equally applicable to CLT-wave propagation as the only difference is in the number of boundary and continuity conditions needed to solve the corresponding system of equations. This difference is addressed later in this section.

To begin, each layer is treated as a single layer, and a time-harmonic steady-state solution to Navier's equation is assumed. As seen in Equation (11.68), the assumed solution is valid for layer m.

$$u_z^{(m)} = \psi^{(m)}(r)e^{i(p\theta - \omega t)}, \qquad r_m \leq r \leq r_{m+1}. \tag{11.68}$$

Using a slightly modified version of Equation (11.12), the expression for $\psi^{(m)}(r)$, in terms of Bessel functions, is

$$\psi^{(m)}(r) = A_1^{(m)} J_p\left(k_s^{(m)}r\right) + A_2^{(m)} Y_p\left(k_s^{(m)}r\right), \tag{11.69}$$

where two unknown coefficients, $A_{1,2}^{(m)}$, are present for each layer, m. The relevant stress equation for layer m is found by using Equation (11.69) in Equation (11.14).

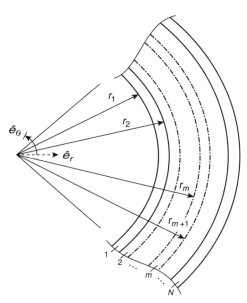

Figure 11.2. Theoretical model used for the development of the governing equations for CSH-wave and CLT-wave propagation in the circumferential direction of a multilayer annulus.

In the CSH case, one displacement-related and one stress-related equation exists for each layer, yielding a total of $2N$ equations ($4N$ for CLT-waves). Therefore, $2N$ boundary/continuity conditions will be required to solve the system of equations. As was done in the single-layer case, the innermost and outermost boundaries are assumed to be traction free. In the case of a multilayered system, some description of the interaction at the interface between two layers must be made. This is done through continuity conditions in which displacement and stress are required to be continuous at the interface. Equation (11.70) summarizes the boundary and interfacial continuity conditions for CSH-wave propagation. For the sake of completeness, the boundary and continuity conditions for the CLT case are also summarized in Equation (11.71).

$$\{\sigma_{rz}\}_{\text{free surface}} = 0, \qquad \left\{ \begin{array}{c} u_z \\ \sigma_{rz} \end{array} \right\}_{\substack{\text{layer } m \\ \text{interface } m+1}} = \left\{ \begin{array}{c} u_z \\ \sigma_{rz} \end{array} \right\}_{\substack{\text{layer } m+1 \\ \text{interface } m+1}}. \tag{11.70}$$

$$\left\{ \begin{array}{c} \sigma_r \\ \sigma_{r\theta} \end{array} \right\}_{\text{free surface}} = 0, \qquad \left\{ \begin{array}{c} u_r \\ u_\theta \\ \sigma_r \\ \sigma_\theta \end{array} \right\}_{\substack{\text{layer } m \\ \text{interface } m+1}} = \left\{ \begin{array}{c} u_r \\ u_\theta \\ \sigma_r \\ \sigma_\theta \end{array} \right\}_{\substack{\text{layer } m+1 \\ \text{interface } m+1}}. \tag{11.71}$$

Next, using Equation (11.18), the layer matrices are formed for the bottom, $D_B^{(m)}$, and top, $D_T^{(m)}$, of each layer. For CSH-waves, the bottom and top layer matrices for layer m are shown in Equation (11.72) and Equation (11.73), respectively.

$$D_B^{(m)} = \left[\begin{array}{cc} J_p(k_s^m r_m) & Y_p(k_s^m r_m) \\ \dfrac{\mu^m k_s^m}{2}\left(J_{p-1}(k_s^m r_m) - J_{p+1}(k_s^m r_m)\right) & \dfrac{\mu^m k_s^m}{2}\left(Y_{p-1}(k_s^m r_m) - Y_{p+1}(k_s^m r_m)\right) \end{array} \right]. \tag{11.72}$$

$$D_T^{(m)} = \begin{bmatrix} J_p(k_s^m r_{m+1}) & Y_p(k_s^m r_{m+1}) \\ \frac{\mu^m k_s^m}{2}\left(J_{p-1}(k_s^m r_{m+1}) - J_{p+1}(k_s^m r_{m+1})\right) & \frac{\mu^m k_s^m}{2}\left(Y_{p-1}(k_s^m r_{m+1}) - Y_{p+1}(k_s^m r_{m+1})\right) \end{bmatrix}.$$

$$(11.73)$$

For the free-surface boundary conditions only the second row of these two matrices are needed, which yields the following for the innermost and outermost surfaces (interfaces):

$$\left[D_B^{(1)}\right]_2 \begin{Bmatrix} A_1^{(1)} \\ A_2^{(1)} \end{Bmatrix} = \left[D_T^{(N)}\right]_2 \begin{Bmatrix} A_1^{(N)} \\ A_2^{(N)} \end{Bmatrix} = \begin{Bmatrix} 0 \\ 0 \end{Bmatrix}, \qquad (11.74)$$

where the subscript "2" after the $D_{B,T}^{(1,N)}$ matrices indicates that only the second row of the corresponding matrix is being considered as this is the row relating to stress. The two equations shown in Equation (11.74) constitute the first and last rows of the global matrix. All rows in between result from the interfacial continuity conditions.

Using the notation introduced in Equation (11.72) and Equation (11.73), the interfacial continuity conditions can be written as follows:

$$\left[D_T^{(m)}\right] \begin{Bmatrix} A_1^{(m)} \\ A_2^{(m)} \end{Bmatrix} = \left[D_B^{(m+1)}\right] \begin{Bmatrix} A_1^{(m+1)} \\ A_2^{(m+1)} \end{Bmatrix}, \qquad (11.75)$$

or, rearranging to obtain a homogeneous form,

$$\left[\begin{bmatrix} D_T^{(m)} \end{bmatrix}\begin{bmatrix} -D_B^{(m+1)} \end{bmatrix}\right] \begin{Bmatrix} A_1^{(m)} \\ A_2^{(m)} \\ A_1^{(m+1)} \\ A_2^{(m+1)} \end{Bmatrix} = \begin{Bmatrix} 0 \\ 0 \\ 0 \\ 0 \end{Bmatrix}. \qquad (11.76)$$

With both the boundary and continuity conditions expressed in homogeneous form, it is now possible to construct the global matrix. This is done by assembling the layer matrices into a single matrix in which the individual matrices are matched according to interface. That is, the top coefficient matrix of layer m should fall into the same rows of the global matrix as the bottom coefficient matrix of layer $m+1$. The global matrix for an N-layered system is shown on the left in Equation (11.77).

$$\begin{bmatrix} \left[D_B^{(1)}\right]_2 & 0 & 0 & 0 & 0 & 0 & 0 & 0 \\ \left[D_T^{(1)}\right] & \left[-D_B^{(2)}\right] & 0 & 0 & 0 & 0 & 0 & 0 \\ 0 & \ddots & \ddots & 0 & 0 & 0 & 0 & 0 \\ 0 & 0 & 0 & \left[D_T^{(m-1)}\right] & \left[-D_B^{(m)}\right] & 0 & 0 & 0 \\ 0 & 0 & 0 & 0 & \left[D_T^{(m)}\right] & \left[-D_B^{(m+1)}\right] & 0 & 0 \\ 0 & 0 & 0 & 0 & 0 & \ddots & \ddots & 0 \\ 0 & 0 & 0 & 0 & 0 & 0 & \left[D_T^{(N-1)}\right] & \left[-D_B^{(N)}\right] \\ 0 & 0 & 0 & 0 & 0 & 0 & 0 & \left[D_T^{(N)}\right]_2 \end{bmatrix} \begin{Bmatrix} \mathbf{A}^{(1)} \\ \mathbf{A}^{(2)} \\ \vdots \\ \mathbf{A}^{(m-1)} \\ \mathbf{A}^{(m)} \\ \vdots \\ \mathbf{A}^{(N-1)} \\ \mathbf{A}^{(N)} \end{Bmatrix} = \begin{Bmatrix} \mathbf{0} \\ \mathbf{0} \\ \vdots \\ \mathbf{0} \\ \mathbf{0} \\ \vdots \\ \mathbf{0} \\ \mathbf{0} \end{Bmatrix},$$

$$(11.77)$$

where:

$$\mathbf{A}^{(m)} = \left\{ \begin{matrix} A_1^{(m)} \\ A_2^{(m)} \end{matrix} \right\}, \text{ and } \mathbf{0} = \left\{ \begin{matrix} 0 \\ 0 \end{matrix} \right\}. \tag{11.78}$$

It is seen from Equation (11.77) that for CSH-waves traveling in a multilayered system, the result will be a $2N$ square global matrix. In the case of CLT-wave propagation in a multilayered system, the global matrix will be a $4N$ square matrix because two stress components must satisfy the boundary conditions and two stress and two displacement components must satisfy the interfacial continuity conditions.

After the formation of the global matrix, the remainder of the solution process is identical to the single-layer cases; the determinant of the global matrix yields the characteristic equation whose roots are the eigenvalues for the structure under study. Once the eigenvalues of the characteristic equation are identified, the unknown constants, $A_{1,2}^{(1)}$ through $A_{1,2}^{N}$, can be determined by arbitrarily setting one to unity and solving for the relative value of the others.

Again, by performing all necessary calculations in ω-p space, the angular phase and group velocities can be determined, from which the linear phase and group velocities can be determined at any arbitrarily chosen radius. Alternatively, one may initially choose some radius, R, for which a solution is desired and then solve in the ω-k space to directly determine the linear phase and group velocities.

This brings the theoretical development of circumferential guided wave propagation to an end. Both the single-layer CSH-wave and CLT-wave cases have been presented and the extension to the multilayer case was illustrated using the CSH-wave case. Though not shown explicitly, the extension to multiple layers for CLT-waves follows the exact same process as that of CSH-waves.

11.4 Numerical Solution of the Governing Wave Equations for Circumferential Guided Waves

While the derivations of the characteristic equations do provide some physical insight, of more practical interest is the resultant propagating wave modes and shapes for a particular structure. These are determined by numerical solution of the characteristic equations as closed-form solutions do not exist.

Following the development of the characteristic equations for CSH-waves and CLT-waves for single-layered and multilayered annuli, presented in Sections 11.1 and 11.2, respectively, it is necessary to verify the validity of the presented equations. In this work, verification is achieved by one or more methods, including direct comparison of numerical results with published results, comparison of numerical results with the limiting case of a plate, and analysis of the eigenfunctions for satisfaction of the appropriate boundary and continuity conditions.

All materials are assumed to be elastic in this chapter and as such, only real angular wavenumbers, p, are considered. Situations in which p would be complex, such as for evanescent modes or viscoelastic materials, are not considered here, but can be found in Van Velsor (2009). The problem of determining the propagating wave modes at some circular frequency, ω, reduces to determining the values of p for which nontrivial solutions, that is, nonzero, roots, of the characteristic equations exist. Any number of root-seeking algorithms may be employed to find the roots of the characteristic equations. Common algorithms include the Bisection method, the Newton-Rhapson method, the Secant method, and the *regula falsi* method.

For real p, the Bisection method is regarded as a more robust routine as it guarantees convergence, though only at the sacrifice of computational speed (Jaluria 1996). For this reason, as well as for its simplistic nature and ease of coding, the Bisection routine is employed in this work.

The process of generating dispersion curves involves determining the roots of the characteristic equation over some frequency range that is of interest, typically determined by the application at hand. For the elastic materials considered here, the angular wavenumber roots, p, are determined at fixed increments of ω, resulting in the frequency-wavenumber (ω-p) dispersion curves from which it is possible to determine the angular or linear phase and group velocity dispersion curves using the relations introduced in Section 11.2.

Sample numerical results are now presented for both CSH-waves and CLT-waves in single-layered and multilayered elastic annuli with the primary goal of proving the validity of the analytical treatment presented in Sections 11.2 and 11.3. Defining characteristics of circumferential guided waves are discussed and the chapter concludes with a discussion on the computational limitations associated with the analytical formulation of the circumferential guided wave problem.

11.4.1 Numerical Results for CSH-Waves

As a first check of the validity of the analytical derivation presented in Section 11.2.1 for CSH-waves, a plate approximation is assumed and the dispersion curves are compared to those obtained using a formulation for SH-waves in a flat plate. As the inner-to-outer radius ratio (or aspect ratio) approaches unity, the CSH-wave solution should approach the solution for the flat plate. This is intuitive from a physical standpoint and is mathematically proven in Zhao and Rose (2004).

Figure 11.3 shows a comparison of the SH-wave dispersion curves for a steel plate and the CSH-wave dispersion curves for a steel annulus with an aspect ratio of 0.984. A bulk shear wave velocity of 3.23 mm/μs was used in the computation and a closed-form solution for SH-waves in a plate, as presented in Rose (1999) and Graff (1991), was used to generate the plate solution shown in the figure. As expected, the plate approximation is a sufficient model for an annulus of this aspect ratio, though small differences in the curves begin to manifest as the frequency and wavenumber increase.

Figure 11.4 depicts the dispersion curve comparisons of a steel plate and annulus with an aspect ratio of 0.5. It is evident from this figure that as the aspect ratio of the annulus decreases, the plate-like assumption is no longer a valid approximation for CSH-waves in an annulus. Several interesting phenomena arise as the aspect ratio decreases. First, the fundamental mode, labeled *SH0* in each plot, becomes increasingly dispersive. Second, the group velocities of the higher order modes can exceed that of the fundamental mode as the aspect ratio decreases. These two phenomena do not hold true for wave propagation in a plate or in the axial direction of a pipe and, therefore, are defining characteristics of CSH-waves in an annulus. Also note that the cutoff frequencies are very close for CSH-waves in an annulus and for SH-waves in a flat plate, even for very small aspect ratios.

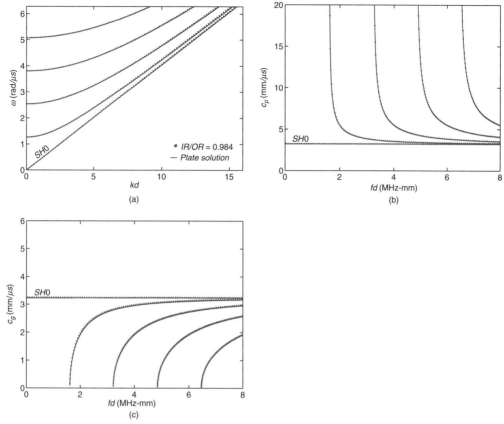

Figure 11.3. CSH **(a)** Frequency-wavenumber (thickness), **(b)** phase velocity, and **(c)** group velocity dispersion curves for a steel plate and an annulus with a 0.984 aspect ratio. The circular wavenumber, k, and the linear phase and group velocity are calculated at the OR of the annulus.

Figure 11.5 shows the dispersion curves for a multilayer annulus with properties as summarized in Table 11.1. In each of the plots, the dispersion curves are shown for the multilayer annulus, for Layer 1 only, and for Layer 2 only. The wavenumber and linear phase and group velocities are calculated at the mid-plane of the multilayer structure and at the top of Layer 1 and the bottom of Layer 2.

It is seen from plots (a) and (b) of Figure 11.5 that any intersection in the solution space for the individual layers also belongs to the solution space of the multilayer system. This is seen more clearly in the magnified view of Figure 11.5(a), shown in Figure 11.6. Physically, these points signify locations in the dispersion space where the boundary and continuity conditions are satisfied for all three annular structures. As seen in the wave structure (or eigenfunction) plots (d), (e), and (f) of Figure 11.7, the intersection points are characterized by zero shear stress on the surfaces of the single-layer annuli and at the surfaces and interface of the multilayer annulus. Simonetti (2004) made similar observations for multilayered plates. The wave structures shown in Figure 11.7 correspond to the points labeled in Figure 11.5(b). Furthermore, these intersections represent "resonance" points of Layer 1. Resonance,

Table 11.1. *Geometric and material parameters used in the generation of the plots shown in Figure 11.5 through Figure 11.7 and Figure 11.10 through Figure 11.13*

	r_{inner} (m)	r_{outer} (m)	ρ (kg/m³)	c_l (m/s)	c_s (m/s)
Layer 1	0.032	0.036	7,850	5,850	3,230
Layer 2	0.036	0.04	1,500	1,700	900

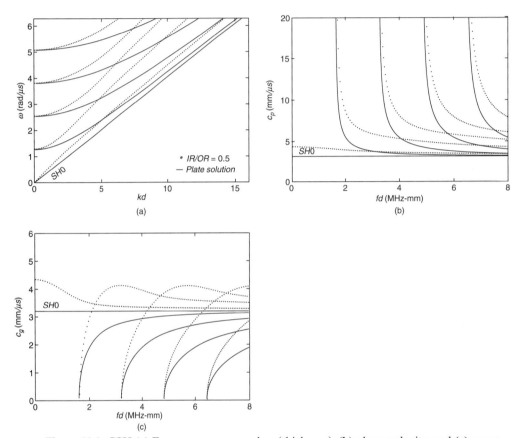

Figure 11.4. CSH **(a)** Frequency-wavenumber (thickness), **(b)** phase velocity, and **(c)** group velocity dispersion curves for a steel plate and an annulus with a 0.5 aspect ratio. The circular wavenumber, k, and the linear phase and group velocity are calculated at the OR of the annulus.

in this context, is taken to mean a mode/frequency combination in which the largest possible displacement is achieved in the layer with the higher acoustic impedance. Figures 11.7(a)–(c) and 11.7(g)–(i) show that points not located near the resonances display dominant particle displacement in Layer 2. From Figure 11.7(c) it is also seen that said resonance points coincide with the peaks of the group velocity dispersion curves.

As observed in Figure 11.5(b),

$$\lim_{f \to \infty} c_p(r) = \frac{r}{r_{N+1}} \min(c_s^m), \qquad (11.79)$$

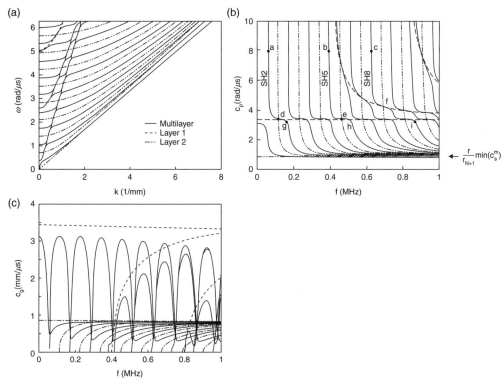

Figure 11.5. CSH **(a)** Frequency-wavenumber, **(b)** phase velocity, and **(c)** group velocity dispersion curves for a multilayered annulus with a 0.8 aspect ratio and properties described in Table 11.1. Each layer is 4 mm thick. Curves are also shown for the individually isolated layers. The circular wavenumber, k, and the linear phase and group velocity are calculated at the common interface of the annuli.

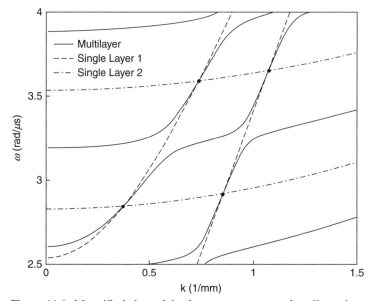

Figure 11.6. Magnified view of the frequency-wavenumber dispersion curve shown in Figure 11.5(a). Black dots denote points of intersection.

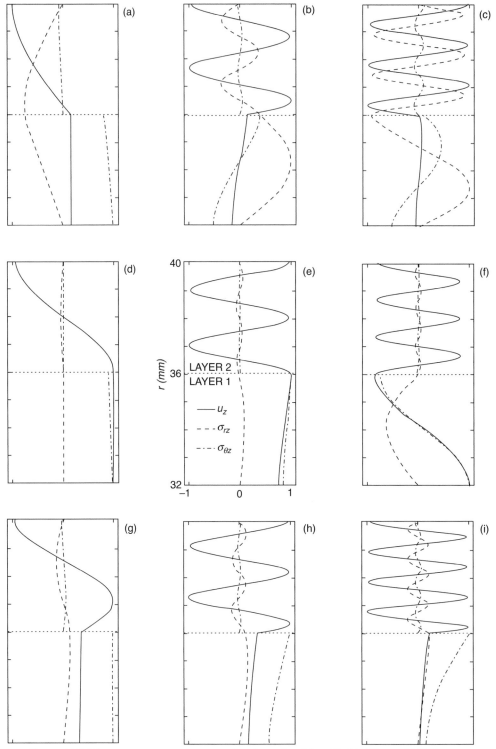

Figure 11.7. Wave structures and stresses for the points identified in Figure 11.5(b).

where r_{N+1} and c_s^m are as seen in Figure 11.2 and r is the arbitrary radius for which the linear phase velocity dispersion curves have been calculated. From Equation (11.79) it is seen that in the limiting case, the phase velocity of all modes approaches the bulk shear wave velocity of the slowest layer with the maximum potential phase velocity occurring at the outer radius of the annulus; all other radii have proportionally slower phase velocities. As previously stated, the variation in linear phase velocity with radius is a necessary requirement to maintain a constant phase front through the thickness of the annulus.

One last interesting observation concerns the linear group velocity dispersion curves seen in Figure 11.5(c). For the multilayered annulus, the peaks of the linear group velocity dispersion curves approximately trace, or approach, the linear group velocity curves of the higher-impedance layer – in this case Layer 1. The actual peak values, though, are always lower than the curves for the isolated high-impedance layer. A practical exploitation of this phenomenon is discussed and demonstrated in Section 11.5.

The next section presents numerical examples of CLT-waves in single-layered and multilayered annuli. The chapter ends with a discussion of the limitations of the numerical methods based on the analytical formulation of the circumferential wave problem.

11.4.2 Numerical Results for CLT-Waves

As was done for the case of CSH-waves, a first check of the analytical solution for CLT-waves will be completed by comparing a circumferential wave plate approximation with the results obtained for a flat plate. Figure 11.8 shows a comparison of the Lamb-wave dispersion curves for a steel plate (solid lines) and the CLT-wave dispersion curves for a steel ($c_s = 3.23$ mm/μs, $c_l = 5.96$ mm/μs, and $\rho = 7850$ kg/m^3) annulus with an aspect ratio of 0.984 (dotted lines). The Lamb-wave dispersion curves were obtained based on the formulation for plates presented in Rose (1999).

As seen in the CSH-wave case, the plate approximation provides an accurate representation of a flat plate with small discrepancies beginning to manifest at high frequencies and large wavenumbers. The first two modes, *LT0* and *LT1*, are labeled in each plot. For the plate approximation, these modes correspond to the fundamental antisymmetric and symmetric modes of the flat plate, respectively. When referring to waves propagating in an annulus, mode shapes are no longer perfectly symmetric or antisymmetric and are therefore referred to as *LT0* through *LTm* in this text, where *m* represents any positive integer extending to infinity.

Several defining characteristics of CLT-waves are seen in Figure 11.8 and Figure 11.9. To start, the mode corresponding to the *S0* plate mode no longer intercepts the vertical axis of the phase velocity dispersion plot at the plate velocity. It instead asymptotically tends toward infinite phase velocity. In the group velocity curves, the same mode tends toward zero as frequency tends toward zero.

Another defining characteristic of CLT-waves is clearly seen for annuli with smaller aspect ratios. A new mode is seen extending from the origin of the frequency-wavenumber dispersion curves, a mode that quickly peaks and then intercepts the kd-axis at the point ($kd = 1$-η), where η represents the aspect ratio of the cylinder. Liu and Qu (1998a) provide a nice discussion on this mode and conclude that it

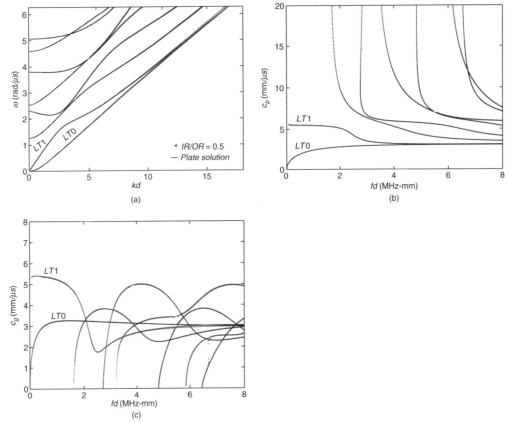

Figure 11.8. CLT **(a)** Frequency-wavenumber (thickness), **(b)** phase velocity, and **(c)** group velocity dispersion curves for a steel plate and an annulus with a 0.984 aspect ratio. The circular wavenumber, k, and the linear phase and group velocity are calculated at the OR of the annulus.

is a result of the inability of the inner surface of the annulus to support a surface wave. As Viktorov (1967) pointed out, convex surfaces may support surface waves while concave surfaces may not. Therefore, any surface wave that forms on the inner surface of the annulus will quickly leak back into the annulus, where it is then reflected off of the outer surface, subsequently forming the new mode. This mode is not seen in plates as they can support surface waves on both sides. The reason that the mode intercepts the kd-axis at $(kd = 1\text{-}\eta)$ is because the wavelength is equal to the outer circumference at this point and no propagation is possible.

Based on the discussion in the previous paragraph, it is possible to conclude that the phase velocity of the CLT-mode that is analogous to the *A0* mode in a plate will converge to the Rayleigh surface wave velocity as frequency tends toward infinity. The Rayleigh wave propagates on the outer surface of the annulus. It is also seen from Figure 11.9 that as the aspect ratio of the annulus begins to decrease, the CLT-mode once analogous to the *S0* plate mode becomes quite dissimilar to the *S0* plate mode. This phenomenon is a manifestation of the inability of small aspect ratio annuli to support symmetric wave displacement profiles.

Figure 11.10 presents the CLT-wave dispersion curves for a multilayered annulus with the properties and geometry previously specified in Table 11.1. In the plots, the

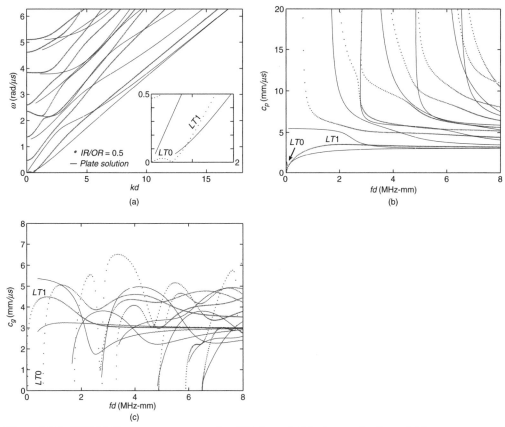

Figure 11.9. CLT **(a)** Frequency-wavenumber (thickness), **(b)** phase velocity, and **(c)** group velocity dispersion curves for a steel plate and an annulus with a 0.5 aspect ratio. The circular wavenumber, k, and the linear phase and group velocity are calculated at the OR of the annulus.

curves corresponding to the multilayered annulus and for Layer 1 are shown. The individual curves for Layer 2 are not shown to prevent the plots from becoming cluttered. Figure 11.11 shows two magnified regions of plot (a) in Figure 11.10 where there appear to be intersecting modes. Examination of the magnified views reveals that the CLT-modes in the multilayered annulus do not actually intersect. Überall and colleagues (1994) and Simonetti (2004) made similar observations for Lamb-waves in multilayered plates.

Figure 11.12 and Figure 11.13 show the wave structures and stress distributions for points a, b, c and d, e, and f in Figure 11.10(b), respectively. It is observed from Figure 11.12 that for regions of the multilayer dispersion curves that do not trace the curves for the high-impedance layer, Layer 1, wave propagation occurs primarily in the lower-impedance layer, Layer 2. As shown in the case of CSH-waves and now for CLT-waves in Figure 11.13, the points at which the multilayer curves intersect the curves for the isolated high-impedance layer constitute resonance points of that layer. Furthermore, Figure 11.10(c) illustrates that the peaks of the group velocity dispersion curves for the multilayer coincide with the resonance points, though the peak values only approach the group velocity of the modes in the isolated high-impedance layer and are always slightly slower. As an example, these resonance

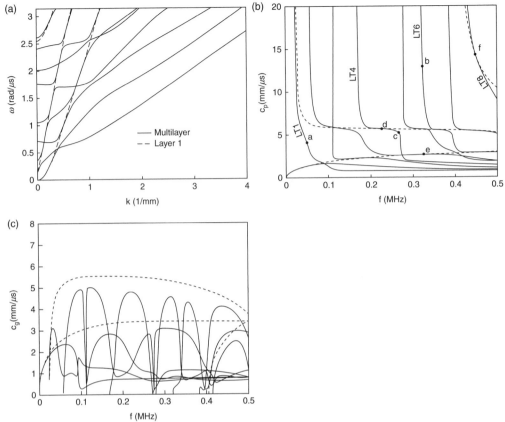

Figure 11.10. CLT **(a)** Frequency-wavenumber, **(b)** phase velocity, and **(c)** group velocity dispersion curves for a multilayered annulus with a 0.8 aspect ratio and properties described in Table 11.1. Each layer is 4 mm thick. Curves are also shown for the isolated Layer 1. The circular wavenumber, k, and the linear phase and group velocity are calculated at the common interface of the annuli.

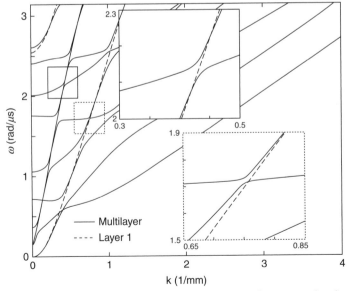

Figure 11.11. CLT Frequency-wavenumber dispersion curves for the multilayered annulus described in Table 11.1. Magnified views illustrate the repulsion of CLT-modes in multilayered annuli.

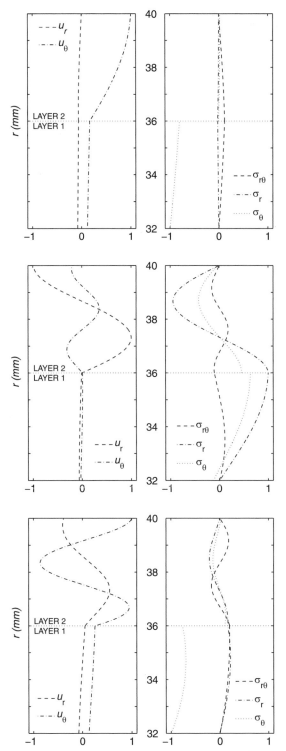

Figure 11.12. Wave structures and stresses for the points *a*, *b*, and *c* identified in Figure 11.10(b).

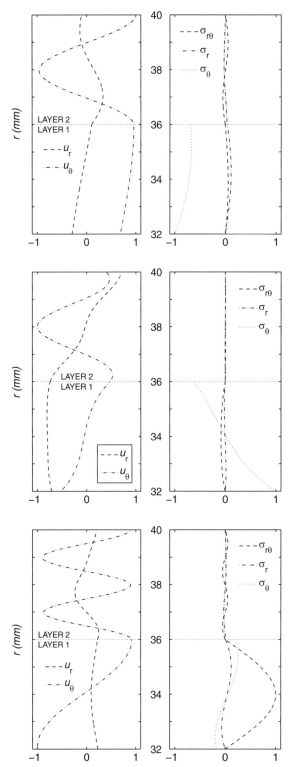

Figure 11.13. Wave structures and stresses for the points d, e, and f identified in Figure 11.10(b).

points are of significant practical importance for the nondestructive evaluation (NDE) of coated metallic annuli.

This concludes the presentation of the numerical results as obtained from the analytical formulation. The next section discusses some of the limitations of the numerical solution of the analytical dispersion equations and presents methods for addressing said limitations.

11.4.3 Computational Limitations of the Analytical Formulation

This section is devoted to identifying the limitations associated with the numerical solution of the analytical formulation of the circumferential wave problem. In fact, there is one primary limitation associated with the numerical computation of the circumferential guided wave dispersion curves and it has to do with the calculations of Bessel functions. For cylindrical geometries, Bessel functions are inevitably encountered. Barshinger (2001) solves the problem of wave propagation in the axial direction of a hollow cylinder with a viscoelastic coating. In the axial formulation, the characteristic equation is composed of 0th- and 1st-order Bessel (or Hankel) functions of the first and second kind (i.e., $J_0(x)$, $J_1(x)$, $Y_0(x)$, and $Y_1(x)$). The expressions for these common functions can be found in any advanced mathematics text or handbook, such as Hayek (2001) or Abramowitz and Stegun (1964).

The necessity of only four Bessel functions is a convenience that is not encountered in the case of circumferential wave propagation. As shown in Equation (11.12) for CSH-waves and in Equations (11.44) and (11.45) for CLT-waves, the order of the Bessel functions is not necessarily an integer and is dependent on wavenumber and radius because $p=kr$. Thus for large radius annuli, the Bessel functions will be of very high order, which will naturally present a problem for small arguments of the Bessel function of the second kind, $Y_p(x)$. To illustrate this, consider the plots of the Bessel functions shown in Figure 11.14. Plots (a) and (b) show the 0th- and 1st-order Bessel functions, respectively, that are used in the calculation of the dispersion curves for wave propagation along the axial direction of a pipe. The Bessel function of the first kind intersects the y-axis for an argument of zero whereas the Bessel function of the second kind asymptotically approaches negative infinity. This typically does not present a problem in computations as values for $Y_0(x)$ and $Y_1(x)$ can typically be calculated for arguments approaching zero.

Now consider plot (c) in Figure 11.14, which has an arbitrarily selected order of *100*. It is immediately seen that computing any Bessel function of the second kind will be very difficult for small arguments. Furthermore, because the argument of the Bessel function involves a product with frequency for circumferential waves, the computation for small arguments is a necessity.

Figure 11.15 shows the frequency-angular wavenumber dispersion curves for the multilayered annulus described in Table 11.2. This plot shows two sets of curves; one set was calculated using double precision and one set was calculated using arbitrary precision arithmetic. A close review of the curves reveals the computation using double precision first fails to find a root near the point $(p, \omega) = (670, 2.4)$ and then continuously fails to find roots for the *SH0* and *SH1* modes.

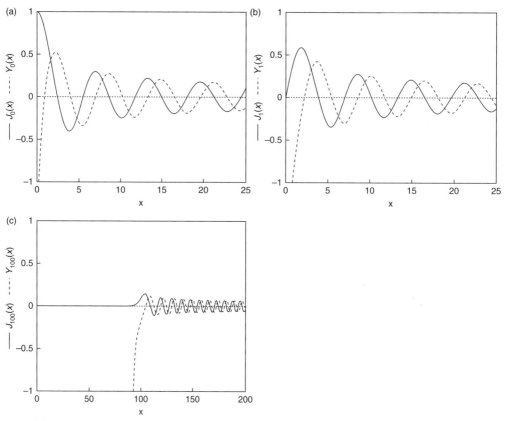

Figure 11.14. Bessel functions of the first and second kind of order **(a)** 0, **(b)** 1, and **(c)** 100.

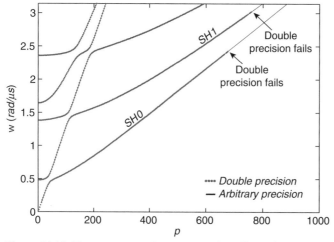

Figure 11.15. Frequency-angular wavenumber dispersion curves for CSH-waves propagating in a two-layer annulus as described in Table 11.2. Double precision eventually fails to find roots at higher frequencies because of the large angular wavenumber.

Table 11.2. Geometric and material parameters used in the generation of Figure 11.15

	r_{inner} (m)	r_{outer} (m)	ρ (kg/m³)	c_l (m/s)	c_s (m/s)
Layer 1	0.24765	0.254	7,850	5,960	3,260
Layer 2	0.254	0.257	1,500	1,400	900

Only frequencies up to 500 kHz are shown in Figure 11.15, and researchers have observed that even more roots are missed at higher frequencies, especially when considering large-radius annuli with low-impedance layers, such as polymers or other viscoelastic materials. The calculation of the CLT-wave dispersion curves for large-radius annuli raises similar problems.

As seen in Figure 11.15, the use of arbitrary precision arithmetic can provide the complete solution for the dispersion space analyzed, albeit with a significantly reduced computational efficiency. In the case of CLT-waves, researchers also found that a larger region of the dispersion space could be computed through the use of arbitrary precision arithmetic, though the algorithm did also eventually fail to find roots. Other problems arising from the numerical computation based on the analytical formulation involved the calculation of wave structures. In general, they could not be computed for very low-phase velocities, such as those encountered with low-impedance materials.

Because the arbitrary precision arithmetic approach still has root-finding limitations, and because the approach increased the calculation time by an approximate factor of ten, it may be desirable to investigate alternative solution methods, even if approximate. Gridin and colleagues (2003) introduced one such approximate approach for CSH-waves and CLT-waves that involves the asymptotic reduction of the exact dispersion relations. The method is tedious, involving five to nine different sets of equations whose use depends on the values of the angular wavenumber and therefore is only applicable to single-layer annuli. Viktorov (1967) also used a similar approximation method when he first addressed the topic of circumferential waves and Rayleigh waves on concave and convex surfaces. To arrive at a first approximation, he replaced the Bessel functions in the characteristic equation with asymptotic representations in terms of semi-convergent Debye series (Watson 1966).

With the anticipation that one may wish to add considerations for viscoelastic layered annuli, an approximate method based on the semi-analytical finite element (SAFE) method would be most appropriate for obtaining complete solutions of the dispersion space. In this manner it is not necessary to calculate the complex arbitrary order Bessel functions that would result from viscoelastic considerations. Additionally, several other authors have shown this method to be sufficiently accurate (Hayashi et al. 2003; Hayashi, Song, and Rose 2003; Bartoli et al. 2006; Mu and Rose 2008). The formulation of the SAFE solution for circumferential waves in multilayered annuli can be found in detail for CLT-waves and CSH-waves in Van Velsor (2009), along with a practical example for CSH-waves in Van Velsor, Rose, and Nestleroth (2009). These works include considerations for both elastic and viscoelastic materials.

This concludes the section on the numerical solution of the governing equations for circumferential guided waves in single-layered and multilayered annuli. The next section presents an example of a practical application of circumferential guided waves.

11.5 The Effects of Protective Coating on Circumferential Wave Propagation in Pipe

The previous chapters dealt with the development and validation of the tools necessary to accurately model circumferential guided waves in single-layered and multilayered elastic and viscoelastic annuli. Such tools have significant potential for applications in nondestructive testing (NDT) using ultrasonic guided waves. There are many practical situations in which the annulus model can accurately represent a structure. For example, the practical analog to the generalized plane-strain assumption used in the development of the circumferential wave model is a hollow cylinder whose boundaries in the axial direction are far removed from the wave field relative to wavelength. Because ultrasonic wavelengths are typically very short, most piping systems fit this criterion. Examples of applications include the NDT of gas transmission line, gas storage well casing, and boiler and heat exchanger tubing.

Researchers designed an experiment to study the effect of increasing coating thickness on the propagation of CSH-waves. As seen in the CSH-wave dispersion examples provided in the previous sections, one effect of the addition of a lower-impedance layer to a single-layer annulus is a decrease in the maximum achievable group velocity for any specific wave mode. Furthermore, if the lower-impedance layer is viscoelastic in nature, wave attenuation will be introduced. Figure 11.16

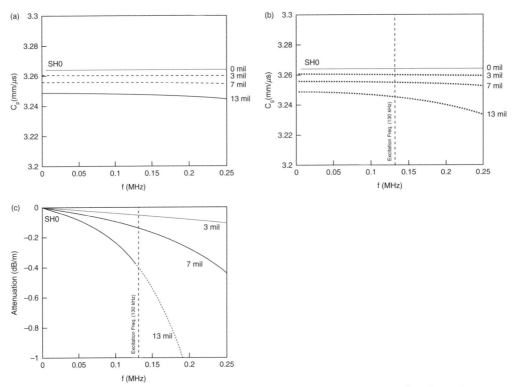

Figure 11.16. CSH **(a)** phase velocity, **(b)** group velocity, and **(c)** attenuation dispersion curves for a twenty-four-inch schedule 10 pipe with various thicknesses of the Bitumastic 50 coating described in Table 11.3. Linear phase and group velocity and attenuation were calculated at the OR.

Table 11.3. Dimensions and properties of pipe specimen used for coating thickness study. Properties for Bitumastic 50 coating from Barshinger and Rose (2004).

	r_{inner} (m)	r_{outer} (m)	ρ (kg/m³)	c_l (m/s)	α_l/ω	c_s (m/s)	α_s/ω
Mild Steel	0.29845	0.3048	7,850	5,850	0	3,230	0
Bitum. 50	0.3048	*varied*	1,500	1,860	0.023	750	0.20

summarizes these effects for a twenty-four-inch-diameter schedule 10 pipe with various thicknesses of a Bitumastic 50 coal-tar mastic coating with the properties published by Barshinger and Rose (2004) and summarized in Table 11.3. The curves were calculated for the outer radius of the pipe.

As seen from Figure 11.16, as the thickness of the coating layer increases, a corresponding decrease in linear phase and group velocity occurs and an increase in attenuation takes place. It should be noted that when dealing with viscoelastic materials, the velocity of the wave group is no longer necessarily the group velocity, but instead is the energy velocity. For regions where attenuation reaches a minimum, there is generally good agreement between the group and energy velocity (Bernard, Lowe, and Deschamps 2001). Furthermore, because Figure 11.16(a) illustrates excellent correspondence between the viscoelastic phase velocity dispersion curves and the equivalent elastic curves, there will be good correspondence between the group velocity curves in the two cases. Considering these points, the curves shown in Figure 11.16(b) are those of the group velocity case.

To verify the trends seen in Figure 11.16, researchers designed an experiment in which two sensors, a transmitter and a receiver, were placed around the circumference of a twenty-four-inch schedule 10 pipe as shown in Figure 11.17. To generate the CSH-waves in the pipe wall, Lorentz-type EMATs were used.

Note that EMATs are not centered at a particular frequency, but rather fixed at a specific wavelength. For this study, an EMAT that generated the *SH0* mode at 130 kHz was used. For this sensor, $\lambda/2$ (and subsequently the magnet width) is approximately equal to 0.5 inches, as determined for a phase velocity of 3.23 mm/µs.

The EMATs were placed on either side of a two-foot-long circumferential section of the pipe, as seen in the photograph in Figure 11.17. Data was collected for the no-coating case and subsequently for coating thicknesses of three mils, seven mils, and thirteen mils. The sensors were not moved at any time during the experiment so that any variation in the amplitude or time of flight of the wave would be attributable to the coating presence.

Figure 11.18 shows the analytic envelopes of the wave packets that have propagated through the bare pipe and pipe with three mils, seven mils, and thirteen mils of coal-tar mastic coating. The envelope amplitudes and changes in time of flight, as compared to the bare pipe case, are summarized in the figure. As the coating thickness increases, a corresponding increase in time of flight and decrease in amplitude occur. The experimentally measured trends agree well with the theoretical predictions. Furthermore, the attenuation is more severe in the case of the thirteen-mil coating, which agrees with the prediction from the attenuation dispersion curves in Figure 11.16(c). The time and amplitude shifts are summarized in Figure 11.19.

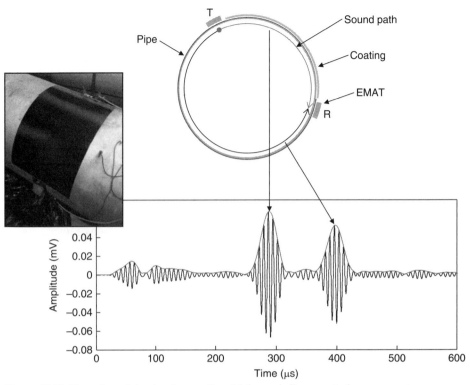

Figure 11.17. Experimental setup for coating thickness influence study.

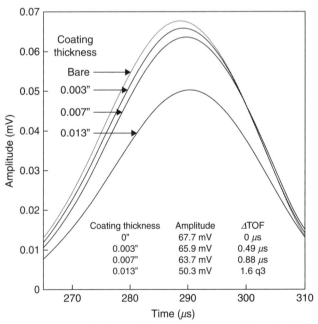

Coating thickness	Amplitude	ΔTOF
0"	67.7 mV	0 μs
0.003"	65.9 mV	0.49 μs
0.007"	63.7 mV	0.88 μs
0.013"	50.3 mV	1.6 q3

Figure 11.18. Analytic envelopes of the wave packets that have traveled through the two-foot-long circumferential section of pipe with different coating thicknesses, showing increased time of flight and decreased amplitude with increasing coating thickness.

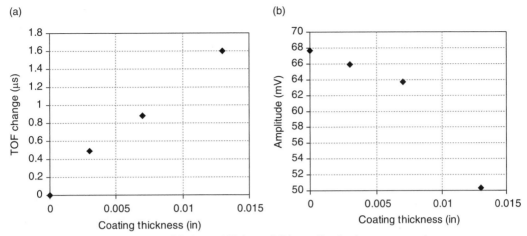

Figure 11.19. Plots showing the (a) time of flight and (b) amplitude change as coating thickness is increased.

According to the theoretical models, the change in time of flight between the bare pipe and thirteen-mil coating cases should be 1.1 μs and the change in amplitude should be approximately 0.24 dB. The actual measured values are larger at 1.6 μs and 2.58 dB, respectively. This is an indication that the actual material properties are slightly different from the published quantities. More important, though, is the fact that the general trends match as the measurement of acoustic properties of coating materials are inherently difficult and can be highly dependent on temperature, age/history of the coating, and variations in the manufacturing process.

Experimentation has verified the general trends of the theoretical models. It has proven that increasing coating thickness causes a reduction in the propagation velocity of a group of waves and that attenuation becomes increasingly severe with coating thickness. As seen from the dispersion curves in Figure 11.16, the increase in time of flight and attenuation could be maximized by moving to a higher frequency, essentially increasing sensitivity to the coating presence.

Practical uses of the circumferential guided wave dispersion problem include the detection and characterization of metal-loss defects in pipe using in-line inspection tools, the detection of disbonded or missing coating on a pipe, the determination of coating material properties, the determination of approximate pipe wall thickness, and the detection of hot-side corrosion or cracking in water-wall and boiler tubing. This concludes the treatment of circumferential guided waves in single-layered and multilayered annuli.

11.6 Exercises

1. Why is wave velocity a function of radius for guided waves propagating in an annulus?
2. If the linear phase velocity dispersion curves have been calculated for the external surface of the thick-walled, small-diameter annulus shown below, by

what factor should the curves be adjusted to determine the linear phase velocity dispersion curves for the internal surface of the annulus?

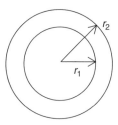

3. What is unique about the Bessel function solutions to the characteristic equations for circumferential guided waves as compared to those for waves propagating in the axial direction of a hollow cylinder?
4. Under what conditions is it acceptable to treat an annulus or other curved surface as a flat plate?
5. What is the fundamental difference between the *SH0* mode in a flat plate and in an annulus?
6. What is the fundamental difference between the *A0* mode in a flat plate and the corresponding wave mode in an annulus?
7. The phase velocity of a 130 kHz circumferential horizontal shear wave propagating in a 0.5-inch thick tube with an outer diameter of 3 inches is determined to be 0.127 in/μs on the inner surface. What is the angular phase velocity of this wave for this particular tube? What is the angular wavenumber of this particular wave?
8. Name several circumferential wave features that could be used to characterize the condition of a protective coating on a pipe and indicate why these features would be relevant based on the theoretical treatment of circumferential waves in a multilayered annulus.
9. Describe a situation in which a numerical solution of the characteristic equation for circumferential guided waves is likely to fail and approximate solution methods may be necessary.
10. When assembling the boundary and continuity conditions for circumferential waves in multilayered structures, what are the relevant stress components for CSH-waves and CLT-waves and why?
11. Why are the multi-layer resonances depicted in Figure 11.10c important in the evaluation of coated metallic annuli?

11.7 REFERENCES

Abramowitz, M., and Stegun, I. A. (1964). *Handbook of Mathematical Functions with Formulas, Graphs, and Mathematical Tables*. New York: Dover Publications.

Barshinger, J. N. (2001). *Guided Wave Propagation in Pipes with Viscoelastic Coatings*. Doctoral Thesis. Department of Engineering Science and Mechanics, Pennsylvania State University.

Barshinger, J. N., and Rose, J. L. (2004). Guided wave propagation in an elastic hollow cylinder coated with a viscoelastic material, *IEEE Transactions on Ultrasonics, Ferroelectrics, and Frequency Control* 51(11): 1547–56.

Bartoli, I., Marzani, A., Lanza di Scalea, F., and Viola, E. (2006). Modeling wave propagation in dampled waveguides of arbitrary cross-section, *J. Sound. Vibr.* 295: 685–707.

Bernard, A., Lowe, M. J. S., and Deschamps, M. (2001). Guided wave energy velocity in absorbing and non-absorbing plates, *The Journal of the Acoustical Society of America* 110(1): 186–96.

Chou, P. C., and Pagano, N. J. (1992). *Elasticity: Tensor, Dyadic, and Engineering Approaches.* New York: Dover Publications.

Graff, K. F. (1991). *Wave Motion in Elastic Solids.* New York: Dover Publications.

Gridin, D., Craster, R. V., Fong, J., Lowe, M. J. S., and Beard, M. (2003). The high-frequency asymptotic analysis of guided waves in a circular elastic annulus, *Wave Motion* 38: 67–90.

Haskell, N. A. (1953). The dispersion of surface waves on multilayered media, *Bulletin of the Seismological Society of America* 43: 17–34.

Hayashi, T., Kawashima, K., Sun, Z., and Rose, J. L. (2003). Analysis of flexural mode focusing by a semi-analytical finite element method, *The Journal of the Acoustical Society of America* 113(3): 1241–8.

Hayashi, T., Song, W.-J., and Rose, J. L. (2003). Guided wave dispersion curves for a bar with an arbitrary cross-section, a rod and rail example, *Ultrasonics* 41: 175–83.

Hayek, S. I. (2001). *Advanced Mathematical Methods in Science and Engineering.* New York: Marcel Dekker.

Jaluria, Y. (1996). *Computer Methods for Engineering.* Bristol, PA: Taylor & Francis.

Knopoff, L. (1964). A matrix method for elastic wave problems, *Bulletin of the Seismological Society of America* 54(1): 431–8.

Liu, G., and Qu, J. (1998a). Guided circumferential waves in a circular annulus, *Journal of Applied Mechanics* 65: 424–30.

(1998b). Transient wave propagation in a circular annulus subjected to transient excitation on its outer surface, *The Journal of the Acoustical Society of America* 104(3): 1210–20.

Lowe, M. J. S. (1995). Matrix techniques for modeling ultrasonic waves in multilayered media, *IEEE Transactions on Ultrasonics, Ferroelectrics, and Frequency Control* 42(4): 525–42.

Morse, P., and Feshbach, H. (1953). *Methods of Theoretical Physics.* New York: McGraw-Hill.

Mu, J., and Rose, J. L. (2008). Guided wave propagation and mode differentiation in hollow cylinders with viscoelastic coatings, *The Journal of the Acoustical Society of America* 124(2): 866–74.

Qu, J., Berthelot, Y. H., and Li, Z. (1996). *Dispersion of Guided Circumferential Waves in a Circular Annulus: Review of Progress in Quantitative NDE.* New York: American Institute of Physics.

Rose, J. L. (1999). *Ultrasonic Waves in Solid Media.* Cambridge, UK: Cambridge University Press.

Sadd, M. H. (2005). *Elasticity: Theory, Applications, and Numerics.* Burlington, MA: Elsevier Butterworth-Heinemann.

Simonetti, F. (2004). Lamb wave propagation in elastic plates coated with viscoelastic materials, *The Journal of the Acoustical Society of America* 115(5): 2041–53.

Thomson, W. T. (1950). Transmission of elastic waves through a stratified solid medium, *Journal of Applied Physics* 21: 89–93.

Überall, H., Hosten, B., Deschamps, M., and Gérard, A. (1994). Repulsion of phase-velocity dispersion curves and the nature of plate vibrations, *The Journal of the Acoustical Society of America* 96(2): 908–17.

Van Velsor, J. K. (2009). *Circumferential Guided Wave in Elastic and Viscoelastic Multilayered Annuli.* Doctoral Thesis. Department of Engineering Science and Mechanics, Pennsylvania State University.

Van Velsor, J. K., Rose, J. L., and Nestleroth, J. B. (2009) Enhanced coating disbond detection capabilities in pipe using circumferential shear horizontal guided waves, *Materials Evaluation* 67(10): 1179–89.

Viktorov, I. A. (1967). *Rayleigh and Lamb Waves, Physical Theory and Applications*. New York: Plenum Press.

Watson, G. N. (1966). *A Treatise on the Theory of Bessel Functions*. London: Cambridge University Press.

Zhao, X., and Rose, J. L. (2004). Guided circumferential shear horizontal waves in an isotropic hollow cylinder, *The Journal of the Acoustical Society of America* 115(5): 1912–16.

12 Guided Waves in Layered Structures

12.1 Introduction

Many engineering structures consist of multiple layers. Examples include plates with coatings, painted structures, diffusion bonded or adhesively bonded structures, ice or contaminant accreted aircraft structures, and laminated composites. To achieve long-distance inspection and monitoring of these structures using ultrasonic guided waves, scholars must study the guided wave propagation characteristics in such structures. This chapter examines the wave propagation problem in layered plate structures consisting of isotropic materials. More advanced studies in complicated structures involving material anisotropy and viscoelasticity are addressed later.

Wave propagation in layered plate structures can be abstracted into several models for different layer thicknesses. When a layer thickness is much larger than the selected wavelength, a half-space model can be used to approximate the thick layer. Interface guided wave modes may exist at the interface between two thick materials. In this case, these two layers are modeled as two half-spaces in the classic problems associated with the Stoneley and Scholte wave solutions. When one of the layers is much thicker than the other layers, guided wave modes exist within the thin layers and the upper region of the thick layer. The thick layer can be modeled as a half-space. One classic problem that falls into this category is the Love wave problem, which studies shear horizontal (SH) guided waves in a layer on a half-space. When all of the layers are of compatible thickness, and the wavelengths of the guided waves are also of a compatible scale, a multilayer model can be established by using finite thicknesses for all of the layers.

Based on studies of classic Stoneley wave, Scholte wave, and Love wave problems, and the numerous references of guided wave literature on layered structures, this chapter is organized into the following sections. Section 12.2 studies interface waves, including waves at a solid–solid interface and waves at a solid–liquid interface. Section 12.3 analyzes guided wave propagation in a layer (layers) on a half-space. Section 12.4 examines the problem of guided wave propagation in multiple layers. All of these sections cover the key topics of governing equations, boundary conditions, characteristic equations, dispersion curves of guided wave propagation, and wave structure analysis. SH wave types and Rayleigh–Lamb type waves are

With significant contribution from Huidong Gao.

studied separately in this chapter. Examples are provided in the text to help students understand the basic concepts. Section 12.5 treats a few special interesting multi-layer situations..

12.2 Interface Waves

This section presents an introduction to two types of interface guided waves: (1) at a solid–solid interface via Stoneley waves, and (2) along a solid–liquid interface via Scholte waves. The basic equations are derived in this section and exercises are provided for students.

12.2.1 Waves at a Solid–Solid Interface: Stoneley Wave

Consider two elastic half-spaces bonded together along the plane $x_3 = 0$ as sketched in Figure 12.1. The terms $\rho^{(m)}$, $\lambda^{(m)}$, and $\mu^{(m)}$, $(m = 1, 2)$ denote the mass density and elastic properties of each half-space. Stoneley first solved the problem of determining whether harmonic wave propagation was possible along the interface in 1924. The solution procedure is described as follows.

The wave motion equation in each material is expressed in Equation (12.1). Here the superscript with parenthesis means material 1 and material 2, respectively.

$$(\lambda^{(1)} + \mu^{(1)})u_{j,ji}^{(1)} + \mu^{(1)}u_{i,jj}^{(1)} = \rho^{(1)}\ddot{u}_i^{(1)} \quad i, j = 1, 2, 3$$

$$(\lambda^{(2)} + \mu^{(2)})u_{j,ji}^{(2)} + \mu^{(2)}u_{i,jj}^{(2)} = \rho^{(2)}\ddot{u}_i^{(2)} \quad i, j = 1, 2, 3$$

$$(12.1)$$

The trial wave solutions in each half-space can be expressed as a combination of partial waves based on the partial wave theory discussed earlier in this book. To study interface Stoneley waves, only those partial waves propagating or decaying from the interface will be involved. In the case of wave propagation mode studies, no incident waves from the half-space toward the interface will be considered. Therefore, only

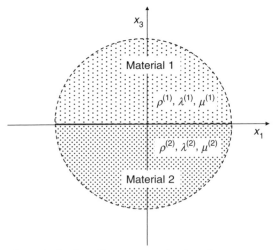

Figure 12.1. Coordinate system for interface wave studies.

partial waves with positive wavenumbers will be considered for material 1, and partial waves with negative wavenumbers will be considered for material 2. The detailed expressions of the partial waves are expressed in Equation (12.2). Here, α_1, α_2, α_3, and α_4 are the ratio of the wavenumbers in the x_3-direction to the wavenumbers in the x_1-direction. These are derived from the Christoffel's equation.

$$\alpha_1 = \sqrt{-1 + \frac{\rho^{(1)}}{\mu^{(1)}}c_p^2}, \ \alpha_2 = \sqrt{-1 + \frac{\rho^{(1)}}{\lambda^{(1)} + 2\mu^{(1)}}c_p^2}, \text{for material 1}$$

$$\alpha_3 = -\sqrt{-1 + \frac{\rho^{(2)}}{\mu^{(2)}}c_p^2}, \ \alpha_4 = -\sqrt{-1 + \frac{\rho^{(2)}}{\lambda^{(2)} + 2\mu^{(2)}}c_p^2}, \text{for material 2}$$

(12.2)

According to the analysis in earlier chapters, the polarization vectors of the partial waves can be expressed as follows.

$$U_{31} = \frac{-1}{\alpha_1}, \ U_{32} = \alpha_2, \ U_{33} = \frac{-1}{\alpha_3}, \ U_{34} = \alpha_4$$

(12.3)

The wave field solutions for displacements are expressed in Equation (12.4) as a combination of partial waves.

$$u_1^{(1)} = \sum_{n=1}^{2} B_n e^{ik(x_1 + \alpha_n x_3 - c_p t)}$$

$$u_3^{(1)} = \sum_{n=1}^{2} B_n U_{3n} e^{ik(x_1 + \alpha_n x_3 - c_p t)}$$

$$u_1^{(2)} = \sum_{n=3}^{4} B_n e^{ik(x_1 + \alpha_n x_3 - c_p t)}$$

$$u_3^{(2)} = \sum_{n=3}^{4} B_n U_{3n} e^{ik(x_1 + \alpha_n x_3 - c_p t)}$$

(12.4)

The stress field can be obtained from elastic theory using the constitutive relation and the strain displacement relation. The detailed expressions are listed in Equation (12.5).

$$\sigma_{33}^{(1)} = \sum_{n=1}^{2} B_n [\lambda^{(1)} + (\lambda^{(1)} + 2\mu^{(1)})\alpha_n U_{3n}](ik)e^{ik(x_1 + \alpha_n x_3 - c_p t)}$$

$$\sigma_{31}^{(1)} = \sum_{n=1}^{2} B_n [\alpha_n + U_{3n}]\mu^{(1)}(ik)e^{ik(x_1 + \alpha_n x_3 - c_p t)}$$

$$\sigma_{33}^{(2)} = \sum_{n=3}^{4} B_n [\lambda^{(2)} + (\lambda^{(2)} + 2\mu^{(2)})\alpha_n U_{3n}](ik)e^{ik(x_1 + \alpha_n x_3 - c_p t)}$$

$$\sigma_{31}^{(2)} = \sum_{n=3}^{4} B_n [\alpha_n + U_{3n}]\mu^{(2)}(ik)e^{ik(x_1 + \alpha_n x_3 - c_p t)}$$

(12.5)

When a perfect bonding condition is considered at the interface of the two half-spaces, the boundary conditions are listed in Equation (12.6).

$$u_1^{(1)} = u_1^{(2)}$$

$$u_3^{(1)} = u_3^{(2)}$$

$$\sigma_{33}^{(1)} = \sigma_{33}^{(2)}$$

$$\sigma_{31}^{(1)} = \sigma_{31}^{(2)}$$

(12.6)

Assembling the four boundary conditions using Equation (12.4) and Equation (12.5), the following characteristic equation can be obtained.

$$
\begin{bmatrix}
1 & 1 & -1 & -1 \\
\frac{-1}{\alpha_1} & \alpha_2 & \frac{1}{\alpha_3} & -\alpha_4 \\
-2\mu^{(1)} & (\lambda^{(1)} + 2\mu^{(1)})\alpha_2^2 + \lambda^{(1)} & 2\mu^{(2)} & -[(\lambda^{(2)} + 2\mu^{(2)})\alpha_4^2 + \lambda^{(2)}] \\
\frac{\alpha_1^2 - 1}{\alpha_1}\mu^{(1)} & 2\alpha_2\mu^{(1)} & -\frac{\alpha_3^2 - 1}{\alpha_3}\mu^{(2)} & -2\alpha_4\mu^{(2)}
\end{bmatrix}
\begin{bmatrix} B_1 \\ B_2 \\ B_3 \\ B_4 \end{bmatrix}
=
\begin{bmatrix} 0 \\ 0 \\ 0 \\ 0 \end{bmatrix}
$$

(12.7)

Similar to the problem studied for Lamb wave propagation for a simple traction-free isotropic plate, this is again a homogeneous linear system of equations. To have nontrivial solutions of B_k, the determinant of the coefficient matrix must be zero. This will lead to a transcendental equation with phase velocity as an unknown parameter when the properties of the two materials are given. Because no frequency term appears in Equation (12.7), the phase velocity solution of Stoneley wave problems does not vary with frequency. Therefore, Stoneley waves are nondispersive.

The wave structure of a Stoneley wave mode can also be obtained by solving Equation (12.7). It is important to note that, although phase velocity is independent of frequency, the wave structure of a Stoneley wave does vary with frequency. The wave structure of a Stoneley wave mode is usually plotted as a function of x_3 / Λ because the distribution of this quantity is frequency independent. Here, Λ is the wavelength. In this example, the Stoneley wave at a tungsten-aluminum interface is studied. The material property of tungsten and aluminum are listed in the following table:

Material	ρ (g/cm³)	λ (GPa)	μ (GPa)
Tungsten	19.25	199.4	158.56
Aluminum	2.7	55.28	25.94

A numerical method is used to search the roots of Equation (12.7). The phase velocity that makes the determinant equal to zero is 2.77 km/s. The corresponding wave structure showing the displacement profile of the Stoneley wave mode is shown in Figure 12.2(a). Using Equation (12.5), the stress distribution profile can also be obtained. The curves plotted in Figure 12.2(b) show that most of the stress is concentrated in the half-space of tungsten.

It is important to note that Equation (12.7) may not have a real solution for phase velocity. Therefore, a Stoneley wave mode may not exist for some material

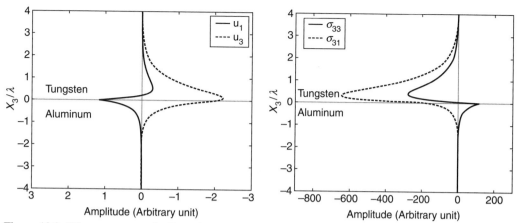

Figure 12.2. Wave structure of a Stoneley wave mode at the interface between tungsten and aluminum. (a) displacement, (b) stress.

combinations. As an example, when aluminum is bonded to titanium, no Stoneley wave mode is observed through numerical simulation. A detailed study of the conditions of the material properties for the existence of a Stoneley wave is beyond the scope of this section. Further information can be found in Auld (1990), Achenbach and Epstein (1967), and Miklowitz (1978).

The generation of Stoneley waves is similar to the case of generating a surface wave or a Lamb wave. According to Snell's law, the incident angle can be calculated as:

$$\frac{c_I}{\sin(\theta)} = c_{stoneley}$$

Note that nonattenuative Stoneley modes exist only for limited material combinations. However, for other material combinations various attenuative mode types *do* exist. There are attenuative interface (Stoneley-type) waves as well as leaky Rayleigh-type waves. See Pilant (1972) for details.

12.2.2 Waves at a Solid–Liquid Interface: Scholte Wave

Wave propagation at a solid–liquid interface is studied in this section. This problem is named after Scholte for his first investigation. This problem can be solved by substituting the upper half-space in the Stoneley wave problem with the material property of the liquid by assuming in this case that $\mu^{(1)} = 0$. The Christoffel equation in the liquid half-space reduces to Equation (12.8) when $\mu^{(1)} = 0$ is substituted into the standard Christoffel equation.

$$\begin{bmatrix} \lambda - \rho c_p^2 & \lambda\alpha \\ \lambda\alpha & \lambda\alpha^2 - \rho c_p^2 \end{bmatrix} \begin{Bmatrix} U_1 \\ U_3 \end{Bmatrix} = \begin{bmatrix} 0 \\ 0 \end{bmatrix} \tag{12.8}$$

Therefore, the characteristic equation for a nontrivial partial wave solution is Equation (12.9).

$$\alpha^2 = -1 + \frac{\rho}{\lambda} c_p^2 \tag{12.9}$$

This means only two partial waves exist in a liquid medium. The corresponding polarization vectors for the partial waves are shown in Equation (12.10).

$$\left\{ \begin{matrix} U_1 \\ U_3 \end{matrix} \right\} \Big|_{1,2} = \left\{ \begin{matrix} 1 \\ \alpha \end{matrix} \right\} \Big|_{1,2} \tag{12.10}$$

Use the same assumption as that used in solving the Stoneley wave problem – that there is no incident wave from the outer layer to the interface. Only the partial waves propagating outward from the interface will be considered here. Therefore, there are three partial waves in total. In addition, at the solid–liquid interface, the continuity of the u_1 displacement is not required. Therefore, the wave fields can be expressed in Equation (12.11).

$$u_3^{(1)} = B_1 U_{31} e^{ik(x_1 + \alpha_1 x_3 - c_p t)}$$

$$u_3^{(2)} = \sum_{n=2}^{3} B_n U_{3n} e^{ik(x_1 + \alpha_n x_3 - c_p t)}$$

$$\sigma_{33}^{(1)} = B_1 [\lambda^{(1)} + (\lambda^{(1)} + 2\mu^{(1)}) \alpha_1 U_{31}](ik) e^{ik(x_1 + \alpha_1 x_3 - c_p t)}$$

$$\sigma_{31}^{(1)} = B_1 [\alpha_1 + U_{31}] \mu^{(1)} (ik) e^{ik(x_1 + \alpha_1 x_3 - c_p t)} \tag{12.11}$$

$$\sigma_{33}^{(2)} = \sum_{n=2}^{3} B_n [\lambda^{(2)} + (\lambda^{(2)} + 2\mu^{(2)}) \alpha_n U_{3n}](ik) e^{ik(x_1 + \alpha_n x_3 - c_p t)}$$

$$\sigma_{31}^{(2)} = \sum_{n=2}^{3} B_n [\alpha_n + U_{3n}] \mu^{(2)} (ik) e^{ik(x_1 + \alpha_n x_3 - c_p t)}$$

Assembling the boundary conditions, the following characteristic equation can be reached.

$$\begin{bmatrix} \alpha_1 & \frac{1}{\alpha_2} & -\alpha_3 \\ (\lambda^{(1)})\alpha_1^2 + \lambda^{(1)} & 2\mu^{(2)} & -[(\lambda^{(2)} + 2\mu^{(2)})\alpha_3^2 + \lambda^{(2)}] \\ 0 & -\frac{\alpha_2^2 - 1}{\alpha_2} \mu^{(2)} & -2\alpha_3 \mu^{(2)} \end{bmatrix} \begin{bmatrix} B_1 \\ B_2 \\ B_3 \end{bmatrix} = \begin{bmatrix} 0 \\ 0 \\ 0 \end{bmatrix} \tag{12.12}$$

The phase velocity of a Scholte wave mode can be obtained by letting the determinant of the coefficient matrix be zero. The wave structure can be calculated after the unknown variable B_k is solved from Equation (12.12). Consider the Scholte wave at the interface of water and aluminum. The material properties are listed in the following table.

Material	ρ (g/cm³)	λ (GPa)	μ (GPa)
Water	1	2.25	0
Aluminum	2.7	55.28	25.94

Using these material properties in Equation (12.8) through Equation (12.12), the phase velocity of the Scholte wave in this case is 1.496 km/s, which is less than

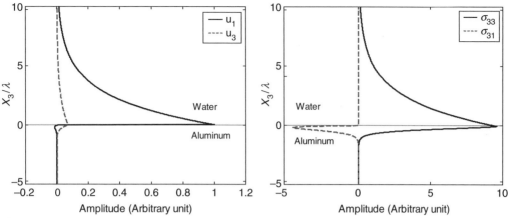

Figure 12.3. Wave structure of a Scholte wave mode at the interface between water and aluminum. (a) displacement, (b) stress.

the bulk wave velocities $(c_L^{(1)}, c_L^{(2)}, c_T^{(2)})$ in the two media. The corresponding wave structure is shown in Figure 12.3. The result indicates that the displacement in the x_1-direction is not continuous at the interface. In addition, the wave field in the water space penetrates deeper than that in the aluminum half-space. The Scholte wave mode carries almost all the energy in the liquid layer rather than in the solid. Hence, Scholte waves are seldom used in nondestructive evaluation (NDE).

12.3 Waves in a Layer on a Half-Space

According to the studies of Lamb and SH-waves, the partial waves polarized in the x_2-direction are totally separable from the partial waves polarized in the (x_1, x_3) plane for an isotropic medium. Therefore, the waves polarized in the (x_1, x_3) plane are generally called Rayleigh–Lamb type guided waves, or R-L type. The waves polarized in the x_2-direction are called SH type guided waves, or SH type. These two types of waves in a layer on a half-space are studied in sections 12.3.1 and 12.3.2, respectively.

12.3.1 Rayleigh–Lamb Type Waves

The sketch of a layer on a half-space is shown in Figure 12.4. Early results for such waves propagating in the x_1-direction were published by Sezawa (1927) and Fu (1946). The thickness of the layer is indicated by h; the density and Lamé constants of these two materials are noted with $\rho^{(1)}, \lambda^{(1)}, \mu^{(1)}$ and $\rho^{(2)}, \lambda^{(2)}, \mu^{(2)}$, respectively.

The R-L type guided wave propagating in this structure can also be solved using partial wave theory and appropriate boundary conditions. The number of partial waves in the layer and the half-space are four and two, respectively. The solutions of the displacement and stress field in the layer are expressed in Equation (12.13). Here, the formulas of α_n and U_{3n} are described in Equation (12.2) and

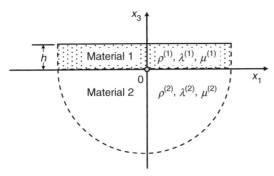

Figure 12.4. Coordinate system for the guided wave propagation study of a layer on a half-space.

Equation (12.3). The wave field solutions in the half-space are expressed in Equation (12.14).

$$u_1^{(1)} = \sum_{n=1}^{4} B_n e^{ik(x_1 + \alpha_n x_3 - c_p t)} \tag{12.13}$$

$$u_3^{(1)} = \sum_{n=1}^{4} B_n U_{3n} e^{ik(x_1 + \alpha_n x_3 - c_p t)}$$

$$\sigma_{33}^{(1)} = \sum_{n=1}^{4} B_n [\lambda^{(1)} + (\lambda^{(1)} + 2\mu^{(1)})\alpha_n U_{3n}](ik) e^{ik(x_1 + \alpha_n x_3 - c_p t)}$$

$$\sigma_{31}^{(1)} = \sum_{n=1}^{4} B_n [\alpha_n + U_{3n}]\mu^{(1)}(ik) e^{ik(x_1 + \alpha_n x_3 - c_p t)}$$

$$u_1^{(2)} = \sum_{n=5}^{6} B_n e^{ik(x_1 + \alpha_n x_3 - c_p t)} \tag{12.14}$$

$$u_3^{(2)} = \sum_{n=5}^{6} B_n U_{3n} e^{ik(x_1 + \alpha_n x_3 - c_p t)}$$

$$\sigma_{33}^{(2)} = \sum_{n=5}^{6} B_n [\lambda^{(2)} + (\lambda^{(2)} + 2\mu^{(2)})\alpha_n U_{3n}](ik) e^{ik(x_1 + \alpha_n x_3 - c_p t)}$$

$$\sigma_{31}^{(2)} = \sum_{n=5}^{6} B_n [\alpha_n + U_{3n}]\mu^{(1)}(ik) e^{ik(x_1 + \alpha_n x_3 - c_p t)}$$

The following boundary conditions must be satisfied for this structure, as demonstrated in Equation (12.15):

$$\begin{aligned}
\sigma_{33}^{(1)} &= 0 &&\text{at} \quad x_3 = h \\
\sigma_{31}^{(1)} &= 0 &&\text{at} \quad x_3 = h \\
u_1^{(1)} &= u_1^{(2)} &&\text{at} \quad x_3 = 0 \\
u_3^{(1)} &= u_3^{(2)} &&\text{at} \quad x_3 = 0 \\
\sigma_{33}^{(1)} &= \sigma_{33}^{(2)} &&\text{at} \quad x_3 = 0 \\
\sigma_{31}^{(1)} &= \sigma_{31}^{(2)} &&\text{at} \quad x_3 = 0
\end{aligned} \tag{12.15}$$

By substituting Equation (12.14) into Equation (12.15), the characteristic equation of the system is obtained and shown in Equation (12.16).

$$\mathbf{DB} = 0 \tag{12.16}$$

Here, $\mathbf{D}$ is a six by six matrix, whose nonzero components are listed in Equation (12.17).

$$
\begin{aligned}
D_{1,n} &= [\lambda^{(1)} + (\lambda^{(1)} + 2\mu^{(1)})\alpha_n U_{3n}](ik)e^{ik\alpha_n h}, \quad n = 1,2,3,4 \\
D_{2,n} &= [\alpha_n + U_{3n}]\mu^{(1)}(ik)e^{ik\alpha_n h}, \quad n = 1,2,3,4 \\
D_{3,n} &= 1, \quad n = 1,2,3,4 \\
D_{3,n} &= -1, \quad n = 5,6 \\
D_{4,n} &= U_{3n}, \quad n = 1,2,3,4 \\
D_{4,n} &= -U_{3n}, \quad n = 5,6 \\
D_{5,n} &= [\lambda^{(1)} + (\lambda^{(1)} + 2\mu^{(1)})\alpha_n U_{3n}](ik), \quad n = 1,2,3,4 \\
D_{5,n} &= -[\lambda^{(2)} + (\lambda^{(2)} + 2\mu^{(2)})\alpha_n U_{3n}](ik), \quad n = 5,6 \\
D_{6,n} &= [\alpha_n + U_{3n}]\mu^{(1)}(ik), \quad n = 1,2,3,4 \\
D_{6,n} &= -[\alpha_n + U_{3n}]\mu^{(2)}(ik), \quad n = 5,6
\end{aligned}
\tag{12.17}
$$

For $\mathbf{B}$ to have nontrivial solutions, the determinant of the coefficient matrix needs to be zero, as in Equation (12.18).

$$|\mathbf{D}| = 0 \tag{12.18}$$

This is a transcendental equation, whose elements depend on both wavenumber and wave velocity. Therefore, the solution of phase velocity is frequency dependent, and wave propagation is dispersive. Similar to the Lamb wave situation, a numerical method is needed to obtain the dispersion curves for this structure. For each wave mode, the wave structures can be calculated from Equation (12.16). This sample problem studies the guided wave propagation characteristics when a thin Plexiglas layer is bonded to a thick aluminum block. Material properties are listed in the following table.

Material	ρ (g/cm³)	λ (GPa)	μ (GPa)	c_L (km/s)	c_T (km/s)	c_R (km/s)
Plexiglas	1.18	5.45	1.48	2.67	1.12	1.05
Aluminum	2.7	55.28	25.94	6.3	3.1	2.89

By using the computational procedure described in Equations (12.13) to (12.18), guided wave dispersion curves can be calculated and are shown in Figure 12.5. The phase velocity values that can lead to guided wave propagation modes are confined over a certain range. When the phase velocity is larger than the shear wave velocity of the half-space, wave energy will leak into the half-space. The guided wave mode with attenuation (complex phase velocity) is not considered here as we are only interested initially in non-leaky wave modes. The upper bound of the phase velocity

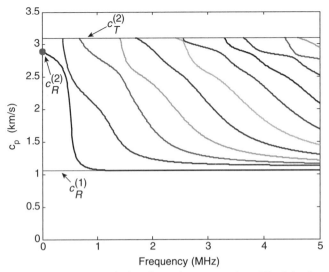

Figure 12.5. Phase velocity dispersion curves for a Plexiglas layer on aluminum half-space. The layer thickness is 1 mm. Material (1) Plexiglas and, (2) Aluminum.

value shown in Figure 12.5 is the shear wave velocity in the aluminum half-space. The lower limit of the phase velocity is the Rayleigh wave velocity in the Plexiglas layer. At the low frequency limit, the wavelength of the guided wave is much larger than the thickness of the Plexiglas layer. The influence of the Plexiglas layer on wave propagation is very small. Therefore, the corresponding wave velocity approaches the Rayleigh surface wave in aluminum.

Similar to the case of Lamb waves in a single-layered plate, the thickness of the layer is always associated with the wavenumber k in Equation (12.16). Therefore, the dispersion curve can also be expressed using $f \cdot h$ as a variable. Curves could be plotted against $f \cdot h$ or f for a particular h.

Because the phase velocity of the guided wave modes changes with frequency, a group velocity dispersion curve will be used to estimate the arrival time of a guided wave package as we did in the case of Lamb waves. Group velocity dispersion curves can be calculated using Equation (12.19):

$$c_g = \frac{d\omega}{d\xi} = \frac{1}{\dfrac{d\left(\dfrac{\omega}{c_p}\right)}{d\omega}} = \frac{c_p^2}{c_p - \omega\dfrac{dc_p}{d\omega}} = \frac{c_p^2}{c_p - f\dfrac{dc_p}{df}} \qquad (12.19)$$

The corresponding group velocity dispersion curves for this example of Plexiglas on aluminum is shown in Figure 12.6.

In this example, we will study the wave propagation in a hard layer on a soft half-space, for example, 1 mm-thick aluminum plate on a Plexiglas block. Using the procedure just discussed, the phase velocity dispersion curve can be calculated and is shown in Figure 12.7. It is evident that the guided wave dispersion curve in this case is dramatically different than it is in the case of a Plexiglas layer on an aluminum half-space. The upper bound of the phase velocity in this case is the shear wave velocity in the Plexiglas. The phase velocity approximates the Rayleigh wave

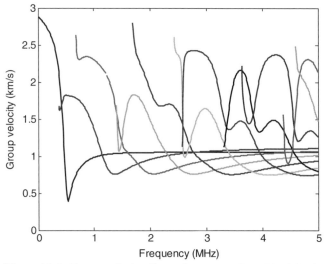

Figure 12.6. Group velocity dispersion curves for a Plexiglas layer on aluminum half-space. The layer thickness is 1 mm.

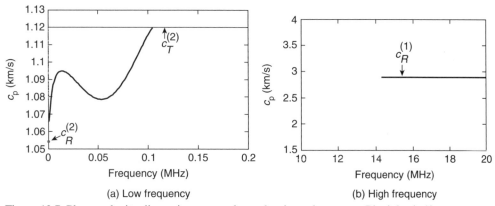

(a) Low frequency   (b) High frequency

Figure 12.7. Phase velocity dispersion curves for a aluminum layer on a Plexiglas half-space. The layer thickness is 1 mm.

velocity in Plexiglas at low frequency. In addition, when the frequency is higher than about 0.1 MHz, no non-leaky guided wave exists in this structure. It is also worthwhile to mention that Rayleigh surface waves on the aluminum surface exist when the frequency is high enough that the wavelength of the mode is much less than the layer thickness. A portion of the dispersion curve in this case is shown in Figure 12.7(b).

12.3.2 Love Waves

The guided wave mode in a layer on a half-space polarized in the x_2-direction is called a Love wave. See Figure 12.4. Based on the partial wave analysis of SH-waves in this book and the discussion of partial waves in a half-space, two partial waves exist in the upper layer and one partial wave exists in the half-space. The expressions for field quantities are listed in Equation (12.20).

$$u_2^{(1)} = \sum_{n=1}^{2} B_n e^{ik(x_1 + \alpha_n x_3 - c_p t)}$$

$$\sigma_{32}^{(1)} = \sum_{n=1}^{2} B_n \alpha_n \mu^{(1)} (ik) e^{ik(x_1 + \alpha_n x_3 - c_p t)} \qquad (12.20)$$

$$u_2^{(2)} = B_3 e^{ik(x_1 + \alpha_3 x_3 - c_p t)}$$

$$\sigma_{32}^{(2)} = B_3 \alpha_3 \mu^{(2)} (ik) e^{ik(x_1 + \alpha_3 x_3 - c_p t)}$$

Detailed expressions of α_n are included in Equation (12.21).

$$\alpha_1 = \sqrt{\frac{\rho^{(1)} c_p^2}{\mu^{(1)}} - 1}$$

$$\alpha_2 = -\sqrt{\frac{\rho^{(1)} c_p^2}{\mu^{(1)}} - 1} \qquad (12.21)$$

$$\alpha_1 = -\sqrt{\frac{\rho^{(2)} c_p^2}{\mu^{(2)}} - 1}$$

The boundary conditions are listed in Equation (12.22).

$$\begin{aligned}
\sigma_{32}^{(1)} &= 0 & \text{at} \quad x_3 &= h \\
u_2^{(1)} &= u_2^{(2)} & \text{at} \quad x_3 &= 0 \qquad (12.22) \\
\sigma_{32}^{(1)} &= \sigma_{32}^{(2)} & \text{at} \quad x_3 &= 0
\end{aligned}$$

Therefore, the characteristic equation of the Love wave problem is expressed in Equation (12.23).

$$\begin{bmatrix} \alpha_1 \mu^{(1)} e^{ik\alpha_1 h} & \alpha_2 \mu^{(1)} e^{ik\alpha_1 h} & 0 \\ 1 & 1 & -1 \\ \alpha_1 \mu^{(1)} & \alpha_2 \mu^{(1)} & -\alpha_3 \mu^{(2)} \end{bmatrix} \begin{bmatrix} B_1 \\ B_2 \\ B_3 \end{bmatrix} = \begin{bmatrix} 0 \\ 0 \\ 0 \end{bmatrix} \qquad (12.23)$$

A numerical solution method is also required to solve this equation for resulting dispersion curves and wave structures. When a 1 mm-thick Plexiglas layer is bonded to an aluminum substrate, the Love wave dispersion curves are shown in Figure 12.8. The phase velocity for a non-leaky Love wave mode with real wavevector is bounded between the shear wave velocity of the layer and that of the half-space. The relation is shown in mathematical form in Equation (12.24).

$$c_T^{(1)} < c_p < c_T^{(2)} \qquad (12.24)$$

If the shear wave velocity in the layer were higher than in the half-space, there would be no real roots for c_p using Equation (12.23). Therefore, a non-leaky Love wave mode would not exist in this structure.

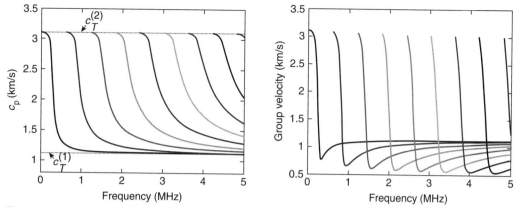

Figure 12.8. Love wave dispersion curve for a 1 mm-thick Plexiglas on an aluminum half-space. (a) phase velocity curves (b) group velocity curves.

12.4 Waves in Multiple Layers

Guided waves in multiple layers can be used in a number of practical situations. Examples include coating problems of plasma spray on a turbine blade, painted structures, aircraft multiple layers (either bonded or with sealant layers), diffusion bonded or adhesively bonded structures in general, and ice or contaminant detection via multiple layers of solid and fluids.

A sketch of a multilayered structure is shown in Figure 12.9. There are N layers in the structure. The thickness, density, and Lamé constants of each layer are denoted by $h^{(n)}$, $\rho^{(n)}$, $\lambda^{(n)}$, and $\mu^{(n)}$, respectively. A Cartesian coordinate system is also shown in the figure, where waves are assumed to propagate along the x_1-direction, and the thickness of the structure is along the x_3-direction. The x_2-direction is perpendicular to the (x_1, x_3) plane. In the case of R-L type wave propagation, the oscillation of the particles is in the (x_1, x_3) plane; for SH wave modes, the oscillation is in the x_2-direction. Two types of boundary condition assembly methods are introduced in this section. One is a global matrix method, and the other is a transfer matrix method.

Scholars often employ two popular approaches for obtaining solutions to wave propagation problems in multilayer structures. A single global matrix can easily be constructed for any N numbers of layers (elastic, viscoelastic, isotropic, anisotropic) by following the technique described here. This method remains numerically robust for any range of fd values, but often involves root extraction from a large-order determinant. Numerical implementation of the transfer matrix may thus be faster, but we prefer the robustness of the global matrix method for a larger class of problem parameters.

The second approach, the transfer matrix method, was introduced in Thomson (1950) and Haskell (1953). The basic idea consists of eliminating all unknowns introduced by the intermediate layers; as a result, the solution to the problem for all layers is expressed in terms of the external boundary conditions. The elimination algorithm can be described by matrix multiplication, where each matrix represents the interface boundary conditions for the intermediate layer. For large fd values, when the waves on the top layers have little influence on those on the bottom layers, the transfer matrix becomes ill conditioned, and so a numerical solution of the

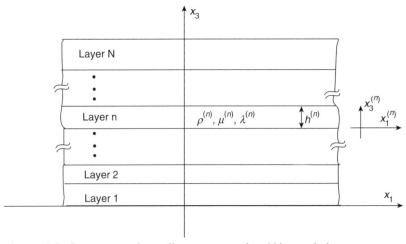

Figure 12.9. Geometry and coordinate system of an *N*-layered plate.

characteristic equations becomes unstable. References to the different approaches for overcoming this numerical problem can be found in Lowe (1995).

12.4.1 The Global Matrix Method

A global matrix method can be derived by two different methods, one associated with satisfying Navier's displacement equations of motion for each layer, utilization of harmonic assumed solutions, application of Helmholtz decomposition useful for isotropic materials, use of elasticity theory and appropriate boundary conditions and continuity of displacement and stress components across each interface, and Snell's law when the four partial waves in each layer have the same value of the projection of the wavevector on the x_1-direction or wavevector direction. This leads to a homogenous system of equations where eigenvalues can be found to eventually produce the phase velocity dispersion curves, and eigenvectors the corresponding wave structures. The four layers lead to a sixteen by sixteen matrix because of sixteen boundary conditions and continuity equations. This approach is presented in Rose (1999).

The second approach uses a partial wave technique that can be used for both isotropic and anisotropic materials. Using the partial wave theory described in Chapter 6, the ultrasonic wave field in an elastic medium can be expressed as a linear combination of partial wave solutions obtained from the Christoffel's equation in Equation (12.25).

$$\begin{bmatrix} (\lambda+2\mu)+\mu\alpha^2-\rho c_p{}^2 & (\lambda+\mu)\alpha \\ (\lambda+\mu)\alpha & (\lambda+2\mu)\alpha^2+\mu-\rho c_p{}^2 \end{bmatrix} \begin{Bmatrix} U_1 \\ U_3 \end{Bmatrix} = 0 \qquad (12.25)$$

The relation of recombination is described in Equation (12.26) for the nth layer.

$$u_1^{(n)} = \sum_{m=4n-3}^{4n} B_m e^{ik(x_1+\alpha_m x_3-c_p t)}$$

$$u_3^{(n)} = \sum_{m=4n-3}^{4n} B_m U_{3m} e^{ik(x_1+\alpha_m x_3-c_p t)}$$

$$\sigma_{33}^{(n)} = \sum_{m=4n-3}^{4n} B_m[\lambda^{(n)} + (\lambda^{(n)} + 2\mu^{(n)})\alpha_m U_{3m}](ik)e^{ik(x_1 + \alpha_m x_3 - c_p t)}$$

$$\sigma_{31}^{(n)} = \sum_{m=4n-3}^{4n} B_m[\alpha_m + U_{3m}]\mu^{(n)}(ik)e^{ik(x_1 + \alpha_m x_3 - c_p t)} \tag{12.26}$$

After the Christoffel's equation is solved for all the N layers, there are $4N$ unknown combination coefficients $B_m, m = 1, 2, \cdots, 4N$ for R-L type wave propagation. Because the thickness of each layer is $h^{(n)}$, it is convenient to present the position of the top surface of layer n as $H^{(n)} = \sum_{m=1}^{n} h^{(m)}$. The boundary conditions for the N-layered structure are listed in Equation (12.27).

$$\sigma_{33}^{(1)} = 0, \ \sigma_{31}^{(1)} = 0, \ \text{at} \ x_3 = 0$$

$$u_3^{(n)} = u_3^{(n-1)}, u_1^{(n)} = u_1^{(n-1)}, \ \text{at} \ x_3 = H^{(n-1)}, \ n = 2, 3, \cdots, N$$

$$\sigma_{33}^{(n)} = \sigma_{33}^{(n-1)}, \ \sigma_{31}^{(n)} = \sigma_{31}^{(n-1)}, \ \text{at} \ x_3 = H^{(n-1)}, \ n = 2, 3, \cdots, N \tag{12.27}$$

$$\sigma_{33}^{(N)} = 0, \ \sigma_{31}^{(N)} = 0, \ \text{at} \ x_3 = H^{(N)}$$

The first line of Equation (12.27) is the stress-free boundary condition at the bottom surface of the structure. The second and third lines are the displacement and the stress continuity relation at the layer interfaces, respectively. The fourth line is the stress-free boundary condition at the top surface of the structure. There are two boundary conditions at the bottom surface, two boundary conditions at the top surface, and four boundary conditions for each interface. Therefore, the total number of boundary condition is $2 + 4(N - 1) + 2 = 4N$.

Several examples of the boundary condition derivation are shown in the following.

σ_{33} *Stress free at bottom surface*

Four partial waves involved in the first layer are represented as: $B_m, m = 1, 2, 3, 4$. At the bottom surface, $x_3 = 0$. Therefore, the boundary condition can be expanded to Equation (12.28).

$$[(\lambda^{(1)} + (\lambda^{(1)} + 2\mu^{(1)})\alpha_1 U_{31})B_1 + (\lambda^{(1)} + (\lambda^{(1)} + 2\mu^{(1)})\alpha_2 U_{32})B_2$$
$$+ (\lambda^{(1)} + (\lambda^{(1)} + 2\mu^{(1)})\alpha_3 U_{33})B_3 + (\lambda^{(1)} + (\lambda^{(1)} + 2\mu^{(1)})\alpha_4 U_{34})B_4](ik)e^{ik(x_1 - c_p t)} = 0$$

$$\tag{12.28}$$

Notice that, for a propagating guided wave solution, $e^{ik(x_1 - c_p t)} \neq 0$. Equation (12.29) can be obtained by neglecting this common term.

$$[(\lambda^{(1)} + (\lambda^{(1)} + 2\mu^{(1)})\alpha_1 U_{31})B_1 + (\lambda^{(1)} + (\lambda^{(1)} + 2\mu^{(1)})\alpha_2 U_{32})B_2$$
$$+ (\lambda^{(1)} + (\lambda^{(1)} + 2\mu^{(1)})\alpha_3 U_{33})B_3 + (\lambda^{(1)} + (\lambda^{(1)} + 2\mu^{(1)})\alpha_4 U_{34})B_4](ik) = 0 \tag{12.29}$$

u_1 *continuity at the interface of layer n and layer n-1.*

Eight partial waves are involved in this equation, four for layer n and four for layer *n-1*. Neglecting the common term $e^{ik(x_1-c_pt)}$, the relation is expressed in Equation (12.30).

$$e^{ik(\alpha_{4n-7}H^{(n-1)})}B_{4n-7} + e^{ik(\alpha_{4n-6}H^{(n-1)})}B_{4n-6} + e^{ik(\alpha_{4n-5}H^{(n-1)})}B_{4n-5} + e^{ik(\alpha_{4n-4}H^{(n-1)})}B_{4n-4}$$
$$-e^{ik(\alpha_{4n-3}H^{(n-1)})}B_{4n-3} - e^{ik(\alpha_{4n-2}H^{(n-1)})}B_{4n-2} - e^{ik(\alpha_{4n-1}H^{(n-1)})}B_{4n-1} - e^{ik(\alpha_{4n}H^{(n-1)})}B_{4n} = 0$$

$$(12.30)$$

With a similar procedure, other boundary conditions listed in Equation (12.27) can all be expressed in terms of the unknown coefficients B_k. The details are reserved for student exercises.

In the global matrix method, all the boundary conditions are assembled into a single linear system of equations, which is also called the characteristic equation. In the characteristic equation, coefficients B_m are the unknown variables, and phase velocity c_p and the wavenumber k are parameters in the matrix **D**.

$$\mathbf{DB} = 0 \qquad (12.31)$$

Example: Specific Configurations of Four-Layer Structures

When the number of layers $N = 4$, the coefficient matrix **D** is a sixteen by sixteen matrix. See Equation (12.32). The components of the matrix are listed in Equation (12.33).

$$\mathbf{D} = \begin{bmatrix}
D_{11} & D_{12} & D_{13} & D_{14} & 0 & 0 & 0 & 0 & 0 & 0 & 0 & 0 & 0 & 0 & 0 & 0 \\
D_{21} & D_{22} & D_{23} & D_{24} & 0 & 0 & 0 & 0 & 0 & 0 & 0 & 0 & 0 & 0 & 0 & 0 \\
D_{31} & D_{32} & D_{33} & D_{34} & D_{35} & D_{36} & D_{37} & D_{38} & 0 & 0 & 0 & 0 & 0 & 0 & 0 & 0 \\
D_{41} & D_{42} & D_{43} & D_{44} & D_{45} & D_{46} & D_{47} & D_{48} & 0 & 0 & 0 & 0 & 0 & 0 & 0 & 0 \\
D_{51} & D_{52} & D_{53} & D_{54} & D_{55} & D_{56} & D_{57} & D_{58} & 0 & 0 & 0 & 0 & 0 & 0 & 0 & 0 \\
D_{61} & D_{62} & D_{63} & D_{64} & D_{65} & D_{66} & D_{67} & D_{68} & 0 & 0 & 0 & 0 & 0 & 0 & 0 & 0 \\
0 & 0 & 0 & 0 & D_{75} & D_{76} & D_{77} & D_{78} & D_{79} & D_{7(10)} & D_{7(11)} & D_{7(12)} & 0 & 0 & 0 & 0 \\
0 & 0 & 0 & 0 & D_{85} & D_{86} & D_{87} & D_{88} & D_{89} & D_{8(10)} & D_{8(11)} & D_{8(12)} & 0 & 0 & 0 & 0 \\
0 & 0 & 0 & 0 & D_{95} & D_{96} & D_{97} & D_{98} & D_{99} & D_{9(10)} & D_{9(11)} & D_{9(12)} & 0 & 0 & 0 & 0 \\
0 & 0 & 0 & 0 & D_{(10)5} & D_{(10)6} & D_{(10)7} & D_{(10)8} & D_{(10)9} & D_{(10)(10)} & D_{(10)(11)} & D_{10(12)} & 0 & 0 & 0 & 0 \\
0 & 0 & 0 & 0 & 0 & 0 & 0 & 0 & D_{(11)9} & D_{(11)(10)} & D_{(11)(11)} & D_{(11)(12)} & D_{(11)(13)} & D_{(11)(14)} & D_{(11)(15)} & D_{(11)(16)} \\
0 & 0 & 0 & 0 & 0 & 0 & 0 & 0 & D_{(12)9} & D_{(12)(10)} & D_{(12)(11)} & D_{(12)(12)} & D_{(12)(13)} & D_{(12)(14)} & D_{(12)(15)} & D_{(12)(16)} \\
0 & 0 & 0 & 0 & 0 & 0 & 0 & 0 & D_{(13)9} & D_{(13)(10)} & D_{(13)(11)} & D_{(13)(12)} & D_{(13)(13)} & D_{(13)(14)} & D_{(13)(15)} & D_{(13)(16)} \\
0 & 0 & 0 & 0 & 0 & 0 & 0 & 0 & D_{(14)9} & D_{(14)(10)} & D_{(14)(11)} & D_{(14)(13)} & D_{(14)(13)} & D_{(14)(14)} & D_{(14)(15)} & D_{(14)(16)} \\
0 & 0 & 0 & 0 & 0 & 0 & 0 & 0 & 0 & 0 & 0 & 0 & D_{(15)(13)} & D_{(15)(14)} & D_{(15)(15)} & D_{(15)(16)} \\
0 & 0 & 0 & 0 & 0 & 0 & 0 & 0 & 0 & 0 & 0 & 0 & D_{(16)(13)} & D_{(16)(13)} & D_{(16)(15)} & D_{(16)(16)} \\
\end{bmatrix}$$

$$(12.32)$$

$$D_{1m} = [\lambda^{(1)} + (\lambda^{(1)} + 2\mu^{(1)})\alpha_m U_{3m}](ik), \quad m = 1,2,3,4$$

$$D_{1m} = 0, \quad m = 5,6,\ldots,16$$

$$D_{2m} = (\alpha_m + U_{3m})\mu^{(1)}(ik), \quad m = 1,2,3,4$$

$$D_{2m} = 0, \quad m = 5,6,\ldots,16$$

$$D_{3m} = e^{ik\alpha_m H^{(1)}}, \quad m = 1,2,3,4$$

$$D_{3m} = -1, \, m = 5,6,7,8$$

$$D_{3m} = 0, \quad m = 9,10,...,16$$

$$D_{4m} = U_{3m} e^{ik\alpha_m H^{(1)}}, \quad m = 1,2,3,4$$

$$D_{4m} = -U_{3m}, \, m = 5,6,7,8$$

$$D_{4m} = 0, \, m = 9,10,...,16$$

$$D_{5m} = [\lambda^{(1)} + (\lambda^{(1)} + 2\mu^{(1)})\alpha_m U_{3m}](ik)e^{ik\alpha_m H^{(1)}}, \, m = 1,2,3,4$$

$$D_{5m} = -[\lambda^{(2)} + (\lambda^{(2)} + 2\mu^{(2)})\alpha_m U_{3m}](ik), \, m = 5,6,7,8$$

$$D_{5m} = 0, \, m = 9,10,...,16$$

$$D_{6m} = (\alpha_m + U_{3m})\mu^{(1)}(ik)e^{ik\alpha_m H^{(1)}}, \, m = 1,2,3,4$$

$$D_{6m} = -(\alpha_m + U_{3m})\mu^{(2)}(ik), \, m = 5,6,7,8$$

$$D_{6m} = 0, \, m = 9,10,\cdots,16$$

$$D_{7m} = 0, \, m = 1,2,3,4$$

$$D_{7m} = e^{ik\alpha m H^{(2)}} \, m = 5,6,7,8$$

$$D_{7m} = -1, \, m = 9,10,11,12$$

$$D_{7m} = 0, \, m = 13,14,15,16$$

$$D_{8m} = 0, \, m = 1,2,3,4$$

$$D_{8m} = U_{3m} e^{ik\alpha_m H^{(2)}}, \, m = 5,6,7,8$$

$$D_{8m} = -U_{3m}, \, m = 9,10,11,12$$

$$D_{8m} = 0, \, m = 13,14,15,16 \tag{12.33}$$

$$D_{9m} = 0, \, m = 1,2,3,4$$

$$D_{9m} = \left[\lambda^{(2)} + (\lambda^{(2)} + 2\mu^{(2)})\alpha_m U_{3m}\right](ik)e^{ik\alpha_m H^{(2)}}, \, m = 5,6,7,8$$

$$D_{9m} = \left[\lambda^{(3)} + (\lambda^{(3)} + 2\mu^{(3)})\alpha_m U_{3m}\right](ik), \, m = 9,10,11,12$$

$$D_{9m} = 0, \, m = 13,14,15,16$$

$$D_{10m} = 0, \, m = 1,2,3,4$$

$$D_{10m} = (\alpha_m + U_{3m})\mu^{(2)}(ik)e^{ik\alpha_m H^{(2)}}, \, m = 5,6,7,8$$

$$D_{10m} = -(\alpha_m + U_{3m})\mu^{(2)}(ik), \, m = 9,10,11,12$$

$$D_{10m} = 0, \, m = 13,14,15,16$$

$$D_{11m} = 0, \quad m = 1,2,\cdots,8$$

$$D_{11m} = e^{ik\alpha_m} H^{(3)}, \quad m = 9,10,11,12$$

$$D_{11m} = -1, \quad m = 13,14,15,16$$

$$D_{12} = 0, \quad m = 1,2,\cdots,8$$

$$D_{12m} = U_{3m} e^{ik\alpha_m} H^{(3)}, \quad m = 9,10,11,12$$

$$D_{12m} = -U_{3m}, \quad m = 13,14,15,16$$

$$D_{13m} = 0, \quad m = 1,2,\cdots,8$$

$$D_{13m} = \left[\lambda^{(3)} + \left(\lambda^{(3)} + 2\mu^{(3)}\right)\alpha_m U_{3m}\right](ik)e^{ik\alpha_m} H^{(3)}, \quad m = 9,10,11,12$$

$$D_{13m} = -\left[\lambda^{(4)} + \left(\lambda^{(4)} + 2\mu^{(4)}\right)\alpha_m U_{3m}\right](ik), \quad m = 13,14,15,16$$

$$D_{14m} = 0, \quad m = 1,2,\cdots,8$$

$$D_{14m} = \left(\alpha_m + U_{3m}\right)\mu^{(3)}(ik)e^{ik\alpha_m} H^{(3)}, \quad m = 9,10,11,12$$

$$D_{14m} = -\left(\alpha_m + U_{3m}\right)\mu^{(4)}(ik), \quad m = 13,14,15,16$$

$$D_{15m} = 0, \quad m = 1,2,\ldots,12$$

$$D_{15m} = \left[\lambda^{(4)} + \left(\lambda^{(4)} + 2\mu^{(4)}\right)\alpha_m U_{3m}\right](ik)e^{ik\alpha_m} H^{(4)}, \quad m = 13,14,15,16$$

$$D_{16m} = 0, \quad m = 1,2,\ldots,12$$

$$D_{16m} = \left(\alpha_m + U_{3m}\right)\mu^{(4)}(ik)e^{ik\alpha_m} H^{(4)}, \quad m = 13,14,15,16$$

After the characteristic equation is obtained for a given structure, the dispersion curves can be derived with numerical methods by satisfying that the determinant of the coefficient matrix has been set to zero.

$$|\mathbf{D}| = 0 \tag{12.34}$$

The corresponding wave structure of a selected wave mode can be calculated by substituting in the values of the (k, c_p) pairs and solving for the unknown vector $\mathbf{B}$ in Equation (12.31). Then, the wave field can be obtained by substituting the results of $\mathbf{B}$ into Equation (12.26).

Note
Before the numerical solutions of Equation (12.31) are obtained, preprocessing of the $\mathbf{D}$ matrix can be carried out to simplify the formulation. For example, the common term (ik) for the first row of Equation (12.32) could be neglected because $k \neq 0$ for a meaningful guided wave mode. Besides the global coordinate system (x_1, x_2, x_3) used earlier, the ultrasonic wave field in each layer can also be expressed in a local coordinate system where the origin is at the bottom of each layer. In this situation, the expression of the $\mathbf{D}$ matrix will also be simplified

to some extent. The details of this method are introduced in the section on the transfer matrix method. The student can also formulate the global matrix method in local coordinate systems as an exercise.

12.4.2 The Transfer Matrix Method

When the number of layers is large, a large $\mathbf{D}$ matrix will be obtained using the global matrix method. Solution of Equation (12.34) with a large $\mathbf{D}$ matrix will be time-consuming. In this case, a transfer matrix method illustrated in this section could be particularly useful.

We will use a local coordinate system for each layer. Figure 12.10 shows an N-layered plate structure. The (x_1, x_2, x_3) coordinate is a global coordinate system; the $(x_1^{(n)}, x_2^{(n)}, x_3^{(n)})$ coordinate system is the local coordinate system for layer n. When the origins of the $x_1^{(n)}$ axis are at the same position, a general coordinate x_1 can be used.

Using the local coordinate systems, the wave fields in each layer are expressed in Equation (12.35) using the partial wave method. Here, the superscript (n) means the material property and field quantities of the nth layer.

$$u_1^{(n)} = \sum_{m=1}^{4} B_m^{(n)} e^{ik(x_1 + \alpha_m^{(n)} x_3^{(n)} - c_p t)}$$

$$u_3^{(n)} = \sum_{m=1}^{4} B_m^{(n)} U_{3m}^{(n)} e^{ik(x_1 + \alpha_m^{(n)} x_3^{(n)} - c_p t)}$$

$$\sigma_{33}^{(n)} = \sum_{m=1}^{4} B_m^{(n)} [\lambda^{(n)} + (\lambda^{(n)} + 2\mu^{(n)}) \alpha_m^{(n)} U_{3m}^{(n)}](ik) e^{ik(x_1 + \alpha_m^{(n)} x_3^{(n)} - c_p t)} \tag{12.35}$$

$$\sigma_{31}^{(n)} = \sum_{m=1}^{4} B_m^{(n)} [\alpha_m^{(n)} + U_{3m}^{(n)}] \mu^{(n)} (ik) e^{ik(x_1 + \alpha_m^{(n)} x_3^{(n)} - c_p t)}$$

When Γ is used to denote the wave field vector $[u_1 \quad u_3 \quad \sigma_{33} \quad \sigma_{31}]^T$ neglecting the harmonic term $e^{ik(x_1 - c_p t)}$, Equation (12.35) can be expressed in a matrix form as in Equation (12.36).

$$\begin{bmatrix} u_1^{(n)} \\ u_3^{(n)} \\ \sigma_{33}^{(n)} \\ \sigma_{31}^{(n)} \end{bmatrix} = \Gamma^{(n)} e^{ik(x_1 - c_p t)}$$

$$\Gamma^{(n)} = \Gamma^{(n)} \mathbf{W}^{(n)} \mathbf{B}^{(n)} \tag{12.36}$$

The details of $\mathbf{X}^{(n)}$, $\mathbf{W}^{(n)}$, and $\mathbf{B}^{(n)}$ are in Equations (12.37), (12.38), and (12.39). $\mathbf{X}^{(n)}$ is a function of wavenumber and wave phase velocity; $\mathbf{W}^{(n)}$ is a function of wavenumber, phase velocity, and position $x_3^{(n)}$; $\mathbf{B}^{(n)}$ is the unknown coefficient vector.

$$\mathbf{X}^{(n)} = \begin{bmatrix} X_{11}^{(n)} & X_{12}^{(n)} & X_{13}^{(n)} & X_{14}^{(n)} \\ X_{21}^{(n)} & X_{22}^{(n)} & X_{23}^{(n)} & X_{24}^{(n)} \\ X_{31}^{(n)} & X_{32}^{(n)} & X_{33}^{(n)} & X_{34}^{(n)} \\ X_{41}^{(n)} & X_{42}^{(n)} & X_{43}^{(n)} & X_{44}^{(n)} \end{bmatrix}$$

$$X_{1m}^{(n)} = 1 \tag{12.37}$$

$$X_{2m}^{(n)} = U_{3m}^{(n)}$$

$$X_{3m}^{(n)} = [\lambda^{(n)} + (\lambda^{(n)} + 2\mu^{(n)})\alpha_m^{(n)}U_{3m}^{(n)}](ik)$$

$$X_{4m}^{(n)} = [\alpha_m^{(n)} + U_{3m}^{(n)}]\mu^{(n)}(ik)$$

$$\mathbf{W}^{(n)} = \begin{bmatrix} e^{ik(\alpha_1^{(n)}x_3^{(n)})} & 0 & 0 & 0 \\ 0 & e^{ik(\alpha_2^{(n)}x_3^{(n)})} & 0 & 0 \\ 0 & 0 & e^{ik(\alpha_3^{(n)}x_3^{(n)})} & 0 \\ 0 & 0 & 0 & e^{ik(\alpha_4^{(n)}x_3^{(n)})} \end{bmatrix} \tag{12.38}$$

$$\mathbf{B}^{(n)} = \begin{bmatrix} B_1^{(n)} \\ B_2^{(n)} \\ B_3^{(n)} \\ B_4^{(n)} \end{bmatrix} \tag{12.39}$$

The wave field at the bottom of each layer is expressed in Equation (12.40) because $x_3^{(n)} = 0$ and $\mathbf{W}^{(n)}(0) = \mathbf{I}$.

$$\Gamma^{(n)-} = \mathbf{X}^{(n)}\mathbf{B}^{(n)} \tag{12.40}$$

The wave field at the top of each layer is expressed in Equation (12.41).

$$\Gamma^{(n)+} = \mathbf{X}^{(n)}(\mathbf{W}^{(n)}\big|_{h^{(n)}})\mathbf{B}^{(n)} \tag{12.41}$$

Therefore, the wave field at the top surface can be expressed as a function of wave field at the bottom surface through the relation in Equation (12.42).

$$\Gamma^{(n)+} = \mathbf{X}^{(n)}(\mathbf{W}^{(n)}\big|_{h^{(n)}})(\mathbf{X}^{(n)})^{-1}\Gamma^{(n)-} = \mathbf{\Psi}^{(n)}\Gamma^{(n)-} \tag{12.42}$$

Here, $\mathbf{\Psi}^{(n)}$ is the transfer matrix of the nth layer. Note that this matrix is only a function of wavenumber and phase velocity.

The rigid bonding boundary condition requires that the displacement and stress field at the top of the nth layer be the same as the bottom surface of the $(n+1)^{th}$ layer (see Equation (12.27)). The abbreviated expression is in Equation (12.43).

$$\Gamma^{(n)+} = \Gamma^{(n+1)-}, \quad n = 1, 2, ..., N - 1 \tag{12.43}$$

Substituting Equation (12.43) into Equation (12.42),

$$\Gamma^{(n+1)-} = \mathbf{\Psi}^{(n)}\Gamma^{(n)-} \tag{12.44}$$

When all of the relations of *N-1* interface conditions are applied together, Equation (12.45) can be obtained.

$$\Gamma^{(N)+} = \Psi^{(n)}\Gamma^{(N)-} = \Psi^{(n)}\Psi^{(n-1)}\dots\Psi^{(1)}\Gamma^{(1)-} = \left(\prod_{n=1}^{N}\Psi^{(n)}\right)\Gamma^{(1)-} \tag{12.45}$$

The dimension of the overall transfer matrix $\Psi = \prod_{n=1}^{N}\Psi^{(n)}$ is always four by four, which is independent of the number of layers. Considering the stress-free condition at the bottom of the first layer and the top of the Nth layer, the expanded format of Equation (12.45) is in Equation (12.46).

$$\begin{bmatrix} u_1^{(N)+} \\ u_3^{(N)+} \\ 0 \\ 0 \end{bmatrix} = \begin{bmatrix} \Psi_{11} & \Psi_{12} & \Psi_{13} & \Psi_{14} \\ \Psi_{21} & \Psi_{22} & \Psi_{23} & \Psi_{24} \\ \Psi_{31} & \Psi_{32} & \Psi_{33} & \Psi_{34} \\ \Psi_{41} & \Psi_{42} & \Psi_{43} & \Psi_{44} \end{bmatrix} \begin{bmatrix} u_1^{(1)-} \\ u_3^{(1)-} \\ 0 \\ 0 \end{bmatrix} \tag{12.46}$$

This equation can be separated into Equation (12.47) and Equation (12.48).

$$\begin{bmatrix} u_1^{(N)+} \\ u_3^{(N)+} \end{bmatrix} = \begin{bmatrix} \Psi_{11} & \Psi_{12} \\ \Psi_{21} & \Psi_{22} \end{bmatrix} \begin{bmatrix} u_1^{(1)-} \\ u_3^{(1)-} \end{bmatrix} \tag{12.47}$$

$$\begin{bmatrix} 0 \\ 0 \end{bmatrix} = \begin{bmatrix} \Psi_{31} & \Psi_{32} \\ \Psi_{41} & \Psi_{42} \end{bmatrix} \begin{bmatrix} u_1^{(1)-} \\ u_3^{(1)-} \end{bmatrix} \tag{12.48}$$

Equation (12.48) is the characteristic equation for guided wave propagation in the multilayered structure. Because the elements in the transfer matrix Ψ are functions of wavenumber and phase velocity, the guided wave dispersion curve can be obtained by searching the appropriate pairs of (k, c_p) that satisfy Equation (12.49).

$$\begin{vmatrix} \Psi_{31} & \Psi_{32} \\ \Psi_{41} & \Psi_{42} \end{vmatrix} = 0 \tag{12.49}$$

After the guided wave dispersion curves are obtained from Equation (12.49) by numerical root searching, the wave displacement at the bottom surface can be calculated using Equation (12.48).

Suppose the solution at the bottom surface is already obtained; the partial wave combination coefficients can be calculated using Equation (12.50):

$$\mathbf{B}^{(1)} = (\mathbf{X}^{(1)})^{-1}\begin{bmatrix} u_1^{(1)-} \\ u_3^{(1)-} \end{bmatrix} \tag{12.50}$$

Then, the wave field at the bottom of the second layer, that is, the field at the top of the first layer, can be calculated using Equation (12.41). By using Equation (12.50) and Equation (12.41) recursively for each layer, the entire wave field can be obtained. The wave structure solution can also be expressed in the global coordinate system (x_1, x_2, x_3).

12.4.3 Examples

There exists a countless number of practical applications of waves in multiple layers that we might consider. Two examples are presented here. The first is ice detection on an aircraft wing. The second is the inspection of diffusion bonds in a three-layer structure for bond integrity.

Multilayer Model for Aircraft Wing Ice Detection

A very light coating of ice could have a deleterious effect on the performance of aircraft. The hazard of icing is not just because of actual weight accumulation, but rather because it destroys the smooth flow of air, increasing drag, degrading control authority, and decreasing the lift ability of an airfoil. In the worst cases, icing has led to aircraft stalls and to the crash of a K-28 aircraft at New York LaGuardia Airport in 1992 (Wikipedia, U.S. Air Flight 405). As a result, early detection and removal of ice accreted on the aircraft surface is very important for aviation safety and efficiency.

To use ultrasonic guided waves for aircraft ice detection, the ultrasonic guided wave propagation characteristics in the aircraft wing skin components have to be understood. As a simplification, the skin of an aircraft is modeled as a single-layer aluminum plate in this example. Figure 12.10 illustrates the conditions before and after an ice layer is accreted on top of the aluminum. The thickness of the aluminum plate is $h^{(1)}$; the thickness of the accreted ice layer is $h^{(2)}$. Table 12.1 lists the material properties and thickness of a glaze ice layer and the aluminum layer.

Researchers have already studied guided wave propagation in the single aluminum plate for Lamb waves and SH waves, respectively. When ice accretes on the aluminum plate, the system becomes a multilayered structure. The global matrix method will be used to generate the dispersion curves. When Lamb type waves are considered, four partial waves exist for each layer. Two stress-free boundary conditions will be applied for each outer surface, and four displacement and stress continuity conditions should be satisfied at the interface when the ice is ideally bonded to the wing skin.

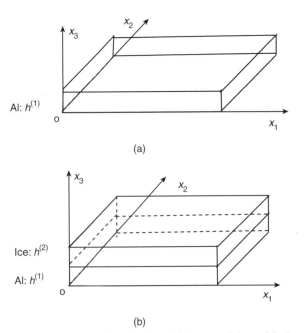

Figure 12.10. Sketches of aircraft icing conditions: (a) clean wing, (b) ice on wing.

Table 12.1 Material properties in an aircraft icing problem

Material	Density (g/cm³)	Young's modulus (GPa)	Poison's ratio	Thickness (mm)
Aluminum	2.7	70.3	0.345	1
Ice	0.9	8.3	0.351	0.2

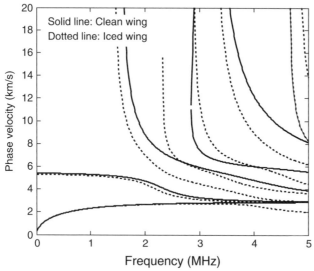

Figure 12.11. Comparison of dispersion curves of an aircraft wing at (a) clean wing condition, (b) aluminum wing with ice.

The components of the **D** matrix for a two-layered structure are part of Equation 12.32. Therefore, we need to substitute the material properties of aluminum and ice into the general formulation. When the eight by eight **D** matrix is formulated, the dispersion curves can be obtained by searching for the wave velocity and frequency pairs that can satisfy $|\mathbf{D}| = 0$. The resulting dispersion curves of the clean wing and the wing with ice are plotted in Figure 12.11.

Figure 12.12 shows that the deposition of a thin ice layer on top of the aluminum plate changes the phase velocity dispersion curve of the structure. The influence is particularly significant for higher order guided wave modes. Therefore, guided waves can be used to detect the ice accretion.

When ice is accreted on top of the aluminum plate, the center of the aluminum plate will no longer be the symmetry plane of the structure. Therefore, the guided wave modes will no longer be exactly classified into groups of symmetric modes and antisymmetric modes. As an example, the wave structure for the third wave mode at 2 MHz is shown in Figure 12.12.

This is only a starting point in solving the ice detection problem. Various thicknesses, properties, patch possibilities, and other parameters should be considered. Also, sliding ice could be modeled over a thin water layer. The dispersion curve differences show promise but selection of the best points on the dispersion curve that could work best is challenging. Transducer design, physically based feature extraction, and establishment of an appropriate ice detection decision algorithm would follow. See for example Gao and Rose (2009) and Rose et al (1997).

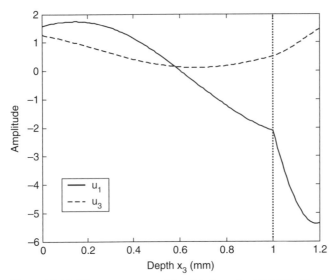

Figure 12.12. Wave structure of the third mode at 2 MHz. The result shows that the wave structure is neither symmetric nor antisymmetric about the mid-plane of the structure.

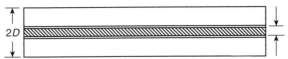

Figure 12.13. Three-layer titanium diffusion bond model.

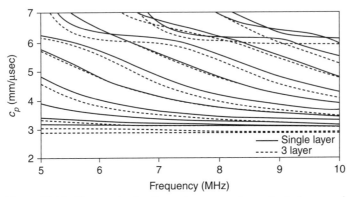

Figure 12.14. Phase velocity dispersion curves for a single-layer and a three-layer titanium model, showing the mode frequency shifts.

Diffusion Bond Inspection

A general sketch of a diffusion bonded interface is shown in Figure 12.13, where a thin layer is used to represent the diffusion bond. A perfect bonding layer will have the same material property as the bulk material, while a poor bond can be modeled as a density and modulus degradation in the interface layer. By doing this, the guided wave dispersion curves for a perfect bond can be compared with degradation models to indicate the bonding quality.

The fine diffusion bonded interface can be viewed as a thin layer; this is shown in Figure 12.13. From the perspective of ultrasonic wave propagation, poor contact can be modeled as a distinct layer with modulus and density degradation values; thus we can depict poor, intermediate, or excellent diffusion bonding (see Rose, Zhu, and

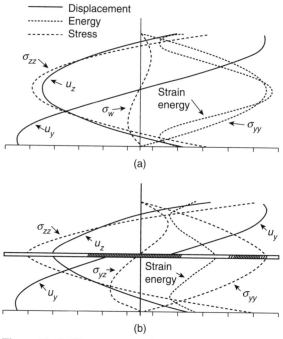

Figure 12.15. Wave structure in titanium plates of (a) the single-layer model and (b) the three-layer model.

Figure 12.16. Guided wave experimental arrangement.

Zaidi 1998 for more details). Following this approach, dispersion curves for a perfect bond can be compared with degradation models. Figure 12.14 shows phase velocity dispersion curves; in general, degradation model dispersion curves shift to the left. Wave structure is illustrated in Figure 12.15 (note that there are reasonable amounts of energy along the interface for the mode studied). An experiment was conducted to see if the frequency shift could be observed; the experimental arrangement is shown in Figure 12.16. A frequency shift experimental result is shown in Figure 12.17. Additional mode features are described in Rose and colleagues (1998).

The solution to this problem also has an interesting twist, as illustrated in Figure 12.18, which depicts two modes. One mode is sensitive to diffusion bond integrity and one mode is not sensitive. Note that amplitude itself is often a poor feature to use in any classification problems but amplitude ratio, here of the two modes, is an excellent feature.

12.5 Fluid-Coupled Elastic Layers

In this section, the influence of a fluid load on an elastic layer is addressed. The problem is tackled in two aspects. One is from an oblique incidence point of view to study the reflection and transmission factors; the other is from the guided wave propagation standpoint to study wave propagation for a solid layer in fluid environments. This problem was studied by Chimenti and Rokhlin (1990).

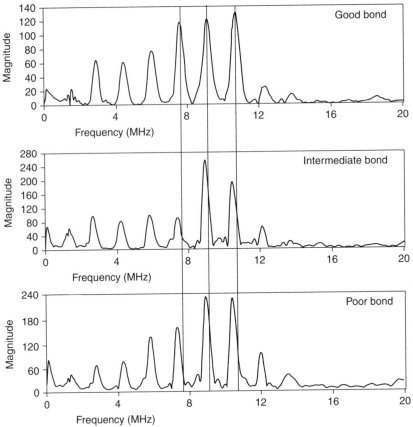

Figure 12.17. Frequency spectrum of the 15-MHz broadband pulse response for 2-mm diffusion bond panels at 60° incidence, showing the mode frequency shifts.

12.5.1 Ultrasonic Wave Reflection and Transmission

When an ultrasonic wave is incident on a plate through a liquid half-space, the entire wave energy will separate into reflected waves, and transmitted waves will separate to the other side of the plate. Let's consider possible relationships with guided wave dispersion curves. Figure 12.19 illustrates the sketch of an oblique wave incidence problem. Here, the elastic plate has thickness h; the density and the Lamé constant are $\rho^{(2)}$, $\lambda^{(2)}$, and $\mu^{(2)}$, respectively. Fluid 1 occupies the half-space with $x_3 > h$, and fluid 2 occupies the half-space with $x_3 < 0$. For a general situation, the second fluid can be different from the first fluid material. In this model, the materials are numbered as (1) and (3), respectively. A plane wave is incident from the liquid layer to the upper surface of the solid plate. Assume the incident wave is a longitudinal wave with frequency f and incident angle θ; the incident wave can be expressed as $B_I e^{i(k_1 x_1 + k_3 x_3 - 2\pi f t)}$. Possible reflected waves and transmitted waves are also plotted in the figure. The longitudinal wave in the incident layer is $C_L^{(1)} = \sqrt{\dfrac{\lambda^{(1)}}{\rho^{(1)}}}$. The wavevector of the incident wave is $\mathbf{k} = k_1 \breve{x}_1 + k_3 \breve{x}_3$, where $k_1 = \dfrac{2\pi f}{\left(C_L^{(1)}\right)} \sin(\theta)$ and $k_3 = \dfrac{2\pi f}{\left(C_L^{(1)}\right)} \cos(\theta)$. According to Snell's law, all the partial waves will have the same wavenumber in

Modes at 7.8 MHz and 70° incidence

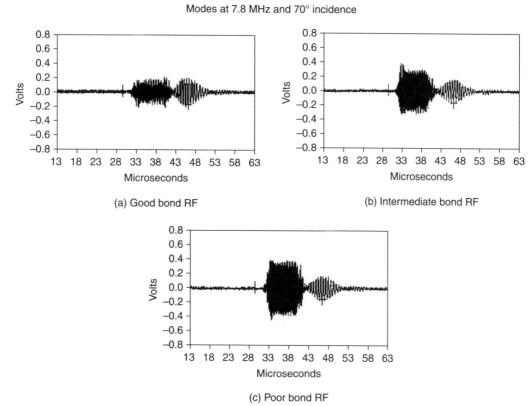

(a) Good bond RF

(b) Intermediate bond RF

(c) Poor bond RF

Figure 12.18. Guided wave experimental results for titanium diffusion bonding inspection.

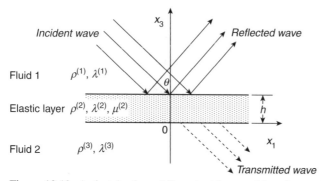

Figure 12.19. A sketch of an oblique incident problem from a fluid half-space to a solid plate.

the x_1-direction. Therefore, the phase velocity values of all the partial waves are

$$c_p = \frac{\omega}{k_1} = \frac{C_L^{(1)}}{\sin(\theta)}.$$ For convenience of expression and to be compatible with the guided

wave analysis, the wavenumber in the x_1-direction k_1 is noted as k in the following.

The partial wave techniques discussed earlier can be used to express the wave fields in the solid and liquid layer, respectively. A list of possible partial waves for these three layers is studied in the following.

In the Incident Layer: Layer 1

Two partial waves exist in the incident layer. One is the incident wave, and the other is a reflected wave. The formal solution of two partial waves in a fluid layer is expressed in Equation (12.51). The incident wave is along the $-x_3$-direction. Therefore, α_1 has a negative value.

$$\text{Incident wave: } \alpha_1 = -\sqrt{-1+\frac{\rho^{(1)}}{\lambda^{(1)}}c_p^2}, \quad U_{31} = \alpha_1$$

$$\text{Reflected wave: } \alpha_2 = \sqrt{-1+\frac{\rho^{(1)}}{\lambda^{(1)}}c_p^2}, \quad U_{32} = \alpha_2 \tag{12.51}$$

The total wave field in the first layer is a summation of the incident wave field and the reflected wave field. Assume that the incident wave field has a given amplitude B_1, and the reflected wave field has an unknown amplitude B_2; hence, the entire wave field can be expressed in Equation (12.52).

$$u_1^{(1)} = \sum_{n=1}^{2} B_n e^{ik(x_1 + \alpha_{n1} x_3 - c_p t)}$$

$$u_3^{(1)} = \sum_{n=1}^{2} B_n U_{3n} e^{ik(x_1 + \alpha_n x_3 - c_p t)} \tag{12.52}$$

$$\sigma_{33}^{(1)} = \sum_{n=1}^{2} B_n [\rho^{(1)} c_p^2](ik) e^{ik(x_1 + \alpha_n x_3 - c_p t)}$$

$$\sigma_{31}^{(1)} = 0$$

In the Solid Layer: Layer 2

As discussed in previous sections, four partial waves exist in a solid layer when R-L types of waves are considered. The results for the wavevector and partial wave polarization can be obtained from the Christoffel's equation. They are also repeated here in Equation (12.53).

$$\alpha_3 = \sqrt{-1+\frac{\rho^{(2)}}{\mu^{(2)}}c_p^2}, \quad U_{33} = \frac{-1}{\alpha_3}$$

$$\alpha_4 = -\sqrt{-1+\frac{\rho^{(2)}}{\mu^{(2)}}c_p^2}, \quad U_{34} = \frac{-1}{\alpha_4}$$

$$\alpha_5 = \sqrt{-1+\frac{\rho^{(2)}}{\lambda^{(2)}+2\mu^{(2)}}c_p^2}, \quad U_{35} = \alpha_5 \tag{12.53}$$

$$\alpha_6 = -\sqrt{-1+\frac{\rho^{(2)}}{\lambda^{(2)}+2\mu^{(2)}}c_p^2}, \quad U_{36} = \alpha_6$$

Here, numbering of the partial waves is continued from the first layer. Partial waves 3 and 4 are shear waves, partial waves 5 and 6 are longitudinal waves; partial waves 3 and 5 propagate in the $+x_3$-direction, and partial waves 4 and 6 propagate in the $-x_3$-direction. The displacement and stress field in the solid layer can be expressed as a linear combination of these four partial waves. Equation (12.54):

$$u_1^{(2)} = \sum_{n=3}^{6} B_n e^{ik(x_1 + \alpha_n x_3 - c_p t)}$$

$$u_3^{(2)} = \sum_{n=3}^{6} B_n U_{3n} e^{ik(x_1 + \alpha_n x_3 - c_p t)}$$

$$\sigma_{33}^{(2)} = \sum_{n=3}^{6} B_n [\lambda^{(2)} + (\lambda^{(2)} + 2\mu^{(2)})\alpha_n U_{3n}](ik) e^{ik(x_1 + \alpha_n x_3 - c_p t)}$$

$$\sigma_{31}^{(2)} = \sum_{n=3}^{6} B_n [\alpha_n + U_{3n}]\mu^{(2)}(ik) e^{ik(x_1 + \alpha_n x_3 - c_p t)}$$

(12.54)

In the Transmitted Layer: Layer 3

Only one partial wave exists in layer 3, which is a longitudinal partial wave propagating toward the $-x_3$-direction. Therefore, the wave field in this layer can be expressed in Equation (12.55).

$$u_1^{(3)} = B_7 e^{ik(x_1 + \alpha_7 x_3 - c_p t)}$$

$$u_3^{(3)} = B_7 U_{37} e^{ik(x_1 + \alpha_7 x_3 - c_p t)}$$

(12.55)

Here, $\alpha_7 = -\sqrt{-1 + \frac{\rho^{(3)}}{\lambda^{(3)}} c_p^2}$. The reason for the negative sign is that the wavevector is along the $-x_3$-direction. Based on partial wave analysis in the liquid, $U_{37} = \alpha_7$, the stress field in this layer can be expressed as in Equation (12.56).

$$\sigma_{33}^{(3)} = B_7 [\rho^{(3)} c_p^2](ik) e^{ik(x_1 + \alpha_7 x_3 - c_p t)}$$

$$\sigma_{31}^{(3)} = 0$$

(12.56)

Boundary Conditions

The boundary conditions in this problem are displacement and stress continuity at the solid–liquid interfaces. Recall the boundary conditions discussed in Section 12.2.2; the conditions for this problem are listed in Equation (12.57).

$$\text{BC1}: u_3^{(2)} - u_3^{(1)} = 0, \text{ at } x_3 = h$$

$$\text{BC2}: u_3^{(2)} - u_3^{(3)} = 0, \text{ at } x_3 = 0$$

$$\text{BC3}: \sigma_{33}^{(2)} - \sigma_{33}^{(1)} = 0, \text{ at } x_3 = h$$

$$\text{BC4}: \sigma_{31}^{(2)} = 0, \text{ at } x_3 = h$$

$$\text{BC5}: \sigma_{33}^{(2)} - \sigma_{33}^{(3)} = 0, \text{ at } x_3 = 0$$

$$\text{BC6}: \sigma_{31}^{(2)} = 0, \text{ at } x_3 = 0$$

(12.57)

Substitute the wave field expressions into the boundary conditions to get Equation (12.58).

$$
\begin{bmatrix}
-U_{32}e^{ik\alpha_2 h} & U_{33}e^{ik\alpha_3 h} & U_{34}e^{ik\alpha_4 h} & U_{35}e^{ik\alpha_5 h} & U_{36}e^{ik\alpha_6 h} & 0 \\
0 & U_{33} & U_{34} & U_{35} & U_{36} & -U_{37} \\
-\rho^{(1)}c_p^2 e^{ik\alpha_2 h} & [-2\mu^{(2)}]e^{ik\alpha_3 h} & [-2\mu^{(2)}]e^{ik\alpha_4 h} & [\rho^{(2)}c_p^2-2\mu^{(2)}]e^{ik\alpha_5 h} & [\rho^{(2)}c_p^2-2\mu^{(2)}]e^{ik\alpha_6 h} & 0 \\
0 & [\alpha_3+U_{33}]\mu^{(2)}e^{ik\alpha_3 h} & [\alpha_4+U_{34}]\mu^{(2)}e^{ik\alpha_4 h} & [\alpha_5+U_{35}]\mu^{(2)}e^{ik\alpha_5 h} & [\alpha_6+U_{36}]\mu^{(2)}e^{ik\alpha_6 h} & 0 \\
0 & -2\mu^{(2)} & -2\mu^{(2)} & \rho^{(2)}c_p^2-2\mu^{(2)} & \rho^{(2)}c_p^2-2\mu^{(2)} & -\rho^{(3)}c_p^2 \\
0 & [\alpha_3+U_{33}]\mu^{(2)} & [\alpha_4+U_{34}]\mu^{(2)} & [\alpha_5+U_{35}]\mu^{(2)} & [\alpha_6+U_{36}]\mu^{(2)} & 0
\end{bmatrix}
\begin{bmatrix} B_2 \\ B_3 \\ B_4 \\ B_5 \\ B_6 \\ B_7 \end{bmatrix}
=
\begin{bmatrix} B_1 U_{31}e^{ik\alpha_1 h} \\ 0 \\ B_1\rho^{(1)}c_p^2 e^{ik\alpha_1 h} \\ 0 \\ 0 \\ 0 \end{bmatrix}
$$

$$(12.58)$$

When the amplitude of the incident wave B_1 is given, the solutions to other partial waves can be obtained by solving this linear system of equations. Therefore, the reflected wave and the transmitted wave, as well as the field in the solid layer, can be calculated. The reflection coefficient and the transmission coefficient can be defined as the amplitude ratio between these partial waves to the incident partial wave Equation (12.59).

$$
R = \frac{B_2}{B_1}
$$
$$
T = \frac{B_7}{B_1}
$$

$$(12.59)$$

In this example, we consider the wave reflection and transmission from an aluminum plate immersed in water. The material properties of the layers are listed in Table 12.2.

In the case of normal incidence, $\theta = 0$. The reflection and transmission coefficients are plotted in Figure 12.20 as a function of frequency.

In the normal incident problem, the incident longitudinal wave only vibrates in the x_3-direction. Therefore, there is also only an x_3 displacement in the aluminum plate and the water layer on the other side. In other words, only longitudinal waves exist in the aluminum plate, and the waves all propagate along the x_3 axis. Therefore, the physical understanding of the transmission and reflection coefficient can be obtained from one-dimensional wave interference analysis. As we know, multiple transmission and reflection occurs at the two boundaries of the aluminum plate. The impedance of the water layer is smaller than the impedance of the aluminum layer. Therefore, there is half wavelength loss, that is, a π phase shift for the reflected waves from the aluminum-water interface to aluminum. Therefore, the phase difference between the second reflected wave TRT and the first reflected wave R is $\Delta\varphi = \dfrac{2hf}{C_L^{(2)}}2\pi + \pi$. When

Table 12.2 *Material and geometric properties of an aluminum plate immersed in water*

	Material	λ (GPa)	μ (GPa)	ρ (g/cm³)	h (mm)
Fluid 1	water	2.19	0	1	---
Solid	aluminum	55.27	25.95	2.7	1
Fluid 2	water	2.19	0	1	---

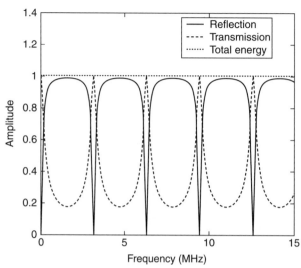

Figure 12.20. Reflection coefficient and transmission coefficient of an ultrasonic wave incident on an aluminum plate immersed in water from zero degree direction.

this phase shift is $(2n+1)\pi$, the reflected wave is under destructive interference. If we express this condition in terms of frequency, we get $f = \dfrac{nC_L^{(2)}}{2h}$. The longitudinal velocity of the aluminum is about 6.3 mm/μs, therefore the frequency corresponds to minimum reflection: $f = 3.15n$ MHz, $n = 0,1,2,\cdots$. For these frequencies, the phase difference of the first transmitted wave (TT) and the second transmitted waves (TRRT) is $\Delta\varphi = \dfrac{2hf}{C_L^{(2)}}2\pi + 2\pi = \dfrac{2h}{C_L^{(2)}}\dfrac{nC_L^{(2)}}{2h}2\pi + 2\pi = 2\pi(n+1)$. This phase difference will lead to constructive interference of the transmitted wave. Therefore, a maximum value of the transmission is obtained at these frequencies. In other words, the waves will be transmitted most efficiently when the aluminum plate is resonance at one of its resonance modes, where $2h = n\lambda$. Because no energy consumption takes place within the aluminum layer for the plane wave case, the total energy of the reflected wave and the transmitted wave will be summed up to the energy of the incident wave, which is also shown in Figure 12.20, when the incident wave amplitude is unity.

Up to now, we have explained the wave transmission and reflection phenomenon for normal incidence from a physical interference point of view. We surmise that when the incident wave frequency corresponds to a thickness vibration mode of the plate, the ultrasonic waves will be transmitted most efficiently. The transmission coefficient will be one at the resonance frequencies and transmission will be small between the vibration mode frequencies.

When the incident wave is at an angle with the surface normal of the plate, the reflection and transmission phenomenon will be much more complicated because both longitudinal and shear waves will be excited in the aluminum layer. Although interference is complicated, the problem can be solved with Equation (12.58). Figure 12.21 shows the reflection and transmission coefficient as a function of incident angle and frequency. It is interesting to see that the reflection coefficient changes with both incident angle and frequency. For certain pairs of (f,θ), the

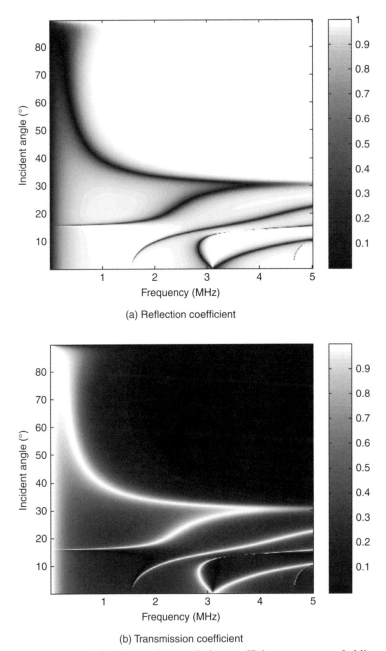

(a) Reflection coefficient

(b) Transmission coefficient

Figure 12.21. Reflection and transmission coefficient spectrum of oblique wave incidence to an aluminum plate immersed in water.

reflection coefficient is zero and the transmission coefficient is one. Figure 12.21 also shows that for most of the (f, θ) combinations, a large portion of the wave energy is reflected back to the "Fluid 1" layer. Despite the energy partition between the reflection and transmission, the energy conservation relation $|R|^2 + |T|^2 = 1$ is always valid for plane harmonic wave incidence.

The pattern of the frequency incident angle spectrum is related to the phase velocity dispersion curve of the aluminum plate. Recall the characteristic matrix for Lamb wave propagation in an isotropic single layer. Figure 12.22 is the Lamb wave

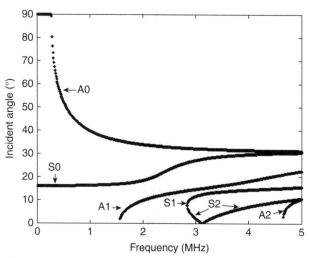

Figure 12.22. Dispersion curve of Lamb type guided waves in a stress free aluminum plate. Thickness of the plate is 1 mm.

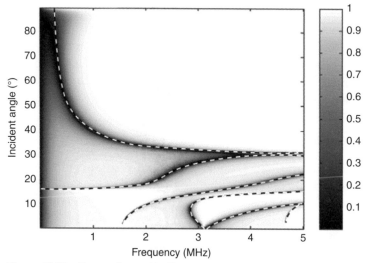

Figure 12.23. Comparison of bulk wave transmission spectrum and guided wave dispersion curve for an 1 mm-thick aluminum plate.

dispersion curves transformed into the frequency incident angle domain using the relation $\theta = \operatorname{asin}\left(\dfrac{C_L^{(1)}}{c_p}\right)$.

Comparing Figure 12.22 to Figure 12.21, there is a correlation between the $|R| = 0$ curves and the dispersion curves for guided wave propagation. The superposition of the guided wave dispersion curve on top of the reflection coefficient spectrum is shown in Figure 12.23. This correlation leads to a very important application of measuring guided wave dispersion curves through the transmission and reflection spectra.

The previous example shows the correlation between the bulk wave reflection spectrum and guided wave propagation dispersion curve. In this example, we are going to study the wave incident on a Plexiglas plate immersed in water. The thickness of the Plexiglas is also 1 mm, but the material properties of the solid layer are now

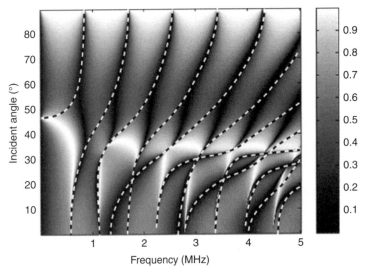

Figure 12.24. Reflection coefficient spectrum of oblique wave incidence to a Plexiglas plate immersed in water. Thickness of the plate is 1 mm.

$\rho^{(2)} = 1.18$ g/cm^3, $\lambda^{(2)} = 5.45$ GPa, and $\mu^{(2)} = 1.48$ GPa. The reflection coefficient spectrum of this system is shown in Figure 12.24.

In this case, the zero in the reflection spectrum is different from that of the dispersion curves for a free plate. This is because the wave velocity in Plexiglas is similar to the longitudinal velocity in water; the loading of the water changes the guided wave propagation mode from that of the free plate.

12.5.2 Leaky Guided Wave Modes

The existence of the fluid layers modifies the guided wave propagation characteristics from the plates with stress-free boundary conditions. The modification is particularly significant when the density of the fluid is comparable to the density of the solid layer. Therefore, there is a need to study the guided wave propagation in the fluid loaded plate. This phenomenon is particularly important when there is energy leakage from the solid layer to the fluid environment.

Theoretical study of the leaky guided wave modes can be carried out in a similar manner to that of the waves in multilayered structures. In particular, for a solid layer immersed in a liquid environment, the formulation of partial waves will be the same as that discussed in Section 12.5.1. When no incident wave exists from the upper half-space, the guided wave propagating in the x_1-direction can be expressed as Equation (12.60).

$$\begin{bmatrix} -U_{32}e^{ik\alpha_2 h} & U_{33}e^{ik\alpha_3 h} & U_{34}e^{ik\alpha_4 h} & U_{35}e^{ik\alpha_5 h} & U_{36}e^{ik\alpha_6 h} & 0 \\ 0 & U_{33} & U_{34} & U_{35} & U_{36} & -U_{37} \\ -\rho^{(1)}c_p^2 e^{ik\alpha_2 h} & [-2\mu^{(2)}]e^{ik\alpha_3 h} & [-2\mu^{(2)}]e^{ik\alpha_4 h} & [\rho^{(2)}c_p^2 - 2\mu^{(2)}]e^{ik\alpha_5 h} & [\rho^{(2)}c_p^2 - 2\mu^{(2)}]e^{ik\alpha_6 h} & 0 \\ 0 & [\alpha_3 + U_{33}]\mu^{(2)}e^{ik\alpha_3 h} & [\alpha_4 + U_{34}]\mu^{(2)}e^{ik\alpha_4 h} & [\alpha_5 + U_{35}]\mu^{(2)}e^{ik\alpha_5 h} & [\alpha_3 + U_{36}]\mu^{(2)}e^{ik\alpha_6 h} & 0 \\ 0 & -2\mu^{(2)} & -2\mu^{(2)} & \rho^{(2)}c_p^2 - 2\mu^{(2)} & \rho^{(2)}c_p^2 - 2\mu^{(2)} & -\rho^{(3)}c_p^2 \\ 0 & [\alpha_3 + U_{33}]\mu^{(2)} & [\alpha_4 + U_{34}]\mu^{(2)} & [\alpha_5 + U_{35}]\mu^{(2)} & [\alpha_6 + U_{36}]\mu^{(2)} & 0 \end{bmatrix} \begin{bmatrix} B_2 \\ B_3 \\ B_4 \\ B_5 \\ B_6 \\ B_7 \end{bmatrix} = \begin{bmatrix} 0 \\ 0 \\ 0 \\ 0 \\ 0 \\ 0 \end{bmatrix}$$

$$(12.60)$$

The dispersion curve of the leaky wave can be obtained by searching the possible pairs of (f, k) that will make the determinant of the coefficient matrix zero. When the frequency of the wave is set to be real, typically, real solution of k exists when no energy is leaking into the fluid layers. This condition can only be true when $c_p < C_L^{(1)}$ and $c_p < C_L^{(3)}$; this means the phase velocity is less than the longitudinal velocity of the fluid material. When the phase velocity of the wave is larger than the longitudinal velocity in the fluid layer, the partial waves in the fluid layer will be propagating waves, which means there will be energy transmission toward $x_3 = \pm\infty$. Therefore, an attenuation of wave amplitude will be involved with the propagation of guided waves in the x_1-direction. The general solution of the wave field can be expressed as in Equation (12.61).

$$u(x_1, x_3, t) = A_o F(x_3) e^{i(kx_1 - \omega t)} = A_o F(x_3) e^{ik'x_1 - i\omega t} e^{-k''x_1} \tag{12.61}$$

Here, k' is the real part of the wavenumber, and k'' is the imaginary part of the wavenumber solution in the dispersion curves.

12.5.3 Nonspecular Reflection and Transmission

When the incident wave is not a continuous harmonic plane wave, the phenomenon of wave reflection and transmission will be much more complicated than the plane wave situation. In general, when the incident wave component does not correspond to a guided wave mode, most of the energy will reflect back to the first fluid layer. When the incident wave component corresponds to a guided wave mode, more energy will be coupled to the plate and potentially radiate into the lower fluid layer and

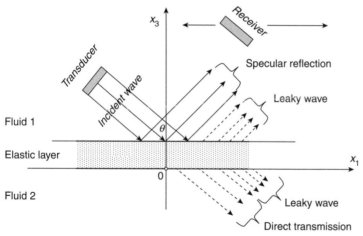

Figure 12.25. Sketch of nonspecular reflection and transmission.

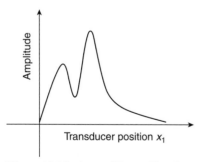

Figure 12.26. A possible profile of nonspecular reflection.

produce a guided wave propagating along the plate. For a guided wave propagating in a plate immersed in water, energy may possibly radiate back to the fluid layers during wave propagation, which forms a leaky Lamb wave mode. A sketch of the physical phenomenon is shown in Figure 12.25.

Because of the existence of the leaky wave produced from guided wave propagation, the receiver can detect reflection signals outside the specular reflection region. The entire reflection wave field is called a nonspecular reflection and transmission. Figure 12.26 shows a sketch of a nonspecular reflection profile. A possible null region is also shown in the sketch.

12.6 Exercises

1. What is the assumption made in the solution of wave propagation along a solid-solid interface?
2. Are Stoneley waves dispersive? Are Scholte waves dispersive? Provide a brief physical explanation for this characteristic.
3. How can we examine the possibility of guided wave propagation in a layer embedded between two half-spaces? How many boundary conditions exist when Rayleigh–Lamb type waves are considered? How many boundary conditions exist when SH type waves are considered?
4. For the case when complex roots are extracted from the Stoneley wave characteristic equation, provide a physical interpretation with respect to wave propagation along the interface.
5. Plot a possible dispersion curve for a layer on a half-space. Consider two cases where the bulk wave velocities of the layer are higher or lower than the velocities of the half-space.
6. How many partial waves exist for R-L type guided waves propagating through two layers on a half-space? How many boundary conditions exist in this system? Please list these boundary conditions.
7. What is the low-frequency limit phase velocity value for a layer on a half-space when R-L type guided waves are considered? Given the bulk wave velocity of the layer is smaller than that of the half-space.
8. For a soft layer on a half-space, what is the high-frequency or small wavelength limit on phase velocity?
9. Show that the upper limit of the Love wave velocity is the shear velocity in the half-space. Provide a physical explanation of this result.
10. Outline a procedure to obtain a dispersion curve for two layers on a half-space.
11. In general, what order determinant must be solved for R-L type guided waves in a ten-layered, traction-free structure? Given all the layers are solid.
12. What order determinant must be solved for R-L type guided waves in a ten-layered solid structure immersed in fluid?
13. What order determinant must be solved for R-L type guided waves in a ten-layered structure? Given the fifth layer is a fluid layer and the outer surfaces of the structure are traction free.
14. Reconsider the SH type guided waves for problems 10, 11, and 12.
15. How might it be possible to generate guided waves across a three-layered structure in such a way that strong modes can be received on a single layer?
16. How would you use wave structure information to locate a defect in a specific interface layer of a four-layer structure?

17. [Program problem] Write a program to generate phase velocity dispersion curves for SH type waves in a two-layered structure. Generate group velocity dispersion curves from the phase velocity dispersion curve and calculate wave structure for a particular guided wave mode.
18. Follow up on the ice detection example. What would the next step be in implementation of an ice detection system on an aircraft?
19. How would you differentiate ice from water on an aircraft with guided wave? Or ice from glycol from water?

12.7 REFERENCES

Achenbach, J. D., and Epstein, H. I., (1967). Dynamic interaction of a layer and a half space, *J. Eng. Mech. Division* 5:27–42.

Auld, B. A. (1990). *Acoustic Fields and Waves in Solids*. Malabar, FL: Krieger Publishing Company.

Chimenti, D. E. (1997). Guided waves in plates and their use in material characterization, *Appl Mech Rev* 50(5): 247–84.

Chimenti, D. E., and Rokhlin, S. I., (1990). Relationship between leaky Lamb modes and reflection coefficient zeroes for a fluid-coupled elastic layer, *J. Acoust. Soc. Am.* 88(3): 1603–11.

Fu, C. Y. (1946). Studies on seismic waves: II. Rayleigh waves in a superficial layer, *Geophys*, 11(1/4): 10–23.

Gao, H. (2007). *Ultrasonic Guided Wave Mechanics for Composite Material Structural Health Monitoring*, PhD thesis, Pennsylvania State University.

Gao, H., Ali, S., and Lopez, B. (2010). Efficient delaminating detection in multilayered composites using ultrasonic guided wave EMATs, *NDT&E International*, 43: 316–22.

Gao, H., Rose, J. L., (2010). Goodness dispersion curves for ultrasonic guided wave based SHM: A sample problem in corrosion monitoring, *The Aeronautical Journal* 114(1151): 797–804.

(2009). Ice detection and classification on an aircraft wing with ultrasonic shear horizontal guided waves. *IEEE Trans. Ultrason. Ferroelectr. Freq. Control.* 56(2): 334–44.

Haskell, N. A. (1953). The dispersion of surface waves on multilayered media, in G. D. Lauderback, H. Benioff, and J. B. Macelwane (Eds.), *Bulletin of the Seismological Society of America*, vol. 43, pp. 17–34. Berkeley: University of California Press.

Lowe, M. J. S. (1995). Matrix techniques for modeling ultrasonic waves in multilayered media, *IEEE Trans. Ultrason. Ferroelec. Freq. Contr.* 42(4): 525–42.

Miklowitz, J. (1978). *The Theory of Elastic Waves and Waveguides*. New York: North-Holland.

Nayfeh, A. H. (1995). *Wave Propagation in Layered Anisotropic Media*. Amsterdam, Lausanne, New York, Oxford, Shannon, & Tokyo: Elsevier.

Pilant, W. L. (1972). Complex roots of the Stoneley-wave equation, *Bull. Seism. Soc. Am.* 62(1): 285–99.

Rose, J. L. (1999). *Ultrasonic Waves in Solid Media*. Cambridge University Press.

Rose, Joseph L., Aleksander B. Pilarski, Jeffrey M. Hammer, Michael T. Peterson, and Philip O. Readio. Contaminant Detection System. The B. F. Goodrich Company, assignee. Patent 5,629,485. 13 May 1997.

Rose, J. L., Zhu, W., and Zaidi, M. (1998). Ultrasonic NDE of titanium diffusion bonding with guided waves, *Mat. Eval.* (56)4: 535–9.

Sezawa, K. (1927). Dispersion of elastic waves propagated on surface of stratified bodies and curved surfaces, *Bull. Earthquake Res. Inst., Univ. Tokyo* 3: 1–8.

Stoneley, R. (1924). Elastic waves at the surface of separation of two solids, *Proc. Roy. Soc. London* 106: 416–28.

Thomson, W. T. (1950). Transmission of elastic waves through a stratified solid medium, *J. Appl. Phys.* 21: 89–93.

USAir Flight 405. (2012, December 25). In *Wikipedia, The Free Encyclopedia*. Retrieved from http://en.wikipedia.org/w/index.php?title=USAir_Flight_405&oldid=529743357.

13 Source Influence on Guided Wave Excitation

13.1 Introduction

Guided wave dispersion curve calculations are based on the assumption of an infinite continuous plane wave excitation producing a set of particular phase velocity values at specific frequencies. However, in real applications, the excitation sources are of a finite size over a finite frequency spectrum. This chapter establishes the guidelines for evaluating the effects of excitation sources on wave excitation. In particular, we address the problem of guided wave excitation and propagation in a traction-free plate.

Many excitation mechanisms can be used to generate ultrasonic waves in a solid medium. The commonly used sources for guided wave excitation are piezoelectric transducers, electromagnetic acoustic transducers (EMAT), magnetostrictive devices, physical impact, and laser ultrasonics. The piezoelectric transducer can be used in a normal incident or oblique incidence situation with an angle wedge. Instead of propagating the waves from the transducer to the structure, the EMAT and laser, for example, generates ultrasound within the structure either as a surface loading or a body loading. Most recently, for purposes of structural health monitoring (SHM), piezoelectric wafer transducers are attached directly to the structure to generate specific kinds of ultrasonic guided waves in the structure.

Besides using the transducer in a single-element fashion, transducers using multiple elements are also used to improve the guided wave mode and frequency selection performance. Recently, ultrasonic phased arrays with various loading patterns have been introduced to achieve beam steering, beam focusing, and guided wave mode tuning.

Because of a variety of available loading sources in wave excitation mechanisms, transducer geometries, and excitation signal configuration, each excitation situation has its own set of design parameters. Some sample problems are discussed extensively in the text and some are left for the students in the exercises. Two theoretical methods are introduced by abstracting the excitation sources into loadings in three coordinate directions. One is an integral transform method; the other is a normal mode expansion (NME) method.

With significant contribution from Huidong Gao.

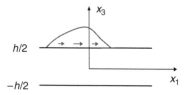

Figure 13.1. A model for wave excitation in a plate with distributed shear loading in the x_1 direction.

13.2 Integral Transform Method

Many integral transforms have been developed in solving partial differential equations. In this section, an example of a Fourier transform method is discussed for guided wave excitation considerations.

13.2.1 A Shear Loading Example

For a wave excitation in a solid layer, the coordinate system used is shown in Figure 13.1. Also shown in the figure is a distributed loading on the surface of the plate in the x_1 direction.

Recall that for ultrasonic wave propagation in an isotropic solid media, Equation (13.1) is the wave equation for Lamb waves at angular frequency ω. Here, ϕ is the scalar potential, ψ is the second component of the vector potential, and c_L and c_T are longitudinal and shear wave velocity respectively.

$$\frac{\partial^2 \phi}{\partial x_1^2} + \frac{\partial^2 \phi}{\partial x_3^2} + \frac{\omega^2}{c_L^2} \phi = 0$$
$$\frac{\partial^2 \psi}{\partial x_1^2} + \frac{\partial^2 \psi}{\partial x_3^2} + \frac{\omega^2}{c_T^2} \psi = 0 \tag{13.1}$$

The loading on the top surface can be expressed as $\tau(x_1)$, which is a shear loading on the surface normal to x_3 and in the direction of x_1. The distributed loading is a function of x_1.

The Fourier transform and inverse Fourier transform for an arbitrary function $f(x_1)$ is in Equation (13.2). Here, k is the corresponding wavenumber.

$$\tilde{f}(k) = \int_{-\infty}^{+\infty} f(x_1) e^{-ikx_1} dx_1$$
$$f(x_1) = \frac{1}{2\pi} \int_{-\infty}^{+\infty} \tilde{f}(\xi) e^{i\xi x_1} d\xi \tag{13.2}$$

Take the Fourier transform of Equation (13.1) from the spatial x_1 domain into the wavenumber domain, thus reading Equation (13.3).

$$\frac{\partial^2 \tilde{\phi}}{\partial x_3^2} + p^2 \tilde{\phi} = 0$$
$$\frac{\partial^2 \tilde{\psi}}{\partial x_3^2} + q^2 \tilde{\psi} = 0 \tag{13.3}$$

Here, the expressions of p and q are illustrated in Equation (13.4).

$$p^2 = \frac{\omega^2}{c_L^2} - k^2$$
$$q^2 = \frac{\omega^2}{c_T^2} - k^2 \qquad (13.4)$$

A general solution to the equations in Equation (13.3) is Equation (13.5):

$$\tilde{\phi} = A_1 \sin(px_3) + A_2 \cos(px_3)$$
$$\tilde{\psi} = B_1 \sin(qx_3) + B_2 \cos(qx_3) \qquad (13.5)$$

Here, A_1, A_2, B_1, and B_2 are four undetermined coefficients.

The displacement field can be expressed as in Equation (13.6) using $\mathbf{u} = \nabla\phi + \nabla \times \boldsymbol{\Psi}$:

$$\tilde{u}_1 = ik\tilde{\phi} + \frac{d\tilde{\psi}}{dx_3}$$
$$\tilde{u}_3 = \frac{d\tilde{\phi}}{dx_3} - i\xi\tilde{\psi} \qquad (13.6)$$

The stress fields in the wavenumber domain are shown in Equation (13.7) using strain displacement and constitutive equation.

$$\tilde{\sigma}_{31} = \mu\tilde{\gamma}_{31} = \mu\left(\frac{d\tilde{u}_1}{dx_3} + ik\tilde{u}_3\right)$$
$$\tilde{\sigma}_{33} = \mu\tilde{\varepsilon}_{11} + (\lambda+2\mu)\tilde{\varepsilon}_{33} = \mu(ik\tilde{u}_1) + (\lambda+2\mu)\frac{d\tilde{u}_3}{dx_3} \qquad (13.7)$$

Notice here, ε_{11}, ε_{33}, and γ_{31} are the engineering strain components, while λ and μ are the elastic constants of the isotropic material.

Taking the Fourier transform of the loading source, the boundary conditions in the wavenumber domain are presented in Equation (13.8):

$$\tilde{\tau}(k) = \int_{-\infty}^{+\infty} \tau(x_1)e^{-ikx_1}dx_1, \quad \text{at } x_3 = h/2 \qquad (13.8)$$

Recall that the solution of Lamb waves in a single-layer isotropic plate can be separated into a symmetric part and an antisymmetric part. In the wavenumber domain, the separation of the two types of solutions is still valid. The graphic sketch of the separation is shown in Figure 13.2. The loading on the top surface is separated into a combination of a symmetric loading and an antisymmetric loading. For the symmetric loading, the entire system is symmetric with respect to $x_3 = 0$; for the antisymmetric loading, the system is antisymmetric with respect to $x_3 = 0$. For these two cases, the response of the structure of the guided waves is expected to be symmetric and antisymmetric with respect to the $x_3 = 0$ plane.

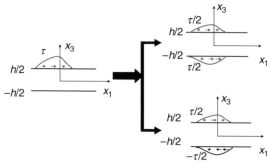

Figure 13.2. Model separation from a single-surface loading to a symmetric and antisymmetric loading.

Solution for the Symmetric Part

To have the wave field symmetric with respect to $x_3 = 0$, the displacement u_1 must be a symmetric function of x_3, and u_3 must be an antisymmetric function of x_3. Therefore, the solutions are in Equation (13.9).

$$\tilde{u}_1 = ikA_2\cos(px_3) + qB_1\cos(qx_3)$$
$$\tilde{u}_3 = pA_2\sin(px_3) - ikB_1\sin(qx_3) \tag{13.9}$$

The boundary conditions are in Equation (13.10).

$$\sigma_{31}\big|_{x_3=\pm h/2} = \frac{\tilde{\tau}}{2}$$
$$\sigma_{33}\big|_{x_3=\pm h/2} = 0 \tag{13.10}$$

These four boundary conditions will reduce to two groups of identical equations when the system symmetry is considered.

$$\begin{bmatrix} -2ikp\sin(ph/2) & (k^2-q^2)\sin(qh/2) \\ (k^2-q^2)\cos(ph/2) & 2ikq\cos(ph/2) \end{bmatrix}\begin{bmatrix} A_2 \\ B_1 \end{bmatrix} = \begin{bmatrix} \tilde{\tau}/2\mu \\ 0 \end{bmatrix} \tag{13.11}$$

Using the Gaussian elimination technique, the solutions of A_2 and B_1 are expressed in Equation (13.12).

$$A_2 = \frac{\tilde{\tau}}{2\mu}\frac{N_{A_2}}{D_s}$$
$$B_1 = \frac{\tilde{\tau}}{2\mu}\frac{N_{B_1}}{D_s} \tag{13.12}$$

The expressions of N_{A_2}, N_{B_1}, and D_s are in Equation (13.13).

$$N_{A_2} = 2ikq\cos(qh/2)$$
$$N_{B_1} = (k^2-q^2)\cos(ph/2)$$
$$D_s = (k^2-q^2)^2\cos(ph/2)\sin(qh/2) + 4k^2pq\sin(ph/2)\cos(qh/2) \tag{13.13}$$

Solution for the Antisymmetric Part

For antisymmetric waves, the displacement u_1 must be an antisymmetric function of x_3, and u_3 must be a symmetric function of x_3. Therefore, the solutions are expressed in Equation (13.14)

$$\tilde{u}_1 = ikA_1 \sin(px_3) - qB_2 \sin(qx_3)$$
$$\tilde{u}_3 = pA_1 \cos(px_3) - ikB_2 \cos(qx_3)$$

(13.14)

The boundary conditions are next in Equation (13.15)

$$\sigma_{31}\big|_{x_3=h/2} = \frac{\tilde{\tau}}{2}$$
$$\sigma_{31}\big|_{x_3=-h/2} = -\frac{\tilde{\tau}}{2}$$
$$\sigma_{33}\big|_{x_3=\pm h/2} = 0$$

(13.15)

These four boundary conditions will reduce to two groups of identical equations when the system symmetry is considered.

$$\begin{bmatrix} 2ikp\cos(ph/2) & (k^2-q^2)\cos(qh/2) \\ (k^2-q^2)\sin(ph/2) & 2ikq\sin(qh/2) \end{bmatrix}\begin{bmatrix} A_1 \\ B_2 \end{bmatrix} = \begin{bmatrix} \tilde{\tau}/2\mu \\ 0 \end{bmatrix}$$

(13.16)

The solution to this linear system of equations is as follows in Equation (13.17).

$$A_1 = \frac{\tilde{\tau}}{2\mu}\frac{N_{A_1}}{D_A}$$
$$B_2 = \frac{\tilde{\tau}}{2\mu}\frac{N_{B_2}}{D_A}$$

(13.17)

The expressions of N_{A_1}, N_{B_2}, and D_A are in Equation (13.18).

$$N_{A_1} = 2ikq\,\sin(qh/2)$$
$$N_{B_2} = -(k^2-q^2)\sin(ph/2)$$
$$D_A = (k^2-q^2)^2\,\sin(ph/2)\cos(qh/2) + 4k^2pq\cos(ph/2)\sin(qh/2)$$

(13.18)

Spatial Domain Solutions

The total solution of the system can be obtained by combining the solutions from the symmetric part and antisymmetric part. Take the displacement component u_1, for example; the formulation is Equation (13.19).

$$\tilde{u}_1 = ikA_2\cos(px_3) + qB_1\cos(qx_3) + ikA_1\sin(px_3) - qB_2\sin(qx_3)$$
$$= \tilde{\tau}\frac{1}{2\mu}[ik\frac{N_{A_2}}{D_s}\cos(px_3) + q\frac{N_{B_1}}{D_s}\cos(qx_3) + ik\frac{N_{A_1}}{D_A}\sin(px_3) - q\frac{N_{B_2}}{D_A}\sin(qx_3)]$$

(13.19)

In Equation (13.19), $\tilde{\tau}$ is related to the source distribution, while the rest is related to the structure. The spatial domain solution of the wave field can be obtained by taking the inverse Fourier transform to $\tilde{u}_1$ using Equation (13.20).

$$u_1(x_1, x_3) = \frac{1}{2\pi} \int_{-\infty}^{+\infty} \tilde{u}_1(k, x_3) e^{ikx_1} dk \tag{13.20}$$

Notice that the integrands in Equation (13.20) can be singular when $D_S = 0$, or $D_A = 0$. Comparing Equation (13.13) and Equation (13.18) with the dispersion curve formulation, the singularity points of the integrand corresponds to the guided wave modes at stress-free boundary conditions. Considering the singularity, the integration in Equation (13.20) can be evaluated using the residue theorem.

$$res(f(z)) = \frac{1}{2\pi} \int_{\Gamma} f(z) dz \tag{13.21}$$

Applying Equation (13.21), the integration is equivalent to Equation (13.22)

$$u_1(x_1, x_3) = (i) \sum_{k=k_A^n, k_s^n} res(\tilde{u}_1(k, x_3) e^{ikx_1}) \tag{13.22}$$

Here, k_A and k_S are the solution of k that satisfies $D_A = 0$ and $D_S = 0$ respectively.

Numerical solutions of these equations have been obtained when we solve for dispersion curves in a single-layered structure.

13.3 Normal Mode Expansion Method

The derivation using the integral transform method indicates that the wave field excited from a finite source is related to the guided wave mode solutions in a stress-free situation. In this section, a NME method is used to study the wave excitation by directly expressing the wave field into combinations of guided wave modal solutions. Before we start the NME for guided waves, some preliminaries are first discussed in the following.

The NME technique of solving forced loading problems is analogous to the so-called eigenfunction expansion methods discussed in most mathematics textbooks. A point on the guided wave dispersion curve serves as an eigenvalue, and the wave structure serves as an eigenfunction. The efficacy of the NME depends directly on two main considerations: completeness and orthogonality of the eigenfunctions. These two aspects are discussed in B. A. Auld's book (1990) in detail from theoretical derivations of reciprocity. A review of the process is also presented here for convenience.

A complete set of guided wave modal solutions includes all possible solutions of the characteristic equation for Lamb waves in a plate. By solving the guided wave dispersion curves, we have obtained the solutions k that are real and positive for a given positive value of frequency. These wave modes correspond to the rightward-propagating waves. It is not difficult to prove that, for a given wave mode (f, k), there is a leftward-propagating mode $(f, -k)$, which also satisfies the condition of wave mode propagation. When the equation is solved in complex space, there exists other solutions of ξ that are either purely imaginary or have a complex value. These solutions are called *nonpropagating modes*, or *evanescent wave modes*, because their amplitude decays exponentially with respect to x_1. The typical effective distance of evanescent waves is within several millimeters. Therefore, only the excitation characteristics of the propagating modes are of critical interest for purposes of nondestructive evaluation (NDE) and SHM.

The proof of orthogonality of guided wave modes is important for the application of the NME theory. Auld (1990) proved the orthogonality of guided wave modes in lossless waveguides through the derivation of the reciprocity relation in piezoelectric media. Equation (13.23) illustrates the approach.

$$\nabla \bullet [-\mathbf{v}_{II}^* \bullet \sigma_I - \mathbf{v}_I \bullet \sigma_{II}^*] = -\frac{\partial}{\partial t}\left([\mathbf{v}_{II}^* \quad \sigma_{II}^*]\begin{bmatrix} \rho & 0 \\ 0 & :s: \end{bmatrix}\begin{bmatrix} \mathbf{v}_I \\ \sigma_I \end{bmatrix}\right) + (\mathbf{v}_{II}^* \bullet \mathbf{F}_I + \mathbf{v}_I \bullet \mathbf{F}_{II}^*) \quad (13.23)$$

Here, the subscripts *I* and *II* denote two wave mode solutions. In addition, the following equation holds when both modes are obtained from stress-free conditions and are of the same frequency ω.

$$\mathbf{v}_I = \mathbf{v}_m(x_3)\exp(i(k_m x_1 - \omega t)$$

$$\mathbf{v}_{II} = \mathbf{v}_n(x_3)\exp(i(k_n x_1 - \omega t)$$

$$\sigma_I = \sigma_m(x_3)\exp(i(k_m x_1 - \omega t)) \quad (13.24)$$

$$\sigma_{II} = \sigma_n(x_3)\exp(i(k_n x_1 - \omega t))$$

$$\mathbf{F}_I = 0$$

$$\mathbf{F}_{II} = 0$$

Therefore, the reciprocity relation reduces to Equation (13.25)

$$\nabla \bullet [-\mathbf{v}_{II}^* \bullet \sigma_I - \mathbf{v}_I \bullet \sigma_{II}^*] = 0 \quad (13.25)$$

Integration of Equation (13.25) over a cross section of the waveguide will lead to Equation (13.26).

$$i(k_m - k_n)4P_{mn} = \{-\mathbf{v}_n^* \bullet \sigma_m - \mathbf{v}_m \bullet \sigma_n^*\} \bullet \hat{x}_3 \Big|_{x_3=0}^{x_3=H} \quad (13.26)$$

where,

$$P_{mn} = \frac{1}{4}\int_0^H \{-\mathbf{v}_n^* \bullet \sigma_m - \mathbf{v}_m \bullet \sigma_n^*\} \bullet \hat{x}_1\, dx_3 \quad (13.27)$$

For the wave mode solutions obtained earlier, stress-free boundary conditions are satisfied, $\sigma_m = 0$ and $\sigma_n^* = 0$. Therefore, the right-hand side of Equation (13.26) is zero. For propagating modes *m* and *n*, $\xi_m \neq \xi_n$, and Equation (13.28) holds.

$$P_{mn} = 0 \text{ for m} \neq \text{n} \quad (13.28)$$

This equation shows that the solutions of two different guided wave modes are orthogonal with respect to Equation (13.27). When the two wave modes are the same, that is, $m = n$, Equation (13.27) reduces to Equation (13.29); this expression is the same as the definition of the complex Poynting's vector.

$$P_{mm} = \frac{1}{2} \int_0^H \left(-\mathbf{v}_m^* \bullet \hat{x}_1 \right) dx_3 \tag{13.29}$$

See further discussion of the Normal Mode Expansion Technique and the completeness and orthogonality of Eigen functions and the complex reciprocity elation associated with the excitability function and source influence development in Rose (1999).

13.3.1 Normal Mode Expansion in Harmonic Loading

Now, the guided wave mode solutions are proved to be an orthogonal set of basic functions for wave propagation in solid media. The NME are then used to express the fields in the loaded waveguide as a combination of the modal solutions. Because only $\mathbf{v}$ and $\sigma \bullet \hat{x}_1$ are involved in Equation (13.29), the expansion equation can be written as Equation (13.30).

$$\mathbf{v}(x_1, x_3) = \sum_n a_n(x_1) \mathbf{v}_n(x_3)$$
$$\sigma(x_1, x_3) \bullet \hat{x}_1 = \sum_n a_n(x_1) \sigma_n(x_3) \bullet \hat{x}_1 \tag{13.30}$$

Here, $a_n(x_1)$ are the combination coefficient functions to be determined for a specific loading situation. In the numerical calculation, the summation covers all the wave modes that make significant contributions to the overall wave field. The common time-dependent factor $\exp(-i\omega t)$ is suppressed. At this point, we make two important notes. (1) The same set of amplitude coefficients is used for the expansion of both the particle velocity and stress field. This is because we assume that $\mathbf{v}_n$ and σ_n are used to represent a single mode component and a_n is a weight coefficient expressing the amplitude of the mode. (2) a_n is a function of x_1. This is because different from the situation of pure wave propagation, the combination of wave mode amplitude changes at different x_1 position because of the effect of excitation sources.

The emphasis of the following work is to determine the coefficient $a_n(x_1)$. Assuming the unknown field is solution I, and the solution II is the wave field of mode n,

$$\mathbf{v}_{II} = \mathbf{v}_n(x_3) e^{i(k_n x_1)}$$
$$\sigma_{II} = \sigma_n(x_3) e^{i(k_n x_1)} \tag{13.31}$$
$$\mathbf{F}_{II} = 0$$

Substituting Equation (13.30) and Equation (13.31) into the reciprocity relation Equation (13.23), and integrating them over the thickness direction of the plate, we obtain Equation (13.32).

$$4 P_{nn} \left(\frac{\partial}{\partial x_1} - i k_n \right) a_n(x_1) = f_{sn}(x_1) + f_{vn}(x_1) \tag{13.32}$$

Here, P_{nn} is the entire power flow in the x_1 direction, f_{sn} is the surface loading, and f_{vn} is the body force loading. In general, the loading terms are functions of x_1. Equation (13.33) and Equation (13.34) are the detailed expressions for the surface loading and the body loading respectively.

$$f_{sn}(x_1) = \{\mathbf{v}_n^*(x_3) \bullet \sigma(x_1, x_3) + \mathbf{v}(x_1, x_3) \bullet \sigma_3^*\} \bullet \hat{x}_3\Big|_{x_3=0}^{H} \tag{13.33}$$

$$f_{vn}(x_1) = \int_0^H \mathbf{v}_n^*(x_3) \bullet \sigma(x_1, x_3) dx_3 \tag{13.34}$$

In general, a surface loading can be either a prescription of particle velocity at the surface ($\mathbf{v}(x_1)$ at the top or bottom surface) or a traction distribution described by $\sigma(x_1)$ at the top or bottom surface. However, because the modal solutions are obtained using stress-free boundary conditions, only traction loading will be effective in Equation (13.33).

Equation (13.32) is a partial differential equation of the function $a_n(x_1)$, that can be solved using a standard integration technique. Assuming the loading area is within $[L_1 \ L_2]$, the wave propagation in the positive x_1 direction must have zero amplitude at the left side of the source. Equation (13.35) is the solution for rightward-propagating wave modes.

$$
\begin{aligned}
a_n(x_1) &= 0, \quad x_1 \le L_1 \\
a_n(x_1) &= \exp(ik_n x_1) \int_{L_1}^{x_1} \frac{f_{sn}(\eta) + f_{vn}(\eta)}{4P_{nn}} \exp(-ik_n \eta) d\eta, \quad L_1 \le x_1 \le L_2 \\
a_n(x_1) &= \exp(ik_n x_1) \int_{L_1}^{L_2} \frac{f_{sn}(\eta) + f_{vn}(\eta)}{4P_{nn}} \exp(-ik_n \eta) d\eta, \quad x_1 \ge L_2
\end{aligned} \tag{13.35}
$$

When the position is outside the source region, $a_n(x_1)$ is a harmonic wave function of x_1 with amplitude as given in Equation (13.36). This means that the amplitude of a given mode is a constant outside the source region.

$$|a_n(x_1)| = \left|\int_{L_1}^{L_2} \frac{f_{sn}(\eta) + f_{vn}(\eta)}{4P_{nn}} \exp(-ik_n \eta) d\eta\right| \tag{13.36}$$

Similarly, for leftward-propagating waves, $n < 0$, the solutions are Equation (13.37):

$$
\begin{aligned}
a_{-n}(x_1) &= 0, \quad x_1 \ge L_2 \\
a_{-n}(x_1) &= \exp(ik_{-n} x_1) \int_{L_2}^{x_1} \frac{f_{s(-n)}(\eta) + f_{v(-n)}(\eta)}{4P_{-n-n}} \exp(-ik_{-n} \eta) d\eta, \quad L_1 \le x_1 \le L_2 \\
a_{-n}(x_1) &= \exp(ik_{-n} x_1) \int_{L_2}^{L_1} \frac{f_{s-n}(\eta) + f_{v-n}(\eta)}{4P_{-n-n}} \exp(-ik_{-n} \eta) d\eta, \quad x_1 \le L_1
\end{aligned} \tag{13.37}
$$

The entire wave field is a summation of all the leftward-propagating waves and the rightward-propagating waves. In a strict sense, a complete solution will include propagating wave modes as well as the evanescent wave modes. The wavenumber of an evanescent mode is either purely imaginary or complex. This imaginary part of the wavenumber will introduce an exponentially decaying factor in the wave field in the propagation direction. Typically, the attenuation factor is large enough that the amplitude of the evanescent wave modes decays around or less than a wavelength of a propagating wave mode at the same frequency. Therefore, for practical purpose, the evanescent waves can be neglected outside the source region.

The NME technique provides a general solution method for guided wave excitation. Because it is directly built upon the knowledge of guided wave propagation mode analysis, it has a more direct physical insight than the integral transform method discussed in Section 13.2. Several examples are discussed in the following.

[Sample Problem] Bonded Wafer Actuator

For the specific problem described in Figure 13.1, only shear loading exists in the upper surface.

$$f_{sn}(x_1) = \left| \{ \mathbf{v}_n^*(x_3) \bullet \mathbf{\sigma}(x_1, x_3) \} \bullet \hat{x}_3 \right|_{x_3=0}^{H}$$
$$= v_{n1}^*(x_3)\sigma_{31}(x_3)\big|_{x_3=H} \tag{13.38}$$
$$f_{vn}(x_1) = \int_0^H \mathbf{v}_n^*(x_3) \bullet \mathbf{\sigma}(x_1, x_3)\, dx_3 = 0$$

In this case, the coefficient of the mode n at the right side of the excitation region is in Equation (13.39).

$$a_n(x_1) = \frac{v_{n1}^*}{4P_{nn}} \left[\int_{L_1}^{L_2} \tau(\eta)\exp(-ik_n\eta)d\eta \right] \exp(ik_n x_1) \tag{13.39}$$

Here, v_{n1}^* is a complex conjugate of the particle velocity component in the x_1 direction at the top surface for the mode n. P_{nn} is the integrated power flow in the x_1 direction. Therefore, the first term is related to the wave mode. The second term in the bracket is related to the excitation source. Because the excitation is only within the range of $[L_1\ L_2]$, then

$$\int_{L_1}^{L_2} \tau(\eta)\exp(-ik_n\eta)d\eta = \int_{-\infty}^{+\infty} \tau(\eta)\exp(-ik_n\eta)d\eta \tag{13.40}$$

This is the value of the Fourier transform of the excitation source at wavenumber k_n. Based on the previous discussion, the guided wave excitation problem can be separated into two key factors. One is an attribute of the guided wave mode in a given structure, which is also called *wave mode excitability*. The other is the wavenumber spectrum of the excitation source in a harmonic loading, which is also referred to as *source influence*.

For the wave excitation with a surface loading, the particle velocity at the excitation surface is the key physical quantity that affects the excitability of the wave mode. Therefore, two wave mode excitability functions can be defined for Lamb wave propagation. For the shear loading case, the wave excitability is defined as $|v_1|$; and for the normal loading situation, the wave excitability function is defined as $|v_3|$. As an example, the mode wave excitability function in a 1 mm-thick aluminum

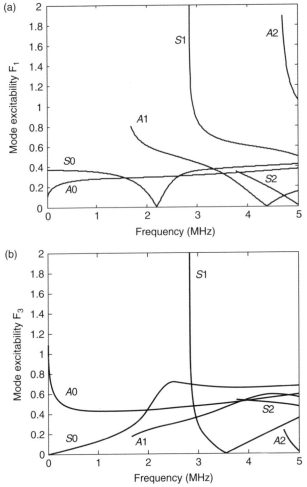

Figure 13.3. Lamb wave mode excitability dispersion curve. (a) shear loading in the x_1 direction. (b) normal loading in the x_3 direction.

plate is plotted in Figure 13.3. This figure indicates that different wave modes have different amplitudes of response. At the low-frequency region, the excitability of the S0 mode is larger than that of the A0 mode using shear loading. However, the excitability of S0 mode is much smaller than that of the A0 mode when normal loading is applied.

According to Equation (13.39) and Equation (13.40), the Fourier transform of the excitation pattern determines the wavenumber spectrum. As an example, when a piston-type normal loading is considered, the corresponding wavenumber spectrum is shown in Figure 13.4. In this example, the loading is evenly distributed within the width of the transducer and the excitation is in the x_3 direction. The figure indicates that for piston-shaped normal loading, the spectrum has large amplitude at low value of wavenumber. In addition, the amplitude oscillates with an increase of wavenumber. Quantitative analysis indicates that the zero values on the spectrum are found when the wavenumber is some integer multiples of $2\pi/w$. Here w is the width of the transducer.

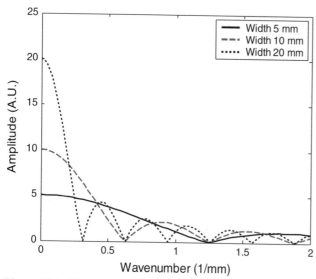

Figure 13.4. Wavenumber spectrum of the piston-type normal loading for the excitation widths of 5 mm, 10 mm, and 20 mm.

When a frequency is given, the wavenumber spectrum can be transformed into a phase velocity spectrum using the relation $c_\rho = \dfrac{\omega}{k}$. Figure 13.5(a) shows the phase velocity spectrum of a transducer with three different widths at a frequency of 1 MHz. These normal loadings have large amplitudes at large phase velocity regions. However, the phase velocity values of a useful guided wave mode are typically within 20 km/s. The truncated spectrum is shown in Figure 13.5(b).

13.3.2 Transient Loading Source Influence

The transient excitation signals are typically used in ultrasonic guided wave applications. Therefore, it is important to study the structural response and source influence of a transient loading source. To solve this problem, we can express the transient excitation signal as a linear combination of time harmonic excitation sources using a Fourier transform. To simplify the formulation, only surface loading at the top surface is considered in the following. The formulation of other types of loadings can be derived using the same method.

$$\mathbf{F}(x_1,\omega) = \int_{-\infty}^{+\infty} \mathbf{F}(x_1,t)e^{-i\omega t}\,dt \tag{13.41}$$

Here, $\mathbf{F}$ is the vector of the surface traction in three directions. Using Equation (13.33), the NME for each frequency component is shown in Equation (13.42).

$$a_n(\omega) = e^{i\xi_n x_1} \int_{L_1}^{L_2} \frac{F_n(x_1,\omega)}{4P_{nn}} e^{-i\xi_n \eta}\,d\eta \tag{13.42}$$

Here, $F_n(x_1,\omega) = \displaystyle\sum_{m=1}^{3} v_{nm}^* F_m(x_1,\omega)$

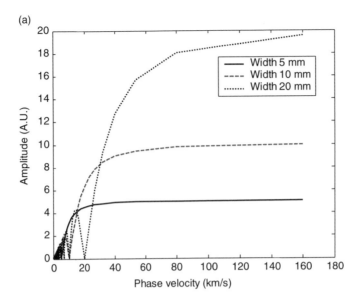

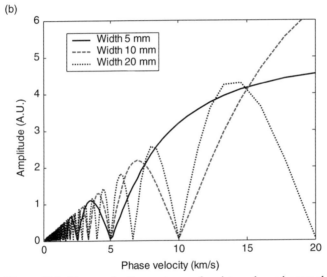

Figure 13.5. Phase velocity spectrum of a piston-shaped normal loading in a width of 5 mm, 10 mm, and 20 mm. (a) Spectrum in a broad phase velocity range, (b) Spectrum of phase velocity below 20 km/s.

Finally, the entire wave field solution is a linear combination of the signals of the entire frequency spectrum.

$$U(x_1, x_3, t) = \frac{1}{2} \int_{-\infty}^{+\infty} \sum_n a_n(\omega) U_n(x_3) e^{i(\xi_n x_1 + \omega t)} d\omega \qquad (13.43)$$

By substituting Equation (13.42) into Equation (13.43), we obtain Equation (13.44).

$$
\begin{aligned}
U(x_1,x_3,t) &= \frac{1}{2\pi}\int_{-\infty}^{+\infty}\sum_n a_n(\omega)U_n(x_3)e^{i(\omega t)}d\omega \\
&= \frac{1}{2\pi}\sum_n\left\{\left[\int_{-\infty}^{+\infty}(e^{i\xi_n x_1}\int_{L_1}^{L_2}F(x_1,\omega)e^{-i\xi_n\eta}d\eta)e^{i\omega t}d\omega\right]\frac{U_n(x_3)}{4P_{nn}}\right\} \\
&= \frac{1}{2\pi}\sum_n\left\{\left[\int_{-\infty}^{+\infty}(\int_{-\infty}^{+\infty}F_n(x_1,\omega)e^{-i\xi_n\eta}d\eta)e^{i\omega t}d\omega\right]\frac{U_n(x_3)}{4P_{nn}}e^{i(\xi_n x_1)}\right\} \\
&= \frac{1}{2\pi}\sum_n\left\{\left[\int_{-\infty}^{+\infty}(\int_{-\infty}^{+\infty}(\int_{-\infty}^{+\infty}F(x_1,t)e^{-\omega t}dt)e^{-i\xi_n\eta}d\eta)e^{i\omega t}d\omega\right]\frac{U_n(x_3)}{4P_{nn}}e^{i(\xi_n x_1)}\right\} \quad (13.44) \\
&= \frac{1}{2\pi}\sum_n\left\{\left[\int_{-\infty}^{+\infty}(\int_{-\infty}^{+\infty}\int_{-\infty}^{+\infty}F(x_1,t)e^{-(\omega t+\xi_n\eta)}dt d\eta)e^{i\omega t}d\omega\right]\frac{U_n(x_3)}{4P_{nn}}e^{i(\xi_n x_1)}\right\} \\
&= \frac{1}{2\pi}\sum_n\left\{\left[\int_{-\infty}^{+\infty}(\tilde{F}(\xi,\omega))e^{i\omega t}d\omega\right]\delta(\xi=\xi_n)\frac{U_n(x_3)}{4P_{nn}}e^{i(\xi_n x_1)}\right\}
\end{aligned}
$$

Besides the phase velocity spectrum, there is usually a frequency spectrum associated with an excitation source because a transient loading is typically used in an ultrasonic NDE test. Commonly used excitation signals are square wave, spike pulse, tone burst, and windowed tone burst. The frequency spectrum of these signals is discussed in detail in many signal processing textbooks. Here, we will just show a few examples.

Source Influence Spectrum

The *source influence spectrum* is a spectrum describing the efficiency of wave excitation due to the pattern of the excitation source. It includes both spatial domain and time domain information. Usually, the dispersion curves are expressed as a relation between phase velocity and frequency in NDE applications. Therefore, in this section we are going to express the source influence spectrum also in the phase velocity and frequency domain.

A Parametric Study of Source Influence Spectrum Possibilities

1. THE COMB TRANSDUCER SITUATION. Comb transducers are commonly used in NDE and SHM applications. This is because these transducers selectively excite those modes with wavelength equal to the period of the transducer. Therefore, an excitation line can be drawn on the phase velocity dispersion curve to show those selected wave modes. However, as we discussed in the previous section, a phase velocity spectrum and a frequency spectrum are associated with a guided wave source. The combined two-dimensional spectrum can be calculated through a two-dimensional Fourier transform.

The design parameters for a comb transducer include the type of loading distribution, number of comb elements, and the period of the comb element. The tone burst excitation signal can also be described with center frequency, number of cycles, and any possible window function. In this section, we will study the influence of these parameters on the source influence spectrum.

The loading distribution from a comb-type transducer is straightforward. For the purpose of simplicity, three approximate models are used here as shown in Figure 13.6. Among these three types of loads, the piston-type loading distribution is studied in detail in the text, and the other two types are reserved for exercise practice.

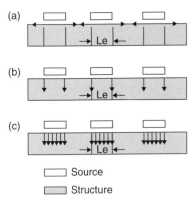

Figure 13.6. Loading distribution models for a comb transducer element.

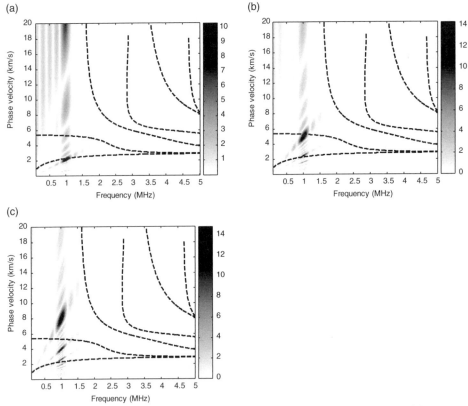

Figure 13.7. Source influence spectrum for the comb transducer with (a) 2.2 mm, (b) 5 mm, and (c) 8 mm period.

Influence of Source Period

In this section, the influence of the source period on the source spectrum will be illustrated. Considering the case of piston-shaped loading and six elements, the source influence spectra for three cases are plotted in Figure 13.7. As is described with the excitation line theory, the most efficient region of the spectrum corresponds to the center frequency and where the wavelength is equal to the period. However, the influence of harmonics in both frequency and wavenumber domain are also

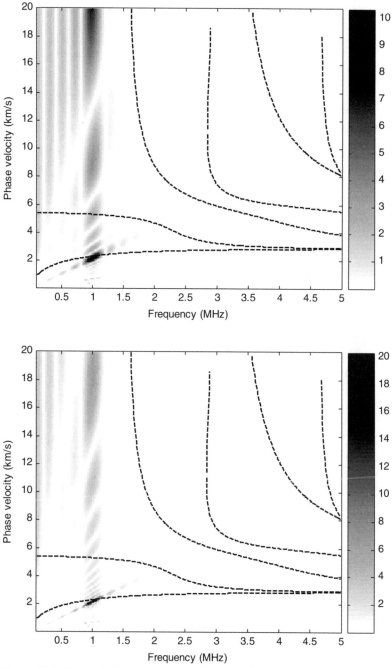

Figure 13.8. Source influence spectrum for a comb transducer with six and twelve periods. The width of the transducer period is 2.2 mm.

shown in the spectrum. If the region happens to be on top of a dispersion curve, the corresponding wave mode will also be efficiently excited. In this figure, when the comb period is 2.2 mm, the A0 mode at 1 MHz is most efficiently excited. When the comb period is 5 mm, the resulting wave mode is S0 at 1 MHz. When the comb period is 8 mm, the optimal point of the source spectrum is not located on any curve; hence, guided wave excitation will not be effective. Observing the spectra, the bandwidth of the spectrum also increases with an increase in the comb period. The frequency domain bandwidth is controlled by the excitation signal. For a piston-type loading, when the period increases, the maximum intensity of the spectrum also increases.

Influence of Number of Periods

Figure 13.8 shows the influence of the number of periods on transducer source influence. With an increase of the number of comb fingers, the phase velocity bandwidth of the source becomes narrower, and the intensity at the maximum spectrum region increases.

Influence of Loading Width

Figure 13.9 shows the influence of loading width on the source spectrum when the period of the loading is kept as 2.2 mm and the number of periods is six. When the width of the transducer finger is 1.1 mm, the intensity at the target spectrum point reaches its maximum. With a further increase of finger width from 1.1 mm to 1.7 mm, the intensity at (1 MHz, Cp 2.2 km/s) reduces. Therefore, maximum intensity is reached when the width of the finger is half the period. Figure 13.9 also shows that

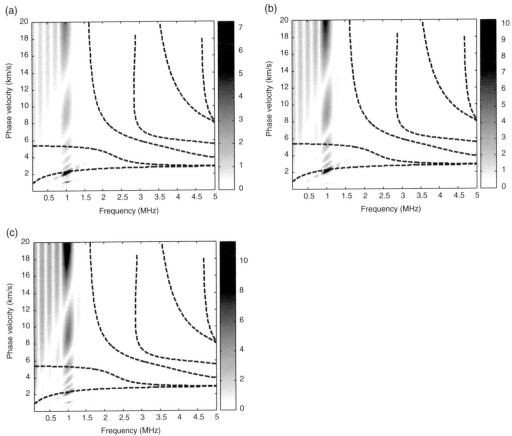

Figure 13.9 Source influence spectrum for a comb transducer by changing the width of the transducer finger from (a) 0.5 mm, (b) 1.1 mm, to (c) 1.7 mm, while keeping the period at 2.2 mm.

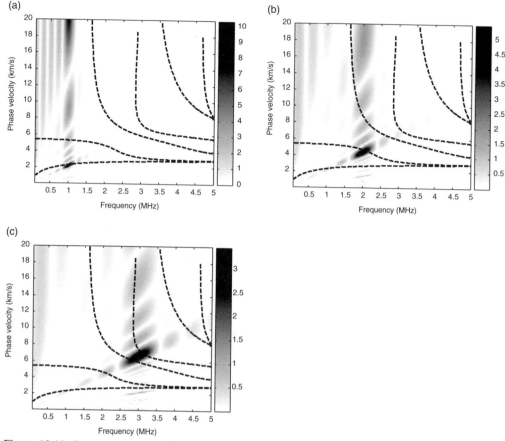

Figure 13.10. Source spectrum of a comb transducer with 2.2 mm and four periods; the excitation frequencies are (a) 1 MHz, (b) 2 MHz, and (c) 3 MHz, respectively.

the intensities of the spectrum at the regions with higher phase velocity increase with an increase of finger width.

Influence of Center Frequency

Figure 13.10 shows an influence of the exciting frequency on the 2-D spectrum of the comb transducer. The most important effect is that the effective region of the excitation moved along the line, on which the wavelength equals the period. When the numbers of cycles are kept as five, the pulse width of the excitation signal decreases with an increase of frequency. Therefore, the bandwidth of the spectrum increases. We also see that the bandwidth of the excitation increases with an increase of excitation frequency. When the input signal is 3 MHz and five cycles, the wave excitation covers both the regions of the A1 and S1 mode lines.

Influence of Window Function

Figure 13.11 shows the influence of different window functions on the source spectrum. When a rectangular windowed tone burst is used, the bandwidth of the main frequency band is narrow. However, multiple side lobes exist together with the main band. When

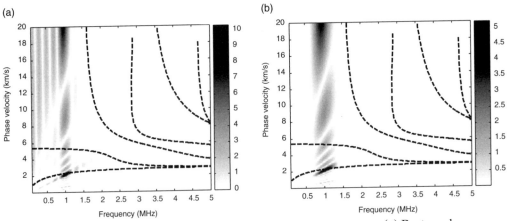

Figure 13.11. Influence of window function on the source spectrum. (a) Rectangular windowed tone burst, (b) Hanning windowed tone burst.

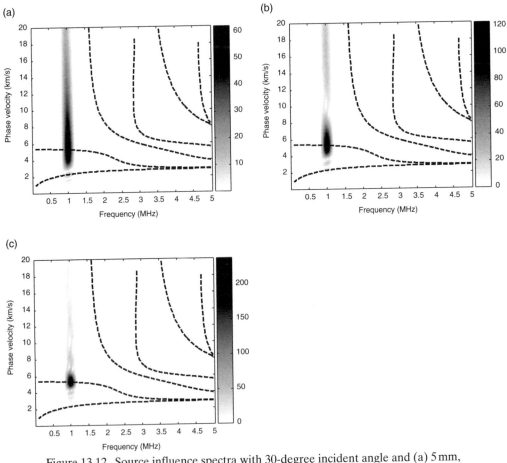

Figure 13.12. Source influence spectra with 30-degree incident angle and (a) 5 mm, (b) 10 mm, and (c) 20 mm transducer width.

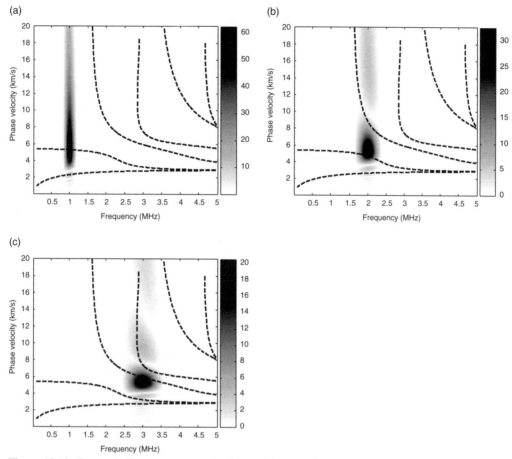

Figure 13.13. Source influence spectra of a 5 mm wide transducer with 30-degree excitation with (a) 1 MHz, (b) 2 MHz, and (c) 3 MHz center frequency. The number of cycles are all five.

a Hanning window is used, the relative intensity of the side lobes are much smaller with the trade-off of increased bandwidth and reduced maximum intensity.

2. THE OBLIQUE INCIDENCE SITUATION. For the case of angle beam incidence, transducer width, excitation frequency, and incident angle are important parameters controlling the wave excitation. Figure 13.12 shows the influence of transducer width on the source influence spectrum. The excitation signal is a 1 MHz tone burst with five cycles. It is shown that with an increase of transducer width, the phase velocity spectrum bandwidth becomes narrower approximating the value given by Snell's law.

Figure 13.13 shows the influence of excitation frequency on the source spectrum. The hot spots move along the frequency axis with constant phase velocity described with Snell's law. At the same time, the bandwidth in phase velocity domain reduces with an increase of frequency.

Figures 13.14 and 13.15 present additional graphs of the source influence results. Several interesting observations can be made from Figure 13.14. As an example, for a specified center frequency of 4.3 MHz and bandwidth of 0.6 MHz for all of

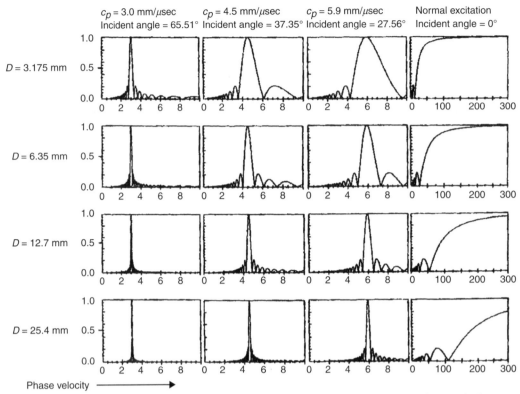

Figure 13.14. Phase velocity spectra showing excitation amplitude versus phase velocity value (frequency = 4.3 MHz, bandwidth = 0.6 MHz).

the curves, note the improvement in the phase velocity spectrum as the transducer diameter is increased – particularly for low-phase velocity values. Phase velocity value is listed with respect to an incident angle using the Cremer hypothesis for wave propagation (oblique incidence) through a Plexiglas wedge. The last column in the diagram is for normal excitation corresponding to a very high phase velocity value, theoretically of infinity. This is the case for normal beam incidence or an incident angle of 0°. Large transducer size helps in most cases, but please note that, with normal beam excitation, the phase velocity spectrum is quite broad (extending from zero to infinity). It is therefore quite difficult to isolate a particular mode and frequency when using a normal beam excitation.

Another interesting presentation is shown in Figure 13.15. A few particular phase velocity values and transducer diameter sizes were selected to illustrate the phase velocity spectrum variations as a function of these input parameters. The results are shown in a gray scale, where the darkest regions indicate the largest amplitude on the phase velocity spectrum. Note that mode and frequency isolation become possible for the smallest excitation zones, which are possible at large transducer diameters and high frequency values. The normal incidence excitation shows side lobes and excitation energy arising from the higher phase velocity values. As the modes become closer together, mode and frequency isolation become more difficult, which summons the need for an even stronger understanding of source influence parameters on the phase velocity spectrum used in guided wave excitation.

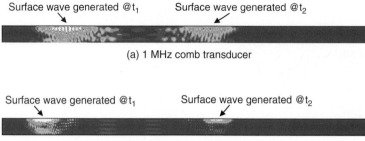

(a) 1 MHz comb transducer

(b) 2 MHz comb transducer

Plate 8.6. Rayleigh surface waves were excited by the 1 MHz and 2 MHz comb transducers on a 1/4″-thick titanium plate.

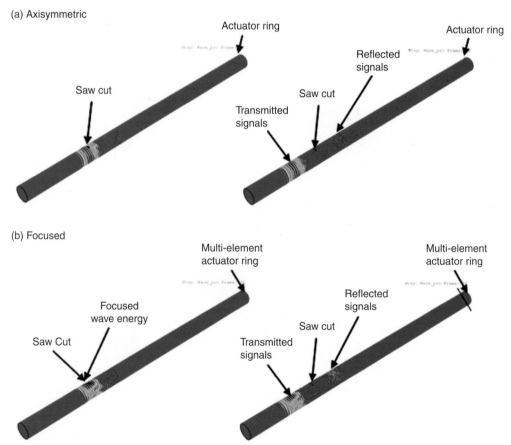

Plate 8.9. FEM simulations of 200 kHz (a) $T(0,1)$ axisymmetric wave and (b) focused $T(m,1)$ wave group reflected from the 9 percent CSA saw cut defect in the two-inch schedule 40 tube. Note that stronger reflected signals were obtained when focusing at the defect.

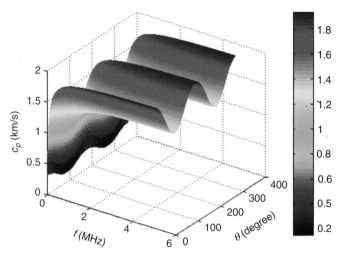

Plate 15.10. Three-dimensional display of the dispersion surface of mode 1 as a function of frequency and wave propagation angle.

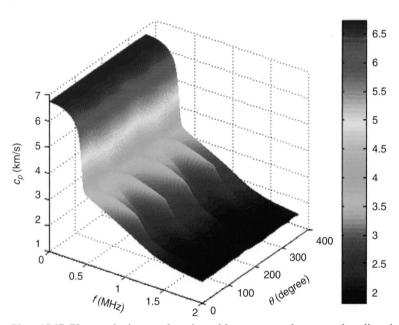

Plate 15.17. Phase velocity as a function of frequency and propagation direction in a quasi-isotropic laminate.

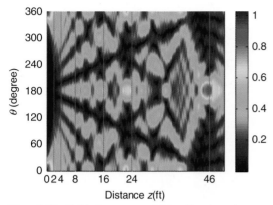

Plate 16.7. Guided wave partial loading interference pattern of the L(n, 2) mode group generated by 45° source loading on a sixteen-inch schedule 30 steel pipe at 50 kHz.

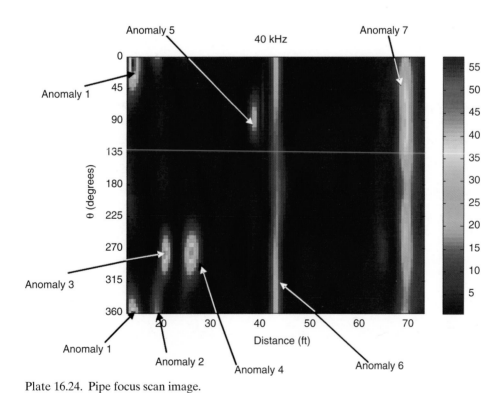

Plate 16.24. Pipe focus scan image.

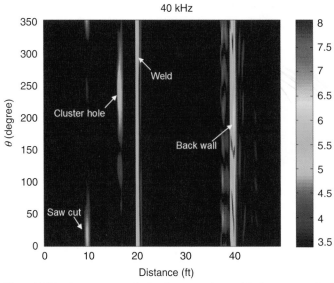

Plate 16.28. Reconstructed pipe image using guided wave synthetic focusing technique.

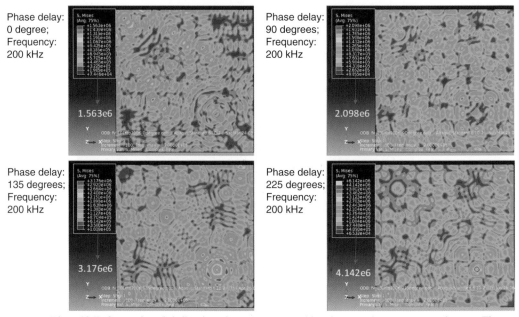

Plate 18.7. Operational deflection shapes generated by the annular array transducer – Fix loading frequency, change phase delay.

Frequency:
101 kHz;
Phase delay:
0 degrees

Frequency:
140 kHz;
Phase delay:
0 degrees

Frequency:
190 kHz;
Phase delay:
0 degrees

Frequency:
200 kHz;
Phase delay:
0 degrees

Plate 18.8. Operational deflection shapes generated by the annular array transducer – Fix phase delay, vary loading frequency.

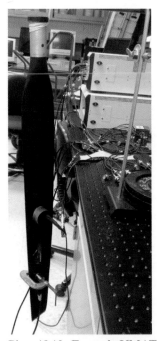

Plate 18.10. Example UMAT Experiment: An Airboat Composite Propeller.

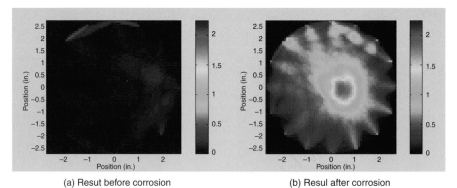

(a) Resut before corrosion   (b) Resul after corrosion

Plate 21.13. Sample ultrasonic guided wave tomographic results for corrosion on the wing of an aircraft.

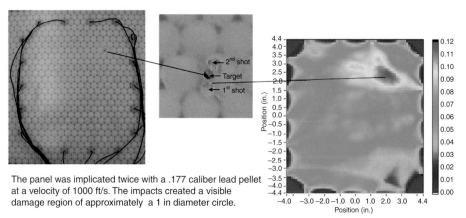

The panel was implicated twice with a .177 caliber lead pellet at a velocity of 1000 ft/s. The impacts created a visible damage region of approximately a 1 in diameter circle.

Plate 21.14. CT testing of ballistic damage to fabricated composite and the resulting guided wave tomogram.

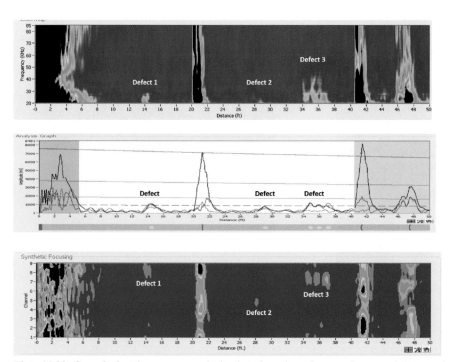

Plate 21.20. Sample fast frequency analysis plot, time domain waveform, and flattened pipe image showing defects and other features.

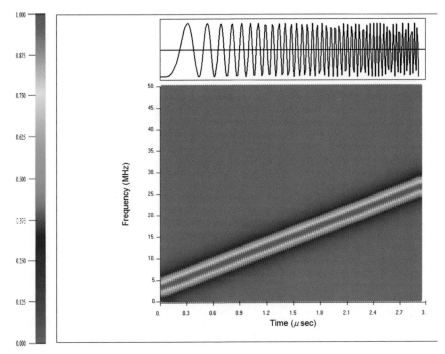

Plate C.10. STFFT spectrogram of a chirp signal showing the spectrogram tracking the frequency change in the signal as a function of time.

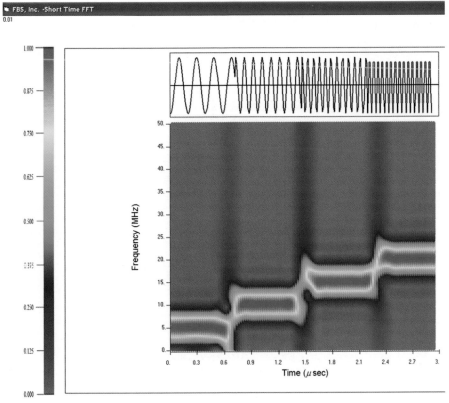

Plate C.11. STFFT spectrogram of a stepped frequency signal showing the spectrogram tracking the frequency change in the signal as a function of time. Window length = 32; Poor frequency resolution, good time resolution.

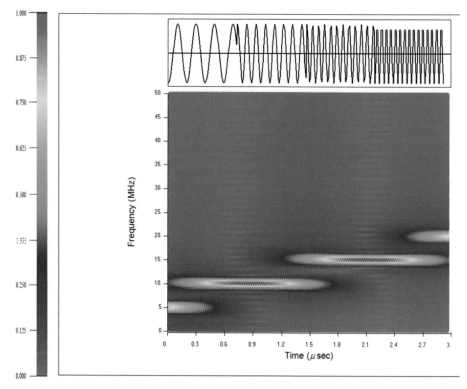

Plate C.12. STFFT spectrogram of a stepped frequency signal showing the spectrogram tracking the frequency change in the signal as a function of time. Window length = 128; Good frequency resolution, poor time resolution.

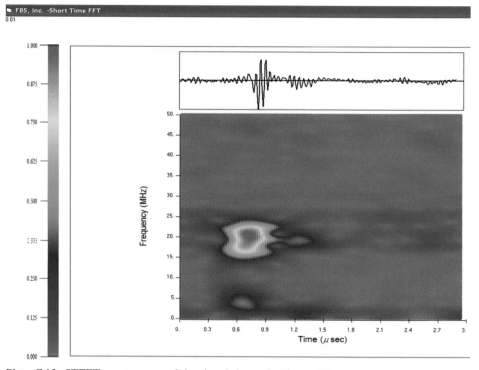

Plate C.13. STFFT spectrogram of the signal shown in Figure C.7.

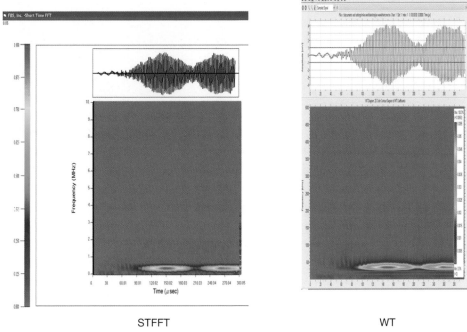

STFFT	WT

Plate C.14. Comparison of the STFFT and WT of the same signal.

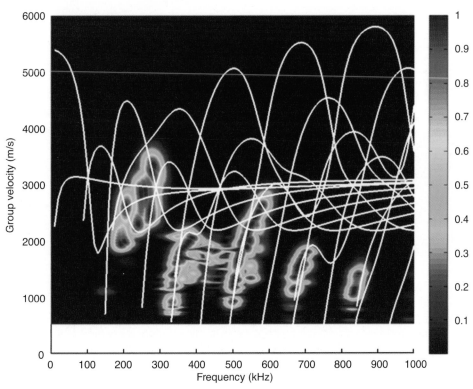

Plate C.16. A compilation of STFFTs created by frequency sweeping a ¾"-thick specimen of aluminum plate.

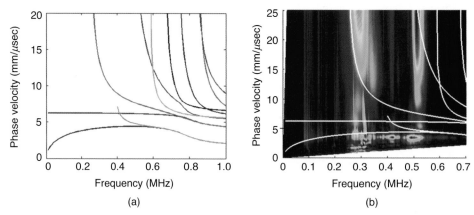

(a)   (b)

Plate C.19. Comparison of a phase velocity dispersion curves with b. 2DFFT results generated from a composite plate.

Software for 2DFFT can be found in Brigham (see references).

MatLab[†] and LabView[‡] also have software that supports the 2DFFT.

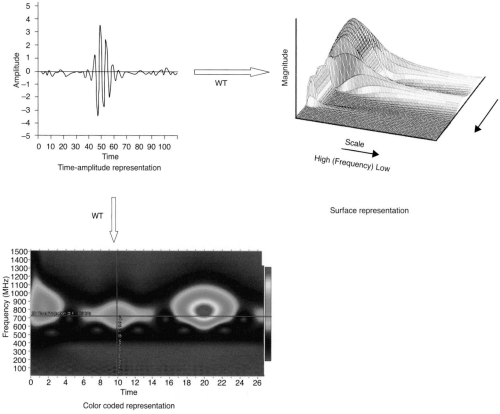

Plate C.20. Illustration of the results of using the WT. The WT takes a function (or waveform) from the time-amplitude domain to the translation-scale domain or to the time-frequency domain. Further, for a fixed distance (the location where the waveform was collected relative to the transmitter), a dispersion curve representation can be obtained.

[†] MatLab is a trademark of The MathWorks, Inc., Natick, Mass. U.S.A.
[‡] LabView is a trademark of National Instruments, Inc., Austin, Texas, U.S.A.

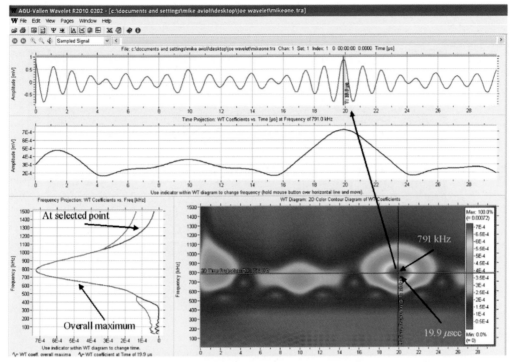

Plate C.23. Example WT with a pair of time-frequency coordinates shown. Using the arrival distance of the selected wave packet, the group velocity, c_g, can be calculated to obtain the dispersion curve coordinates (f, c_g). [AGU-Vallen Wavelet free software.]

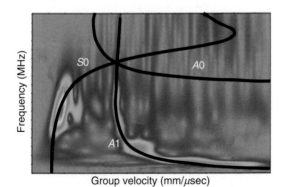

Plate C.25. Illustration of rotated group velocity dispersion curves superimposed on a WT. The red hot spot shows that the **S0** mode is the strongest propagating mode with **A1** as a weaker propagating mode.

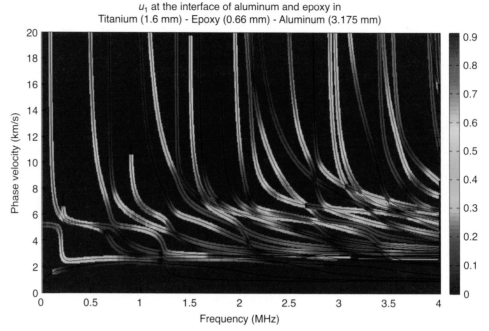

Plate D.9. Interfacial displacement (in-plane) at Al-epoxy interface in Ti-epoxy-Al specimen.

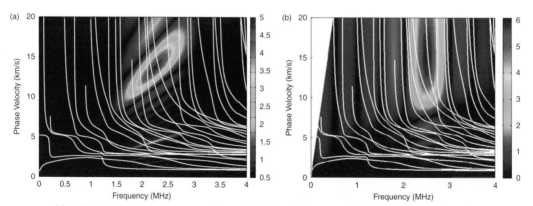

Plate D.10. The white lines are the guided Lamb-type wave phase velocity dispersion curves for the repair patch. (a) Source influence of $\lambda = 6.36$ mm (0.25") comb loading using four elements each 1.58 mm wide and supplied with 2.5 MHz tone burst voltage for 3 cycles on the range of phase velocities and frequencies excited. (b) Source influence of a 6 mm diameter transducer mounted on a 10° acrylic angle beam wedge and supplied with a 2.5 MHz tone burst input voltage for 5 cycles on the range of phase velocities and frequencies excited.

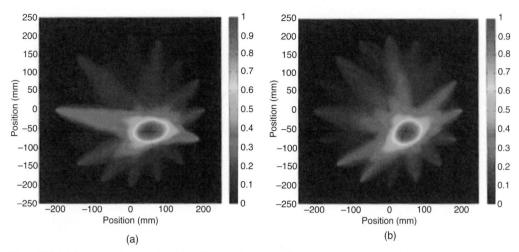

Plate D.11. Tomography result of the dry steel plate with a 10 percent wall thickness loss corrosion defect (a) A1 mode at 2.5 MHz (b) S1 mode at 2.8 MHz.

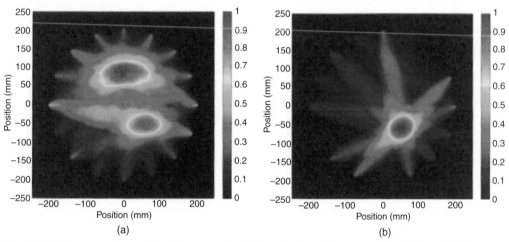

Plate D.12. Tomography result of the water-loaded steel plate with a corrosion defect (a) A1 mode at 2.5 MHz (b) S1 mode at 2.8 MHz.

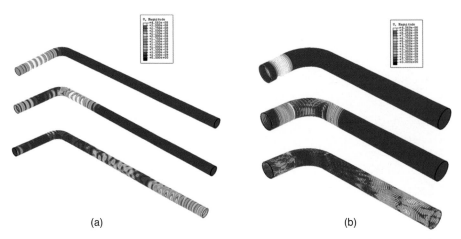

(a)

(b)

75 kHz L(m, 2) wave group propagating in a 2" schedule 40 elbowed steel pipe.

75 kHz L(m, 2) wave group propagating in a 16" schedule 30 elbowed steel pipe.

Plate D.14. Guided wave propagation in elbowed pipes with different sizes.

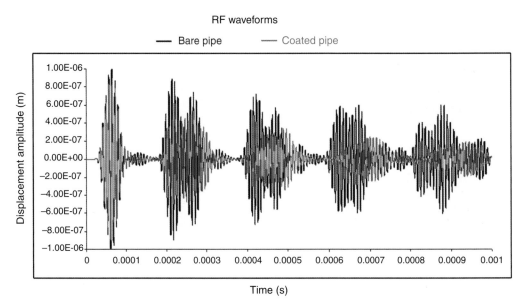

Plate D.20. Waveforms obtained from numerical modeling – SH circumferential wave propagation in an eight-inch diameter ¼″-thick bare pipe and the same pipe with a 3 mm coal tar enamel coating.

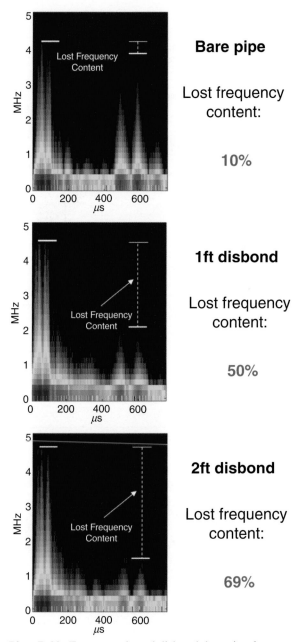

Bare pipe

Lost frequency
content:

10%

1ft disbond

Lost frequency
content:

50%

2ft disbond

Lost frequency
content:

69%

Plate D.22. Frequency-based disbond detection feature.

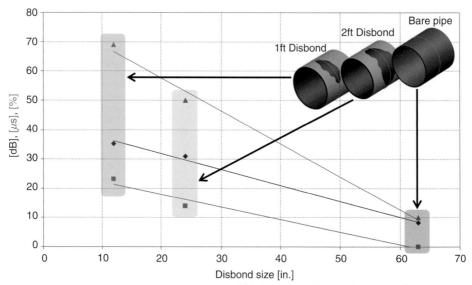

Plate D.23. Coating disbond detection using circumferential SH guided waves (Mode: SH0, Frequency: 130 kHz, Source: EMAT, Pipe: 20in S10 with coal tar coating).

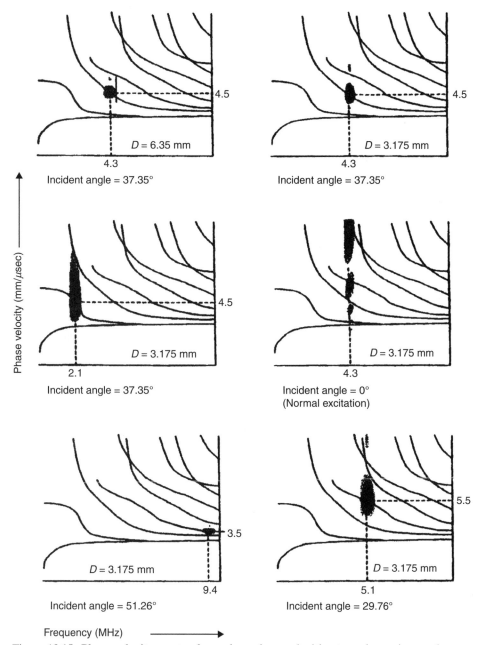

Figure 13.15. Phase velocity spectra for various phase velocities, transducer sizes, and frequencies for a frequency bandwidth of 0.6 MHz.

13.4 Exercises

1. What impact does transducer size have on the phase velocity spectrum in a waveguide?
2. What effect does frequency bandwidth have on the two-dimensional source spectrum?
3. Define the wave excitability in the case of a point force loading at the middle of a plate in the x_1 direction.

4. Determine the phase velocity spectrum for a normal beam incident on a flat plate. How can you achieve a smaller phase velocity bandwidth in the spectrum?

5. Calculate the phase velocity spectrum for piston pressure distribution for the following cases: (1) Incident angle $\alpha = 32.68°$, transducer width $D = 2\,mm$, five-cycled tone burst signals for center frequency 1.725 MHz, 5.765 MHz, and 9.725 MHz. (2) Desired spectrum maximum 2 MHz and $c_p = 5$ km/s with a comb transducer.

6. Design a program to calculate the 2-D source influence spectrum of an oblique incidence onto an aluminum plate using a parabolic transducer model.

7. Design a program to calculate the 2-D source influence spectrum of a phased array system for guided wave mode tuning in an aluminum plate.

13.5 REFERENCES

Auld, B. A. (1990). *Acoustic Fields and Waves in Solids*. Malabar, FL: Krieger Publishing Company.

Gao, H. (2007). *Ultrasonic Guided Wave Mechanics for Composite Material Structural Health Monitoring*. PhD Thesis, Pennsylvania State University.

Gao, H., and Rose, J. L. (2010). Goodness dispersion curves for ultrasonic guided wave based SHM: A sample problem in corrosion monitoring, *The Aeronautical Journal* 114 (1151): 797–804.

Rose, J. L. (1999). *Ultrasonic Waves in Solid Media*. Cambridge University Press.

14 Horizontal Shear

14.1 Introduction

Many aspects of horizontal shear wave propagation are intriguing and quite valuable for applications involving wave propagation, including ultrasonic NDT. Traditionally, the longitudinal and vertical shear modes of wave propagation have been the most commonly used – probably because they are simple to understand and to generate. Yet horizontal shear waves can also be generated quite easily through a variety of different transducers. This chapter covers the fundamental concepts of such propagation.

14.2 Dispersion Curves

In addition to the Lamb wave modes that exist in flat layers, there also exists a set of time-harmonic wave motions known as shear horizontal (SH) modes. The term "horizontal shear" means that the particle vibrations (displacements and velocities) caused by any of the SH modes are in a plane that is parallel to the surfaces of the layer. This is depicted in Figure 14.1, where the wave propagates in the x_1 direction and the particle displacements are in the x_3 direction.

Physically, any mode in the SH family can be considered as the superposition of up- and down-reflecting bulk shear waves, polarized along x_3, with wavevectors lying in the (x_1, x_2)-plane and inclined at such an angle that the system of waves satisfies traction-free boundary conditions on the surfaces of the layer.

The dispersion equation governing the SH modes can be derived in several ways, including the use of Helmholtz potentials, partial wave analysis, or transverse resonance (Auld 1990). Because of the simple physical nature of the SH modes, the most straightforward way to solve the problem is to deal directly with the displacement equations of motion. This is the approach taken here; for more discussion of this technique, see Achenbach (1984).

For any isotropic medium, the particle displacement field $u(x, t)$ must satisfy Navier's displacement equations of motion:

$$\mu \nabla^2 u(x,t) + (\lambda + \mu) \nabla \nabla \bullet u(x,t) = \rho \frac{\partial^2 u(x,t)}{\partial t^2} \tag{14.1}$$

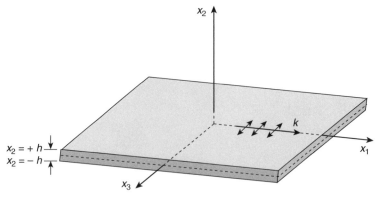

Figure 14.1. SH wave mode propagation, where the propagation is along x_1 and particle displacements are along x_3.

For the SH modes, we consider particle displacement vectors that have only an x_3 component, that is, $u_1(x, t) = u_2(x, t) = 0$ (Malvern 1969). Furthermore, for the x_3 displacement component, we specify a variation of the form at the outset:

$$u_3(x_1, x_2, t) = f(x_2)e^{i(kx_1 - \omega t)}, \tag{14.2}$$

where k is the wavenumber of the mode ($k = \omega/c_p = 2\pi/\lambda$) and ω represents circular frequency. Notice that u_3 is independent of x_3 and hence the wavefronts are infinitely extended in the x_3 direction.

This form of the solution is chosen because it represents a wave motion that propagates along the x_1-coordinate direction (due to the exponential term) and has a fixed distribution in the x_2 direction given by $f(x_2)$. As usual, the actual physical displacement vector field is the real part of the expression on the right-hand side of Equation (14.2).

If only the u_3 component of the particle displacement field is nonzero and if u_3 is independent of x_3, then (14.1) reduces to

$$\frac{\partial^2 u_3}{\partial x_1^2} + \frac{\partial^2 u_3}{\partial x_2^2} = \frac{1}{c_T^2}\frac{\partial^2 u_3}{\partial t^2}, \tag{14.3}$$

where $c_T^2 = \mu / \rho$.

Substituting the assumed solution (Equation (14.2)) into (14.3) results in

$$\frac{\partial^2 f(x_2)}{\partial x_2^2} + \left(\frac{\omega^2}{c_T^2} - k^2\right)f(x_2) = 0. \tag{14.4}$$

This equation has the general solution

$$f(x_2) = A\sin(qx_2) + B\cos(qx_2), \tag{14.5}$$

where q is defined as

$$q = \sqrt{\frac{\omega^2}{c_T^2} - k^2} \tag{14.6}$$

and A and B are arbitrary constants. The general form of the displacement field is therefore

$$u_3(x_1, x_2, t) = [A\sin(qx_2) + B\cos(qx_2)]e^{i(kx_1 - \omega t)}. \tag{14.7}$$

At this point it is advantageous (though hardly necessary) to separate the total displacement field, Equation (14.7), into symmetric and antisymmetric components (with respect to x_2). The parts of the total displacement field that represent symmetric and antisymmetric motions are the $\cos(qx_2)$ and $\sin(qx_2)$ terms, respectively. We thus consider two separate displacement fields:

$$\begin{aligned} u_3^s(x_1, x_2, t) &= B\cos(qx_2)e^{i(kx_1 - \omega t)}, \\ u_3^a(x_1, x_2, t) &= A\sin(qx_2)e^{i(kx_1 - \omega t)}. \end{aligned} \tag{14.8}$$

The superscript s denotes a symmetric mode and a denotes an antisymmetric mode.

The boundary conditions imposed on either type of mode are that the surfaces of the layer ($x_2 = \pm h$) be free of tractions:

$$\sigma_{22}(x_1, x_2, t)\big|_{x_2 = \pm h} = \tau_{12}(x_1, x_2, t)\big|_{x_2 = \pm h} = \tau_{23}(x_1, x_2, t)\big|_{x_2 = \pm h} = 0.$$

However, for a displacement field of either form in (14.8), the stresses σ_{22} and τ_{12} vanish identically. Hence, the only remaining nontrivial boundary conditions are

$$\tau_{23}(x_1, x_2, t)\big|_{x_2 = \pm h} = 0. \tag{14.9}$$

The strain field associated with either displacement field in (14.8) has only two nonzero components, $\varepsilon_{13} = (\partial u_3/\partial x_1)/2$ and $\varepsilon_{23} = (\partial u_3/\partial x_2)/2$. The traction component τ_{23} is therefore given by

$$\tau_{23} = 2\mu\varepsilon_{23} = \mu\frac{\partial u_3}{\partial x_2}, \tag{14.10}$$

the form of which will depend on whether the symmetric or antisymmetric displacement field of Equation (14.8) is used. Calculating τ_{23} yields

$$\tau_{23}(x_1, x_2, t) = \begin{cases} -\mu\, Bq\sin(qx_2)e^{i(kx_1 - \omega t)} & \text{for symmetric modes,} \\ \mu\, Aq\cos(qx_2)e^{i(kx_1 - \omega t)} & \text{for antisymmetric modes.} \end{cases} \tag{14.11}$$

Finally, imposing the boundary conditions (14.9) yields the dispersion equations

$$\sin(qh) = 0 \tag{14.12a}$$

and

$$\cos(qh) = 0, \tag{14.12b}$$

for the traction due to symmetric and antisymmetric displacements, respectively. Owing to the simplicity of these equations, we can obtain explicit solutions. Since $\sin(x) = 0$ when $x = n\pi$ ($n \in \{0, 1, 2, \ldots\}$) and $\cos(x) = 0$ when $x = n\pi/2$ ($n \in \{1, 3, 5, \ldots\}$), the solutions to (14.12a) and (14.12b) can be written as

$$qh = n\pi/2, \tag{14.13}$$

where $n \in \{0, 2, 4, \ldots\}$ for symmetric SH modes and $n \in \{1, 3, 5, \ldots\}$ for antisymmetric ones. Thus, the dispersion equation has an infinite number of solutions for either SH mode. The individual SH modes are specified by the integer n, which should be even for symmetric modes and odd for antisymmetric modes.

Using the form of q given in (14.13), denoting the thickness of the plate by $d = 2h$, and taking the real parts of Equation (14.8), the displacement fields of the symmetric and antisymmetric SH modes can be written as

$$u_3^s(x_1, x_2, t) = B\cos(n\pi x_2/d)\cos(kx_1 - \omega t),$$
$$u_3^a(x_1, x_2, t) = A\sin(n\pi x_2/d)\cos(kx_1 - \omega t).$$

Note that, since the arguments of the sine and cosine terms are independent of frequency and wavenumber, the variation of the fields of the SH modes across the thickness of the plate do not vary along any mode's dispersion curve. This is in sharp contrast to Lamb wave behavior, where the fields are actually functions of the position of the dispersion curves.

14.3 Phase Velocities and Cutoff Frequencies

We have found explicit solutions to the dispersion equation. Hence we can also construct explicit solutions for the phase velocity versus fd curves of any mode specified by the integer n.

With q as defined in (14.6) and using (14.13) and the definition of wavenumber, $k = \omega/c_p$, the dispersion equation can be written as

$$\frac{\omega^2}{c_T^2} - \frac{\omega^2}{c_p^2} = \left(\frac{n\pi}{2h}\right)^2. \qquad (14.14a)$$

This equation can easily be solved for the phase velocity c_p in terms of the frequency thickness product fd (where $d = 2h$ and $\omega = 2\pi f$). The result is

$$c_p(fd) = \pm 2c_T \left\{\frac{fd}{\sqrt{4(fd)^2 - n^2 c_T^2}}\right\}. \qquad (14.14b)$$

When $n = 0$ (corresponding to the zeroth-order symmetric SH mode) we have $c_p = c_T$, a dispersionless wave propagating at the shear wave speed c_T. All other SH modes (i.e., for all $n \neq 0$) are dispersive. Figure 14.2 plots the phase velocity dispersion curves for the first eight SH modes over a frequency thickness range of 0–15 MHz-mm. The (solid) even-numbered curves represent symmetric modes and the (dashed) odd-numbered curves represent antisymmetric modes.

The cutoff frequencies of the SH modes can be found by setting the denominator in (14.14b) equal to zero. This corresponds to infinite phase velocities and (as will be seen later) zero group velocities. The nth cutoff frequency is therefore given by:

$$(fd)_n = \frac{nc_T}{2}; \qquad (14.15)$$

once again, even integer n represents symmetric modes and odd integer n represents antisymmetric modes.

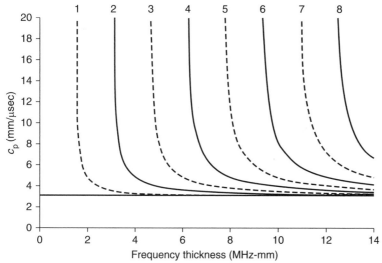

Figure 14.2. SH mode phase velocity dispersion curves for an aluminum layer ($c_T =$ 3.1 mm/μs): solid curves denote symmetric modes; dashed curves denote antisymmetric modes.

14.4 Group Velocity

Having explicit solutions for the dispersion equation also enables us to determine explicit expressions for the group velocity of any given SH mode. From the dispersion equation,

$$\frac{\omega^2}{c_T^2} - k^2 = \left(\frac{n\pi}{2h}\right)^2, \tag{14.16}$$

we take the differential of both sides (using that the right-hand side is a constant for any n). The result is

$$\frac{2\omega d\omega}{c_T^2} - 2k\, dk = 0. \tag{14.17}$$

Solving this equation for the quantity $d\omega/dk$ (by definition, the group velocity) yields

$$\frac{d\omega}{dk} = \frac{kc_T^2}{\omega} = \frac{c_T^2}{c_p} \tag{14.18}$$

Solving (14.16) for k, substituting the result into (14.18), and simplifying, we have

$$c_g(fd) = c_T \sqrt{1 - \frac{(n/2)^2}{(fd/c_T)^2}} \quad (fd \geq (fd)_n). \tag{14.19}$$

From this expression for the group velocity, we can see that – at the cutoff frequencies (given by (14.15)) – the group velocity of any mode is zero. As fd approaches infinity for any given fixed n, the group velocity of any SH mode approaches that of bulk shear waves, c_T.

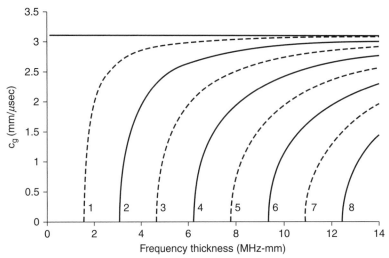

Figure 14.3. SH mode group velocity dispersion curves for the aluminum layer of Figure 14.2.

Figure 14.3 plots the SH mode group velocity curves for the first eight SH modes in an aluminum layer with shear wave speed $c_T = 3.1$ mm/μs. As in Figure 14.2, the solid curves (even n) represent symmetric modes and the dashed curves (odd n) represent antisymmetric modes.

14.5 Summary

The main results concerning the propagation of SH modes in flat layers are summarized here for easy reference. First, recall the definitions

$$q^2 = \frac{\omega^2}{c_T^2} - k^2, \quad d = 2h = \text{plate thickness}, \quad c_T = \text{bulk shear wave speed},$$

and that the wave is assumed to propagate in the x_1 direction with polarization entirely in the x_3 direction. We then have the following equations, where $n = 0, 2, 4, \ldots$ for symmetric modes and $n = 1, 3, 5, \ldots$ for antisymmetric modes.

Dispersion
$$\sin(qh) = 0 \ (\text{symmetric modes})$$
$$\cos(qh) = 0 \ (\text{antisymmetric modes})$$

Cutoff frequency
$$(fd)_n = \frac{nc_T}{2}$$

Phase velocity
$$c_p(fd) = 2c_T \left\{ \frac{fd}{\sqrt{4(fd)^2 - n^2 c_T^2}} \right\}$$

Group velocity
$$c_g(fd) = c_T \left\{ \sqrt{1 - \frac{(n/2)^2}{(fd/c_T)^2}} \right\}$$

Displacement

$$u_3(x_1, x_2, t) = \begin{cases} B\cos(n\pi x_2/d)e^{i(kx_1-\omega t)} & \text{(symmetric modes)} \\ A\cos(n\pi x_2/d)e^{i(kx_1-\omega t)} & \text{(antisymmetric modes)} \end{cases}$$

Stress

$$\tau_{23}(x_1, x_2, t) = \begin{cases} -B\mu q\,\sin(n\pi x_2/d)e^{i(kx_1-\omega t)} & \text{(symmetric modes)} \\ A\mu q\,\cos(n\pi x_2/d)e^{i(kx_1-\omega t)} & \text{(antisymmetric modes)} \end{cases}$$

$$\tau_{13}(x_1, x_2, t) = \begin{cases} -ik\mu B\cos(n\pi x_2/d)e^{i(kx_1-\omega t)} & \text{(symmetric modes)} \\ ik\mu A\sin(n\pi x_2/d)e^{i(kx_1-\omega t)} & \text{(antisymmetric modes)} \end{cases}$$

14.6 Exercises

1. Calculate cutoff frequencies for horizontal shear waves in an aluminum plate of thickness 1 mm.
2. Plot horizontal shear phase and group velocity dispersion curves for an aluminum plate of thickness 1 mm.
3. Discuss a method of generating both vertical and horizontal shear waves in a structure.
4. Define "excitability" with respect to the capability of producing horizontal shear waves in an aluminum plate.
5. Are horizontal shear waves all nondispersive in character? Explain.
6. Describe the displacement field for a nondispersive ($n = 0$) SH plate mode.
7. Why can't the phase velocities of the SH modes be smaller than the bulk shear velocity c_T?
8. Discuss several applications where it may be beneficial to use SH-type guided waves for inspection, as opposed to Rayleigh–Lamb type.
9. Does the wave structure for the $n = 0$ SH mode change with frequency? What about the order modes? Explain your answer.
10. Plot the displacement wave structure of SH modes $n = 1, 2$, and 3, at values of $c_p = 10, 5$, and 3.2 mm/μs for an aluminum plate ($c_T = 3.1$ mm/μs, $d = 1$ mm). Compare and contrast these wave structures with the $n = 0$ mode.

14.7 REFERENCES

Achenbach, J. D. (1984). *Wave Propagation in Elastic Solids*. New York: North-Holland.

Auld, B. A. (1990). *Acoustic Fields and Waves in Solids*, 2nd ed., vols. 1 and 2. Malabar, FL: Kreiger.

Malvern, L. E. (1969). *Introduction to the Mechanics of a Continuous Medium*. Englewood Cliffs, NJ: Prentice-Hall.

15 Guided Waves in Anisotropic Media

15.1 Introduction

The problem of elastic wave propagation in anisotropic layers has received a fair amount of attention in the literature during the past several decades, and recent interest in this subject has increased even more. This is undoubtedly due, at least in part, to the increased use of composite materials in many new facets of structure design. Composite materials that are mechanically anisotropic offer many benefits over more conventional material – a higher stiffness-to-weight ratio, for example. This advantage of composites is in turn due to the fact that their mechanical properties, such as elastic moduli, can be tailored to be high in the directions that are expected to see high loads while remaining considerably lower in other directions. This directional dependence of the mechanical properties of composites classifies them as anisotropic media.

The benefits of using composites come at the cost of a more complicated mechanical response to applied loads, static or dynamic. The anisotropic nature of the solid introduces many interesting wave phenomena not observed in isotropic bodies: a directional dependence of wave speed, a difference between phase and group velocity of the waves, wave skewing, three wave velocities instead of two, and many somewhat more subtle differences. An understanding of the nature of waves in plates made of anisotropic materials is certainly required if one wants to use these materials effectively in structure design or if one wants to inspect them using ultrasonic methods.

In this chapter, we investigate the key points in the propagation of elastic waves in free anisotropic plates. At the outset, we refer the reader to Auld (1990), which contains an excellent exposition of the general methods of analyzing such wave motions. We will see that, even though the anisotropic case is somewhat more involved than its isotropic plate counterpart, many of the final results will be qualitatively the same. For example, we will still find that the waves are governed by a dispersion equation whose roots define the possible modes in the structure. The difference is that in the case of generally anisotropic media, the dispersion curve is much more complicated than in the case of an isotropic layer.

With significant contribution from Huidong Gao.

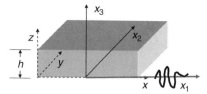

Figure 15.1. Coordinate system for the study of guided waves in an anisotropic plate.

An excellent review and state-of-the-art report that exceeds the basic introductory material presented here can be found in the textbook by Nayfeh (1995). Other related readings can be found in Auld (1990) and in Datta and colleagues (1988).

15.2 Phase Velocity Dispersion

This section presents the derivation of guided wave dispersion relations in an anisotropic layer. We will build our derivation on the knowledge of bulk wave propagation in anisotropic media covered earlier and on the knowledge of the partial wave method.

Figure 15.1 shows an illustration of an anisotropic plate with a coordinate system used to study guided wave propagation in this structure. Different from an isotropic medium, two sets of coordinate systems are needed to describe wave propagation in an anisotropic medium. One is the (x, y, z) system, which is a coordinate system where the nominal material properties are defined. The other is a (x_1, x_2, x_3) system, which is the coordinate system used for wave propagation studies. To remain consistent with the previous study of isotropic media, the waves are still assumed to propagate along the x_1-direction. Figure 15.1 shows that the two coordinate systems can be matched without rotation. Therefore, the material properties can be used directly without future coordinate system rotations.

Recall the governing equations of elastic waves in anisotropic media discussed earlier in the text; they are repeated here in equation (15.1), Equation (15.2), and Equation (15.3).

$$\rho \frac{\partial^2 u_i}{\partial t^2} = c_{ijkl} \frac{\partial^2 u_l}{\partial x_j \partial x_k} \tag{15.1}$$

$$\sigma_{ij} = c_{ijkl} s_{kl} \tag{15.2}$$

$$s_{kl} = \frac{1}{2} \left(\frac{\partial u_l}{\partial x_k} + \frac{\partial u_k}{\partial x_l} \right) \tag{15.3}$$

Here, u_i is the displacement vector, s_{kl} is the strain tensor, and σ_{kl} is the stress tensor. The partial wave technique in anisotropic media assumes that the partial waves have the solutions in the format expressed in Equation (15.4).

$$u_1 = U_1 e^{ik(x_1 + \alpha x_3 - c_p t)}$$

$$u_2 = U_2 e^{ik(x_1 + \alpha x_3 - c_p t)} \tag{15.4}$$

$$u_3 = U_3 e^{ik(x_1 + \alpha x_3 - c_p t)}$$

Substituting Equation (15.4) into Equation (15.1), the Christoffel's equation can be reached as in Equation (15.5).

$$\begin{bmatrix} A_{11} & A_{12} & A_{13} \\ A_{12} & A_{22} & A_{23} \\ A_{13} & A_{23} & A_{33} \end{bmatrix} \begin{bmatrix} U_1 \\ U_2 \\ U_3 \end{bmatrix} = \begin{bmatrix} 0 \\ 0 \\ 0 \end{bmatrix} \tag{15.5}$$

Detailed expressions of the components of $\mathbf{A}$ are listed in Equation (15.6).

$$\begin{aligned}
A_{11} &= C_{11} + 2C_{15}\alpha + C_{55}\alpha^2 - \rho c_p^2 \\
A_{12} &= C_{16} + (C_{14} + C_{56})\alpha + C_{45}\alpha^2 \\
A_{13} &= C_{15} + (C_{13} + C_{55})\alpha + C_{35}\alpha^2 \\
A_{22} &= C_{66} + 2C_{46}\alpha + C_{44}\alpha^2 - \rho c_p^2 \\
A_{23} &= C_{56} + (C_{36} + C_{45})\alpha + C_{34}\alpha^2 \\
A_{33} &= C_{55} + 2C_{35}\alpha + C_{33}\alpha^2 - \rho c_p^2
\end{aligned} \tag{15.6}$$

This is a general expression for an anisotropic material with triclinic symmetry. When all of the terms of the $\mathbf{A}$ matrix exist, the displacements in the three directions are coupled together. Therefore, instead of getting a pure longitudinal wave or a pure shear wave, we obtain quasi-longitudinal and quasi-shear partial waves.

Following the procedure used in solving Christoffel's equation for isotropic media, the determinant of the $\mathbf{A}$ matrix is set to zero to have nontrivial solutions of $[U_1, U_2, U_3]$. The determinant of the $\mathbf{A}$ matrix is a sixth-order equation of α when a value of phase velocity c_p is given. Therefore, for anisotropic media, we get six partial wave solutions. The isotropic counterpart also has six partial waves but is separated into four partial waves in Lamb wave considerations and into two partial waves in shear horizontal (SH) wave studies. The expression of the solution for α is very complicated and will not be covered here. After the solution of α is obtained, the corresponding value of $[U_1, U_2, U_3]$ can also be obtained by solving this linear system of equations.

After the solutions of the partials are obtained, the ultrasonic wave field within the plate can be expressed as a linear combination of the six partial waves.

$$u_l = \left[\sum_{m=1}^{6} B_m U_{lm} \exp(ik(\alpha_m x_3)) \right] \exp(i(kx_1 - \omega t)), \quad l = 1, 2, 3 \tag{15.7}$$

Here, α_m consists of the six solutions obtained from $|\mathbf{A}| = 0$, and U_{lm} represents the corresponding solutions of $[U_1, U_2, U_3]$ for a selected α_m. B_m is six combination coefficients to be determined through an evaluation of the boundary conditions.

The boundary conditions for the stress-free plate are listed in Equation (15.8).

$$\sigma_{3l} = 0, l = 1, 2, 3, \text{ at } x_3 = 0 \text{ and } x_3 = h \tag{15.8}$$

Therefore, to solve the boundary value problem, we need to express the stress components in terms of the partial wave solutions, which can be obtained using Equations (15.2) and (15.3).

$$\sigma_I = \left[\sum_{m=1}^{6} M_{Im} B_m \exp(ik\alpha_m x_3) \right] (ik) \exp(i(kx_1 - \omega t)), \quad I = 1, 2, \ldots, 6 \tag{15.9}$$

The detailed expressions of M_l are listed in Equation (15.10).

$$M_{1m} = C_{11}U_{1m} + C_{13}U_{3m}\alpha_m + C_{14}U_{2m}\alpha_m + C_{15}(U_{1m}\alpha_m + U_{3m}) + C_{16}U_{2m}$$
$$M_{2m} = C_{12}U_{1m} + C_{23}U_{3m}\alpha_m + C_{24}U_{2m}\alpha_m + C_{25}(U_{1m}\alpha_m + U_{3m}) + C_{26}U_{2m}$$
$$M_{3m} = C_{13}U_{1m} + C_{33}U_{3m}\alpha_m + C_{34}U_{2m}\alpha_m + C_{35}(U_{1m}\alpha_m + U_{3m}) + C_{36}U_{2m}$$
$$M_{4m} = C_{14}U_{1m} + C_{34}U_{3m}\alpha_m + C_{44}U_{2m}\alpha_m + C_{45}(U_{1m}\alpha_m + U_{3m}) + C_{46}U_{2m}$$
$$M_{5m} = C_{15}U_{1m} + C_{35}U_{3m}\alpha_m + C_{45}U_{2m}\alpha_m + C_{55}(U_{1m}\alpha_m + U_{3m}) + C_{56}U_{2m}$$
$$M_{6m} = C_{16}U_{1m} + C_{35}U_{3m}\alpha_m + C_{46}U_{2m}\alpha_m + C_{56}(U_{1m}\alpha_m + U_{3m}) + C_{66}U_{2m} \quad (15.10)$$

Substituting the expressions of σ_3, σ_4, and σ_5, that is, σ_{33}, σ_{32}, σ_{31} into the six boundary conditions listed in Equation (15.8), we get Equation (15.11).

$$\begin{bmatrix} M_{31} & M_{32} & M_{33} & M_{34} & M_{35} & M_{36} \\ M_{41} & M_{42} & M_{43} & M_{44} & M_{45} & M_{46} \\ M_{51} & M_{52} & M_{53} & M_{54} & M_{55} & M_{56} \\ M_{31}\exp(ik\alpha_1 h) & M_{32}\exp(ik\alpha_2 h) & M_{33}\exp(ik\alpha_3 h) & M_{34}\exp(ik\alpha_4 h) & M_{35}\exp(ik\alpha_5 h) & M_{36}\exp(ik\alpha_6 h) \\ M_{41}\exp(ik\alpha_1 h) & M_{42}\exp(ik\alpha_2 h) & M_{43}\exp(ik\alpha_3 h) & M_{44}\exp(ik\alpha_4 h) & M_{45}\exp(ik\alpha_5 h) & M_{46}\exp(ik\alpha_6 h) \\ M_{51}\exp(ik\alpha_1 h) & M_{52}\exp(ik\alpha_2 h) & M_{53}\exp(ik\alpha_3 h) & M_{54}\exp(ik\alpha_4 h) & M_{55}\exp(ik\alpha_5 h) & M_{56}\exp(ik\alpha_6 h) \end{bmatrix} \begin{bmatrix} B_1 \\ B_2 \\ B_3 \\ B_4 \\ B_5 \\ B_6 \end{bmatrix} = \begin{bmatrix} 0 \\ 0 \\ 0 \\ 0 \\ 0 \\ 0 \end{bmatrix}$$

$$(15.11)$$

Letting the determinant of the characteristic equation equal zero, we can get the dispersion curves of the structure as we did in the case of an isotropic plate. The solution to this equation usually requires a utilization of numerical methods.

Wave Propagation in the Fiber Direction of a Unidirectional Composite
A unidirectional, fiber-reinforced, composite material is usually modeled as a homogeneous anisotropic medium. The homogeneity assumption is valid when the wavelength of the ultrasound is much larger than the size of the fibers, which are typically several microns for carbon and glass fibers. When the fibers are randomly arranged in the matrix material, the mechanical properties are usually identical along all the directions perpendicular to the fiber direction. Therefore, a transversely isotropic material model is commonly used for a unidirectional composite lamina. The elastic constant matrix has the form of Equation (15.12).

$$C = \begin{bmatrix} C_{11} & C_{13} & C_{13} & 0 & 0 & 0 \\ C_{13} & C_{33} & C_{23} & 0 & 0 & 0 \\ C_{13} & C_{23} & C_{33} & 0 & 0 & 0 \\ 0 & 0 & 0 & \frac{1}{2}(C_{33} - C_{13}) & 0 & 0 \\ 0 & 0 & 0 & 0 & C_{55} & 0 \\ 0 & 0 & 0 & 0 & 0 & C_{55} \end{bmatrix} \quad (15.12)$$

In this example, we consider the IM7/977–3 lamina, whose engineering constants and material density are listed in Table 15.1. The elastic stiffness matrix can be derived from the engineering constants, shown in Equation (15.13). See Hahn and Tsai (1980).

Table 15.1. Material properties of an IM7/977–3 composite lamina

Density (g/cm³)*	E_1 (GPa)	E_2 (GPa)	G12 (GPa)	G23 (GPa)	v12	v23
1.608	172	9.8	6.1	3.2	0.37	0.55

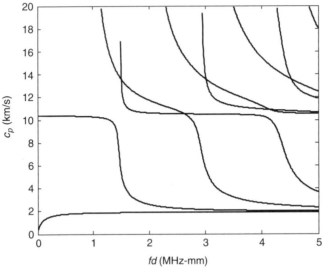

Figure 15.2. Lamb type guided wave dispersion curve for IM7/977–3 composite unidirectional laminate.

$$C = \begin{bmatrix} 178.2 & 8.347 & 8.347 & 0 & 0 & 0 \\ 8.347 & 14.44 & 8.119 & 0 & 0 & 0 \\ 8.347 & 8.119 & 14.44 & 0 & 0 & 0 \\ 0 & 0 & 0 & 3.161 & 0 & 0 \\ 0 & 0 & 0 & 0 & 6.1 & 0 \\ 0 & 0 & 0 & 0 & 0 & 6.1 \end{bmatrix} \tag{15.13}$$

When the material symmetry is considered, the general format of Christoffel's Equation (15.5) reduces to Equation (15.14).

$$\begin{bmatrix} A_{11} & 0 & A_{13} \\ 0 & A_{22} & 0 \\ A_{13} & 0 & A_{33} \end{bmatrix} \begin{bmatrix} U_1 \\ U_2 \\ U_3 \end{bmatrix} = \begin{bmatrix} 0 \\ 0 \\ 0 \end{bmatrix} \tag{15.14}$$

The nonzero terms of the **A** matrix are listed in Equation (15.15).

$$\begin{aligned} A_{11} &= C_{11} + C_{55}\alpha^2 - \rho c_p^2 \\ A_{13} &= (C_{13} + C_{55})\alpha \\ A_{22} &= C_{66} + C_{44}\alpha^2 - \rho c_p^2 \\ A_{33} &= C_{55} + C_{33}\alpha^2 - \rho c_p^2 \end{aligned} \tag{15.15}$$

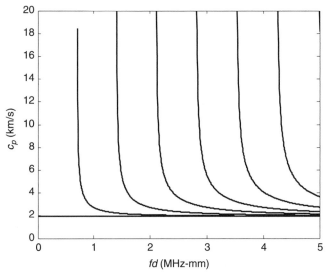

Figure 15.3. SH type guided wave dispersion curve for IM7/977–3 composite unidirectional laminate.

Therefore, Equation (15.14) can be decoupled into two sets of Christoffel's equation. One of them only has displacements in the sagittal plane, and the other only has displacement in the x_2-direction. This means that for the wave propagating in the fiber direction of a unidirectional composite, the Lamb type wave modes can be separated from the SH waves. The dispersion curves for the Lamb type and SH type waves are shown in Figures 15.2 and 15.3.

15.3 Guided Wave Directional Dependency

Besides propagating in the x-direction, ultrasonic guided waves can propagate in any of the other directions within the (x, y) plane of Figure 15.1. As a general model, when the angle between x and x_1 is θ, an illustration is shown in Figure 15.4.

Two methods can be used to study wave propagation in the x_1-direction. One is to build the model in the (x, y, z) coordinate system. In this method, instead of assuming that the plane wave is independent of y, we will have the wavevector in the θ direction. Therefore, the general solution will be $u_l = U_l e^{ik(\cos(\theta)x + \sin(\theta)y + \alpha z - c_p t)}$ and the material properties will still be provided as in the nominal value. The second solution method is built on the (x_1, x_2, x_3) coordinate system. Here, we still assume that the wave propagation is independent of x_2. Therefore, the same partial wave formulation can be used as described in Equation (15.2). However, the material properties need also to be expressed in the (x_1, x_2, x_3) coordinate system, and thus a tensor rotation is required for the material properties. The density of the material is a scalar, therefore, it is an invariant before and after the rotation. The elastic constant is a fourth-order tensor; the rotation formula is expressed in Equation (15.16).

$$c_{mnop} = \beta_{mi}\beta_{nj}\beta_{ok}\beta_{pl}c'_{ijkl}$$

(15.16)

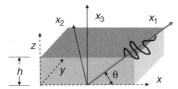

Figure 15.4. Coordinate system to study guided wave propagation in an arbitrary direction in an anisotropic medium.

Here β is the transformation matrix.

$$\beta_{ij} = \begin{bmatrix} \cos\theta & \sin\theta & 0 \\ -\sin\theta & \cos\theta & 0 \\ 0 & 0 & 1 \end{bmatrix} \tag{15.17}$$

With the material properties defined in the (x_1, x_2, x_3) direction, we can now proceed with the procedure discussed in Section 15.2 for dispersion curve analysis. The directional dependence of the wave propagation can be illustrated after we obtain the wave propagation characteristics in all directions.

Wave Propagation in the 90° Direction

When $\theta = 90^0, \beta = \begin{bmatrix} 0 & 1 & 0 \\ -1 & 0 & 0 \\ 0 & 0 & 1 \end{bmatrix}$. Therefore, the elastic stiffness tensor can be calculated

through Equation (15.16). The matrix format elastic stiffness is expressed in Equation (15.18).

$$C = \begin{bmatrix} 14.44 & 8.347 & 8.119 & 0 & 0 & 0 \\ 8.347 & 178.2 & 8.347 & 0 & 0 & 0 \\ 8.119 & 8.347 & 14.44 & 0 & 0 & 0 \\ 0 & 0 & 0 & 6.1 & 0 & 0 \\ 0 & 0 & 0 & 0 & 3.161 & 0 \\ 0 & 0 & 0 & 0 & 0 & 6.1 \end{bmatrix} \tag{15.18}$$

In this coordinate system, the following components of the stiffness matrix remain zeros.

$$C_{16}, C_{14}, C_{56}, C_{45}, C_{36}, C_{34} \tag{15.19}$$

Therefore, using Equation (15.6), we still have $A_{12} = 0$ and $A_{23} = 0$. Therefore, the decoupling of Lamb type and SH type waves is still valid. The phase velocity dispersion curves of these types of waves are shown in Figure 15.5 and Figure 15.6, respectively.

Comparing these dispersion curves with those for wave propagation along the fiber direction, significant differences in the dispersion curves are observed.

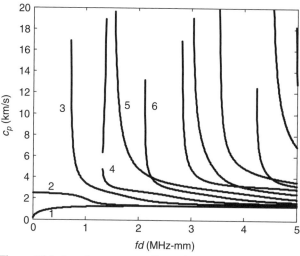

Figure 15.5. Lamb type wave dispersion curves for wave propagation in the 90° direction.

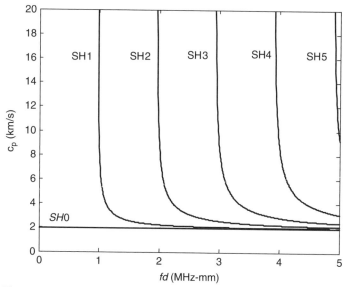

Figure 15.6. SH type wave dispersion curves for wave propagation in the 90° direction.

Therefore, in anisotropic media, the wave propagation characteristics are also anisotropic.

Wave Propagation in the 45° Direction

When $\theta = 45^0$, $\beta = \begin{bmatrix} \frac{\sqrt{2}}{2} & \frac{\sqrt{2}}{2} & 0 \\ -\frac{\sqrt{2}}{2} & \frac{\sqrt{2}}{2} & 0 \\ 0 & 0 & 1 \end{bmatrix}$. The transformed stiffness matrix is shown in

Equation (15.20).

$$C = \begin{bmatrix} 58.43 & 46.23 & 8.233 & 0 & 0 & -40.93 \\ 46.23 & 58.43 & 8.233 & 0 & 0 & -40.93 \\ 8.233 & 8.233 & 14.44 & 0 & 0 & -0.1143 \\ 0 & 0 & 0 & 4.631 & -1.469 & 0 \\ 0 & 0 & 0 & -1.469 & 4.631 & 0 \\ -40.93 & -40.93 & -0.1143 & 0 & 0 & 43.98 \end{bmatrix} \qquad (15.20)$$

Substituting this stiffness constant matrix into Equation (15.5) and Equation (15.6), we can get the Christoffel's equation, which involves all three displacements together. Therefore, for the guided wave propagation in the 45° direction, there is no decoupling of Lamb type and SH type waves. All the dispersion curves must be plotted in a single figure. See Figure 15.7.

The wave structure has displacements in all three directions. As an example, the displacement profiles of the three wave modes at 500 kHz are plotted in Figure 15.8. The amplitudes in the x_1-direction are similar to those in the x_2-direction for all three modes. If we classify the x_1 and the x_2 as in-plane direction, and the x_3 as out-of-plane direction, then mode 2 and mode 3 are still dominated by in-plane displacements, while mode 1 is dominated by the out-of-plane displacement. This qualitative conclusion is compatible to the statement: for an isotropic traction-free plate, the *S0* and *SH0* modes are in-plane dominant, while the *A0* mode is out-of-plane displacement dominant for lower-frequency values.

Note that a typical phase velocity variation as a function of propagation angle for a given mode and frequency is illustrated in Figure 15.9.

The directional dependency of ultrasonic guided waves can also be expressed with the concept of a dispersion surface, where the dispersion curves in all the wave propagation directions are plotted over a three-dimensional surface as shown in Figure 15.10.

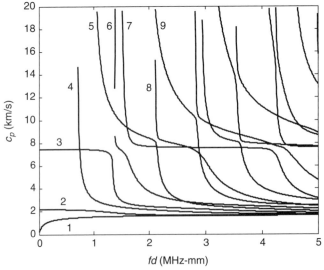

Figure 15.7. Guided wave dispersion curves for wave propagation in the 45° direction of the unidirectional composite laminate.

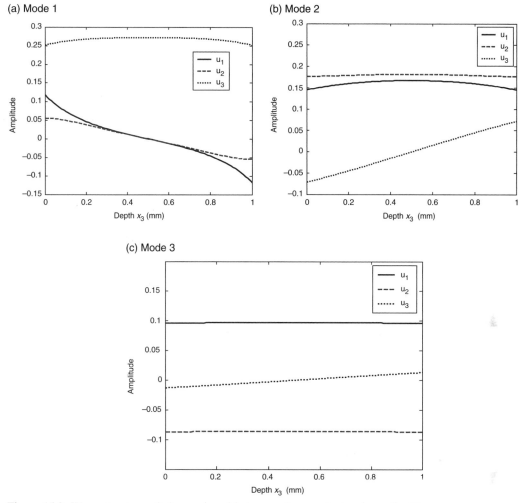

Figure 15.8. Wave structure of ultrasonic guided waves in an anisotropic media. Wave propagation direction is at 45° from the fiber direction.

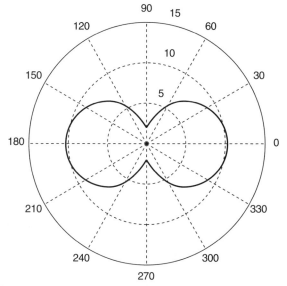

Figure 15.9. Phase velocity variation as a function of propagation angle for a given mode and frequency.

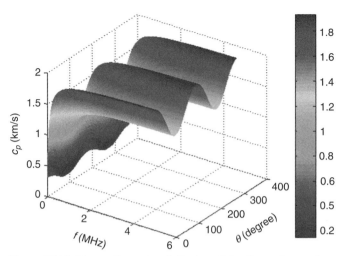

Figure 15.10. Three-dimensional display of the dispersion surface of mode 1 as a function of frequency and wave propagation angle. See plates section for color version.

15.4 Guided Wave Skew Angle

One other unique feature of guided wave propagation in an anisotropic plate is the possibility of having a skew angle for a particular wave mode and frequency, which is the angle between the energy propagation direction and the wavevector direction. The skew angle can be calculated in two ways. One is from wave structure analysis. The other is from the slowness profile.

As in the case of an isotropic plate, the wave structure of a particular guided wave mode can be calculated after the dispersion curve is obtained. The power flow carried in the wave mode can be evaluated with Equation (15.21).

$$\mathbf{P} = \frac{-\mathbf{v}^* \bullet \boldsymbol{\sigma}}{2} \tag{15.21}$$

This is a vector describing the power flow at a particular point in the wave field. If we integrate the power flow over the plate thickness, we can get the entire power transmission rate of the wave mode in both x_1- and x_2-directions. The skew angle Φ of the wave mode can then be defined using Equation (15.22).

$$\tan(\Phi) = \frac{\int_0^H P_{x_2} dx_3}{\int_0^H P_{x_1} dx_3} \tag{15.22}$$

The second method of defining skew angle is from a spatial variation of a phase velocity dispersion curve. A slowness value is defined for each wave mode by taking the reciprocal of the phase velocity value according to Equation (15.23).

$$Slowness = \frac{1}{c_p} \tag{15.23}$$

For a given frequency and mode index, the value of phase velocity varies with the wave propagation direction. Therefore, the slowness also varies with direction.

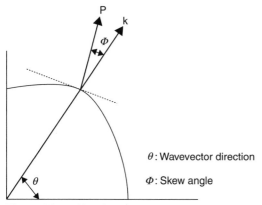

Figure 15.11. Sketch of slowness profile and skew angle.

Figure 15.11 shows the relation between the direction of phase velocity, the direction of power flow, and the skew angle in a slowness profile. In this figure, the curve is the slowness profile with respect to the wave propagation angle (θ). The dashed line is the tangent of the profile for direction (θ). The surface normal direction is the actual power flow direction. The angle between the power flow direction and the wavevector direction is defined as the skew angle of the guided wave mode.

15.5 Guided Waves in Composites with Multiple Layers

Composite materials used in most applications are commonly fabricated by stacking the unidirectional composite lamina into laminate structures. The orientation of each layer can be engineered to achieve a desired mechanical response. Therefore, to study the problem of ultrasonic guided wave propagation in composite laminates, we must combine the knowledge of wave propagation in anisotropic media with the knowledge of wave propagation in multiple layers. These two topics are covered in this chapter and in Chapter 12.

Figure 15.12 is a sketch of the multiple layers and the coordinate systems. As an example, a five-layered structure is shown in this figure. The thickness of the layers are denoted with $h^{(n)}$. The waves are propagating in the x_1-direction of the (x_1, x_2, x_3) coordinate system. The material properties are defined in each of its local coordinate systems. As an example, the coordinate systems for the first two layers are shown on the right side of the figure.

Following the procedure described in this chapter, the ultrasonic wave field in each layer can be expressed as a linear combination of six partial wave solutions as demonstrated in Equation (15.24).

$$u_l^{(n)} = \left[\sum_{m=6(n-1)+1}^{6n} B_m U_{lm} \exp(ik(\alpha_m x_3)) \right] \exp(i(kx_1 - \omega t)), \quad l = 1, 2, 3$$

$$\sigma_I^{(n)} = \left[(ik) \sum_{m=6(n-1)+1}^{6n} M_{Im} B_m \exp(ik\alpha_m x_3) \right] \exp(i(kx_1 - \omega t)), \tag{15.24}$$

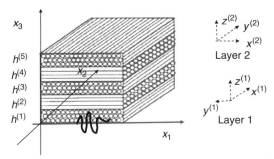

Figure 15.12. A sketch of a composite laminate and the coordinate systems for wave propagation.

Following the discussion in Chapter 12 on layered structures, the boundary conditions will include the interface continuity conditions at the interfaces and the stress-free boundary conditions at the top and the bottom surfaces.

$$\sigma_{3l}^{(1)} = 0,\, l = 1,2,3,\text{ at } x_3 = 0$$

$$u_l^{(n)} = u_l^{(n-1)},\, l = 1,2,3,\text{ at } x_3 = H^{(n-1)}, n = 2,3,\cdots,N$$

$$\sigma_{3l}^{(n)} = \sigma_{3l}^{(n-1)},\, l = 1,2,3,\text{ at } x_3 = H^{(n-1)},\, n = 2,3,\cdots,N$$

$$\sigma_{3l}^{(N)} = 0,\, l = 1,2,3,\text{ at } x_3 = H^{(N)}$$

$$(15.25)$$

Following the procedure of assembling the boundary conditions into a global matrix, a characteristic equation of guided wave propagation can be obtained. It is expressed in Equation (15.26) in a contracted format.

$$\mathbf{D} \bullet \mathbf{B} = 0 \qquad (15.26)$$

Here, the $\mathbf{D}$ matrix is in the dimension of $6N$ by $6N$, and the dimension of the unknown vector $\mathbf{B}$ is $6N$. Although the general expressions of the elements in the $\mathbf{D}$ matrix are not elaborated here, the reader can derive these elements by following the procedure discussed in the layered structure chapter.

After the characteristic equation is obtained, the process of calculating dispersion curves and wave structures is discussed in detail in previous chapters and sections. Just as a brief summary, the dispersion curves can be calculated by searching the appropriate pairs of frequency and wavenumber (or phase velocity) values that make the determinant of the $\mathbf{D}$ matrix equal to zero. The wave structure solution corresponding to a particular wave mode can be obtained by substituting the pairs of frequency and phase velocity and calculating the solutions for the linear system of equations.

Phase velocity dispersion curves for several laminated composite structures are now illustrated in a few examples.

Cross Ply Laminate

Cross ply laminates are commonly used in engineering structures. An example of guided wave propagation in a cross ply laminate is studied here. This cross ply laminated sample composite consists of eight layers of unidirectional composite lamina. The lamina properties are listed in Table 15.1. The lay-up sequence is

$[(0/90)_s]_2$; if we elaborate the stack sequence, it is [0/90/90/0/0/90/90/0]. The thickness of each ply is 0.2 mm. Therefore, the total thickness is 1.6 mm.

For cross ply laminates, we can prove that the Lamb type waves are still separable from the SH type waves when the wave propagates along the $0°$ or $90°$ directions. In addition, the structure still has a symmetric middle plane between layers 4 and 5. Therefore, the guided wave modes can still be classified into categories of symmetric and antisymmetric waves.

When the wave propagation is in the $0°$ direction, the Lamb type and SH type wave dispersion curves are plotted in Figures 15.13 and 15.14, respectively.

In the $45°$ direction, the guided wave modes are not separable. The total set of dispersion curves is plotted in Figure 15.15.

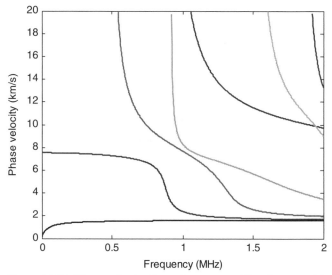

Figure 15.13. Phase velocity dispersion curves of Lamb type guided wave propagation in the $0°$ direction of the $[(0/90)_s]_2$ cross ply laminate. Note: material lamina properties are in Table 15.1.

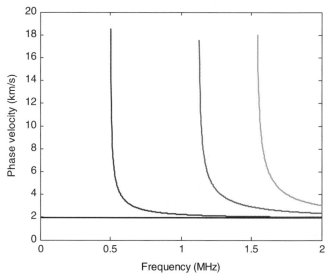

Figure 15.14. Phase velocity dispersion curves of SH type guided wave propagation in the $0°$ direction of the $[(0/90)_s]_2$ cross ply laminate.

Quasi-Isotropic Laminate

An example of a quasi-isotropic composite laminate is studied in this section, whose lay-up sequence is $[(0/45/90/-45)_s]_2$. The material used is still an IM7/977–3 laminate, whose properties are listed in Table 15.1. Notice that in this quasi-isotropic

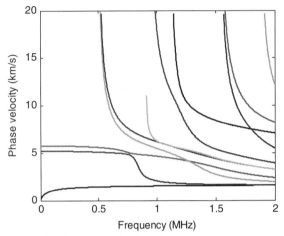

Figure 15.15. Phase velocity dispersion curves of guided wave propagation in the 45° direction of the $[(0/90)_s]_2$ cross ply laminate.

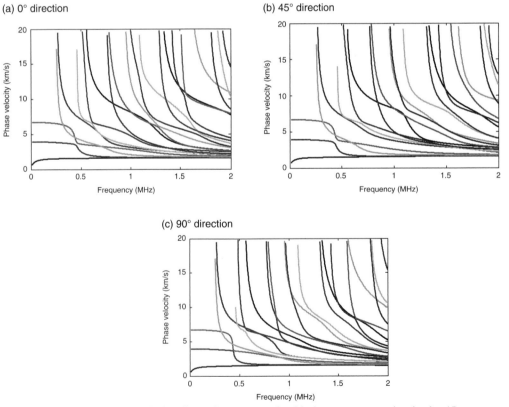

(a) 0° direction

(b) 45° direction

(c) 90° direction

Figure 15.16. Phase velocity dispersion curves of guided wave propagation in the 45° direction of the $[(0/45/90/-45)_s]_2$ quasi-isotropic laminate.

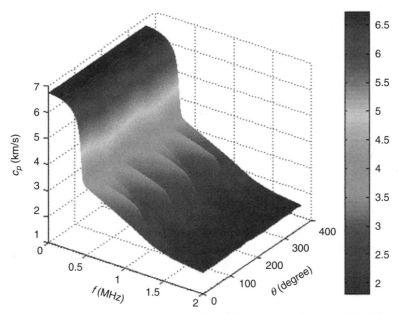

Figure 15.17. Phase velocity as a function of frequency and propagation direction in a quasi-isotropic laminate. See plates section for color version.

composite laminate, there are no wave propagation directions where Lamb type and SH type waves can be separated. Therefore, all of the guided wave modes are plotted together in a single set of dispersion curves. Figure 15.16 shows the dispersion curves for the guided wave modes propagating in the 0°, 45°, and 90° directions, respectively.

In the low-frequency region, the second and third mode dispersion curves are approximately independent of wave propagation angle. The low-frequency extremes of these wave modes are in plane tension and in plane shear of the laminate. Quasi-isotropic laminates exhibit quasi-isotropic characteristics under these loads. However, for the wave modes at high-frequency and higher-order wave modes, significant anisotropy of wave propagation can be observed. As an example, Figure 15.17 shows the variation of the third mode phase velocity as a function of both frequency and propagation direction. Strong angle dependence is shown in the frequency range of 500 kHz to 1 MHz in Figure 15.17.

Skew angle in a quasi-isotropic laminate could be very large. Figure 15.18 shows the skew angle dispersion curves for guided waves propagating along the 0° direction of the $[(0/45/90/-45)_s]_2$ laminate. Skew angle may reach 40° and −40°, depending on the mode and frequency selected. Therefore, detailed skew angle analysis must be studied to obtain an accurate prediction of guided wave propagation.

It is possible to plot such special characteristics of points on a dispersion curve as skew angle, upper surface in-plane displacement, or any other feature of interest superimposed on top of the phase velocity dispersion curve with a suitable legend following methods outlined by Gao and Rose (2010).

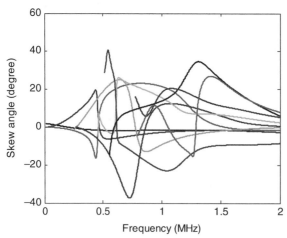

Figure 15.18. Skew angle of guided wave modes propagating along the 0° direction of a quasi-isotropic laminate. (Mode color code following Figure 15.16(a)).

15.6 Exercises

1. Derive the Rayleigh–Lamb dispersion equation for a unidirectional composite for planes of symmetry.
2. Develop a general procedure to solve a problem of guided waves in a multilayer composite plate.
3. Develop an equation for group velocity computation from the phase velocity results in a composite laminate.
4. Determine the Rayleigh wave velocity for an anisotropic half-space for different directions of propagation.
5. What is the assumption to model a composite lamina as a homogeneous anisotropic layer?
6. What are the limitations of considering a multilayered composite laminate as a homogeneous anisotropic plate?
7. Describe the relation between the skew angle, wavevector direction, and wave energy propagation direction. How would you derive the skew angle of a wave mode based on phase velocity data?
8. Outline an experimental procedure to produce dispersion curves for an anisotropic structure.
9. a. Prove that mechanical response of quasi-isotropic composite laminates under static in-plane tension is independent of loading direction using theory of mechanics in composite materials. (Optional). b. Explain why the third guided wave mode phase velocity is independent of propagation angle in quasi-isotropic composite laminates.
10. Calculate skew angle as a function of launch angle for a sample composite structure. Use two methods.
11. For a cross ply laminate, how would you show that the Lamb type waves are still separable from the SH type waves when the waves propagate along the 0° or 90° directions?

12. At what angle are higher phase velocities achieved for the results shown in Figure 15.10?
13. For what frequencies shown in Figure 15.10 can higher phase velocities be achieved?

15.7 REFERENCES

Auld, B. A. (1990). *Acoustic Fields and Waves in Solids*. Malabar, FL: Krieger Publishing Company.

Datta, S. K., Shah, A. K., Bratton, R. L., and Chakraborty, T. (1988). Wave propagation in laminated composite plates, *J. Acoust. Soc. Am.* 83(6):2020–6.

Gao, H. (2007). *Ultrasonic Guided Wave Mechanics for Composite Material Structural Health Monitoring*, PhD thesis, Pennsylvania State University.

Gao, H., and Rose, J. L. (2010). Goodness dispersion curves for ultrasonic guided wave based SHM: A sample problem in corrosion monitoring, *The Aeronautical Journal of the Royal Aeronautical Society* 114(1151): 797–804.

Hahn, H. T., and Tsai, S. W. (1980). *Introduction to Composite Materials*. Lancaster, PA:, Technomic Publishing Company.

Nayfeh, A. H. (1995). *Wave Propagation in Layered Anisotropic Media*. Amsterdam, Lausanne, New York, Oxford, Shannon, & Tokyo: Elsevier.

Rose, J. L. (1999). *Ultrasonic Waves in Solid Media*. Cambridge University Press.

16 Guided Wave Phased Arrays in Piping

16.1 Introduction

Phased array technology has taken the world by storm in the inspection ability to scan structures and to achieve dynamic beam focusing and beam steering without mechanically moving actuators over the part under examination. Modern-day electronics and cost reduction have inspired activity in this area. The first real-time phased array multi-element transducer system used in medical ultrasound was introduced in the '70s with applications in solid materials following in the '80s. In the case of using bulk wave ultrasound, only straight line computation was needed to control wave interference phenomena at selected points inside the human body or inside a specific solid structure. In addition to real-time phased array systems, synthetic focusing was introduced whereby enough information was transmitted from a multi-element array and subsequently received by the individual elements, where software was used to simulate the real-time phased array approach. Both methods provide special advantages.

During the '90s, real-time phased array and synthetic focusing methods were introduced for cylindrical structures, primarily pipelines. The time delay computations to achieve this are more complex. This chapter primarily focuses on principles for cylindrical piping structures.

Although the phased array focusing technique has been widely used in such fields as electromagnetic waves and radar and conventional ultrasonic bulk waves, it was rarely seen in guided wave applications because of the complexity of guided waves. In 2002, Li and Rose reported that if the angular profile for a single-channel loading was known, the total guided wave angular profile could be controlled and thus focused at any predetermined circumferential location by a circumferentially placed phased array with appropriate voltage amplitude and time-delay inputs. With the guided wave phased array focusing technique in pipes, a stronger and more narrowly focused beam could be achieved. Better circumferential resolution and higher penetration power than axisymmetric type waves in defect detection became possible. As a result, the circumferential location of a defect can be determined. It also provided better potential in defect sizing. Sun and colleagues

With significant contribution from Jing Mu.

294

(2003) extended the guided wave phased array focusing technique further by utilizing torsional waves.

Another focusing technique receiving more attention is the time-reversal focusing technique (Ing and Fink 1998). This technique reverses the received signals in the time domain and utilizes these time-reversed signals as transmitting signals to focus at the largest reflector. The advantage of this technique is that there is no computation complexity. Only hardware with arbitrary function generators is needed. Another advantage is that this technique is applicable in various media, such as inhomogenous material and the human body, and in complex waveguide structures like pipe elbows. Although seemingly straightforward, the technique presents many challenges when trying to implement such an approach, primarily because of the signal-to-noise ratio requirement of the signals utilized in the process.

The focusing techniques just discussed are usually referred to as *real-time focusing*, where the wave energy is actually tuned to focus at a predetermined location by sending signals from the phased array transducers with different amplitude factors and time delays. In applications, the hardware that is capable of accurately controlling time delays is often expensive. A phased array with a large number of channels or transducers can be costly as well. These are the cases where synthetic focusing techniques could be employed instead of real-time focusing. In synthetic focusing, signals are sent and received by different transducer pairs and the focal waveforms are obtained by signal processing in a post-processing procedure. Multiple ways exist of achieving synthetic focusing using guided waves. Excellent work can be found in Wilcox (2003) and Giurgiutiu and Bao (2004) for plate structures, and in Hayashi and Murase (2005) and Mu and Rose (2010) for pipes. Different from the synthetic focusing conventionally used in bulk waves, dispersion associated with multiple mode propagation has to be taken into account in guided wave synthetic focusing. Compared to real-time focusing, synthetic focusing may offer less penetration power because of the existence of noise in each signal utilized in the computation, but it is more cost-effective in some cases. Synthetic focusing also has the disadvantage of possible confusion in the case of multiple reflections and defects.

This chapter discusses the active real-time guided wave phased array focusing technique (Li and Rose 2002) in detail and also introduces the synthetic guided wave focusing technique that uses back propagation based on normal mode decomposition (NMD) (Hayashi and Murase 2005).

16.2 Guided Wave Phased Array Focus Theory

Guided waves propagating in a pipe can be classified into a doubly infinite number of propagation modes. The doubly infinite propagation modes have proven orthogonal to each other, therefore they are called *normal modes*. The ultrasonic field of a normal mode can be represented by two indices n and M, where M is circumferential order and n is the index of mode group (Ditri and Rose 1992; Li and Rose 2002):

$$\mathbf{v}_n^M e^{i(\omega t - k_n^M z)} = \sum_{\xi = r, \theta, z} R_{n\xi}^M(r) \Theta_\xi^M(M\theta) e^{i(\omega t - k_n^M z)} \hat{\mathbf{e}}_\xi \tag{16.1}$$

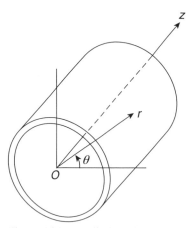

Figure 16.1. A cylinder with a cylindrical coordinate system.

where ω and k are angular frequency and wavenumber respectively and $\mathbf{v}_n^M$ is the particle displacement distribution function, which is independent of the cylinder axial direction z and time t. The particle displacement is decomposed into three directions denoted by unit vectors $\hat{\mathbf{e}}_r$, $\hat{\mathbf{e}}_\theta$, and $\hat{\mathbf{e}}_z$ in cylindrical coordinates shown in Figure 16.1. Function $R_{n\xi}^M(r)$ denotes the distribution of the particle displacement produced by mode (M, n) in the ξ-direction ($\xi = r, \theta, z$). Function $\Theta_\xi^M(M\theta)$ is the angular distribution function ($\cos(M\theta)$ or $\sin(M\theta)$) of the particle displacement produced by mode (M, n) in the ξ-direction. For $M = 0$, the modes are axisymmetric longitudinal or torsional, according to the nature of sinusoidal functions. For $M > 0$, the mode is longitudinal flexural or torsional flexural. The dispersion curve for all of the guided wave modes propagating in an elastic hollow cylinder can be obtained by solving a boundary value problem (Gazis 1959).

The longitudinal axisymmetric ($m = 0$) and flexural modes ($m > 0$) are represented as L(M, n). The phase velocity of the longitudinal modes and flexural modes depends on pipe diameter and wall thickness. The phase velocities of the various modes are different. The modes with the same n value are called a *mode group*. For the same group of modes L(M, n), longitudinal mode L$(0, n)$ has the lowest phase velocity. The phase velocity gets higher as the circumferential order M increases. The phase velocity difference between a group of modes decreases with frequency increase, which is also the feature indicated in Figure 16.2. This is also the case with the torsional type of guided wave modes. The phase velocity dispersion curve of torsional-type modes T(M, n) in a sixteen-inch schedule 30 steel pipe is shown in Figure 16.3.

For a normal beam source loading on the outer surface of the elastic isotropic hollow cylinder shown in Figure 16.4, the loading condition can be described as:

$$\hat{T} \cdot \vec{n} = \begin{cases} -p_1(\theta)p_2(z)\vec{e}_{\xi_0}, & |z| \le L, |\theta| \le \alpha, r = b \\ 0, & |z| > L, \text{ or } |\theta| > \alpha, \text{ or } r \ne b \end{cases} \tag{16.2}$$

where $\vec{e}_{\xi_0}$ ($\xi_0 = r, \theta, \text{ or } z$) represents the loading vibration direction. The transducer covers an axial length of $|z| \le L$ and a circumferential angle of $|\theta| \le \alpha$. a and b denote the inner and outer surfaces of the cylinder, respectively. The normal beam

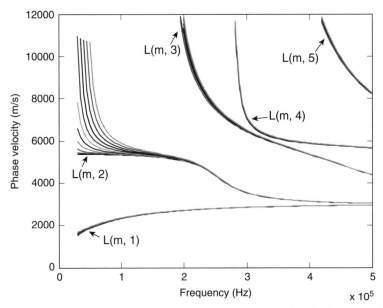

Figure 16.2. The phase velocity dispersion curve of longitudinal type modes $L(M, n)$ in a sixteen-inch schedule 30 steel pipe including axisymmetric modes $m = 0$ and flexural modes ($M = 1, 2, 3, \ldots$).

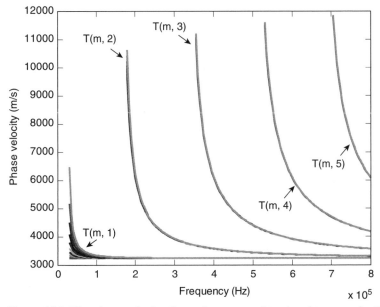

Figure 16.3. The phase velocity dispersion curve of torsional type modes $T(M, n)$ in a sixteen-inch schedule 30 steel pipe including axisymmetric modes $m = 0$ and flexural modes ($M = 1, 2, 3, \ldots$).

transducer loading is assumed to produce a traction loading in the z-direction only. Functions $p_1(\theta)$ and $p_2(z)$ describe the traction variation in the θ- and z-directions, respectively.

As an example, such a partial loading can generate a group of longitudinal and flexural modes resulting from the match between the loading condition and modal wave structures. The amplitude factor of each generated mode propagating in the

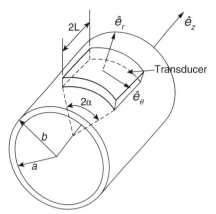

Figure 16.4. An elastic isotropic hollow cylinder loaded by an ultrasonic transducer.

+z-direction A_{+n}^M can be calculated using a normal mode expansion (NME) method as described in Ditri and Rose (1992):

$$A_{+n}^M = -\frac{R_{n\xi_0}^{M*}(b)e^{-ik_n^M z}}{4P_{nn}^{MM}} \int_{-\alpha}^{-\alpha+2\pi} \Theta_{\xi_0}^M(M\theta)p_1(\theta)\,d\theta \int_{-\infty}^{\infty} p_2(z)e^{ik_n^M z}dz, \quad z \geq L \quad (16.3)$$

where the superscript * denotes a complex conjugate. The term P_{nn}^{MM} is the power flow carried by the mode L(M, n) across the cross section D of the cylinder. It is defined in Equation (16.4).

$$P_{nn}^{MM} = -\frac{1}{4}\iint_D (\mathbf{V}_n^{M*} \cdot \mathbf{T}_n^M + \mathbf{V}_n^M \cdot \mathbf{T}_n^{M*}) \cdot \hat{\mathbf{e}}_\xi \, d\sigma, \quad (16.4)$$

where $\mathbf{V}_n^M$ and $\mathbf{T}_n^M$ are the particle velocity vector and stress tensor of mode L(M,n), respectively. If $\xi_0 = \theta$, Equation (16.3) can also be used to describe the amplitude factor of a torsional guided wave mode.

It can be clearly seen from Equation (16.3) that the amplitude factor of a generated guided wave mode is determined by three parts: $\dfrac{R_{n\xi_0}^{M*}(b)e^{-ik_n^M z}}{4P_{nn}^{MM}}$, $\displaystyle\int_{-\alpha}^{-\alpha+2\pi}\Theta_{\xi_0}^M(M\theta)p_1(\theta)d\theta$,

and $\int_{-\infty}^{\infty}p_2(z)e^{ik_n^M z}dz$. $R_{n\xi_0}^M(b)$ is the wave structure of mode L(M, n) in the dominant particle vibration direction ξ_0 ($\xi_0 = r, \theta, or z$) at the outer surface of the cylinder $r = b$. The term $R_{n\xi_0}^M(b)$ exists because the traction loading in Equation (16.2) is in the ξ_0-direction. In other words, if the traction is loaded in the r-, θ- or z-direction, the corresponding function will be $R_{nr}^M(b), R_{n\theta}^M(b)$ or $R_{nz}^M(b)$. The term $\dfrac{R_{n\xi_0}^{M*}(b)}{4P_{nn}^{MM}}$ is the ratio between the outer surface wave structure value and the power flow of mode L(M,n). The higher the value of $R_{n\xi_0}^M(b)$ compared to P_{nn}^{MM}, the larger the amplitude factor and the better the mode L(M,n) can be generated. The term $\dfrac{R_{n\xi_0}^{M*}(b)}{4P_{nn}^{MM}}$ therefore describes the excitability of mode L(M, n). For example, torsional modes have minimal displacements in the r-direction, so they will not be generated in this case. The second part, $\int_{-\alpha}^{-\alpha+2\pi}\Theta_{\xi_0}^M(M\theta)p_1(\theta)d\theta$, is the inner product of the circumferential distribution of mode L(M, n) and the loading,

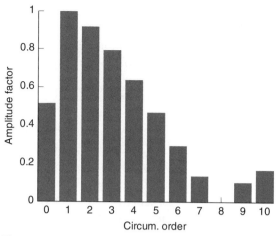

Figure 16.5. Normalized amplitude factors of the L(m, 2) mode group when 45° normal loading is applied to a sixteen-inch schedule 30 steel pipe at a frequency of 50 kHz.

which represents the likelihood of the loading and mode $L(M,n)$ in the circumferential direction. The better these two distributions match each other, the greater the potential that the mode $L(M,n)$ will be accurately generated. Notice that when the source loading is axisymmetric, which can be represented by $p_1(\theta) = 1$ for $0 \le \theta < 2\pi$, the inner product in the circumferential direction vanishes except for M = 0. This explains quite well the fact that only axisymmetric modes will be generated in the case of axisymmetric source loading. Likewise, $\int_{-\infty}^{\infty} p_2(z)e^{ik_n^M z}dz$ describes how similar the traction distribution in the z-direction is compared to that of mode $L(M, n)$. The more similar they are to each other, the greater chance that the mode will be accurately generated. This is also a manifestation of Snell's law.

To summarize, Equation (16.3) describes the similarity between distributions of the source and the mode $L(M, n)$. The better they match each other, the greater chance that the mode will be generated. Figure 16.5, as an example, shows the numeric calculated amplitude factor distribution with respect to the circumferential order for 45° normal loading on a sixteen-inch schedule 30 carbon steel pipe at 50 kHz. Notice that the first several modes are generated with significant amplitude factors and the other higher order modes have much smaller amplitude.

Summing up the generated modes weighted by their corresponding amplitude factors and phases yields the guided wave displacement distribution or interference pattern in specific pipes. We use the function $\mathbf{v}(r,\theta,z)$ to denote the displacement field created by a partial loading inside a pipe. The term $\mathbf{v}(r,\theta,z)$ is the summation of all generated guided wave modes weighed by their corresponding amplitude factors as described in Equation 16.5.

$$\mathbf{v}(r,\theta,z) = \sum_{\gamma \in (T \cup L)} \left\{ \sum_{n=1}^{\infty} \sum_{M=0}^{\infty} \left\{ A_{+n\gamma}^{M} \sum_{\xi=r,\theta,z} [R_{n\xi}^{M}(r)\Theta_{\xi}^{M}(M\theta)\hat{\mathbf{e}}_{\xi}]e^{i\omega t} \right\} \right\} \tag{16.5}$$

A polar plot illustrates the displacement distribution around the pipe circumference at a certain axial distance from the transducer location; this distribution is called an *angular profile*. Because the total field in the pipe is the superposition of all of the generated modes with different phase velocities, the slight difference in phase velocities leads to a variation of the phase match between the different modes, therefore resulting in variations of angular profiles along the propagating direction. The phase velocity difference plays an important role in the development of angular profiles.

The angular profiles of the L(m, 2) mode group at several axial distances for the 45° source loading on the sixteen-inch schedule 30 steel pipe at 50 kHz are demonstrated in Figure 16.6. Figure 16.6 shows that the energy is concentrated on the top of the pipe when the axial distance is close to the transducer (Figure 16.6(a)). Then, the energy spreads out as the axial distance increases (Figure 16.6(b)–(f)).

Figure 16.7 illustrates a guided wave partial loading interference pattern (unwrapped) sample result of the L(m,2) mode group generated by 45° source loading on a sixteen-inch schedule 30 steel pipe at 50 kHz. The horizontal axis represents the axial distance and the vertical axis is the circumferential angle. The source loading is centered at 0° as can be seen in Figure 16.7. The propagation distances used in the angular profile calculation in Figure 16.6 are also marked in Figure 16.7 by vertical magenta lines. It can be clearly seen that Figure 16.6 and Figure 16.7 agree quite well. The interesting phenomenon of guided wave natural focusing is clearly shown in Figure 16.7 at about forty-six feet in axial distance. The first natural focusing happens at an angle of 180°, opposite to the source loading position in the circumference. It agrees with the angular profile shown in Figure 16.6(f).

Note that the term *natural focusing* is introduced here. Natural focusing points are points in a structure where strong constructive interference of many waveforms occurs as a result of partial loading around the circumference of a pipe or after a series of waves have passed an elbow region. The natural focal points change with frequency and degree of partial loading around the pipe circumference.

Figure 16.8 illustrates the schematic of a phased array mounted around a pipe circumference. The phased array is driven by a multichannel (N) time-delay system. Each channel corresponds to a normal beam transducer element mounted partially on the pipe outer surface. Note that the closer the elements are in filling the circumference of the pipe, the closer the wave is to being axisymmetric in character when all elements are pulsed in phase.

Suppose we are tuning the angular profile at a certain axial distance, z_0, from the transducers on a pipe. The angular profile at the pipe outer surface for element 0, the one placed on top of the pipe, is denoted by a complex function, $H(\theta)$, at the distance z_0. The term $H(\theta)$, based on the previous derivation, can be expressed as:

$$H(\theta) = \mathbf{v}(r = b, \theta, z = z_0), \tag{16.6}$$

If the responses of the elements are identical, the angular profile of element i with the same input will be the shifted function of $H(\theta)$; that is, $H(\theta - \theta_i)$, where θ_i is the circumferential position of element i and θ is the target position of the field. The total angular profile $G(\theta)$ at distance z_0 is the superposition of all of the elements' angular profiles at that distance, expressed by the convolution of the contributions from each element as shown in Equation (16.7).

$$G(\theta) = \sum_{i=0}^{N-1} A(\theta_i)H(\theta - \theta_i) = A \otimes H, \tag{16.7}$$

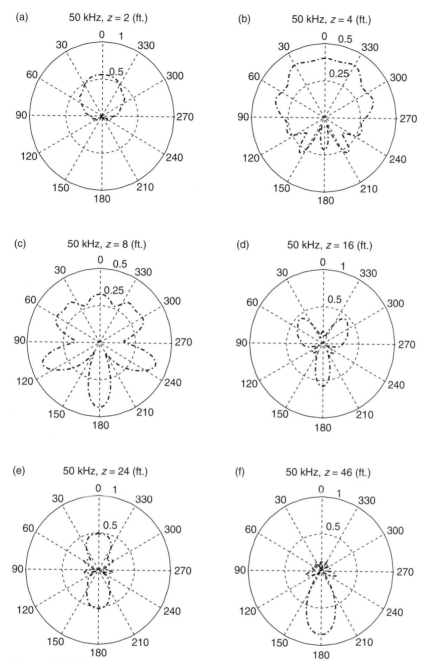

Figure 16.6. The angular profiles of the L(m, 2) mode group at different axial distances for 45° source loading on a sixteen-inch schedule 30 steel pipe at 50 kHz.

where $A(\theta_i)$ is the discrete weighting function for element i in the summed angular profile. $A(\theta_i)$ is a complex function, where the norm represents the voltage level applied to element i and the phase of $A(\theta_i)$ corresponds to the time delay to be applied to the tone burst signal for channel i. The relation between the time delay input of each channel and phase of the function $A(\theta_i)$ can be represented as

$$t_i = -\phi_i / (2\pi f), \tag{16.8}$$

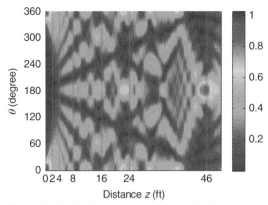

Figure 16.7. Guided wave partial loading interference pattern of the L(m, 2) mode group generated by 45° source loading on a sixteen-inch schedule 30 steel pipe at 50 kHz. See plates section for color version.

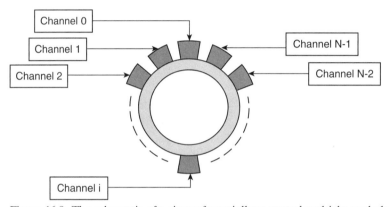

Figure 16.8. The schematic of a circumferentially mounted multichannel phased array.

where f is frequency and ϕ_i is the phase of function $A(\theta_i)$ for element i. ϕ_i is in the range from -2π to 0. t_i is a positive value representing the physical time delay applied to channel i to generate the tone burst with a phase output ϕ_i.

Equation (16.7) shows that a total angular profile is the circular convolution between the weighting function $A(\theta_i)$ and the single-element angular profile function $H(\theta)$. Therefore, given the voltage amplitudes and time delays applied to the elements, the total angular profile can be calculated based on the knowledge of the single-element angular profile.

For an inverse problem (i.e., the angular profile tuning procedure), the desired total angular profile $G(\theta)$ is given. We need to calculate the discrete weight function $A(\theta_i)$ to generate the total field $G(\theta)$. Equation (16.7) makes the calculation procedure quite straightforward, which is a deconvolution algorithm as shown in the following:

$$A(\theta) = G \otimes^{-1} H \qquad (16.9)$$

$$A(\theta) = \text{FFT}^{-1}[\mathbf{G}(\omega) / \mathbf{H}(\omega)] \qquad (16.10)$$

where $\otimes$ and $\otimes^{-1}$ are convolution and deconvolution operators, respectively.

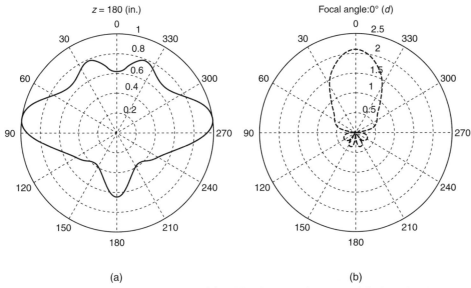

Figure 16.9. Sample angular profiles of (a) 90°-loaded transducer and (b) phased array focused result at 0° in a sixteen-inch schedule 80 steel pipe at z = 18 in. The excited wave mode group is the 35 kHz T(m, 1) group.

If a desired angular profile is given, the voltage level and time delays can be calculated by the deconvolution algorithm based on the knowledge of a single-element angular profile. For the procedure of guided wave beam focusing, the total angular profile is a spike function versus the circumferential location. The Fourier transform of the spike function is a constant function. Therefore, to focus a guided wave beam onto the top of the pipe (zero-degree circumferential location), the voltage level and time delay can be calculated based on the following equation:

$$A(\theta) = 1 \otimes^{-1} H \qquad (16.11)$$

$$A(\theta) = \text{FFT}^{-1}[1 / \mathbf{H}(\omega)]. \qquad (16.12)$$

Although Equation (16.12) is for focusing the guided wave beam to a zero-degree circumferential location, the focused point can be adjusted to other locations by circularly shifting the $A(\theta)$ function, that is, the voltage levels and time delays.

In addition, by changing the target function, G, the circumferential focal position will change. For instance, if we let G_i become one at 90° and be zero at any other angle, the focal beam in Figure 16.9(b) will rotate 90°. By turning the focal point along the circumference of the cylinder, one can determine the circumferential location of a reflector. If the focal beam is narrow enough, the circumferential length of the defect can be measured by turning the focal point around the circumference.

16.3 Numerical Calculations

As an example, let us investigate a circumferential phased array with eight normal beam transducer segments placed around a sixteen-inch schedule 30 steel pipe. Each segment can be composed of multiple normal beam transducer elements, which covers a 45° circumferential angle and 20 mm axial length. The transducers produce traction in the pipe axial direction z; therefore, the major generated modes are L(m, 2) at a frequency of 50 kHz.

Table 16.1. Amplitude factors and time delays for obtaining the focused profiles in Figure 16.10(e) and (f)

Channel no.		1	2	3	4	5	6	7	8
Figure 10(e)	Amplitude factor	0.72	0.40	0.16	0.28	1.00	0.28	0.16	0.40
	Time delay (μs)	4.1	19.2	6.9	11.9	17.9	11.9	6.9	19.2
Figure 10(f)	Amplitude factor	0.20	0.11	0.09	0.11	1.00	0.11	0.09	0.11
	Time delay (μs)	4.8	14.0	17.5	5.4	12.5	5.4	17.5	14.0

Suppose we are concerned with the partial loading angular profiles at the distances listed in Figure 16.6. If we need to construct the angular profiles with energy focused at the zero-degree circumferential position at these distances, the voltage levels (amplitude factors) and time delays can be calculated from Equations (16.8) and (16.12). The resulting focused angular profiles are plotted in Figure 16.10, showing that the guided wave beam is effectively focused at the desired locations. The amplitude factors and time delays used in producing the focused profiles in Figure 16.10(e) and (f) are listed in Table 16.1.

For a partial loading source on a pipe, the angular profiles of the generated guided waves are symmetric about the center line crossing the center of the pipe and source location. Therefore, the derived amplitude factor and time delay values are also symmetric about the center line crossing the center of the pipe and the focused circumferential location. This feature is indicated in Table 16.1. The amplitude factor and time delay values for channels 2, 3, and 4 are identical to those for channels 8, 7, and 6, respectively.

Therefore, based on the deconvolution technique, we can calculate the voltage levels and time delays to obtain desired angular profiles. As an example, we could turn the angular profile to focus the beam at a 0° circumferential position based on Equation (16.12). If the voltage level and time delay on each channel are shifted to the channel right next to it, the focus position of the total guided wave beam will shift to the 45° circumferential location. If all eight positions are tried for focusing, a circumferential scan is implemented with guided waves.

Note that the existence of side lobes may cause artifacts in pipe inspection. The focused amplitude and the amplitudes of side lobes vary with frequency and focal distance. Numerical calculations should be performed as part of the design process to allow the selection of a good frequency having high focused amplitude and small side lobes at a predetermined focal distance.

An important facet of the focused guided wave beam is the circumferential beam width. A narrower beam width is usually more favorable for NDE tests because of stronger focused energy and better lateral resolution for guided wave beam scanning. Similar to the ultrasonic bulk wave array, the number of array elements plays an important role in the focused beam width of the guided waves. Generally, the more channel numbers, the narrower the guided wave beam width. As an illustrative example, the focused angular profiles for four, eight, and sixteen channels are shown in Figure 16.11. Torsional waves T(m, 1) created by a traction loading source in the circumferential direction at 40 kHz in a sixteen-inch schedule 30 steel pipe are used in this example. The size of a focused beam is reduced with an increase in the number of channels. Calculations have also shown that the six dB down point of the focused beam width is almost the same as the angular coverage

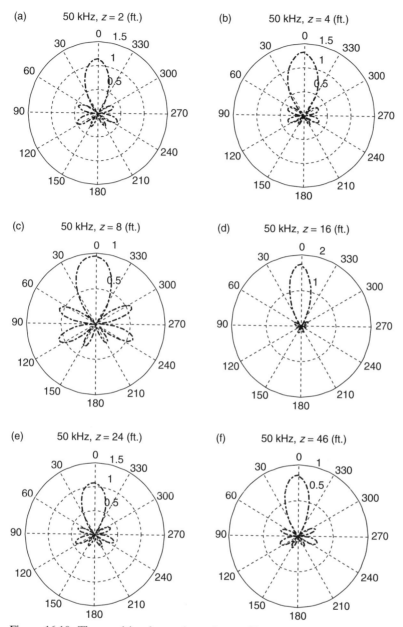

Figure 16.10. The resulting focused angular profiles of the L(m, 2) mode group at different axial distances for an eight-channel phased array mounted on a sixteen-inch schedule 30 steel pipe at 50 kHz, when using appropriate time delays and amplitude factors.

of each channel. A narrower focused beam width is good because it will give higher circumferential resolution in defect sizing. However, the cost also increases with an increase in channel numbers and scanning completely around the circumference becomes more time consuming. Beam overlap in fixed point scanning around the circumference is recommended.

It is also instructive to investigate the focal zones in the axial direction of the pipe. Figure 16.12, as an example, shows the displacement amplitude distributions of four-, eight-, and sixteen-channel focusing and their corresponding partial loading

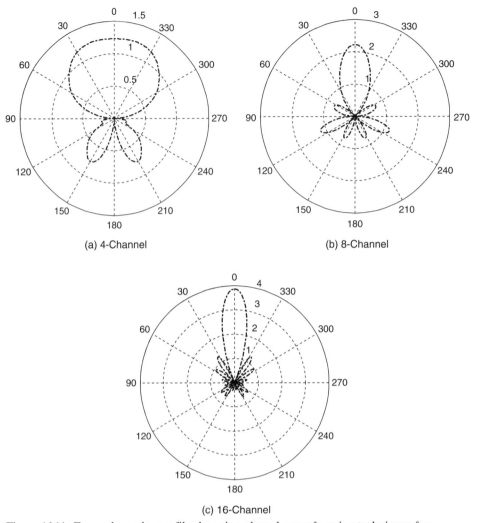

(a) 4-Channel

(b) 8-Channel

(c) 16-Channel

Figure 16.11. Focused angular profiles by using phased-array focusing techniques for different channel numbers. Torsional modes T(m, 1) at 40 kHz are generated to focus at eleven feet axial distance in a sixteen-inch schedule 30 steel pipe.

amplitude distributions at $0°$. The designed focal distance in this example is 3.35m. (11 ft.) in a sixteen-inch schedule 30 steel pipe. In Figure 16.12, the horizontal axis is the axial distance from the transducer array along a pipe; the vertical axis is the displacement amplitude normalized by that of the axisymmetric wave $L(0, 2)$ in dB. The amplitude of the axisymmetric $L(0, 2)$ wave is at 0 dB. The focal zone is defined as the axial area where focusing provides higher displacement amplitude than the axisymmetric wave $L(0, 2)$. The length of the focal zone decreases with an increase in the number of channels. When more channels are used, the angular profile varies more rapidly with axial distance. The shorter focal zone will result in smaller step sizes in scanning a pipe.

The computational process of achieving phased array focusing in pipes is summarized in Figure 16.13. The problem starts from a separation of variables in the governing equations. The normal modes in the dispersion curves are obtained by applying the gauge invariance and boundary conditions. The waves generated

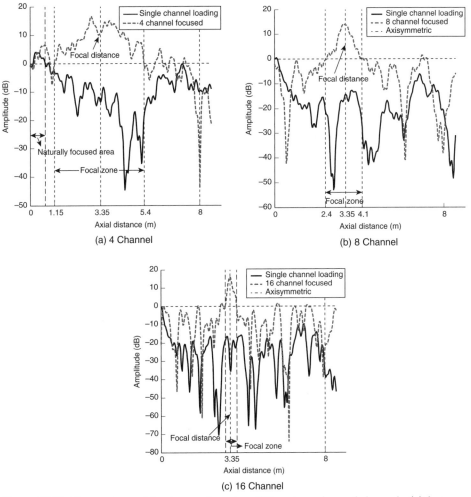

Figure 16.12. Comparison of focal zones by using different numbers of channels: (a) four channels; (b) eight channels; (c) sixteen channels. L(m, 2) waves at a frequency of 50 kHz are focused at 3.35 m (11 ft.) in a sixteen-inch schedule 30 steel pipe.

by a certain source loading are represented in terms of these normal modes. A deconvolution algorithm is then involved to calculate the time delays and amplitude factors in order to focus at a specific location in a pipe.

16.4 Finite Element Simulation of Guided Wave Focusing

Various fields increasingly use FE modeling because of its power, simplicity, and flexibility. It is suitable for problems with complex structures and can also be used as a tool to verify theories and to visualize physical phenomena. Many easy-to-use commercial software codes are currently available. In this study, ABAQUS is adopted to simulate wave propagation in pipes. The maximum length of the elements is chosen to be no more than one-eighth of a wavelength. The time steps in the calculation are automatically decided by ABAQUS. If chosen manually, the time steps should be at least several times less than the period of the input signal and less than the time the ultrasonic wave propagates from one end of the element to the other end.

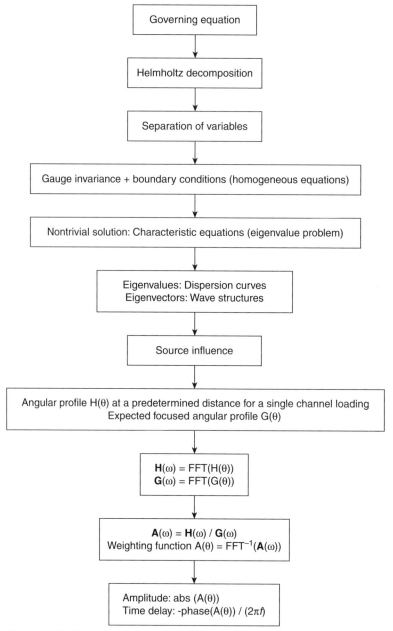

Figure 16.13. The phased array focusing technique in pipes.

Torsional wave focusing by using four channels is simulated in ABAQUS. The results are shown in Figure 16.14. The phased array excitation is simulated by applying boundary conditions with calculated time delay and amplitude values at the left end of a sixteen-inch schedule 30 steel pipe in Figure 16.14(a). The predetermined focal point is 3.35m (11 ft.) away from the phased array. Figure 16.14(a) shows that the generated guided waves are focused at the designated location in the pipe. Figure 16.14(b) shows the focused angular profile comparison at the focal distance in the pipe. The dashed line is the focused angular profile from theoretical

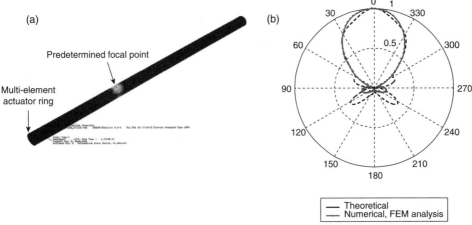

Figure 16.14. Torsional waves $T(m, 1)$ at 35 kHz are focused at 3.35 m axial distance in a sixteen-inch schedule 30 steel pipe. Four channels are used to achieve focusing.

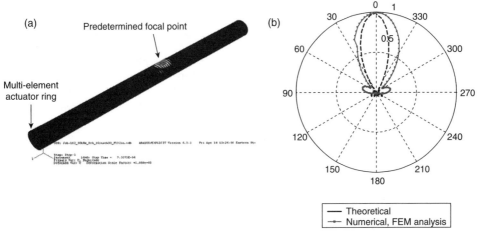

Figure 16.15. Longitudinal waves $L(m, 2)$ at 50 kHz are focused at 3.35 m axial distance in a sixteen-inch schedule 30 steel pipe. Eight channels are used to achieve focusing.

calculation and the solid line is the focused angular profile obtained from the FE simulation. The focusing results from theory and FE simulation match quite well. Figure 16.15(a) provides a static shot of eight-channel longitudinal waves focused at 3.35m (11 ft.) in a sixteen-inch schedule 30 steel pipe. The static shot in the FE simulation shows a well-focused guided wave beam. Figure 16.15(b) illustrates the corresponding focused profiles obtained from theoretical calculation and FEM simulation. In this case, the FE-simulated focused profile is slightly wider than the theoretical prediction. Compared to the four-channel focusing in Figure 16.13(a), it is also clear from the focusing simulations and focused angular profiles that eight-channel focusing provides a much narrower focal beam than four-channel focusing. The focal beam width is usually about the same as the circumferential length of each channel in the phased array. Therefore, the greater the number of channels one uses, the narrower the focal beam one can obtain.

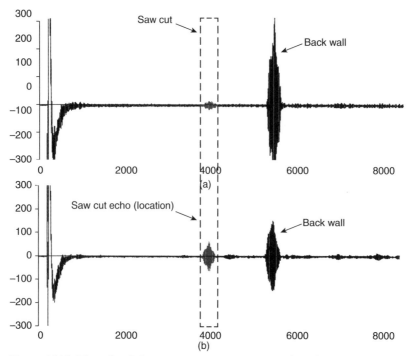

Figure 16.16. There is a 3.6 percent cross-sectional area (CSA) saw cut located at z = 240" in a sixteen-inch schedule 30 steel pipe. Sample pipeline inspection signals by employing the 55 kHz (a) axisymmetrically excited $T(0, 1)$ mode and (b) $T(m, 1)$ group, which is focused at the defect location. Each segment in the transducer array has a 90° circumferential loading length.

16.5 Active Focusing Experiment

The active focusing technique effectively enhances guided wave penetration power in pipe. Sample experimental results in Figure 16.16 show that the active focusing technique led to significant improvement for defect detections in a sixteen-inch schedule 30 steel pipe. The reflected signal from the defect is much higher when active focusing is used compared to when axisymmetric is excited for pipe inspection.

The direct application of the guided wave focusing technique in pipes is to use a combination of axial and circumferential scans at selected increments with focused guided wave beams. This technique provides the information on the defect axial and circumferential distribution of a pipe at a certain distance from the transducer. Another benefit of the focused guided wave is that stronger focused wave energy helps detect smaller defects in pipes as well as having increased penetration power. In this section, experiments are discussed to demonstrate the beam focusing, circumferential scan, and complete pipe scan concepts in terms of practical pipe tests.

Figure 16.17 shows the schematic of a guided wave focus scan experiment conducted on a twenty-inch schedule 40 eight-foot-long steel cased example pipe like those commonly used in road crossings (Mu and Rose 2008). A sketch of a multichannel guided wave phased array system is shown in Figure 16.18. The system

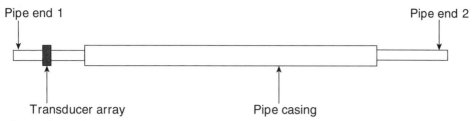

Figure 16.17. Schematic of the experiment setup on a twenty-inch schedule 40 cased steel pipe.

Figure 16.18. An example of guided wave phased array tool mounted on a pipe.

has elements wrapped around the pipe circumference that can be segmented into quadrants or octants, for example. The phased array can contain more or fewer elements depending on the pipe diameter. The closer the elements are together, the better chance of more closely approximating an axisymmetric wave when all elements are pulsed simultaneously. Time delays and amplitude factors can be input to the system through a control from a laptop to tune the guided wave energy to focus at desired positions in a pipe. Each element may contain more than one sensor in the axial direction. Multiple rings of sensors can form similar to comb-type transducers for mode control and/or wave suppression in the pipe axial direction. For normal beam transducers loaded on a pipe, the generated guided waves will propagate in both directions, forward and backward from the transducers. This can make defining the defect locations difficult. To eliminate this problem, a separate set of time delays can be applied to different rings to alter the wave interference patterns, allowing most of the ultrasonic energy to travel in one direction.

Active phased array systems are currently available in the marketplace.

Let's consider a sample phased array experiment. In the experiment, the phased array was segmented into four channels. Theoretically, the four-channel focused beam covers approximately 90° in circumference. However, to ensure full coverage of energy on the circumference, eight equally spaced positions are scanned at each focal distance with the four-channel focusing. In a circumferential scan, the time delays and amplitude factors are calculated for eight focal positions around the circumference to spin the focal beam around the pipe at each focal distance. A full scan of the pipe is achieved by carrying out circumferential scans at consecutive

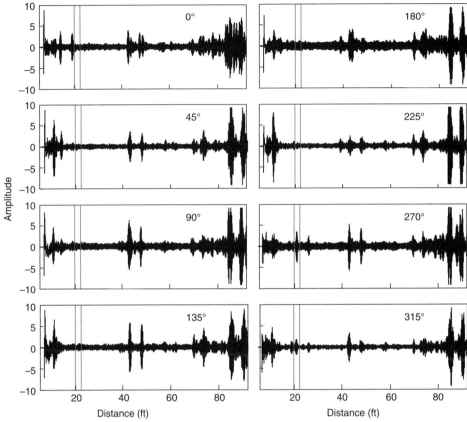

Figure 16.19. Waveforms received by torsional waves T(m, 1) focusing at eight different angles around the circumference at a focal distance of 20'9" from pipe end 1. The vertical lines gate the effective focal range. Note a defect echo is located at 270°.

axial distances. Longitudinal L(m, 2) and torsional T(m, 1) wave modes can be generated from both sides of the pipe with the transducer array placed 4'9" from each pipe end. Figure 16.19 shows the waveforms received by the phased array when torsional waves T(m, 1) were focused at 20'9" from pipe end 1. Each of the eight plots corresponds to a focal angle as noted in the figure. The waveforms are displayed in such a way that the zero distance in each plot represents pipe end 1. The focal zone in the axial direction is the region indicated between the two vertical lines in each waveform. The biggest reflection inside the focal zone is received when the guided waves are focused at 270°. The axial location can also be precisely determined by the arrival time of the biggest reflection. Notice that there is an echo in front of the gate in the 0° focused waveform. This echo is caused by a defect and it is picked up in the 0° focused waveform because the focal zone is actually longer than the distance indicated by the two vertical lines in the waveform.

The maximum amplitude inside each focal zone is then recorded and plotted with respect to the focal angle in constructing a circumferential profile as shown in Figure 16.20. This process decreases any false alarm calls because the defect grows upon focusing as seen in Figure 16.20. Artifacts via any mode conversion or noise will not grow or give rise to non-axisymmetric profiles as shown in Figure 16.20. The circumferential profile illustrates the circumferential distribution of the anomaly

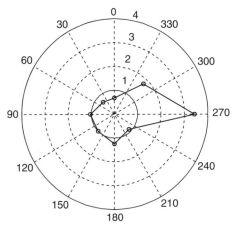

Figure 16.20. The circumferential profile of the defect located at twenty feet and 270° in the pipe. The unit of the amplitude is mV.

located inside the focal zone. For an anomaly with a non-axisymmetric feature, for example, a defect, the circumferential profile will be non-axisymmetric. However, if the anomaly has an axisymmetric feature, such as a weld, the circumferential profile is approximately axisymmetric. In this way, for example, defects can be differentiated from welds quite easily.

In the cased pipe example, spacers are used inside the casing to separate the pipe from the spacing. To illustrate the difference in reflection from different anomalies, Figure 16.21 provides the waveforms from spacers located at eleven feet from pipe end 1. Spacers are distributed around the pipe circumference to support the casing. The waveforms are also taken from torsional wave focus inspection from pipe end 1. Different from defects, spacers are distributed around the pipe circumference. Therefore, evident reflections may be expected at multiple focal angles when the focal beam moves around the pipe. This is well demonstrated in Figure 16.21. The corresponding circumferential profile is shown in Figure 16.22. It can be observed from Figure 16.22 that the amplitudes of the echoes reflected by the spacers are in general larger than the amplitudes of defect reflections. Possible explanations are as follows: first, multiple spacers may be located within the focal beam width; second, the spacers are located closer to the transducer array, thus less attenuation of the focal beam when focusing at the spacers compared to focusing at the defect shown in Figure 16.20.

Another interesting anomaly to investigate is a weld. A good weld can demonstrate a fairly axisymmetric feature in the reflected ultrasonic waves. Figure 16.23 shows the circumferential profile reflected by a weld located at 41'11" inspected by longitudinal waves L(m, 2) focusing from pipe end 2. The circumferential profile is almost axisymmetric in this case. Thus, the defect can be well separated from spacers and welds by the circumferential profiles of reflection in a focal scan.

An image of the pipe can be obtained by combining the circumferential plots with focal distance. Figure 16.24 displays the image of the unwrapped pipe. The vertical axis in Figure 16.24 denotes the circumferential direction of the pipe and the horizontal axis denotes focal distance. The reflection amplitude is displayed in color as a third dimension. Interpolation is used in both the circumferential and axial directions to smooth the image. Because the pipe is inspected from both ends, the

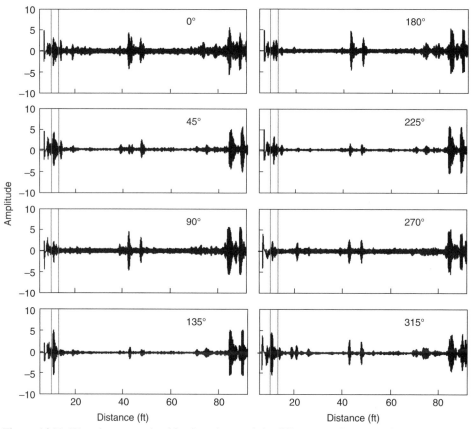

Figure 16.21. Waveforms received by focusing at eight different angles around the circumference at a focal distance of 12.75 ft. from the pipe end.

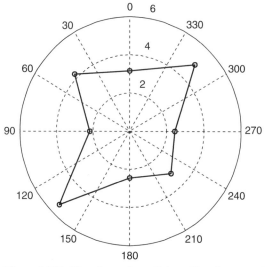

Figure 16.22. The circumferential profile of spacers in the cased pipe located at eleven feet.

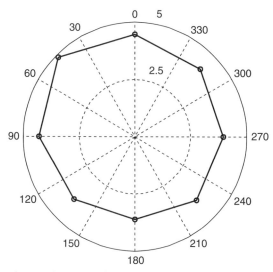

Figure 16.23. The circumferential profile of the weld located at 41'11".

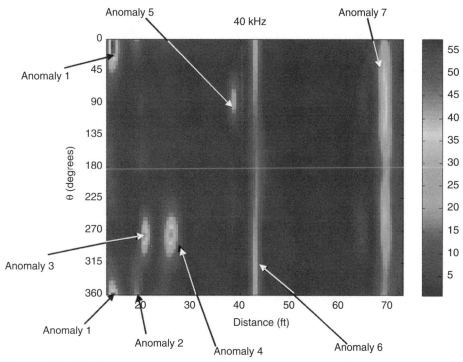

Figure 16.24. Pipe focus scan image. See plates section for color version.

image is formed by superposition in combining the results from the two pipe ends. Imaging from both pipe ends produces a higher signal-to-noise ratio than using the inspection results from a single pipe end. Also, because of the fact that guided waves attenuate as they propagate along the pipe, the guided wave attenuation along the pipe is compensated by $e^{2\alpha z}$, where α is the attenuation factor and z is the focal distance. The attenuation factor can be measured experimentally from pipe end reflection or from the reflections of multiple welds.

Several anomalies are shown and marked in the pipe image illustrated in Figure 16.24. From the focused scan image, it is clear that anomalies 1 to 5 are defects. Anomaly 6 can be distinguished as a weld because of its relatively uniform distribution in the circumferential direction as discussed earlier. Anomaly 7 consists of spacers as revealed by their large and nonuniform reflections around the circumference.

16.6 Guided Wave Synthetic Focus

Pipe defects are usually non-axisymmetric features. Therefore, even though axisymmetric type waves are often excited to inspect a pipe, the reflected waves can contain both axisymmetric waves and a number of flexural (non-axisymmetric) modes. These modes contain the information on defect characteristics and are essential for accurate defect localization and sizing. This section describes the synthetic focus approach based on a combination of normal mode decomposition concepts with time-reversed back propagation in constructing a defect image.

A hollow cylindrical guided mode can be represented as in Equation (16.1). The displacement received from a transducer located at the surface of a pipe is the combination of a number of guided wave modes emitted from a defect. The displacement is a function of θ, z, and t that can be denoted by Equation (16.13).

$$u(\theta,z,t) = \sum_{n=1}^{+\infty} \sum_{m=-\infty}^{+\infty} A_{mn}(\omega)e^{i(m\theta+k_n^M z+\omega t)}. \tag{16.13}$$

where A_{mn} is the amplitude of the defect-emitted guided wave. Here m represents the mth circumferential order and equals to M that was used in previous equations. In general cases, the reflected guided waves can contain a doubly infinite number of modes as in Equation (16.13). Nevertheless, Equation (16.13) can be simplified at certain circumstances. For example, guided waves at relatively low frequencies (≤ 100 kHz) are commonly used in NDT for their low attenuation and long inspection range. Also, the torsional axisymmetric wave T(0, 1) over such a frequency range is usually chosen for inspection because of its nondispersive property. This frequency range is usually below the cutoff frequency of the second torsional mode group, T(m, 2), as shown in Figure 16.3. Therefore, the defect-reflected guided waves will contain primarily the guided wave modes T(m, 1). Using this fact to simplify Equation (16.13) results in

$$u(\theta,z,t) = \sum_{m=-\infty}^{+\infty} A_m(\omega)e^{i(m\theta+k_m z+\omega t)}, \tag{16.14}$$

where M = 1 and therefore n is eliminated from Equation (16.13).

Guided wave synthetic focusing utilizes axisymmetric type excitation and multichannel receiving. Suppose N is the total number of transducer segments used for guided wave reception. Then the ith ($i = 0, 1, \ldots, N-1$) transducer segment centers at (θ_i, z_R) and has a circumferential coverage from $\theta_i - \dfrac{\pi}{N}$ to $\theta_i + \dfrac{\pi}{N}$. The signal received from the ith transducer segment $u_R(\theta_i, t)$ can be represented by the following integral.

$$u_R(\theta_i,t) = \int_{\theta_i-\frac{\pi}{N}}^{\theta_i+\frac{\pi}{N}} u(\theta, z = z_R, t)b\,d\theta \tag{16.15}$$

where b is the outer radius of the pipe as shown in Figures 16.1 and 16.4. Substituting Equation (16.14) into Equation (16.15) yields

$$u_R(\theta_i,t) = b \int_{\theta_i-\frac{\pi}{N}}^{\theta_i+\frac{\pi}{N}} \sum_{m=-\infty}^{+\infty} A_m(\omega)e^{i(m\theta+k_m z_R+\omega t)}d\theta$$

$$= b \sum_{m=-\infty}^{+\infty} A_m(\omega)e^{i(k_m z_R+\omega t)} \int_{\theta_i-\frac{\pi}{N}}^{\theta_i+\frac{\pi}{N}} e^{im\theta}d\theta, \tag{16.16}$$

$$= b \sum_{m=-\infty}^{+\infty} A_m(\omega)\frac{2\pi}{N}\mathrm{Sa}\left(\frac{m}{N}\right)e^{i(m\theta_i+k_m z_R+\omega t)}$$

where $\dfrac{2\pi}{N}$ is the circumferential coverage of each transducer segment and function $\mathrm{Sa}(x)$ is defined as

$$\mathrm{Sa}(x) = \frac{\sin(\pi x)}{\pi x}. \tag{16.17}$$

It can be seen from Equation (16.16) that there are N received signals $u_R(\theta_i, t)$ (i = 0, 1, 2, ...N-1) in total. They are functions of transducer segment position θ_i and time t. Because of the sinusoidal nature $e^{im\theta_i}$ in Equation (16.16), the amplitude A_m of a reflected guided wave mode can be extracted using a decomposition scheme based on a two-dimensional Fourier transform in time t and circumferential position θ_i (or i) domain. The 2-D Fourier transform (2DFFT) process is illustrated in Figure 16.25. Figure 16.25(a) shows the layout of the received guided wave signals in numerical computation as a 2-D matrix in time t and circumferential position i domain. After the 2DFFT, time t is converted to frequency ω domain and circumferential position i is converted to the circumferential order m of the guided wave axisymmetric ($m = 0$) and flexural modes ($m > 0$). The resulting matrix can be converted to the amplitudes of the guided wave modes $A_m(\omega)$ generated from pipe features, such as defects, welds, and back wall.

Based on Equation (16.16) and Figure 16.25, the relation between the decomposed guided wave modal amplitude $A_m(\omega)$ and the received signal $u_R(\theta_i, t)$ can be described using Equation (16.18).

$$A(m,\omega) = A_m(\omega) = \frac{2\mathrm{D\ FFT}[u_R(\theta_i,t)]}{b\dfrac{2\pi}{N}\mathrm{Sa}\left(\dfrac{m}{N}\right)e^{ik_m z_R}}, \quad m = 0,\ 1,\ 2,\ ...,\ N-1. \tag{16.18}$$

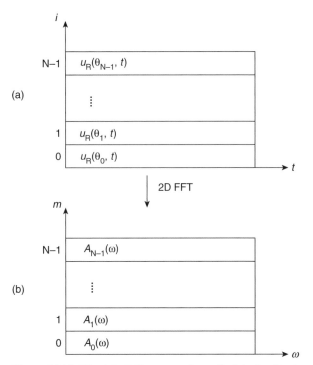

Figure 16.25. The 2-D FFT process for cylindrical guided wave mode decomposition.

If the axial location of the transducers are assumed to be zero, that is, $z_R = 0$, Equation (16.18) can be further simplified to

$$A(m,\omega) = A_m(\omega) = \frac{\text{2D FFT}[u_R(\theta_i,t)]}{b\dfrac{2\pi}{N}\text{Sa}\left(\dfrac{m}{N}\right)}, \quad m = 0, 1, 2, ..., N-1. \quad (16.19)$$

It should be noted that according to the properties of a Fast Fourier Transform (FFT), the decomposed guided wave modal amplitudes $A(m, \omega)$ has only N values, that is, $m = 0, 1, 2, ..., $ N–1. Because of symmetry, $A(m, \omega)$ for $m = (N/2) + 1,...,$ N–1 corresponds to the amplitude of the flexural mode with circumferential order $m = -(N/2) +1, ..., -1$. The defect-emitted guided waves can contain an infinite number of modes $m \in (-\infty,\infty)$ as described in Equation (16.14). However, the number of circumferential modes that can be resolved in guided wave synthetic focus equals the number of receiving transducer segments N in pipe circumference. The circumferential order of the resolved flexural mode is:

$$m = -\frac{N}{2}+1,..., -1, 0, 1,..., \frac{N}{2}. \quad (16.20)$$

The other guided wave modes with circumferential order $m < -N/2 + 1$ or $m > N/2$ will be superimposed to the range described in Equation (16.20) because of aliasing. Therefore, the resulting synthetic focus pipe image is highly dependent on the number of receiving segments N. Better image quality and sizing ability can be achieved by segmenting the transducer array into more receiving channels.

$A(m, \omega)$ is a 2-D matrix as shown in Figure 16.25(b). For each given frequency ω, the wavenumber k_m of mode T(m, 1) can be calculated theoretically if the pipe size and material properties are known. By multiplying the back propagation factor $e^{i(k_m z + m\theta)}$ with the corresponding modal amplitude factor $A(m, \omega)$ and superimposing all of the decomposed modes, the originally received guided wave signals are mapped onto a pipe image. As a result, the synthesized pipe image $u_s(\theta, z)$ is a function of pipe circumferential angle θ and axial position z:

$$u_s(\theta, z) = \sum_{\omega=\omega_1}^{\omega_l} \sum_{m=-\frac{N}{2}+1}^{\frac{N}{2}+1} A(m, \omega) e^{i(k_m z + m\theta)}, \tag{16.21}$$

where $\omega_1, \omega_2, \dots, \omega_l$ represents the discrete frequency contents within the effective frequency bandwidth. From this analysis, we can conclude that the circumferential resolution of the synthetic focus scheme equals the aperture of the receiving segments: $2\pi/N$ and the axial resolution equals approximately the pulse length of the excitation signal.

16.7 Synthetic Focusing Experiment

A synthetic focus sample experiment is conducted on an eight-inch schedule 40 steel pipe. The schematic of the pipe is shown in Figure 16.26. The pipe is forty feet long with a saw cut at nine feet, a cluster hole at fifteen feet, and a weld at twenty feet from one pipe end. The saw cut is 50 percent through wall in depth and is approximately 140° apart from the saw cut in circumference. An eight-channel phased array collar is mounted at the end of the pipe. The transducers are designed to produce shear traction in the pipe circumferential direction to generate and receive torsional type guided waves. The pulse echo mode is used in the pipe inspection. A torsional axisymmetric type wave T(0, 1) is used in excitation and eight channels are used for reception. According to the synthetic focusing discussion, reflected guided wave modes T(m, 1) with m = −3, −2, …, 4 can be resolved from the synthetic focusing technique. The steel properties used in the synthetic focus calculation are density $\rho = 7860$ kg/m³, longitudinal velocity $c_L = 5850$ m/s, and shear velocity $c_S = 3230$ m/s.

The circumferential locations of channels 1 to 8 are 0°, 45°, 90°, 135°, 180°, 225°, 270°, and 315°, respectively. The circumferential location of the saw cut is close to that of channel 1. The synthetic focusing experiment is first conducted from pipe end A (illustrated in Figure 16.26). An 40 kHz axisymmetric mode T(0, 1), as an example, is excited by the transducers. The ultrasonic waveforms received from each channel are plotted in Figure 16.27. The axial locations of the pipe features: saw cut, cluster hole, weld, and back wall, are all marked on the figure. The reflection from the weld can be seen on all eight waveforms. The reflection from the saw cut can only be observed clearly in some of the waveforms. The amplitudes of reflections are quite different for the eight channels, which indicate that the defect reflection is non-axisymmetric in nature. Because of the wraparound and interference characteristics of the guided waves, the maximum received amplitude of the saw cut reflected echo is at channel 6 (225°), which is different from the real location of the defect, 0°. Similarly, the maximum reflection from the cluster hole is received at channel 2 (45°), which is different from the

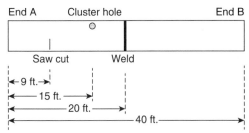

Figure 16.26. Experiment pipe schematic.

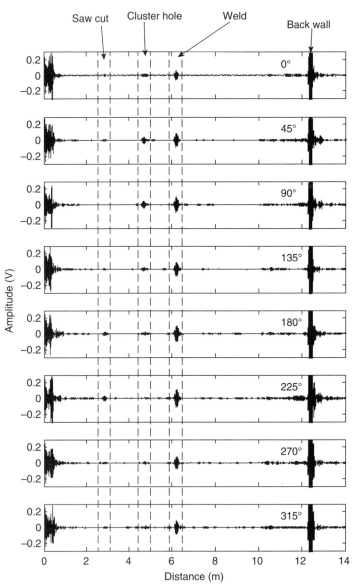

Figure 16.27. Waveforms received from eight channels for a torsional axisymmetric 40 kHz input with the defect and weld locations indicated.

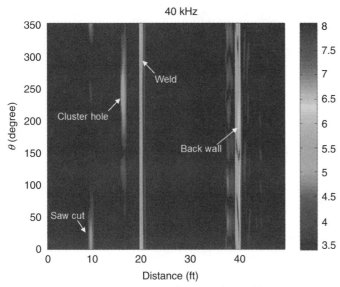

Figure 16.28. Reconstructed pipe image using guided wave synthetic focusing technique. See plates section for color version.

actual circumferential location, 225°. These results show that the circumferential location of the maximum reflected amplitude should not be directly used to denote the circumferential location of the defect.

Pipe image reconstruction from the waveforms is shown in Figure 16.28 using the guided wave synthetic focusing technique. The logarithm of the Hilbert envelope of the reconstructed pipe image is shown in Figure 16.28. The high-amplitude areas represent the features in the pipe. They are labeled in the figure. Both the saw cut and the cluster hole can be clearly identified in Figure 16.28 as non-axisymmetric features. Their axial and circumferential locations are very accurate. On the other hand, the weld and the back wall in the image are axisymmetric features. In this way, defects can be easily differentiated from welds in the pipe image.

16.8 Summary

This chapter discusses the theory and application of the guided wave active phased array focusing and synthetic focusing techniques. Active phased array focusing physically forms a focused guided wave beam at the targeted area inside a pipe during excitation, while synthetic focusing decomposes the received signals into individual guided wave modes and recombines them to construct a pipe image upon reception. Active phased array can produce higher power and better detection sensitivity at the focal spot, but can be time consuming as a full scan of the pipe is required. Synthetic focusing can construct the pipe image through a single axisymmetric shot. These two techniques can be complementary to each other in pipe testing because each has its own advantages.

16.9 Exercises

1. What factors will affect the focused beam amplitude?
2. What factors will affect the size of the side lobes in active phased array focusing?

3. How would you estimate the circumferential length of a defect using guided wave focusing techniques?

4. What is the axial resolution for active and synthetic guided wave focusing in pipe inspection?

5. How far can you focus?

6. How would you choose active versus synthetic focusing in application?

7. What are the advantages of using eight channels versus four channels in phased array focusing?

8. List several advantages and disadvantages to using phased arrays to focus guided waves in pipes.

9. List several advantages and disadvantages to using active focusing instead of synthetic focusing when performing guided wave pipe inspection in a field situation.

10. Find natural focal point in Figure 16.7 and explain diagram.

11. How are the focusing concepts discussed in this chapter applicable to focusing guided waves in a plate and in other structures?

12. For the natural focal point in red in Figure 16.7, what could be done to move the natural focal point to a new location?

13. Is defect circumferential sizing possible from the image in Figure 16.24?

14. Approximately how long is the circumferential saw cut length shown in Figure 16.28?

16.10 REFERENCES

Ditri, J. J., and Rose, J. L. (1992). Excitation of guided elastic wave modes in hollow cylinders by applied surface tractions, *J. Appl. Phys.* 72: 2589–97.

Gazis, D. C. (1959). Three dimensional investigation of the propagation of waves in hollow circular cylinders. I. Analytical foundation, *J. Acoust. Soc. Am.* 31: 568–73.

Giurgiutiu, V., and Bao, J. (2004). Embedded-ultrasonics structural radar for in-situ structural health monitoring of thin-wall structures, *Structural Health Monitoring – an International Journal* 3(2): 121–40.

Hayashi, T., and Murase, M. (2005). Defect imaging with guided waves in a pipe, *J. Acoust. Soc. Am.* 117: 2134–40.

Ing, R. K., and Fink, M. (1998). Time-reversed Lamb waves, *IEEE Transactions on Ultrasonics, Ferroelectrics, and Frequency Control* 45(4): 1032–43.

Li, J., and Rose, J. L. (2001). Excitation and propagation of non-axisymmetric guided waves in a hollow cylinder, *J. Acoust. Soc. Am.* 109(2): 457–64.

(2002). Angular-profile tuning of guided waves in hollow cylinders using a circumferential phased array, *IEEE Transactions on Ultrasonics, Ferroelectrics and Frequency* 49(12): 1720–10.

Mu, J., and Rose, J. L. (2008). Long range pipe imaging with a guided wave focal scan, *Materials Evaluation* 66(6): 663–6.

Mu, J., Zhang, L., Hua, J., and Rose, J. L. (2010). Pipe testing with ultrasonic guided wave synthetic focusing techniques, *Materials Evaluation* 68(10): 1171–6.

Sun, Z., Zhang, L., Gavigan, B. J., Hayashi, T., and Rose, J. L. (2003). Ultrasonic flexural torsional guided wave pipe inspection potential, *ASME Proceedings of Pressure Vessel and Piping Division Conference*, PVP-456, 29–34.

Wilcox, P. D. (2003). A rapid signal processing technique to remove the effect of dispersion from guided wave signals, *IEEE Transactions on Ultrasonics, Ferroelectrics, and Frequency Control* 50(4): 419.

17   Guided Waves in Viscoelastic Media

17.1 Introduction

This chapter outlines basic concepts and analysis of viscoelasticity and its impact on wave propagation. Even though the attenuation due to viscoelastic effects has plagued investigation in ultrasonic NDE for years, limited progress has been made in the study of viscoelasticity, especially in guided wave analysis. However, the reality calls for an understanding of attenuation principles as a function of material properties, wave modes, and frequency. In this chapter, we present possible approaches and discuss a few sample problems.

In many guided wave problems, material viscoelasticity may be ignored in the calculations and analysis of mode selection and wave propagation. However, in this chapter, we examine situations where a consideration of viscoelasticity is important. Significant advances have been made recently in the ability to efficiently model and solve for guided wave modes in viscoelastic waveguides that has aided our understanding of wave propagation.

General elastic theory assumes that, during deformation, a material stores energy with no dissipation. This is accurate for most metals, ceramics, and some other materials. However, many modern artificial materials, including polymers and composites, dissipate a great deal of energy during deformation. The behavior of these materials combines the energy-storing features of elastic media and the dissipating features of viscous liquids; such materials are called *viscoelastic*. Stresses for viscoelastic materials are functions of strains and derivatives of strains over time. If the stresses and strains and their derivation over time are related linearly, then the material has properties of linear viscoelasticity. It is important to note that viscoelastic material properties are typically very sensitive to temperature changes.

There are several ways to model viscoelasticity. The Maxwell, Kelvin-Voight, and Hysteretic models, for example, are discussed in Section 17.2. There are many different viscoelastic and inelastic models in use to describe the behavior of solid materials. As with the discovery of the basic guided wave types in the early twentieth century, viscoelastic models were first developed in the geophysics community to describe the propagation of waves (earthquakes) traveling in layers of the earth's crust. One example from the literature is a paper by Sivaselvan and Reinhorn (2000)

With significant contribution from Jason Philtron

that describes several different hysteretic models that references relevant historical papers describing model development.

Beyond needing the bulk wave speeds (longitudinal and transverse) and density of a material, we also need the bulk longitudinal and transverse wave attenuation values to fully characterize an isotropic viscoelastic material and to solve for the dispersion curves. A short discussion about measuring these properties is given in Section 17.3. Section 17.4 discusses how dispersion curves and mode behavior changes when solving for guided waves in waveguides of increasing viscoelasticity. Section 17.5 discusses a problem on horizontal shear waves in a viscoelastic orthotropic plate. Section 17.6 outlines a sample problem of Lamb type waves in a viscoelastic layer. Section 17.7 examines the effect of including viscoelasticity when calculating guided wave modes in a composite laminate. In Section 17.8, we briefly discuss guided wave propagation in pipes with viscoelastic coatings.

We find that the attenuation in a mode generally increases with frequency, but that there are mode frequency regions where the opposite is true. One universal concept when using guided waves to inspect viscoelastic material systems is that the user will want to choose a mode with low attenuation if a long propagation distance is desired.

17.2 Viscoelastic Models

Described next will be two well-known viscoelastic models for uniaxial stress: Maxwell and Kelvin-Voight. Further aspects of a Hysteretic and Kelvin-Voight model are also discussed.

17.2.1 Material Viscoelastic Models

Usually, the elastic effect is modeled with a spring using the Hooke's Law, while the viscous effect is modeled with a dashpot model using Newton's Law. The numerical expressions of these two material characteristics are in Equation (17.1) and Equation (17.2) for a one-dimensional problem.

$$\text{Hooke's Law: } \sigma = C\varepsilon \tag{17.1}$$

$$\text{Newton's Law: } \sigma = \eta \frac{d\varepsilon}{dt} \tag{17.2}$$

The constitutive relations for viscoelastic material are commonly formulated in two ways. One is a Maxwell model, where a spring is in series with a dashpot. The other is a Kelvin-Voigt model, where a spring is in parallel with a dashpot. The graphical illustrations of these two models are shown in Figure 17.1.

17.2.2 Kelvin-Voight Model

In the Kelvin-Voight model shown in Figure 17.1(b), the spring is in parallel with the dashpot. Therefore, the strain of the two components is the same and the total stress is the summation of the stress, as shown in Equation (17.3).

$$\sigma = \sigma_1 + \sigma_2 = C\varepsilon + \eta \frac{d\varepsilon}{dt} \tag{17.3}$$

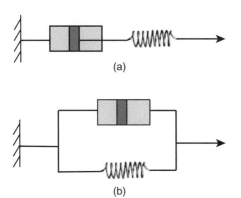

Figure 17.1. Viscoelastic material model (a) Maxwell model (b) Kelvin-Voight model.

For harmonic waves described with Equation (17.4), Equation (17.5) is as follows.

$$\varepsilon = \hat{\varepsilon}e^{i\omega t} \text{, and } \sigma = \hat{\sigma}e^{i\omega t} \tag{17.4}$$

$$\text{and } \hat{\sigma} = C\hat{\varepsilon} + i\omega\eta\hat{\varepsilon} = (C + i\omega\eta)\hat{\varepsilon} \tag{17.5}$$

Therefore, for the viscoelastic material, the constitutive relation can still be expressed in the same form as in the elastic theory. The relation of the viscoelastic solution with the solution of the corresponding elastic problem is known as the *correspondence principle*. The difference is that now the stiffness constant has a complex value and depends on frequency. The real part of the complex stiffness constant is the elastic stiffness constant, which represents the stored energy. The imaginary part of the complex value represents the loss of energy.

The correspondence principle, Auld (1990), of the one-dimensional expression can be generalized for three-dimensional anisotropic materials in three-dimensional wave mechanics. In Equation (17.6), the elements of stiffness matrix C are complex and frequency dependent.

$$\sigma_{ij} = C_{ijkl}\varepsilon_{kl} \tag{17.6}$$

17.2.3 Maxwell Model

As shown in Figure 17.1(a), the spring and the dashpot are in series in the Maxwell model. The total strain is the sum of the strains from two components as shown in Equation (17.7).

$$\varepsilon = \varepsilon_1 + \varepsilon_2 \tag{17.7}$$

Taking the derivatives with respect to time, we have Equation (17.8).

$$\dot{\varepsilon} = \dot{\varepsilon}_1 + \dot{\varepsilon}_2 \tag{17.8}$$

Using the relations in Equation (17.1) and Equation (17.2), the strain-stress relation is in Equation (17.9).

$$\dot{\varepsilon} = \frac{\dot{\sigma}}{C} + \frac{\sigma}{\eta} \tag{17.9}$$

In wave propagation problems, all field variables are expressed as a harmonic function of time, as in Equation (17.4).

Substituting these expressions into Equation (17.9), the following relation holds.

$$\hat{\sigma} = \hat{C}\hat{\varepsilon} = (\hat{C}_1 + \hat{C}_2)\hat{\varepsilon} \tag{17.10}$$

$$\hat{C}_1 = \frac{C\omega^2}{C\eta^2 + \omega^2} \tag{17.11}$$

$$\hat{C}_2 = \frac{C^2\omega\eta}{C\eta^2 + \omega^2} \tag{17.12}$$

17.2.4 Further Aspects of the Hysteretic and Kelvin-Voight Models

In this section, we will now consider further aspects of the Hysteretic and Kelvin-Voight models for viscoelasticity in a solid material. The hysteretic model is so named because it expresses the dynamic behavior of the hysteresis-shaped stress-strain relationship using complex stiffness coefficients. Lake (1999) describes a basic model where the complex stiffness coefficients are each given by a real and imaginary component. In this model, the imaginary part of the stiffness that corresponds to the viscoelasticity is independent of frequency. This is in contrast to the Kelvin-Voigt model, where the viscoelastic stiffness is dependent on frequency. For more general references on viscoelastic solids, see Lake (1999) or Haddad (1995). In both models, viscoelasticity is included by using complex stiffness values in the governing wave equations. The same governing equations can be used as for the elastic case according to the correspondence principle (Auld 1990). However, instead of real valued moduli, we use complex values where the imaginary component represents the viscoelastic loss. For example, consider the stiffness variable $C*$. The real part of the stiffness value, C', corresponds to energy storage while the complex part, C'', corresponds to the damping introduced by material viscoelasticity. The stiffness can be expressed as

$$C* = C' - iC''. \tag{17.13}$$

The Hysteretic model assumes that C'' is independent (and constant, regardless) of frequency, so that this equation may be rewritten using the loss, η, as

$$C* = C' - i\eta. \tag{17.14}$$

The Kelvin-Voight model assumes that C'' is a linear function of frequency. η is measured at a reference frequency, f_0, and the relationship is

$$C* = C' - i\frac{f}{f_0}\eta. \tag{17.15}$$

Both of these models give the same result at the characterization frequency, f_0, but give differing results the further the working frequency, f, is from the characterization frequency. The Kelvin-Voight model results in a smaller attenuation below f_0, and a larger attenuation above f_0.

We solve the governing equations for the guided wave mode solutions using a real valued frequency, so each solution is a complex wavenumber, $k*$. From the complex wavenumber we may calculate the phase velocity and attenuation shown on typical dispersion curves. For a mode with a complex wavenumber

$$k* = k' + ik'',$$ (17.16)

we can write the complex phase velocity as

$$c_p*(\omega) = \left(\frac{1}{c_p(\omega)} - i\frac{\alpha(\omega)}{\omega} \right)^{-1}.$$ (17.17)

From Equation (17.17) we have the real phase velocity, c_p, given as

$$c_p(\omega) = \left(\frac{k'}{\omega} \right)^{-1},$$ (17.18)

and the attenuation, α, given as

$$\alpha(\omega) = -\omega \mathrm{Im} \left\{ \frac{1}{c*(\omega)} \right\} = -k'',$$ (17.19)

where both variables are functions of the circular frequency, ω.

The units of α are Nepers per meter (Np/m). This is because the unit amplitude was defined as the decay of $1/e$ after traveling one wavelength (e.g., see Lowe 1995). However, more often we use dB/m or Np/wavelength because it is more intuitive. The conversion between Nepers and decibels is

$$1 \text{ neper} = 20\log_{10}(e) \text{ dB} = 8.69 \text{ dB}.$$

Additionally, because we now have complex wavenumbers and complex velocities for damped modes, we need to calculate the energy (not group) velocity to find the speed of the guided wave. This is because taking the derivative of a complex value is ambiguous, so the equation for group velocity from an earlier chapter cannot be used directly. However, we will find that using just the real part of the phase velocity to calculate the group velocity provides us with a good approximation for lightly damped modes. For more information on how to calculate the energy velocity see Chapter 9 or Bartoli and colleagues (2006).

17.3 Measuring Viscoelastic Parameters

When finding GW mode solutions in damped waveguides using analytical or numerical methods, complex material properties are used. Accurate material properties are necessary for accurate models of real structures. Although measured material properties may be easily found in look-up tables for elastic materials, viscoelastic material properties are not as easy to find. Therefore, it is necessary to measure the attenuative properties of materials so that we may calculate the guided wave modes more accurately.

One challenge in modeling viscoelastic materials is deciding what values to use for the complex material properties. There are several methods for determining these values, all of which must be performed experimentally. An example of a relatively simple through-thickness measurement is given by Barshinger and Rose (2004). In this method, a sample is placed in a water bath between two longitudinal transducers to characterize the bulk longitudinal and shear wave viscoelastic behavior. Baudouin and Hosten (1996) and Castaings and colleagues (2000) use immersion methods to infer anisotropic attenuation in composite plates by changing the angle between the source, sample, and receive transducer. Van Velsor (2009) developed a pulse-echo method based off of reflection coefficients that can be used to measure material properties in situ. Van Velsor's method was developed to measure coating properties on gas pipeline in the field.

When modeling viscoelastic materials, one important consideration is how material attenuation varies with frequency. If we assume that the complex moduli used in calculations are not frequency dependent, numerical calculations are easier. Several papers (Baudouin and Hosten 1996; Chan and Cawley 1998; Barshinger and Rose 2004) have found that for polymer-based materials this assumption is true in the 0.3 to 5 MHz frequency range. This is likely because the attenuation of longitudinal and shear bulk waves in this type of material varies linearly with respect to frequency.

17.4 Viscoelastic Isotropic Plate

A variety of papers in the literature address problems where viscoelasticity is a concern. Based on the correspondence theory, ultrasonic guided waves in a viscoelastic media can be analyzed by substituting a complex stiffness constant matrix into the governing equation. The partial theory, the Christoffel's equation, the global matrix method, and the transfer matrix method can still be used for viscoelastic media. The details of these theories are discussed in earlier chapters. Some examples will be illustrated to show the influence of material viscoelasticity on guided wave propagation.

We will now briefly summarize one of the more fundamental studies. A fundamental study of a viscoelastic isotropic plate by Chan and Cawley (1998) addresses guided wave propagation in a plate waveguide as the severity of the viscoelasticity is increased. They find that, with increasing attenuation, both the phase velocity dispersion curves and attenuation of the modes change. Additionally, it is found that attenuation in a mode generally increases with frequency, but that there are mode regions where the opposite is true. From examination of the dispersion curves, we find that these locations are typically in mode regions where the bulk longitudinal wave is dominant.

At high levels of attenuation, some plate modes become asymptotic to the bulk longitudinal velocity, as opposed to the bulk shear velocity, at high frequencies. Chan and Cawley find that, in general, modes dominated by the longitudinal partial wave have less attenuation than the shear-dominated modes. Typically, bulk shear wave attenuation is three to ten times higher than bulk longitudinal wave attenuation, which likely contributes to this effect. This result suggests that for guided wave propagation in highly attenuative materials we should look for modes dominated by the longitudinal partial wave because they will have less attenuation, and therefore a longer useful inspection range.

17.5 Viscoelastic Orthotropic Plate

17.5.1 Problem Formulation and Solution

We shall now discuss a viscoelastic orthotropic plate (occupying the region $|z| \leq d/2$. $|x| < \infty$. $|y| < \infty$) with a time-harmonic shear stress loading $p(x)e^{i\omega t}$ on the upper surface in the finite area $|x| \leq b$. The lower surface of the plate is stress free, and $\omega = 2\pi f$ is the angular frequency of vibration. The equation of motion for the viscoelastic case is found from the elastic case when we replace the real elastic moduli with a complex frequency dependence $C_{kl}(\omega)$. For SH waves, we have thefollowing governing equation (the time factor $e^{i\omega t}$ has been suppressed for ease of discourse):

$$C_{66}(\omega)\frac{\partial^2 u_y}{\partial x^2} + C_{44}(\omega)\frac{\partial^2 u_y}{\partial z^2} + \rho\omega^2 u_y = 0, \tag{17.20}$$

$$C_{kl}(\omega) = C_{kl}^r(\omega) + iC_{kl}^i(\omega), \tag{17.21}$$

where $u_y = u_y(x, z)$ (see Auld 1990).

The appropriate boundary conditions for the plate are

$$\sigma_{yz} = p(x)e^{i\omega t} \text{ at } z = d/2, |x| \leq b, \tag{17.22a}$$

$$\sigma_{yz} = 0 \text{ at } z = d/2, |x| > b, \tag{17.22b}$$

$$\sigma_{yz} = 0 \text{ at } z = -d/2, |x| \geq 0. \tag{17.22c}$$

The complex moduli represent the capability of storing energy under elastic deformation, where $C_{kl}^r(\omega)$ is the real part and energy damping $iC_{kl}^i(\omega)$ is the imaginary part. We now introduce the following dimensionless variables:

$$\zeta = z/d, \xi = x/d, a = b/d. \tag{17.23}$$

Applying a Fourier transform to Equation (17.20) with respect to the x-axis and satisfying boundary conditions (17.22) (see Achenbach 1984), one discovers an expression for the displacement field in the form

$$u_y(\xi, \zeta, \omega) = \frac{d}{2\pi C_{44}(\omega)}\int_{-\infty}^{\infty}\left[\frac{\sinh(\lambda\zeta)}{\lambda\cosh(\lambda/2)} + \frac{\cosh(\lambda\zeta)}{\lambda\sinh(\lambda/2)}\right]T(\alpha)e^{-i\alpha\xi}d\alpha, \tag{17.24a}$$

where

$$\lambda^2 = (C_{66}(\omega)\alpha^2 - s^2)\,C_{44}^{-1}(\omega) \quad (s^2 = \rho(\omega d)^2), \tag{17.24b}$$

$$T(\alpha) = \frac{1}{2}\int_{-a}^{a}p(\gamma)e^{i\alpha\gamma}d\gamma. \tag{17.24c}$$

The formal solution to (17.24a) is the same as in the elastic case. When evaluating the integrals, one must take into account the poles of the integrands. Poles for the symmetric part of the solution (first term in (17.24a)) are

$$\alpha_n = \sqrt{\{s^2 - C_{44}(\omega)(2\pi n)^2\}C_{66}^{-1}(\omega)} \quad (n = 0, 1, 2, \ldots); \tag{17.25a}$$

for the antisymmetric part (second term in (17.24a)), the poles are

$$\beta_n = \sqrt{\{s^2 - C_{44}(\omega)[\pi(2n+1)^2]\}C_{66}^{-1}(\omega)} \quad (n = 0, 1, 2, \ldots). \tag{17.25b}$$

In an elastic solid, for each particular value of ω, one calculates from Equation (17.25) a finite number of real poles and an infinite number of imaginary poles that are symmetric to the coordinate directions. For the viscoelastic case, α_n and β_n are always complex wavenumbers.

Waves in a viscoelastic material decay in the direction of propagation. Accordingly, we attach signs to the imaginary parts of the wavenumbers α_n and β_n, where the imaginary part of a wavenumber is the attenuation coefficient. Applying the residue theory to (17.24a) yields

$$
\begin{aligned}
u_y(\xi, \zeta, \omega) = &-\frac{id}{C_{66}(\omega)} \\
&\times \left\{ \frac{T(\alpha_0)}{2\alpha_0} e^{-i\alpha_0 \xi} + \sum_{n=1}^{\infty} \left[\frac{\cos(2\pi n\zeta)}{\cos(\pi n)} \frac{T(\alpha_n)}{2\alpha_n} e^{-i\alpha_n \xi} \right. \right. \\
&\left. \left. + \frac{\sin(\pi(2n-1)\zeta)}{\cos\pi(n-1)} \frac{T(\beta_n)}{2\beta_n} e^{-i\beta_n \xi} \right] \right\}.
\end{aligned}
\tag{17.26}
$$

where $|\xi| > a$. The summation in (17.26) is over poles with negative imaginary parts for positive ξ and over poles with positive imaginary parts for negative ξ.

All calculations are carried out for $p(x) = 1$ and for two cases of viscoelastic solids. The first case is the Voight solid, which has an elastic moduli given by Coquin (1964) and Auld (1990):

$$C_{kl}(\omega) = C_{kl}^r + i\omega C_{kl}^i. \tag{17.27}$$

For the second case, the real and imaginary parts of the moduli are independent of the frequency. This assumption approximates the response of some materials over a fairly wide frequency range.

17.5.2 Numerical Results

The units of C_{kl} are given in gigapascals (GPa); data are taken from experimental results in Deschamps and Hosten (1992). For the Voight solid:

$$C_{44}^r = 6.15, \quad C_{44}^i = 0.02, \quad C_{66}^r = 3.32, \quad C_{66}^i = 0.009; \quad \rho = 1.5 \text{ g/cm}^3, \quad a = 1.5.$$

For the elastic solid, $C_{44}^i = C_{66}^i = 0$.

The phase velocity dispersion curves for the Voight material are plotted in Figure 17.2. Figure 17.3 graphs phase velocity when the elastic moduli are independent of frequency. In part (b) of Figure 17.3, dispersion phase velocity curves overlap the dispersion curves for the elastic plate.

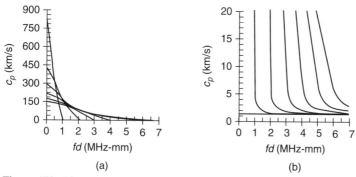

Figure 17.2. Phase velocity curves for a Voight viscoelastic solid, where (a) and (b) show the same curves using different scales.

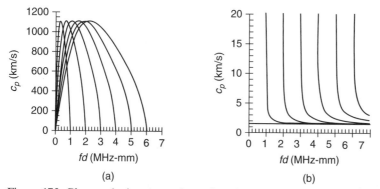

Figure 17.3. Phase velocity curves for a viscoelastic solid when the complex elastic moduli are independent of frequency, where (a) and (b) show the same curves using different scales.

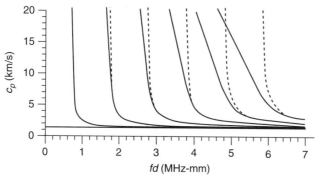

Figure 17.4. Phase velocity curves for the Voight model (solid lines) and the elastic solid (dashed lines), where the first two modes overlap each other for viscoelastic and elastic cases.

The character of the viscosity effect is illustrated in Figure 17.4, whose graphed curves show the change in phase velocity from the Voight material (solid curves) to the elastic solid (dashed curves). The first two modes overlap each other for the viscoelastic and elastic cases.

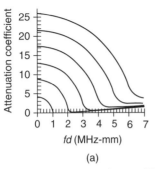
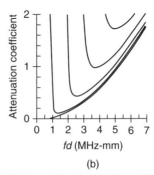

Figure 17.5. Attenuation coefficient for Voight solid, where (a) and (b) show the same curves using different scales.

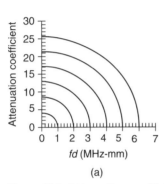
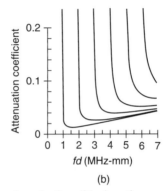

Figure 17.6. Attenuation coefficient for a viscoelastic solid when the complex elastic moduli are independent of frequency, where (a) and (b) show the same curves using different scales.

We now consider the wave field representation, using Equation (17.26). We can decompose the complex wavenumbers α_n and β_n as $k = k_{Re} + i k_{Im}$. Every term of (17.26) can thus be written as

$$u_y \sim A e^{ik\xi} = A e^{ik_{Re}\xi} e^{-k_{Im}\xi},$$

where $e^{ik_{Re}\xi}$ is a typical wave propagation term for an elastic material and $e^{-ik_{Im}\xi}$ introduces attenuation while distance ξ increases. The attenuation coefficients so derived are shown for the Voight solid in Figure 17.5 and the viscoelastic material in Figure 17.6. The graphs in Figure 17.6 show the significant effect of elastic constants depending on frequency: the phase velocity and attenuation coefficients are sensitive to elastic moduli changes.

In the practical excitation of SH waves, the energy $E(f)$ that the transducer radiates will spread in the frequency domain – assuming a Gaussian distribution for energy in that domain. Hence, for any excitation frequency f_0, there exists a frequency *spread area* $(f_0 - 3\sigma, f_0 + 3\sigma)$, where σ is the standard deviation. To model a practical situation, one must express the resultant amplitude for any f_0 as follows:

$$\bar{u}_y(\xi,\ \zeta, f_0) = \int_{f_0-3\sigma}^{f_0+3\sigma} u_y(\xi,\ \zeta,\ 2\pi f) E(f)\, df. \qquad (17.28)$$

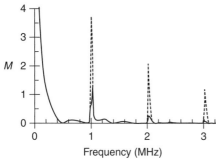

Figure 17.7. Magnitude-frequency relationship for an independent viscoelastic solid, showing frequency-elastic moduli (solid line) and elastic material (dotted line).

Figure 17.7 graphs the resulting dimensionless magnitude $M = |\bar{u}_y(\xi, d/2, f)|$ in the frequency domain (where $d = 1, x = 7, a = 1.5, \sigma = 0.03$). The peaks in this figure are associated with cut-off frequencies for phase velocity curves from Figure 17.2 (elastic solid). Viscosity makes the peaks much smoother for the viscoelastic solid (with frequency-independent moduli); for the Voight material, there is no such relation because $C_{kl}(\omega)$ elevates with increasing frequency.

17.5.3 Summary

In this section we have examined the viscoelastic effects of guided SH waves; our results can be summarized as follows. The phase velocity curves of viscoelastic material approach the corresponding curves for the elastic case for a small imaginary part of the complex elastic moduli. For an elastic plate, attenuation coefficients are smallest in the neighborhood of the cut-off frequency. The amplitude peaks reflect clear cut-off frequencies for the elastic case; the peaks for viscoelastic material are smaller. We have demonstrated also the source influence on wave excitation.

The practical significance of this work is demonstrated, for example, in powder metal injection molding, where transducers can be used to excite guided waves in the part being manufactured. Selecting points of minimum attenuation can be achieved experimentally by frequency and angle tuning.

17.6 Lamb Waves in a Viscoelastic Layer

In this section, we evaluate waves in a two-dimensional viscoelastic isotropic layer with traction-free boundary conditions.

We assume that x is in the direction of wave propagation and a layer thickness d. Just as for the viscoelastic case, we will use complex wavenumbers and wave velocities. The real part of the wavenumber represents the propagation of the wave, and the imaginary part describes the wave attenuation. According to equations for the Kelvin-Voight model, we assume that the attenuation of bulk waves varies linearly with frequency and hence defines the two damping parameters (per unit frequency) η_L and η_T for longitudinal and shear (transverse) bulk waves, respectively. Complex wavenumbers and bulk velocities can then be introduced as follows (see Chimenti and Nayfeh 1989):

$$k_L^* = \omega/c_L + i\eta_L\omega/(2\pi), \quad k_T^* = \omega/c_T + i\eta_T\omega/(2\pi)$$
$$(c_L^* = \omega/k_L^*, \ c_T^* = \omega/k_T^*).$$

$$(17.29)$$

As before, c_L and c_T are longitudinal and shear (transverse) wave velocities in the elastic media.

The analytical solution of this wave propagation problem for a viscoelastic material can be found in the same fashion as that used for the elastic case: simply replace real wavenumbers and bulk velocities with complex wavenumbers and velocities, according to Equation (17.29). We may begin by stating the stress-free boundary conditions as

$$\sigma_x = \sigma_{xy} = 0 \quad \text{at } y = \pm d/2.$$

$$(17.30)$$

This leads to the complex Rayleigh–Lamb dispersion equation found earlier in Equations (6.31) and (6.32) and rewritten here.

$$\frac{\tan(qd/2)}{\tan(pd/2)} + \left\{\frac{4pqk^2}{(q^2-k^2)^2}\right\}^{\pm1} = 0$$
$$(p^2 = (\omega/c_L^*)^2 - k^2, \ q^2 = (\omega/c_T^*)^2 - k^2)$$

$$(17.31)$$

Now consider the root $k = \alpha + i\beta$ of Equation (17.31). For each root, the components of the displacement field can be written as

$$u_k(x, y) = U_k(y, k)e^{i(kx-\omega t)} = U_k(y, k)e^{i(\alpha x-\omega t)}e^{-\beta x},$$

$$(17.32)$$

where $u_1 = u_x$ and $u_2 = u_y$. As shown in Equation (17.32), the amplitude of the guided wave will decay as $e^{-\beta x}$ after propagating a distance x. Then, phase velocity values may be obtained as $c_p = \omega/\alpha$ for an appropriate value k.

Because Equation (17.31) is itself complex, it does not have complex conjugate roots. This leads to a violation of the symmetry of the frequency spectrum with respect to the system of coordinates. The elastic spectrum is evaluated as asymptotic for a viscoelastic material when both η_L and η_T approach zero. The concept of a cut-off frequency does not have meaning for viscoelastic material, because $k = 0$ is not a root of the Equation (17.31) for $\omega > 0$.

As we can see from Figure 17.8, the phase velocity and attenuation dispersion curves for the viscoelastic case have maximum curvature near the cut-off frequencies of the elastic spectrum. One can therefore generate a number of low-attenuation guided wave modes for long-distance propagation by using minimum attenuation values. Two cases of viscoelastic materials are graphed in Figure 17.8 and Figure 17.9, which show that only the A0 mode attenuation varies linearly with the frequency change.

17.7 Viscoelastic Composite Plate

In this section, we calculate the solutions (roots) to the guided wave equation using the Semi-Analytical Finite Element Method (SAFEM). For more information on the SAFEM, see Chapter 9. Because the formulation of the eigenvalue problem in

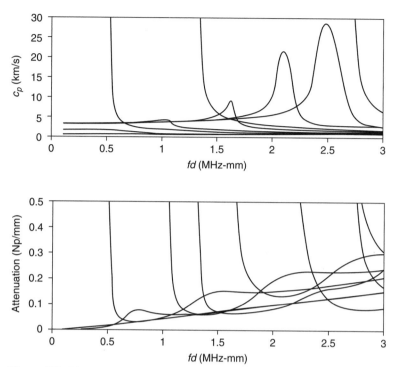

Figure 17.8. Phase velocity and attenuation dispersion curves for a viscoelastic plate with bulk damping parameters $\eta_L = 0.02$ Np/MHz-mm and $\eta_T = 0.07$ Np/MHz-mm; bulk velocity for the appropriate elastic media are $c_L = 2.7$ km/s and $c_T = 1.1$ km/s.

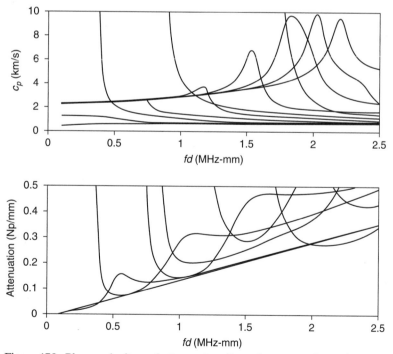

Figure 17.9. Phase velocity and attenuation dispersion curves for a viscoelastic plate with bulk damping parameters $\eta_L = 0.1$ Np/MHz-mm and $\eta_T = 0.2$ Np/MHz-mm; $c_L = 1.8$ km/s and $c_T = 0.8$ km/s.

Table 17.1. *Lamina properties of the viscoelastic composite used in simulation. The measurement frequency is 2 MHz and the density is 1.6 g/cm³*

Stiffness component	C_{11}	C_{12}	C_{13}	C_{22}	C_{23}	C_{33}	C_{44}	C_{55}	C_{66}
Real part (GPa)	178	8.35	8.35	14.4	8.12	14.4	3.16	6.10	6.10
Imaginary part (GPa)	8.23	0.65	0.60	0.34	0.25	0.65	0.24	0.28	0.25

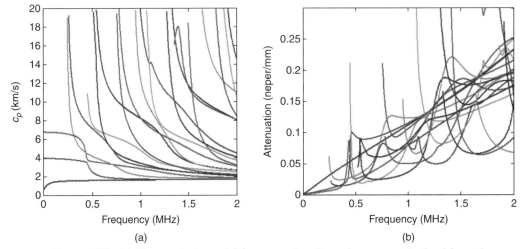

Figure 17.10. (a) Phase velocity and (b) attenuation dispersion curves obtained from the Hysteretic model.

the SAFEM is independent of whether real or complex stiffness values are used, the method takes no additional computation time or considerations over the elastic problem. The SAFEM provides an advantage over matrix-based (e.g., the global matrix method) methods used for solving for the mode solutions because the matrix-based methods need to perform root searching in the complex domain (in the real and imaginary dimensions). This matrix-based calculation ends up being more computationally expensive, and root-finding algorithms may miss certain roots.

Table 17.1 displays the properties of the viscoelastic composite plate used in the calculations. It is a sixteen-layer quasi-isotropic composite made of IM7/977-3 prepreg. The layup sequence is $[(0/45/90/-45)_s]_2$ and the total thickness of sixteen layers is 3.2 mm. In this section, we generally follow the analysis of this composite plate given by Gao (2007).

Figure 17.10 shows the dispersion curves for this plate using the Hysteretic model. Figure 17.11 shows the dispersion curves for this plate using the Kelvin-Voight model. Only modes with an attenuation of less than 1 Np/m are plotted. Although the phase velocity dispersion curves of these two figures appear very similar, the attenuation dispersion curves show significant differences. In Figure 17.11b, we see a relatively higher (lower) attenuation at frequencies lower (higher) than the characterization frequency compared to Figure 17.11b, as discussed in Section 17.2. The attenuation characteristics of a given mode, however, follow similar trends of increasing or decreasing attenuation with changes in frequency.

Figure 17.12 shows the result from Figure 17.10a overlaid with the same composite plate with elastic material properties (i.e., the complex stiffness coefficients were set

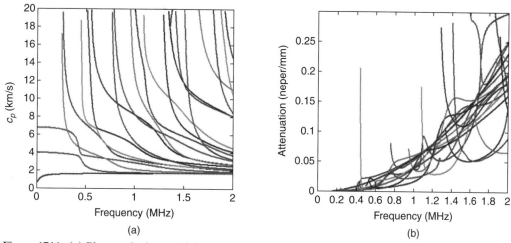

Figure 17.11. (a) Phase velocity and (b) attenuation dispersion curves obtained from the Kelvin-Voigt model.

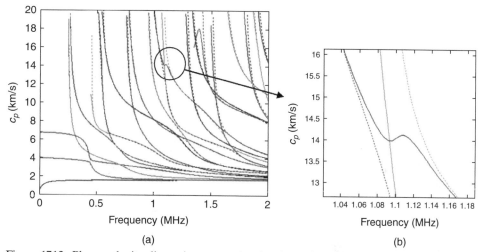

Figure 17.12. Phase velocity dispersion curves for the elastic (dotted) and Hysteretic viscoelastic (solid) models. (a) Full set of dispersion curves. (b) Magnified curve in a mode interaction region.

to zero). When considering a viscoelastic material throughout the rest of this chapter, we use the Hysteretic model. We find that the phase velocity dispersion curves for this viscoelastic composite plate are quite similar to the elastic case, and in many places the curves overlap. This is in general true for lightly damped structures. In fact, as we continue our analysis throughout this section, we will find many similarities between the elastic case and the lightly damped composite.

One of the key differences between the dispersion curves for elastic and viscoelastic plates is in the way nearby modes interact. As shown in Figure 17.12, for the elastic case nearby modes repel each other. However, for the viscoelastic case we see mode intersection, or crossing. Distinguishing which mode is which may seem a somewhat academic distinction, as opposed to a practical one, but one very important consideration of a mode is its wave structure. Indeed, in determining which mode is which in mode interaction regions, researchers often look at a continuity in wave

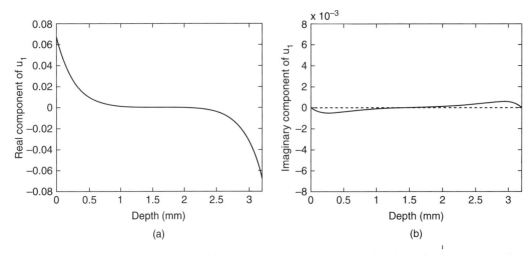

Figure 17.13. The (a) real and (b) imaginary components of the in-plane displacement, u_1, for the first mode at 200 kHz for the elastic (dashed) and viscoelastic (solid) models.

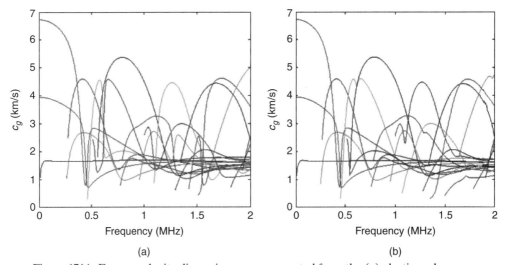

Figure 17.14. Energy velocity dispersion curve generated from the (a) elastic and (b) viscoelastic models.

structure to indicate continuity in mode. That is, if a certain mode solution has a more similar wave structure at a lower and higher frequency around the crossing point, as compared to other solutions of similar phase velocity, then those mode solution points with a similar wave structure are considered to track the same mode.

Beyond the similarity in the phase velocity dispersion curves of the elastic and viscoelastic cases, we also see many regions that have similar wave structures. For example, Figure 17.13 shows the first wave mode at 200 kHz. This mode shows an almost identical real component of the wave structure for the two cases. We can deduce from this result, and other similar modes, that for modes with low attenuation the wave structure obtained from the elastic model is still a good approximation.

Next we compare the speed at which a guided wave mode travels along the waveguide. Figure 17.14 shows the group velocity for the elastic and viscoelastic cases. Again we find very similar curves. It has been shown that for minimum attenuation

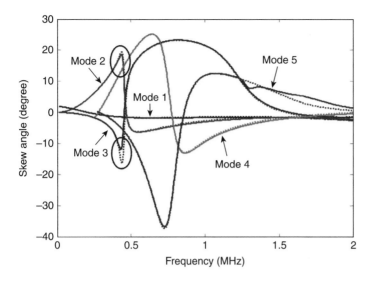

Figure 17.15. Skew angle dispersion curves for the elastic (dotted) and viscoelastic (solid) models. Two ovals indicate a mode interaction region near 0.5 MHz.

regions, there is generally good agreement between the group and energy velocities (Bernard et al. 2001). This is clear from Figure 17.14, where the group velocities are very similar between the two cases. The locations where differences occur are in the mode interaction regions. For example, one is at approximately 1.1 MHz and 2.5 km/s. In these mode interaction regions it is necessary to calculate the energy velocity to have an accurate description of the speed with which a mode will propagate in this damped waveguide. It should be noted that when using the SAFEM to calculate the energy velocity, it is necessary to use smaller elements than for root finding or the wave structure calculation to have the same accuracy. This is because the energy velocity is a function of stress and therefore the displacement derivatives (Bartoli et al. 2006).

Figure 17.15 shows the skew angle dispersion curves for the elastic and viscoelastic models. Again, the elastic and viscoelastic models predict the same result for most of the mode solutions. However, at mode interaction regions, a significant difference is shown between the two models. One of these mode interaction regions occurs near 0.5 MHz, where modes 2 and 3 interact. In this region, we find that the viscoelastic model predicts a lower skew angle for both of the modes. This is not always the case in mode interaction regions, as shown by mode 5 above 1.5 MHz.

An interesting question to ask is: What modes have the least attenuation at a given frequency? These might be the modes that we wish to use if we are performing a guided wave inspection over a limited frequency range. Another question is: Which modes will arrive first at, or travel fastest to, a receiver location? These will be the modes with the fastest energy (group) velocity for the viscoelastic (elastic) case. Performing an inspection with a mode that arrives first at a sensor may simplify signal processing when multiple modes are generated, which happens in cases of high modal density due to source influence.

Figure 17.16 highlights the modes with the lowest attenuation and largest group velocity for a given frequency in the viscoelastic plate. This figure shows that the same modes often fit both characteristics. This is useful for guided wave inspection

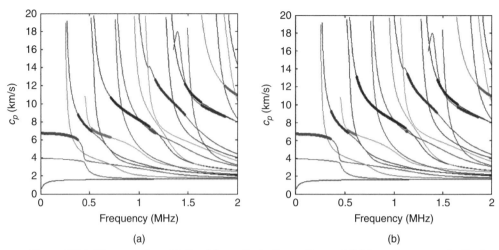

Figure 17.16. Wave modes with the (a) smallest attenuation and (b) largest group velocity for a given frequency.

because a mode with the highest group velocity (at a given frequency) may be chosen; we will inherently have chosen a mode with relatively low attenuation.

In summary, we have shown that many guided wave mode characteristics are similar for an elastic and a lightly damped composite plate. In particular, the phase velocity, group velocity, and skew angle dispersion curves are very similar. Differences in these variables occur, however, in mode interaction regions. In these regions it is necessary to consider the viscoelastic mode solution to obtain an accurate calculation of all mode properties. Additionally, we have found that the modes with the largest group velocity at a given frequency are typically the modes that have the least attenuation.

17.8 Pipes with Viscoelastic Coatings

In many applications, pipelines are coated with viscoelastic materials for the purpose of protection against corrosion or other damage. The existence of the coating will affect guided wave propagation. It is therefore important to understand the coating effects to predict penetration power and to select proper modes for defect detection and characterization.

The SAFE method can be used to calculate guided wave dispersion curves in a coated pipe for faster computation compared to analytical matrix methods (Mu et al. 2008). An attenuation dispersion curve will be associated with the phase velocity dispersion curve for each coated pipe configuration. Each guided wave mode propagating in a coated pipe will have a specific phase velocity and attenuation based on the frequency. Based on the phase velocity and attenuation dispersion curves, coated pipe focusing can be calculated similarly as that is done in a bare pipe (Mu 2008).

Figure 17.17 shows the phase velocity dispersion curves for guided wave modes in a four-inch schedule 40 hollow cylinder (pipe) coated with 0.02 inch-thick Bitumastic 50. Figure 17.18 shows the attenuation dispersion curves for the same modes shown in Figure 17.17. The material properties used for these two figures are given in Table 17.2.

Table 17.2. Properties used for coated pipe calculation

Material	c_L (mm/μs)	α_L/ω (μs/mm)	c_T (mm/μs)	α_L/ω (μs/mm)	ρ (g/cm³)
Steel	5.85	-	3.23	-	7.86
Bitumastic 50	1.86	0.023	0.75	0.24	1.5

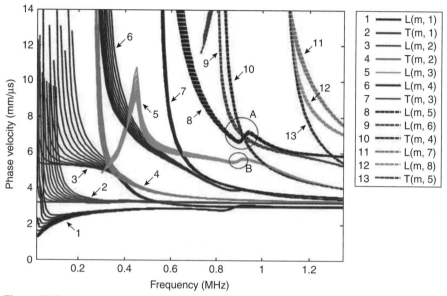

Figure 17.17. Phase velocity dispersion curves for modes with circumferential order m from zero to ten in a coated steel pipe. Circled regions A and B indicate the regions for modes L(m, 5) and L(m, 3) that experience non-monotonic change with frequency.

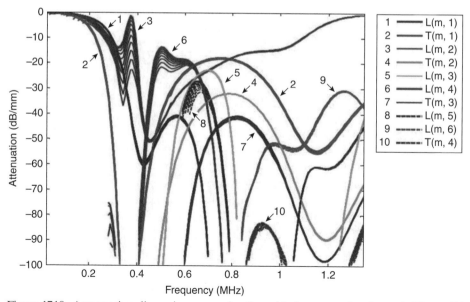

Figure 17.18. Attenuation dispersion curves for the guided wave modes shown in Figure 17.17.

Figure 17.17 shows many similar characteristics to the single-layer elastic pipe dispersion curves calculated in Chapter 10. However, there are several key differences, most notably that the modes all have some finite amount of attenuation and the shape of the dispersion curves in some mode-crossing regions. From Figure 17.18, we find that at most frequencies the mode group with the lowest attenuation is $L(m, 2)$, with mode groups $L(m, 1)$, $L(m, 4)$, and $T(m, 1)$ each having a lower attenuation region for a frequency region near 0.3, 0.5 and 0.7 MHz, respectively.

One particularly interesting region in Figure 17.17 is near 0.9 MHz, where mode groups $L(m, 5)$ and $L(m, 3)$ experience a non-monotonic change with frequency (i.e., an increasing phase velocity with increasing frequency). This is indicated with a circle and the labels A and B, respectively. In these two regions we see a very high level of attenuation. In fact, the attenuation for these modes is greater than 100 dB/m. However, we find that the attenuation of the other two modes in the vicinity of region A, $L(m, 6)$ and $T(m, 4)$, have a much lower attenuation, of only about 60 and 85 dB/m, respectively.

17.9 Exercises

1. Name several applications where consideration of viscoelastic materials is important for guided wave propagation.
2. Would you prefer to calculate dispersion curves for a viscoelastic plate using the GMM or SAFEM? Why?
3. What are some of the disadvantages when using the GMM with viscoelastic materials? How about the SAFEM?
4. Consider an unknown material. Describe an experiment to test whether the hysteretic or Kelvin-Voight viscoelastic model is more accurate for the material.
5. Assume the following material properties for a viscoelastic isotropic plate: c_L=1.8 km/s, c_T=0.75 km/s, a_L/ω=0.03 s/km, and a_T/ω=0.24 s/km. What is the attenuation (in dB/m) for bulk longitudinal and transverse waves at 1 MHz? At 0.5 MHz? Without performing the calculation, what would you expect the attenuation to be at 2 MHz and why? Perform the calculation for 2 MHz. Were you correct?
6. A particular guided wave mode has a frequency of 0.4 MHz and a complex phase velocity of $1.57 + i*0.0467$ km/s. What is the attenuation for this mode in Np/m, Np/wavelength, and dB/m?
7. Consider a cube of unknown material with a side length of 4 inches (10 cm). Describe how you would measure c_L, c_T, and density. (You may cut the block into smaller pieces if you desire.) What in your experiment or result would suggest needing to measure the viscoelastic properties of the material?
8. Consider the unknown material from Exercise 7. Describe how you would measure the bulk longitudinal and transverse attenuation.
9. In Exercises 7 and 8, how would you quantify the error in your measurement?
10. Describe the difference in the behavior of phase and group velocity dispersion curves for guided waves in elastic versus viscoelastic plates.
11. How does energy loss arise for harmonic vibration in viscoelastic media?

12. Choose a mode and frequency from the dispersion curves in Figure 17.10. Describe an experiment where you would confirm the attenuation and velocity calculated using the SAFEM. How would you quantify the error in your measurement?

13. How would you calculate the group velocity dispersion curve for a viscoelastic plate?

14. Consider the dispersion curves shown in Figure 17.10. You'd like to inspect this plate using guided waves. What modes would you wish to use to inspect for internal voids? How does your answer change if you wish to inspect for surface defects?

15. For the plate in the previous problem, what changes, if any, would you need to make to your excitation technique when compared to exciting guided wave modes in an elastic plate? Select a specific mode (frequency and phase velocity) and describe two (or more) possible excitation methods.

16. At a given frequency in a viscoelastic plate, why do the fastest traveling modes tend to have the least attenuation?

17. Let's say you calculate the guided wave modes in a composite laminate while assuming elastic material properties. At what phase velocity, frequency, and modes would you expect to have significant errors in the result? For which mode results would you expect the calculations to be acceptable?

18. How would your previous answers to Exercises 7 and 8 change if you realized that c_L and c_T were a function of direction in your sample?

19. Describe how the addition of viscoelastic coatings on pipeline affect guided wave propagation.

20. You are approached by a pipe inspection company. Currently they can inspect bare pipe at distances up to 100 feet (30 m). However, there is a new VE coating that will be applied to the outside of the pipe to protect the pipe from corrosion. The company gives you a sample of the coating material. How would you estimate the new inspection distance for the coated pipe?

17.10 REFERENCES

Achenbach, J. D. (1984). *Wave Propagation in Elastic Solids*. New York: North-Holland.

Auld, B. A. (1990). *Acoustic Fields and Waves in Solids,* 2nd ed., vols. 1 and 2. Malabar, FL: Kreiger.

Barshinger, J. N., and Rose, J. L. (2004). Guided wave propagation in an elastic hollow cylinder coated with a viscoelastic material, *IEEE Trans. Ultrason. Ferroelectr. Freq. Control* 51(11): 1547–56.

Bartoli, I., et al. (2006). Modeling wave propagation in damped waveguides of arbitrary cross-section, *J. Sound and Vib.* 295(3–5): 685–707.

Baudouin, S., and Hosten, B. (1996). Immersion ultrasonic method to measure elastic constants and anisotropic attenuation in polymer-matrix and fiber-reinforced composite materials, *Ultrasonics* 34: 379–82.

Bernard, A., Lowe, M. J. S., and Deschamps, M. (2001). Guided wave energy velocity in absorbing and non-absorbing plates, *J. Acoust. Soc. Am.* 110(1): 186–96.

Bouc, R. (1967). Forced vibration of mechanical systems with hysteresis, *Proceedings of the Fourth Conference on Nonlinear Oscillation*. Prague, Czechoslovakia, 315.

Castaings, M. et al. (2000). Inversion of ultrasonic, plane-wave transmission data in composite plates to infer viscoelastic material properties, *NDT and E International* 33(6): 377–92.

Chan, C. W., and Cawley, P. (1998). Lamb waves in highly attenuative plastic plates, *J. Acoust. Soc. Am.* 104(2): 874–81.

Chervinko, O. P., and Savchenkov, I. K. (1986). Harmonic viscoelastic waves in a layer and in an infinite cylinder, *Sov. Appl. Mech.* 22: 1136.

Chimenti, D. E., and Nayfeh, A. H. (1989). Ultrasonic leaky waves in a solid plate separating a fluid and vacuum, *J. Acoust. Soc. Am.* 85(2): 555–60.

Coquin, G. A. (1964). Attenuation of guided waves in isotropic viscoelastic materials, *J. Acoust. Soc. Am.* 36: 1074.

Deschamps, M., and Hosten, B. (1992). The effects of viscoelasticity on the reflection and transmission of ultrasonic waves by an orthotropic plate, *J. Acoust. Soc. Am.* 91: 2007–15.

Eder, J. E., and Rose, J. L. (1995). Composite cure evaluation using obliquely incident ultrasonic guided waves, in D. O. Thompson and D. E. Chimenti (Eds.), *Review of Progress in Quantitative Nondestructive Evaluation*, vol. 14, p. 1279. New York: Plenum.

Gao, H. (2007). *Ultrasonic Guided Wave Mechanics for Composite Material Structural Health Monitoring*, PhD Thesis, The Pennsylvania State University.

Haddad, Y. M. (1995). *Viscoelasticity of Engineering Materials*. London: Chapman and Hall.

Lake, R. S. (1999). *Viscoelastic Solids*. Boca Raton, FL: CRC Press.

Lowe, M. J. S. (1995). Matrix techniques for modeling ultrasonic waves in multilayered media, *IEEE Trans. Ultrason. Ferroelectr. Freq. Control* 42(4): 525–42.

Mu, J., (2008). *Guided Wave Propagation and Focusing in Viscoelastic Multilayered Hollow Cylinders*, PhD Thesis, The Pennsylvania State University.

Mu, J., and Rose, J. L. (2008). Guided wave propagation and mode differentiation in hollow cylinders with viscoelastic coatings, *J. Acoust. Soc. Am.* 124(2): 866–74.

Rokhlin, S. I., Lewis, D. K., Graff, K. F., and Adler, L. (1986b). Real-time study of frequency dependence of attenuation and velocity of ultrasonic guided waves during the curing reaction of epoxy resin, *J. Acoust. Soc. Am.* 79: 1786–93.

Sivaselvan, M., and Reinhorn, A. (2000). Hysteretic models for deteriorating inelastic structures, *J. Eng. Mechanics* 126(6): 633–40.

Tamm, K., and Weiss, O. (1961). Wellenausbreitung in unbegrenzten Scheiben und in Scheibenstreifen, *Acustica* 11: 8–17.

Van Velsor, J. (2009). *Circumferential Guided Waves in Elastic and Viscoelastic Multilayered Annuli*, PhD Thesis, The Pennsylvania State University.

Weiss, O. (1959). Uber die Schallausbreitung in verlustbehafteten Median mit komplexem Schub und kompressions Modul, *Acustica* 9: 387–99.

Wen, Y.-K. (1976). Method for random vibration of hysteretic systems. *J. Eng. Mech. Div., ASCE*, 102(2): 249–63.

18 Ultrasonic Vibrations

18.1 Introduction

Ultrasonic vibrations have often been used in the past on various structures without any real understanding of the impact of a loading function on structural resonances or resulting vibrational patterns at different frequencies. The method was used to obtain an ultrasonic signal signature of a part being inspected, most often in quality control after manufacture. A purpose of this chapter is therefore to establish an understanding of the ultrasonic vibration as a superposition of guided wave modes traveling in a structure and ways to optimize sensitivity to certain defects by way of loading function choice. In more traditional low-frequency modal analysis, the loading function choice is not so critical; results depend primarily on excitation frequency. With ultrasonic vibration, we will see that the choice of a loading function plays a major role in the design of an inspection system for developing various quality control and in-service inspection solutions.

This chapter now examines the subject of ultrasonic vibrations, the topic of which is a logical extension to ultrasonic guided waves. A few applications are considered.

Let's consider the long time solution to a wave propagation problem. In the bulk wave case, because waves are traveling in infinite space, there is no vibration aspect of the problem to be considered as there are no wave reflection and transmission factors. On the other hand, for guided wave propagation, the long time solution in many cases leads to a vibrations problem. This may not occur if no boundary exists in a particular direction as the wave is transmitted to infinity. In examining a closed structure, such as a finite plate or tube, the reflection and transmission factors for each entry of a wave onto a boundary leads to a variety of constructive and destructive interference phenomena. The long time solution therefore leads to a vibrations or modal analysis problem. There will be specific resonant frequency values for a structure as well as specific vibrational patterns at on and off resonance. Researchers are carrying out work examining the transition from the initial transient response to a long time vibrations solution. As a consequence, ultrasonic nondestructive evaluation (NDE) and structural health monitoring (SHM) is being developed by considering many aspects of ultrasonic bulk wave analysis, ultrasonic guided wave analysis, and modal vibration analysis.

Ultrasonic vibration is the steady-state vibration of a structure at ultrasonic frequencies. Of interest to the work discussed here, it specifically refers to such vibration in waveguide-like structures such as plates, pipes, shells, and so forth. While more common low-frequency vibration often assumes simplified field variation through the thickness of the structure, at ultrasonic frequencies such assumptions are often inaccurate. In fact, significant variations can exist in displacement, stress, and so forth through the thickness of a waveguide-like structure under ultrasonic vibration. It can be shown that such variations through the thickness for orthogonal natural modes of vibration of such structures are in some ways related to the wave structures of guided waves in that structure at the resonant frequencies of the plate. In fact, for some structures it can be shown that the orthogonal natural modes of vibration of the structure are comprised of a combination of guided wave modes propagating in the structure and creating a resonance (Borigo et al. 2011; Rose et al. 2012). This relationship between ultrasonic vibration and guided waves can be exploited for purposes of establishing a useful ultrasonic vibrations signature of a test part to determine if defects are present.

The simplest case of such ultrasonic guided wave vibration can be developed for a semi-infinite platelike structure, although this can be expanded to higher order structures, as shown for the bar case. Scholars will focus work efforts on randomly shaped objects with various boundary conditions.

The elastic field that satisfies the harmonic equilibrium condition of a semi-infinite platelike structure in steady-state vibration can be expanded as a superposition of propagating and evanescent guided wave modes traveling in the positive and negative x-directions. This can be written in terms of stresses σ as:

$$\sigma(x,z) = \sum_{\mu=1}^{\infty} \alpha_{\mu} \bar{\sigma}_{\mu}(z) e^{i(k_{\mu}x - \omega t)} + \sum_{\mu=1}^{\infty} \alpha'_{\mu} \bar{\sigma}_{\mu}(z) e^{i(k_{\mu}(x-L) - \omega t)}, \qquad (18.1)$$

in which μ signifies the guided wave mode number, α and α' are the relative complex amplitude coefficients for each mode in the $+x$- and $-x$-directions, k is the wavenumber, L is the length of the plate, ω is the angular frequency, and $\bar{\sigma}$ is the transverse stress field (i.e., wave structure) solution for each guided wave mode. The first term corresponds to waves traveling in the $+x$-direction and the second term to waves traveling in the $-x$-direction.

The plate was then discretized with N nodes along the thickness direction, as shown in Figure 18.1, and the semi-analytical finite element (SAFE) method (Gavric 1995) was utilized to develop the dispersion curves and wave structure solutions

Figure 18.1. An example of a possible discretization of the semi-infinite plate in the thickness dimension, in which five nodes were used.

for the desired frequency range in the structure. The result of a SAFE dispersion analysis with N nodes is a set of $3N$ guided wave solutions in each x-direction.

All six possible combinations of the three classical boundary conditions (free, simply supported, and clamped) for the semi-infinite plate were successfully considered. The free-free boundary conditions, given as one example, are:

$$\begin{cases} \sigma_{xx}\left(x=0,z\right)=\sigma_{xx}\left(x=L,z\right)=0 \\ \sigma_{xz}\left(x=0,z\right)=\sigma_{xz}\left(x=L,z\right)=0. \\ \sigma_{xy}\left(x=0,z\right)=\sigma_{xy}\left(x=L,z\right)=0 \end{cases} \tag{18.2}$$

Upon applying the boundary conditions Equation (18.2), by the collocation principle, to the discretized Equation (18.1), a matrix equation is developed as a function of frequency:

$$\begin{bmatrix} \sigma_{xx}^1\left(z_1\right) & \cdots & \sigma_{xx}^{3N}\left(z_1\right) & \sigma_{xx}^1\left(z_1\right)e^{-ik_1L} & \cdots & \sigma_{xx}^{3N}\left(z_1\right)e^{-ik_{3N}L} \\ \sigma_{xz}^1\left(z_1\right) & \cdots & \sigma_{xz}^{3N}\left(z_1\right) & \sigma_{xz}^1\left(z_1\right)e^{-ik_1L} & \cdots & \sigma_{xz}^{3N}\left(z_1\right)e^{-ik_{3N}L} \\ \vdots & \vdots & \vdots & \vdots & \vdots & \vdots \\ \sigma_{xy}^1\left(z_N\right) & \cdots & \sigma_{xy}^{3N}\left(z_N\right) & \sigma_{xy}^1\left(z_N\right)e^{-ik_1L} & \cdots & \sigma_{xy}^{3N}\left(z_N\right)e^{-ik_{3N}L} \\ \sigma_{xx}^1\left(z_1\right)e^{ik_1L} & \cdots & \sigma_{xx}^{3N}\left(z_1\right)e^{ik_{3N}L} & \sigma_{xx}^1\left(z_1\right) & \cdots & \sigma_{xx}^{3N}\left(z_1\right) \\ \sigma_{xz}^1\left(z_1\right)e^{ik_1L} & \cdots & \sigma_{xz}^{3N}\left(z_1\right)e^{ik_{3N}L} & \sigma_{xz}^1\left(z_1\right) & \cdots & \sigma_{xz}^{3N}\left(z_1\right) \\ \vdots & \vdots & \vdots & \vdots & \vdots & \vdots \\ \sigma_{xy}^1\left(z_N\right)e^{ik_1L} & \cdots & \sigma_{xy}^{3N}\left(z_N\right)e^{ik_{3N}L} & \sigma_{xy}^1\left(z_N\right) & \cdots & \sigma_{xy}^{3N}\left(z_N\right) \end{bmatrix} \begin{Bmatrix} \alpha_1 \\ \alpha_2 \\ \alpha_3 \\ \vdots \\ \alpha_{6N} \end{Bmatrix} = \begin{Bmatrix} 0 \\ 0 \\ 0 \\ 0 \\ \vdots \\ 0 \end{Bmatrix}$$

$$\tag{18.3}$$

Here the vector of complex amplitudes α is the only unknown and is a function of frequency. This can be written in a simpler manner by representing the large matrix as Λ:

$$[\Lambda]\{\alpha\} = \{0\} \tag{18.4}$$

Thus all nontrivial solutions to this matrix equation must correspond to the frequencies at which the determinant of Λ is equal to zero:

$$|\Lambda| = 0 \tag{18.5}$$

This determinant was plotted as a function of ω and an iterative root-finding algorithm was used to locate the frequencies at which Equation (18.5) is approximately satisfied. These frequencies are the natural resonant frequencies of the structure.

To determine the complex amplitude vector α associated with each natural resonant frequency, the nullspace of Λ at each resonant frequency must be determined. However, because discrete values of ω and a finite number of guided wave modes were used in the superposition formulation, the matrix is not an exact representation of the system. Thus it is possible that all rows of the matrix can be linearly independent, which leads to a situation in which the rank of matrix Λ is $6N$ and the nullspace of the matrix subsequently does not exist. To overcome this, an artificial eigenvalue problem can be developed as in Equation (18.6).

$$\Lambda\alpha - \lambda\alpha = 0; \quad \lambda = 0 \tag{18.6}$$

$$\lambda = \begin{Bmatrix} \lambda_1 \\ \lambda_2 \\ \vdots \\ \lambda_{6N} \end{Bmatrix}; \quad \underline{\Phi} = \begin{bmatrix} \alpha_1^{(1)} & \alpha_1^{(2)} & \cdots & \alpha_1^{(6N)} \\ \alpha_2^{(1)} & \alpha_2^{(2)} & \cdots & \alpha_2^{(6N)} \\ \vdots & \vdots & \cdots & \vdots \\ \alpha_{6N}^{(1)} & \alpha_{6N}^{(2)} & \cdots & \alpha_{6N}^{(6N)} \end{bmatrix} = \begin{bmatrix} \alpha^{(1)} & \alpha^{(2)} & \cdots & \alpha^{(6N)} \end{bmatrix} \tag{18.7}$$

This equation is identical to Equation (18.4) as long as the eigenvalue Λ is equal to zero. To approximate this case, the eigenvalue problem Equation (18.6) was solved and the smallest eigenvalue and corresponding eigenvector solution were considered. As long as this minimum eigenvalue is close to zero, the solution is a good approximation of Equation (18.4) and of Equation (18.5) as well. Thus the complex amplitude vector for each resonant frequency is the eigenvector associated with the minimum eigenvalue, as shown in Equations (18.8) and (18.9).

$$\lambda^{\min} = \min(\lambda) \approx 0 \tag{18.8}$$

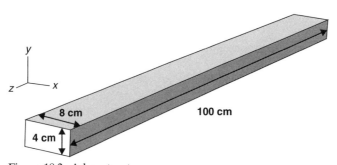

Figure 18.2. A bar structure.

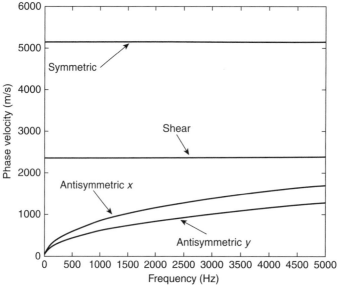

Figure 18.3. Dispersion curve for the bar structure.

Table 18.1. *Resonant frequencies for a bar calculated by finite-element eigenvalue solver and by the guided wave vibration method*

Steady-State Vibration		Resonant Frequencies				Guided Wave Amplitudes				
Vib. Mode Type	Vib. Mode Order	FEM_lim	FEM_mesh	GW Vib	% error	GW1(A_y)	GW2(A_x)	GW3(SH)	GW4 (S)	Match?
Flexural	1_y	**211**	209	212	0.47%	1	0.01	0.01	0	✓
Flexural	1_k	**408**	405	409	0.25%	0	1	0	0	✓
Flexural	2_y	**573**	569	575	0.35%	1	0	0	0	✓
Flexural	2_x	**1,069**	1,062	1,071	0.19%	0.01	1	0	0	✓
Flexural	3_y	**1,103**	1,097	1,107	0.36%	1	0	0.01	0	✓
Torsional	1	**1,189**	1,167	1,197	0.67%	0	0.02	1	0	✓
Flexural	4_y	**1,785**	1,774	1,792	0.39%	1	0.03	0.01	0	✓
Flexural	3_x	**1,976**	1,965	1,981	0.25%	0	1	0	0	✓
Torsional	2	**2,382**	2,337	2,398	0.67%	0.01	0.01	1	0.01	✓
Compressional	1	**2,583**	2,583	2,584	0.04%	0	0	0	1	✓
Flexural	5_y	**2,602**	2,586	2,611	0.35%	1	0	0	0	✓
Flexural	4_x	**3,063**	3,047	3,071	0.26%	0	1	0	0	✓
Flexural	6_y	**3,536**	3,513	3,550	0.40%	1	0	0	0	✓
Torsional	3	**3,581**	3,516	3,605	0.67%	0.01	0.01	1	0	✓
Flexural	5_x	**4,281**	4,260	4,294	0.30%	0.01	1	0.01	0.01	✓
Flexural	7_y	**4,573**	4,548	4,590	0.37%	1	0	0	0	✓
Torsional	4	**4,788**	4,694	4,821	0.69%	0	0.01	1	0	✓

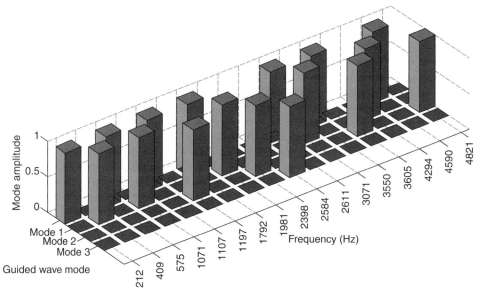

Figure 18.4. Plot of the relative amplitudes of various guided wave modes associated with each natural resonance of the bar structure.

$$\alpha = \alpha^{(\lambda_{\min})} = \begin{Bmatrix} \alpha_1 \\ \alpha_2 \\ \alpha_3 \\ \vdots \\ \alpha_{6N} \end{Bmatrix} \tag{18.9}$$

These solutions were then used to reconstruct the full stress or displacement vibration fields associated with each resonant frequency according to Equations (18.10) and (18.11), respectively.

$$\sigma(x,z) = \sum_{\mu=1}^{3N} \alpha_\mu \bar{\sigma}_\mu(z) e^{i(k_\mu x - \omega t)} + \sum_{\mu=3N+1}^{6N} \alpha_\mu \bar{\sigma}_\mu(z) e^{i(k_\mu(x-L)-\omega t)} \tag{18.10}$$

$$u(x,z) = \sum_{\mu=1}^{3N} \alpha_\mu \bar{u}_\mu(z) e^{i(k_\mu x - \omega t)} + \sum_{\mu=3N+1}^{6N} \alpha_\mu \bar{u}_\mu(z) e^{i(k_\mu(x-L)-\omega t)} \tag{18.11}$$

Results for the guided wave vibration case of an aluminum bar with a rectangular cross section are shown in the following figures and Table 18.1.

Practical considerations of the ultrasonic vibrations problem are discussed in the next section.

18.2 Practical Insights into the Ultrasonic Vibrations Problem

Let's now consider some practical aspects of ultrasonic vibrations, an extension to guided wave analysis. Investigators have tried for decades to inspect components just manufactured as well as those in service from an NDT and SHM point of view by a single activation source and a single receiver to provide an ultrasonic signature

Table 18.2. *A brief comparison of vibration modal analysis, ultrasonics, and UMAT*

	Detected defect size	Actuator position	Testing speed
Vibration Modal Analysis	Large	One	Fast
Transient Guided Wave	Small	Multiple	Slow
UMAT	Small to intermediate	One	Fast

▪ advantage
☐ disadvantage

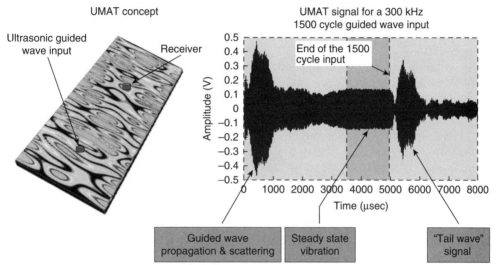

Figure 18.5. Ultrasonic vibration concept.

of the part being tested. In many cases this worked well, primarily because of the large number of parts under inspection and the statistical basis decision algorithm established for a long series of trial-and-error experiments. The approach was somewhat magical, with little understanding of wave mechanics. The work presented here provides a scientific basis for such a signature approach, called an ultrasonic modal analysis technique (UMAT). The answer lies in the understanding of ultrasonic guided waves and ultrasonic vibrations.

A motivation for developing such an approach compared to more standard vibration studies in guided waves is summarized in Table 18.2. Note the advantages of the UMAT with respect to finding intermediate-sized defects compared to those found by the other two methods and the benefit of simplicity and speed to run a test.

Implementation of the UMAT concept is depicted in Figure 18.5. A sample ultrasonic vibration field is illustrated on the left for a single exciter and single receiver. A pulse of 1,500 cycles is used to excite the part. Note that this leads to a wave propagation problem, but because of the large number of cycles, multiple reflections take place in the part leading to all sorts of interference results. An examination of the RF waveform shows an initial transient guided wave propagation and scattering followed by a steady state vibration with many oscillations, and finally a "tail wave" signal of the transient response again, this time as the wave reduces its energy to zero.

- The long time structural response depends on the initial ultrasonic loading functions (guided wave mode-frequency combinations) and the structure geometry and boundary conditions.

To obtain a receiver signal that reliably provides an indication of whether there is a defect in the structure depends on many variables; of utmost importance will be the starting point in selecting an appropriate mode and frequency on the phase velocity dispersion curve with expected sensitivity to the kind of defects expected in the structure. Quite often, this is difficult to do, so a special actuator is recommended that can be used to search the phase velocity dispersion curve space for the optimal point in providing the most useful signature of the component test condition.

A multi-element annular array transducer was selected to do this because when used in a phased array mode with frequency tuning, provided that frequency bandwidth is sufficient, one could traverse a large area of the phase velocity dispersion curve space. The starting point of the search should use physical insight on a wave parameter that has a good chance of being sensitive to the defect expected. See Appendix D for details on this mode and frequency selection process. Note, however, that even if the wave structure selected at the particular mode and frequency is appropriate for a transient guided wave test, because of mode conversion that occurs at each boundary of the test component, the wave structure may or may not be preserved. For example, wave structure is preserved for a normal incidence onto a boundary when impinging below the first cutoff frequency. Oblique incidence situations should be studied. Hence, the principal reason for the annular array tool is to search for an adequate solution.

A few useful comments are presented here. Vibrational patterns for any waveguide-type structure can be calculated using finite element computer codes in a fashion similar to that used to calculate responses associated with a transient wave propagation problem. So some effort is required to learn how to do this. Second, a receiver at one point could possibly work well, but a series of different transducers at different locations might be more beneficial, say, in a grid of three by three or four by four. Also, a laser vibrometer could be used to capture a full field displacement image. Note, however, that the practical aspects of using a laser vibrometer could be limited. Next, note that resonant frequencies are so close together at higher ultrasonic frequencies compared to normal low-frequency vibration methods that they cannot generally be used to indicate a defect presence in a structure. Therefore, the operational deflection shapes (ODS) can provide us with a suitable approach for tackling this problem. The ODS changes over the structure quite significantly for small defect changes. Note that the ODS also changes as a result of boundary condition variations that could be good or bad depending on the problem definition. If a weakened boundary condition is an indication of a structural defect, then that detection is valuable.

See Figure 18.6 for the loading concept for a simple structure. Receivers can be placed anywhere on the structure, but if possible, especially close to the defect regions if their whereabouts might be known. Let's examine two interesting modeling results shown in Figures 18.7 and 18.8. Note that in Figure 18.7, even though the excitation frequency is fixed at 200 kHz, the ODS changes. The ODS changes because the exciting transducer is shifting to a different mode on the phase velocity dispersion

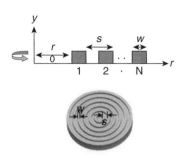

FEM setup to study the ODS's of the plate excited by the annular array transducer

Plate dimension: 0.15m*0.15m *3e-3m

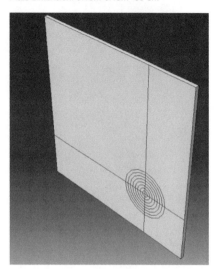

(a) Annular array transducer schematic, in which *N* is the number of elements, r_0 is the inner radius, *w* is the element width, and *s* is the spacing.

(b) Another view of an annular array.

Size of the annular array in FEM

Annular array element No.	Inner radius (mm)	Outer radius (mm)
1	2.625	5.250
2	7.875	10.500
3	13.125	15.175
4	18.375	21.000

Figure 18.6. Plate excitation by an annular array actuator. (Note that the receiver can be placed anywhere on the structure.)

Phase delay: 0 degree; Frequency: 200 kHz

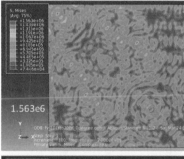

Phase delay: 90 degrees; Frequency: 200 kHz

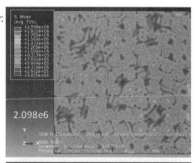

Phase delay: 135 degrees; Frequency: 200 kHz

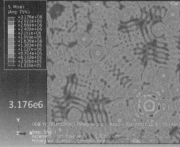

Phase delay: 225 degrees; Frequency: 200 kHz

Figure 18.7. Operational deflection shapes generated by the annular array transducer – Fix loading frequency, change phase delay. See plates section for color version.

curve as a result of the phase delay between annular array elements, which are related to the time-delay profiles applied to the elements of an annular array.

On the other hand, if no time delays are applied to the annular array elements, the activation line is fixed close to slope Λ in the phase velocity dispersion curve, and

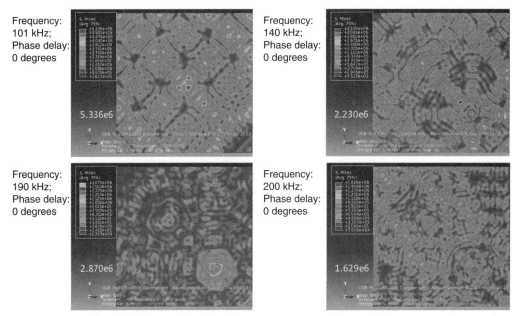

Figure 18.8. Operational deflection shapes generated by the annular array transducer – Fix phase delay, vary loading frequency. See plates section for color version.

hence mode change occurs as a result of frequency shifting as it is tuned over the range of interest as demonstrated in Figure 18.8.

We now see the challenge and opportunity to apply this method to experiments and actual inspection.

An experimental arrangement is illustrated in Figure 18.9. Multichannel continuous sinusoidal signals with varying frequencies and phase delays among different channels are applied to annular array actuators for mode and frequency tuning. Distributed receivers are placed on the structure for measuring forced vibration amplitudes under different loading functions. Vibration amplitude maps are plotted in a 2-D frequency-phase delay space. A sample problem is outlined here. Suppose you want to inspect the composite propeller shown in Figure 18.10. The propeller was freely suspended as shown to eliminate boundary condition variations from the inspection result. We will now place the annular array activation source on the blade initially to establish a baseline. Several baselines were taken on different days before inducing any damage into the composite propeller blade. Several receiving transducers were then placed on the test part. Data were then taken after damage was initiated, in this case by a hammer impact situation. Data were obtained by plotting amplitude as a function of phase angle and frequency for some of the receiver locations. See Figure 18.11 for a plot of phase angle from 0° to 360° versus frequency from 20 kHz to 210 kHz in a test bed data acquisition approach that included more than 4,775 different loading functions. The most sensitive areas were selected for establishing a decision algorithm as shown in the zoomed-in profile in Figure 18.11. Alternatively, specific points can be selected. As an example, see the results in Table 18.3. Clearly there is an indication of change taking place in the propeller blade by looking either at Table 18.3 or Figure 18.11.

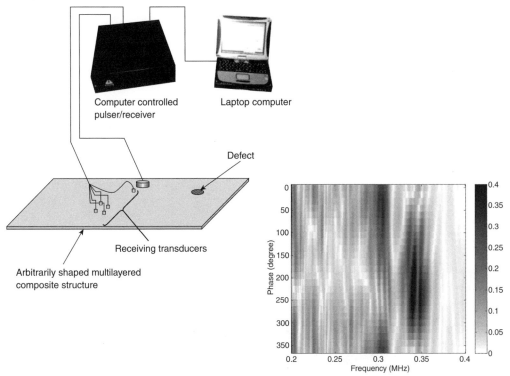

Figure 18.9. Ultrasonic vibration tests with guided wave mode and frequency tuning.

Figure 18.10. Example UMAT Experiment: An Airboat Composite Propeller. See plates section for color version.

Table 18.3. Ten loading functions most sensitive to the delamination defect

Frequency (MHz)	Phase delay (degree)
0.1	90
0.128	240
0.165	150
0.1	105
0.099	135
0.167	135
0.128	225
0.167	150
0.099	255
0.168	135

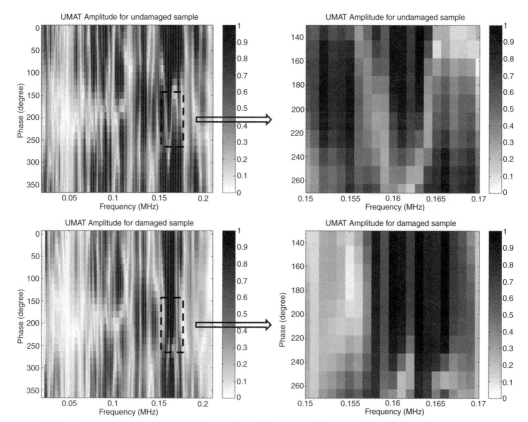

Figure 18.11. Example UMAT experiment I – composite propeller impact damage.

At this point, pattern recognition could be used to establish a defect decision algorithm utilizing either the total phase velocity versus frequency space or the zoomed-in space. The solution is certainly not unique. Many other pseudo-image methods could be developed to solve this problem. The ideas of ultrasonic vibration are presented here that could be useful for addressing a variety of problems.

18.3 Concluding Remarks

A new concept of ultrasonic vibration for damage detection was introduced here. A theoretical foundation for ultrasonic vibration was presented through the development of a guided wave mode decomposition method for investigating the connections between guided wave propagations and steady-state vibrations. It was demonstrated that, via selections of different guided wave loading functions, specific vibrations associated with the guided waves can be selectively excited to produce high sensitivity to specific types of damage. Annular array actuators were introduced as a powerful tool for tuning guided wave loading functions. The actuators were designed based on the theories for guided wave excitations by annular arrays. A unique UMAT system was developed and demonstrated on various test samples. The UMAT technique developed can serve as a good alternative damage detection method to traditional vibration measurements and transient guided wave techniques. The damage detection sensitivity of UMAT is superior compared to traditional vibration measurements. Compared to transient guided wave tests, UMAT provides better coverage and faster inspection speed for inspecting a large area and requires less accessible locations.

Note also that ultrasonic guided wave vibration could also be useful in de-icing applications by finding vibration modes with strong shear stresses at the ice-substrate interface.

18.4 Exercises

1. What kind of mode conversion occurs as an ultrasonic guided wave impinges onto a boundary of a test object?
2. Name several applications where UMAT may be applicable. For each application, describe the practical advantages and disadvantages to using UMAT over other methods currently available.
3. Assume you are performing UMAT on a test object with a single, fixed excitation probe position and one receiving probe position. What factors define the limits on the size of an object you could test?
4. One important factor when testing an object using UMAT is the boundary condition applied to the structure's edges. What practical techniques might you employ to enforce consistent boundary conditions in multiple or repeated laboratory tests?
5. What are the differences between "traditional" ultrasonic guided wave inspection and an inspection performed with UMAT? What are the similarities?
6. What are the differences between traditional vibration modal analysis and UMAT? What are the similarities?
7. Why is an annular array transducer advantageous for testing using UMAT? List several other transducer types and make an argument for why they would or would not be advantageous for use with UMAT.
8. See Figures 18.7 and 18.8. High frequency ultrasonic vibrational resonances are very close to each other and therefore not as useful as those considered in low frequency vibrational modal analysis. ODS shapes however are quite useful. How would you design an ultrasonic vibration quality control inspection for an

unusually shaped part? How would you select an appropriate phase delay and frequency?

9. Discuss how ultrasonic guided wave vibration could be used to de-ice a complete waveguide or airfoil.

18.5 REFERENCES

Borigo, C., Yan, F., Liang, Y., and Rose, J. L. (2011). Ultrasonic guided wave vibration formulation, *Quantitative Nondestructive Evaluation Conference*, Burlington, VT.

Gavric, L. (1995). Computation of propagative waves in free rail using a finite element technique, *J. Sound Vib.* 185(3): 531–43.

Liew, K. M., Hung, K. C., and Lim, M. K. (1993). A continuum three-dimensional vibration analysis of thick rectangular plates, *Int. J. Solids Structures* 30(24): 3357–79.

Rose, J. L., Yan, F., Liang, Y., and Borigo, C. (2012). Ultrasonic vibration method for damage detection in composite aircraft components, *IMAC XXX Conference*, Jacksonville, FL.

Salas, K. I., and Cesnik, C. E. S. (2010). Guided wave structural health monitoring using CLoVER transducers in composite materials, *Smart Materials and Structures* 19: 015014.

Srinivas, S., Rao, C. V., and Rao, A. K. (1970). An exact analysis for vibration of simply-supported homogeneous and laminated thick rectangular plates, *J. Sound Vib.* 12(2): 187–99.

Wilcox, P. D. (2003). Omni-directional guided wave transducer arrays for the rapid inspection of large areas of plate structures, *IEEE Trans. Ultrason., Ferroelect., Freq.* 50(6): 699–709.

Yan, F., and Rose, J. L. (2009). Time delay comb transducers for aircraft inspection, *The Aeronautical Journal* 113(1144): 417–27.

19 Guided Wave Array Transducers

19.1 Introduction

Multiple element array transducers are extremely useful and popular in today's inspection environment. The first applications were to carry out electronic B and C scans opposed to earlier developed mechanical scans. This was followed by phasing of the elements in a bulk wave problem where beam steering and focusing were possible during the electronic scanning process. Today, array transducers are being used in guided wave inspection. Linear comb and annular array sensors are two possibilities. Beyond electronic scanning and focusing it is also now possible to select time delay profiles for guided wave mode and frequency selection to optimize sensitivity to certain defects and penetration power in special situations. Thus, time delays for mode selection and electronic scanning can be superimposed for rapid and efficient NDT and SHM. As a consequence, such items as the excitation spectrum and the mode excitability function will be studied along with phasing principles for linear combs and annular arrays.

To employ guided waves for nondestructive evaluation (NDE) or structural health monitoring (SHM) purposes, these waves must first be generated in the structure of interest. Accordingly, to fully reap the benefits guided waves can offer, such wave generation should be performed in a well-designed, highly controlled manner, which is only possible through deliberate transducer design. Proper guided wave mode control can provide distinct advantages in terms of sensitivity to particular defects, sensitivity to environmental variables, penetration power, and other factors in guided wave inspection. Additionally, the suppression of spurious modes and/or the excitation of a particularly nondispersive mode can greatly enhance the potential signal analysis of gathered data by simplifying the waveforms.

In Chapter 13, a detailed analysis of guided wave mode control with a finite-sized transducer was performed by applying the normal mode expansion (NME) technique. This analysis was then applied to the case of an angle beam transducer, and it was shown that the design of the angle beam transducer (such as the incident angle and the shape, size, and distribution of the beam profile) heavily influences the guided waves that will be generated. This influence can be expressed mathematically in the form of the source excitation spectrum. In reality, a single transducer is not always the best approach to

With significant contribution from Cody Borigo.

guided wave excitation. Rather, an array of transducers distributed in some spatial pattern can provide more precise mode control and significant electronic mode control flexibility without the need to physically manipulate the transducer (as would be required to excite a number of modes with an angle beam wedge). Additionally, array transducers can also be designed to be much more compact than angle beam wedges and can provide multidirectional mode control if designed to do so. However, to realize the benefits of array transducers, it is imperative that we first develop analytical expressions of the excitation spectra that describe the mode control properties of any particular array.

19.2 Analytical Development

The source influence analysis of array transducers follows a common logic with that derived in Chapter 13. Recall that the NME technique or integral transform techniques can be employed to solve the boundary value problem of a forced loading of a structure, the solutions of which must obey the governing wave equations in the structure. The application of either of these techniques will eventually lead to an identical solution, such as that outlined in Chapter 13 and repeated here as Equation (19.1). See Rose (1999) for details.

$$A_{\pm v}(x) = \frac{e^{\pm ik_v x}}{4P_{vv}} \tilde{\bar{v}}_{\pm v}\left(\frac{b}{2}\right) \cdot \int_{-\infty}^{\infty} e^{\pm ik_v \eta} \bar{t}(\eta) d\eta \qquad (19.1)$$

Recall that $A_v(x)$ indicates the amplitude of a particular guided wave mode v as a function of distance in the wave propagation direction x. Pvv is the time-averaged power flow in the wave propagation direction; k_v is the wavenumber of mode v; $\bar{v}_v$ $(b/2)$ is the normalized surface velocity of mode v; $\bar{t}(\eta)$ describes the surface traction applied to the structure over the domain η.

This expression can be reformulated by separating various terms based on the parameters that govern them. This reformulation in Equation (19.2) allows us to more clearly understand the influence of the individual terms.

$$A_{\pm v}(x) = GFE_v e^{\mp ik_v x} \qquad (19.2)$$

If, for simplicity's sake, we assume out-of-plane (y-direction) loading over a distribution $p(\eta)$, as described in Equation (19.3), we can then define the three individual source influence terms from Equation (19.2) as given in Equation (19.4)

$$\bar{t}(\eta) = \rho \cdot p(\eta)\hat{e}_y \qquad (19.3)$$

where ρ is a traction density factor proportional to the transducer load.

$$G \equiv \frac{\rho}{4}, \qquad (19.4a)$$

$$E_v \equiv \frac{\tilde{v}_{\pm v_y}(b/2)}{P_{vv}}, \qquad (19.4b)$$

$$F \equiv \int_{-\infty}^{\infty} e^{\pm ik_v \eta} p(\eta) d\eta, \qquad (19.4c)$$

Here G simply conveys an amplitude factor relative to the loading amplitude of the transducer; this is insignificant for our analysis and will thus be disregarded during further development. E_v is the *mode excitability function* of mode v, which describes the compatibility of guided wave mode v with the surface traction applied for the transducer. Although this factor is related to the transducer, it is governed only by the type of loading applied and not the distribution of that loading, which is the factor that we seek to analyze for array transducer development. Finally, the factor F is the *excitation spectrum* of the transducer and depends solely on the distribution of the loading source over the domain η. The excitation spectrum uniquely describes the relative excitation of various guided wave modes in the k, or conversely the λ or c_p–f, domains.

At this point in the derivation, it is critical that we notice the form of F as described in Equation (19.4c). Clearly this is simply the spatial Fourier expansion of the loading distribution $p(\eta)$ over the spatial domain η. Therefore the excitation spectrum is simply the wavenumber (inverse spatial) spectrum of the source loading distribution. This realization will allow us to directly analyze any distributed transducer loading, such as those of a variety of array transducers, by simply taking the spatial Fourier transform of their distribution functions.

19.2.1 Linear Comb Array Solution

One of the most commonly employed array transducers is the *linear comb* (or *comb*) array. A comb array is nothing more than a number of parallel elements spaced at a set distance apart from each other; typically the length of the elements (perpendicular to the wave propagation direction) will be much greater than the width of the elements (parallel to the wave propagation direction). As one might intuitively infer, the spacing of these elements will greatly determine the wavelength of the guided wave modes that the comb array will generate. To further understand the parameters that dictate the source influence of comb arrays, the analytical description of the excitation spectrum must be derived.

The linear comb array is assumed to have N elements with a pitch (spacing) denoted by s and an element width denoted by w, as detailed in Figure 19.1. The analysis will be performed with the assumption that the comb elements extend to $z = \pm\infty$. This assumption is valid if the length of the elements is sufficiently greater than their width, as is typically the case.

To simplify the problem, we will assume the case of even pressure loading across the face of each transducer element. Therefore the geometry of the comb array can be defined in terms of a sum of rectangle functions, each denoted by $\prod_a(x)$. This rectangle function is defined in Equation (19.5).

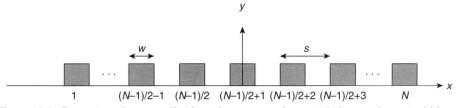

Figure 19.1. Geometry of a generalized comb array; s = element pitch; w = element width; N = number of elements.

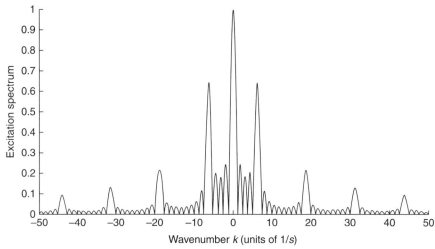

Figure 19.2. Excitation spectrum in the wavenumber domain for a five-element comb array with width-spacing ratio $w = s/2$.

$$\prod_a (x) = \begin{cases} 1, & -a \leq x \leq a \\ 0, & x < -a \; or \; x > a \end{cases} \tag{19.5}$$

Using such functions, the entire comb array geometry can be described, with full generality, by Equation (19.6):

$$f^{comb}(x) = \sum_{n=0}^{N-1} \prod_{w/2} (x - ns) \tag{19.6}$$

Applying the spatial Fourier transform to a single element (or single term in Equation (19.6)), the excitation spectrum of an individual rectangular element is given by Equation (19.7):

$$F_n^{comb}(k) = w \cdot \sin c \left(\frac{kw}{2} \right) \tag{19.7}$$

and utilizing the modulation property of the Fourier transform allows us to translate and sum any number of elements along the x-axis, which, after some manipulation, yields the excitation spectrum of a complete comb array in Equation (19.8). Here $\sin c(x)$ is the unnormalized sinc function $\sin c(x) = \sin(x)/x$.

$$F^{comb}(k) = w \cdot \sin c \left(\frac{kw}{2} \right) \cdot \frac{\sin(ksN/2)}{\sin(ks/2)} \cdot e^{i\frac{ks}{2}(N-1)} \tag{19.8}$$

Although the excitation spectrum described in Equation (19.8) is in fact a complex function of k, we are interested only in the magnitude of the spectrum. We are also not particularly interested in the explicit magnitude of the spectrum, as it is dictated by the amplitude of the loading and not the distribution. Thus we will normalize the spectra by the peak value in the wavenumber domain (which occurs at $k = 0$). Figures 19.2 and 19.3 show a sample excitation spectrum for a

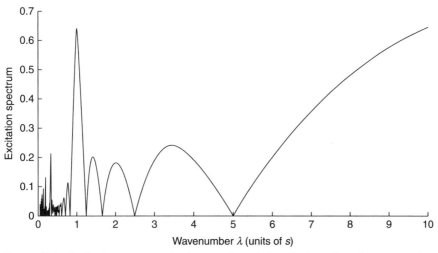

Figure 19.3. Excitation spectrum in the wavelength domain for a five-element comb array with width-spacing ratio $w = s/2$.

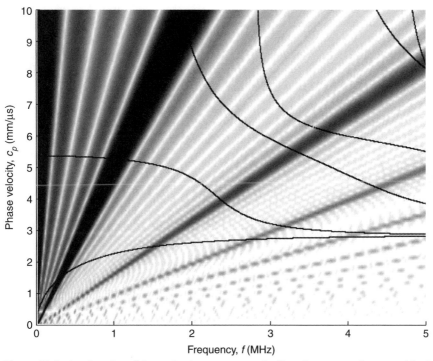

Figure 19.4. A color plot of the excitation spectrum of a five-element comb array with pitch $s = 5$ mm and element width $w = 2.5$ mm, in which the darker shades represent greater excitation. The dispersion curve for a 1 mm aluminum plate is superimposed in black as a reference.

five-element comb array with $w = s/2$ in the wavenumber (k) and wavelength (λ) domains, respectively.

Another way the excitation spectrum of a comb array can be displayed is in the phase velocity versus frequency dispersion curve space, as done in Figure 19.4. The advantage of plotting the data in this manner is that it can be easily understood in relation to mode selection on a dispersion curve. However, for a more thorough

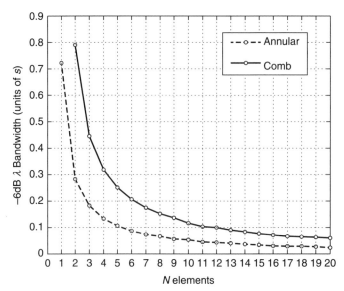

Figure 19.5. The effect of the number of elements in a comb or annular array is to reduce the bandwidth of the primary peak, as shown here. Note that the bandwidth of the secondary peaks is also reduced as N increases.

analysis, it is best to plot the excitation spectrum in the wavenumber of wavelength domains because of the inherent simplification and the fact that much data is easily lost when viewed within the normal boundaries of the dispersion curve space.

Note the symmetry in Figure 19.2 about $k = 0$; this means that the guided waves generated by the comb transducer will be identical in the forward ($+x$) and in the backward ($-x$) directions. Also note that the maximum of the excitation spectrum occurs at $k = 0$ (i.e., $\lambda \to \infty$); such a condition can only occur at cutoff frequencies or at zero frequency for some wave modes, and is thus not of interest in most applications. The peaks of primary importance are the fundamental peak, which occurs near $\lambda = 1s$ in Figure 19.3, and the secondary peaks, which decay in magnitude as λ decreases (i.e., as $|k|$ increases). The location of the primary peak in the spectrum tells us that to optimize the design of a comb transducer (with $w = s/2$) is to set the comb pitch to be equal to the wavelength of the guided wave mode that it is intended to generate at a particular frequency.

Further analysis of the effects of various comb parameters can be accomplished by studying Equation (19.8). Although a complete parametric study is beyond the scope of this chapter, some general relations can be examined. For instance, increasing the number of elements in the array narrows the peaks but negligibly affects their magnitudes and locations; this is analogous to the narrowing of the frequency bandwidth of a time-domain signal by increasing the number of cycles. Therefore the more elements in an array, the more precise the mode selection will be. This relationship is shown in Figure 19.5. Increasing or decreasing the width of the elements, w, with respect to the pitch, generally leads to the deleterious effects of an increase in the number and relative magnitude (with respect to the primary peak) of the secondary peaks in the spectrum.

Note that the assumption of infinite element length L (in the $\pm z$-direction) imposes some limitations on the accuracy of the comb transducer excitation spectrum

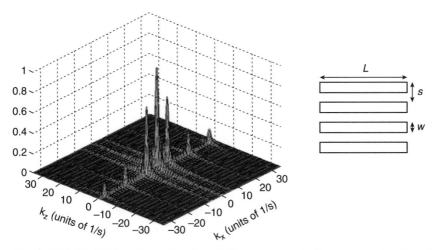

Figure 19.6. (Left) Two-dimensional excitation spectrum in the wavenumber domain for a four-element comb transducer with element $w = s/2$ and $L = 10w$. (Right) Loading profile of the comb transducer with $L = 10w$.

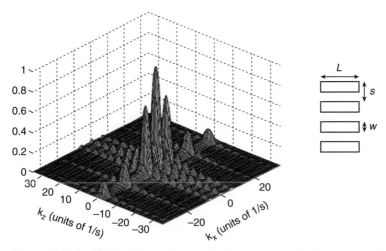

Figure 19.7. (Left) Two-dimensional excitation spectrum in the wavenumber domain for a four-element comb transducer with element $w = s/2$ and $L = 3w$. (Right) Loading profile of the comb transducer with $L = 3w$.

analysis. Our analysis assumed that the length of these elements is much greater than their width, yet even when this condition holds, the accuracy of the spectrum derived here is applicable to small angles from the x-axis.

A complete comb transducer excitation spectrum analysis can be performed by calculating the Fourier transform in Equation (19.4c) over the two-dimensional spatial $(x$-$y)$ domain using the complete two-dimensional loading profile as done in Equations (19.9) and (19.10). This yields an excitation spectrum in the $(k_x$-$k_z)$ domain such as those shown in Figures 19.6 and 19.7 for comb arrays with $L = 10w$ and $L = 3w$, respectively.

$$f^{2Dcomb}(x,z) = \Pi_{L/2}(z) \cdot \sum_{n=0}^{N-1} \Pi_{w/2}(x-ns) \tag{19.9}$$

$$F^{2Dcomb}(k_x,k_z) = wL \cdot \text{sin} c\left(\frac{k_z L}{2}\right) \cdot \text{sin} c\left(\frac{k_x w}{2}\right) \cdot \frac{\sin(k_x sN/2)}{\sin(k_x s/2)} \cdot e^{i\frac{ks}{2}(N-1)} \quad (19.10)$$

A narrower comb array leads to increased excitation along the z-axis and a broader beam with more angular variation in the spectrum along the x-direction; both of these are undesirable characteristics as comb transducers are generally designed to excite guided waves solely in the x-direction with a narrow wavelength spectrum while suppressing wave generation in all other directions.

19.2.2 Annular Array Solution

Another commonly employed array transducer is the *annular array*. An annular array is a set of concentric annular transducer sections positioned coaxially. The concept behind the annular array is to generate guided waves with a controlled wavelength based on the spacing and width of the individual array elements, just as in the comb case; however, the fundamental difference between the comb and annular arrays is the multidirectional nature of the annular array due to its axial symmetry. This symmetry allows the array to generate a circular-crested guided wave packet with equivalent mode control in all directions (for an isotropic material). This can be quite beneficial in a number of cases including guided wave computed tomography, phased array beam steering, time reversal, and ultrasonic vibrations. However, with the benefits of radial symmetry come additional complexities due to the interference of the inward- and outward-propagating waves emitted from the array.

The geometry of an annular array is described in Equation (19.11) using rectangle functions in cylindrical coordinates, as defined in Equation (19.12) (Figure 19.8). Here w and s are the element width and spacing, just as with the comb array, and r_0 is the inner radius of the central annular element.

$$f^{ann}(r) = \sum_{n-0}^{N-1}\left[\Pi_{w/2}(r - ns - r_o)\right] \quad (19.11)$$

$$\Pi_a(r) = \begin{cases} 1, & 0 \leq r \leq a \\ 0, & r > a \end{cases} \quad (19.12)$$

Because the annular array exhibits loading over a distributed pattern in the x-z plane, we must utilize a two-dimensional spatial Fourier transform in Equation (19.13) to develop the excitation spectrum solution. Because of the radial symmetry

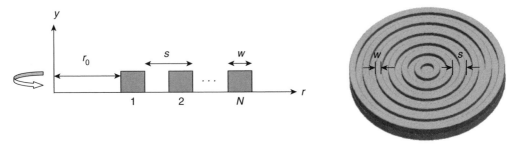

Figure 19.8. Geometry of a generalized annular array; s = element spacing; w = element width; r_0 = inner radius; N = number of elements.

of the geometry function in Equation (19.11), this is equivalent to the 0th-order Hankel transform in Equation (19.14).

$$F^{ann}\left(k_x, k_z\right) = \int\limits_{-\infty}^{\infty} \int\limits_{-\infty}^{\infty} f^{ann}\left(x,z\right) e^{-i\left(k_x x + k_z z\right)} dx\, dz \qquad (19.13)$$

$$F^{ann}\left(k_r\right) = \int\limits_{0}^{\infty} f^{ann}\left(r\right) r J_0\left(k_r r\right) dr \qquad (19.14)$$

Inserting the geometry function Equation (19.11) into the Hankel transform integral in Equation (19.14) yields the excitation spectrum solution Equation (19.15) after some manipulation. Here k is assumed to be equivalent to k_r, $J_1(x)$ is the first-order Bessel function of the first kind, and the substitution $\alpha_n = r_0 + sn$ is utilized.

$$F^{ann}\left(k\right) = \frac{1}{k}\sum_{n=0}^{N-1}\left\{\left(\alpha_n + w\right) J_1\left[\left(\alpha_n + w\right)k\right] - \alpha_n J_1\left[\alpha_n k\right]\right\} \qquad (19.15)$$

The excitation spectrum for the annular array Equation (19.15) is equivalent for all radial directions originating at the center of the array; a sample spectrum for a five-element annular array with $w = s/2$ and $r_0 = s/2$ is given in Figures 19.9 and 19.10 in the wavenumber and wavelength domains, respectively. Note that the primary peak in the comb spectrum is replaced with a double peak near $\lambda = s$. This double peak is due to the interference of the inward- and outward-propagating waves emitted from the annular array. The excitation spectrum for each of these waves can be individually derived from Equation (19.15). By substituting the Bessel functions for Hankel functions using the identity Equation (19.16),

$$J_v\left(x\right) = \frac{1}{2}\left[H_v^{(1)} + H_v^{(2)}\right] \qquad (19.16)$$

the spectrum of the outward-propagating waves is given by Equation (19.17a) and that of the inward-propagating waves is given by Equation (19.17b).

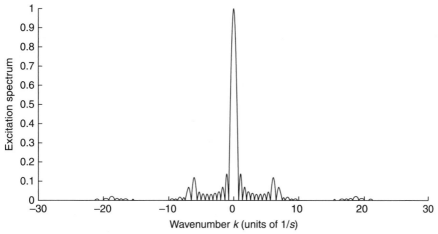

Figure 19.9. Excitation spectrum in the wave length domain for a five-element annular array with element width $w = s/2$ and inner radius $r_0 = s/2$.

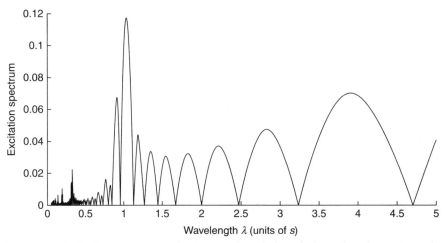

Figure 19.10. Excitation spectrum in the wavelength domain for a five-element annular array with element width $w = s/2$ and inner radius $r_0 = s/2$.

$$F_O^{ann}(k) = \frac{1}{k} \sum_{n=0}^{N-1} \left\{ (\alpha_n + w) H_1^{(1)} \left[(\alpha_n + w)k \right] - \alpha_n H_1^{(1)} \left[\alpha_n k \right] \right\} \qquad (19.17\text{a})$$

$$F_I^{ann}(k) = \frac{1}{k} \sum_{n=0}^{N-1} \left\{ (\alpha_n + w) H_1^{(2)} \left[(\alpha_n + w)k \right] - \alpha_n H_1^{(2)} \left[\alpha_n k \right] \right\} \qquad (19.17\text{b})$$

This conclusion is reached based on the nature of the Hankel functions $Hv^{(1)}$ and $Hv^{(2)}$, which are the solutions to the cylindrical Helmholtz wave equation that correspond to outward- and inward-propagating waves, respectively. Further analysis can be performed using these separated equations, and optimization of the array design can be accomplished by using the inner radius to modulate the interaction of these two waves. For simplicity's sake, the remaining analysis of the annular array transducer will assume an inner radius of $r_0 = w$ is utilized. The other array parameters such as the element width-to-spacing ratio and the total number of elements have a very similar effect on the annular array spectrum as they did on the comb array spectrum.

We can see in Figure 19.10 that for the annular array, the maximum excitation point in the spectrum does not occur precisely at $\lambda = s$, but rather at a slightly longer wavelength. Because of the frequency bandwidth that may typically be used to excite a particular guided wave mode and frequency, one might think that the relatively small shift in the primary excitation spectrum peak away from $\lambda = s$ is negligible and that the annular array can simply be designed under the assumption that the generated wavelength will be equal to the element spacing. However, because of the substantially narrower primary peak that occurs in the annular array spectrum, as well as the complex relationship between the point of maximum excitation and the array parameters such as the number of elements and the inner radius, this assumption can lead to poor excitation.

Figure 19.12 shows a correction factor $\alpha = s/\lambda_{peak}$ for the spacing-wavelength ratio that can be used in the design of an annular array with any particular number of

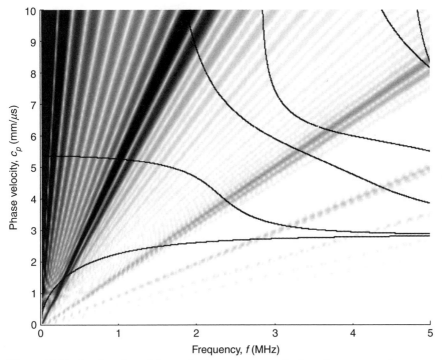

Figure 19.11. A color plot of the excitation spectrum of a five-element annular array with pitch $s = 5\,\mathrm{mm}$, element width $w = 2.5\,\mathrm{mm}$, and inner radius $r_0 = 2.5\,\mathrm{mm}$, in which the darker shades represent greater excitation. The dispersion curve for a 1 mm aluminum plate is superimposed in black as a reference.

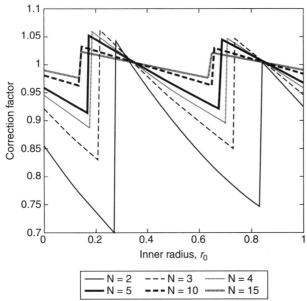

Figure 19.12. A correction factor $\alpha = s/\lambda_{peak}$ for the design of annular arrays with a variety of N and r_0 values.

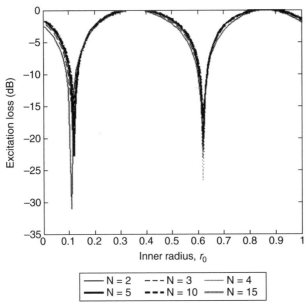

Figure 19.13. The excitation loss, that is, the decrease in the spectrum between its peak amplitude and the amplitude at $\lambda = s$, for annular arrays with a variety of N and r_0 values.

elements and inner radius. Any value on this chart can be directly calculated for a particular array based on the excitation spectrum, but a range of values is shown here for comparison. The difference (in dB) of the excitation spectrum at $\lambda = s$ and at the actual peak in the spectrum, termed the *excitation loss*, is proportional to the loss in generated wave amplitude that would occur if an annular array is designed without the correction factor (for a narrowband excitation). The excitation loss is shown in Figure 19.13 as a function of r_0 and N.

19.3 Phased Transducer Arrays for Mode Selection

As mentioned at the beginning of the chapter, one of the extraordinary benefits of array transducers is their dynamic mode selection capability. By appropriately delaying the electronic signals sent to individual elements in an array transducer, the excitation spectrum can be altered without the need for replacing or physically manipulating the transducer in any way; this technique is known as *phasing*. In this section, we further investigate the effects of phasing on mode control for comb and annular array transducers.

19.3.1 Phased Array Analytical Development

To derive the excitation spectrum of a phased array transducer, the complex exponential phasing term $e^{-i\varphi_n}$ must be included in the calculation for each transducer element n with phase delay φ_n. Mathematically, this term accounts for the delay in phase of the waves generated by each element; physically, the effect of phasing is to essentially delay or accelerate the arrival of wave fronts from an element at a particular point beyond the array. This either "stretches" or "compresses" the successive wave crests, leading to a guided wave with either a longer or shorter wavelength, respectively. One may intuitively consider this as analogous to the Doppler effect in terms of spatial frequency.

In reality, any phase delay can be applied across the various elements in an array, and in some cases nonlinear phase delays may provide optimal mode selection (and conversely, mode suppression). However, nonlinear phasing requires more complex optimization and is beyond the scope of this chapter. Therefore *linear phase delays* will be considered; the excitation spectra for comb and annular arrays with linear phase delays are given in Equations (19.18) and (19.19), respectively.

$$F^{comb}(k) = w \cdot \sin c\left(\frac{kw}{2}\right) \cdot \frac{\sin\left[\frac{1}{2}N(ks - \varphi)\right]}{\sin\left[\frac{1}{2}(ks - \varphi)\right]} e^{i\frac{ks-\varphi}{2}(N-1)}$$

(19.18)

$$F^{ann}(k) = \frac{1}{k}\sum_{n=0}^{N-1}\left\{(\alpha_n + w)J_1\left[(\alpha_n + w)k\right] - \alpha_n J_1\left[\alpha_n k\right]\right\}e^{-in\varphi}$$

(19.19)

Here $\alpha_n = r_0 + sn$ and φ is the phase delay in radians. Note that each element n will be subject to a phase delay $(n\text{-}1)\varphi$, based on the linear phase delay approach.

19.3.2 Phased Array Analysis

Examination of the excitation spectra calculated by these solutions yields interesting observations. Figure 19.14 shows the wavenumber-domain excitation spectrum for a five-element comb array with 90° phase delays applied. Note that the peaks in the spectrum have shifted in the $+k$-direction and spectrum symmetry about $k = 0$ has been broken. In the wavelength domain, this manifests itself as a decrease in primary wavelength for the forward-propagating $(+x)$ wave and an increase in primary wavelength for the backward-propagating $(-x)$ wave, as shown in Figure 19.15.

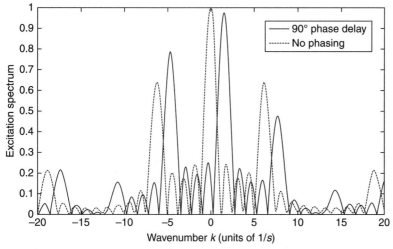

Figure 19.14. Excitation spectrum in the wavenumber domain for a five-element comb array with width-spacing ratio $w = s/2$ and 90° phase delays. The dashed line shows the original unphased spectrum.

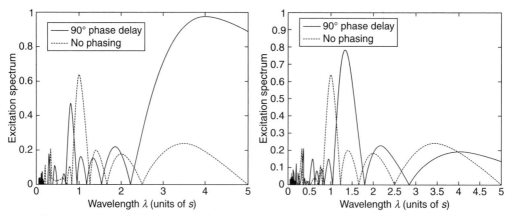

Figure 19.15. Excitation spectrum in the wavelength domain for the forward-propagating (left) and backward-propagating (right) waves from a five-element comb array with width-spacing ratio $w = s/2$ and 90° phase delays. The dashed lines show the original unphased spectrum.

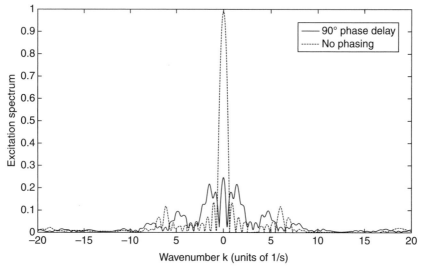

Figure 19.16. Excitation spectrum in the wavenumber domain for a five-element annular array with width-spacing ratio $w = s/2$, inner radius $r_0 = s/2$, and 90° phase delays. The dashed line shows the original unphased spectrum.

In the case of the phased annular array, on the other hand, the symmetry in the k-domain is retained because of the axisymmetric nature of the annular array, as seen in Figure 19.16. Although the outward- and inward-propagating waves from the annular array, which are analogous to the forward- and backward-propagating waves from the comb array, are uniquely affected by the phasing, the resulting excitation spectrum for the array is a combination of these two individual spectra. This is, of course, because the inward-propagating wave passes through the central point of the array and subsequently propagates outward along with the outward-propagating wave. Note that in Figure 19.17, we can see that this interference phenomenon distorts the excitation spectrum when phasing is applied, which is not the case for the comb array.

For annular arrays with a large inner radius, a physical distinct separation in these waves may occur, particularly if the group velocity of the two waves is substantially

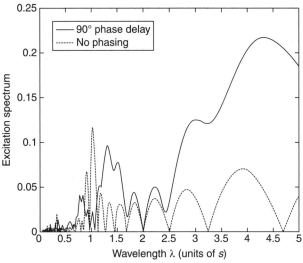

Figure 19.17. Excitation spectrum in the wavelength domain for a five-element annular array with width-spacing ratio $w = s/2$, inner radius $r_0 = s/2$, and 90° phase delays. The dashed line shows the original unphased spectrum.

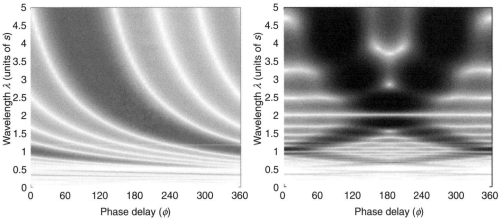

Figure 19.18. Excitation spectrum in the wavelength versus phase delay space for the forward wave spectrum of a five-element comb array (left) with width-spacing ratio $w = s/2$, as well as the spectrum of a five-element annular array (right) with width-spacing ratio $w = s/2$ and inner radius $r_0 = s/2$. Note that the backward wave spectra for the comb array would be identical to the forward spectra (left) reversed about $\varphi = 180°$. The interference of two similar spectra can be seen in the annular array spectrum, with additional effects due to the axisymmetric geometry and the interference of the inward and outward waves based on the inner radius.

different. In such cases, it may be beneficial to study the phased excitation spectra of the outward- and inward-propagating waves individually. Analysis of these spectra can be exploited with a combination of linear or nonlinear phase delays to suppress one of the spectra and thus enhance the guided wave mode control of the annular array; however, this is not discussed here and is left to the reader to explore.

The effect of linear phase delays on the comb and annular array spectra can be most easily interpreted with a multidimensional plot of excitation spectrum amplitude versus wavelength λ and phase delay φ, as shown in Figure 19.18. Note that the backward wave spectra for the comb array would be identical to the forward spectra (left side of Figure 19.18) reversed about $\varphi = 180°$.

In other words, the backward comb spectrum is described by switching the sign of the phase delay, which makes physical sense. The order in which the elements are phased is essentially switched, that is, from element 1 to element N versus element N to element 1 (see Figure 19.1). Therefore it is critical to carefully determine what element numbering system is used for a given array to avoid phase delay sign confusion. Also note that the interference of two spectra similar to those of the forward and backward comb waves can be seen in the annular array spectrum, with additional effects due to the axisymmetric geometry and the interference of the inward and outward waves based on the inner radius.

Although the same cannot be said for the phased annular array, the simplicity of the phased comb array spectrum allows for an analytical solution to the wavelength control problem for both the forward and backward waves using linear phase delays. Recall the solution for the linearly phased comb array spectrum in Equation (19.18); because the backward wave is described by reversing the sign of the phase delays in the forward wave equation, we will rewrite Equation (19.18) as:

$$\left|F_{\pm}^{comb}(k)\right| = w \cdot \sin c\left(\frac{kw}{2}\right) \cdot \frac{\sin\left[\frac{1}{2}N(ks \mp \varphi)\right]}{\sin\left[\frac{1}{2}(ks \mp \varphi)\right]}, \tag{19.20}$$

in which F_+ represents the spectrum for the forward-propagating comb waves and F_- represents the spectrum of the backward-propagating comb waves. To maximize the magnitude of the excitation spectrum in Equation (19.20), we simply need to maximize the denominator by finding conditions in which $\frac{1}{2}(ks \mp \varphi) = m\pi$, in which m is an integer.

By considering time delays instead of phase delays, and after some manipulation, we arrive at the relationship for the optimum linear time delays required to excite a desired wavelength λ in the forward or backward waves from a comb transducer.

$$\tau_{\pm} = \pm\frac{1}{f}\left(\frac{s}{\lambda} - m\right) \tag{19.21}$$

Here $\tau\pm$ represents the optimum time delays for the forward (+) and backward (−) waves, f represents the frequency of excitation, λ is the wavelength of the desired guided wave, and m is an integer. Note that this integer m corresponds to the various peaks in the spectrum (see Figure 19.15). The maximum excitation point on the dispersion curve for any solution corresponding to a particular value of m and a particular time delay can be plotted in the c_p–f dispersion curve space as in Figure 19.19. Here the dashed lines represent the forward-propagating wave, the dotted lines represent the backward-traveling wave, and the solid lines are the dispersion curves for a 2 mm-thick aluminum plate, included for reference.

Figures 19.20 and 19.21 show finite element results for eight-element phased annular array guided wave mode control simulations on a 1 mm aluminum plate. The annular array had element spacing $s = 5$ mm, element width $w = 2.5$ mm, and inner

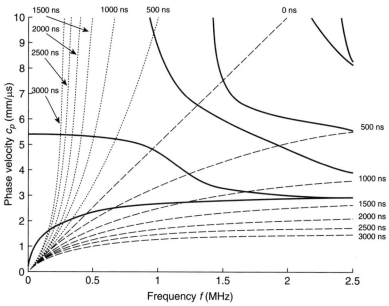

Figure 19.19. Lines of maximum excitation for the forward (dashed) and backward (dotted) waves from a comb transducer with spacing $s = 5$ mm for the $m = 1$ peak for various linear time delays from 0 – 3000 nanoseconds. The dispersion curves for a 2 mm-thick aluminum plate are included (solid) for reference.

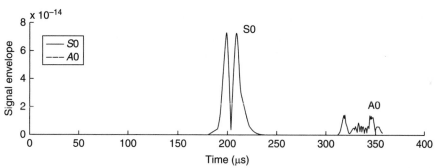

Figure 19.20. Finite element simulation results showing S0 mode enhancement and A0 mode suppression at 1.14 MHz in a 1 mm aluminum plate using an eight-element annular array transducer with $s = 5$ mm, $w = 2.5$ mm, $r_0 = 2.5$ mm, and no phase delays.

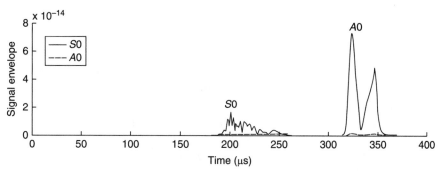

Figure 19.21. Finite element simulation results showing A0 mode enhancement and S0 mode suppression at 1.14 MHz in a 1 mm aluminum plate using an eight-element annular array transducer with $s = 5$ mm, $w = 2.5$ mm, $r_0 = 2.5$ mm, and 329-ns linear phase delays.

radius $r_0 = 2.5$ mm. In Figure 19.20, the transducer array is driven at 1.14 MHz with no phase delays applied; the S0 mode is enhanced and the A0 mode is suppressed. In Figure 19.21, the array is driven at the same frequency, but 329-ns (135°) linear phase delays are applied; the A0 mode is enhanced and the S0 mode is suppressed. In addition to the effectiveness of the mode control, note the clear separation of the inward- and outward-propagating wave packets for each mode.

19.4 Concluding Remarks

Comb and annular array transducers are excellent tools in guided wave analysis, particularly for mode control applications. Our ability to dynamically alter the guided wave excitation spectrum of these arrays with electronic phasing greatly increases their usefulness in many applications. However, to fully exploit these benefits, we must first carefully analyze the design of these array transducers.

In this chapter we developed analytical expressions for the excitation spectra of phased comb and annular array transducers by applying spatial Fourier transforms to the transducer array loading distributions. We briefly discussed the design parameters for such arrays and how those parameters may influence the excitation spectrum of the array. Finally, we discussed the effects of applying linear phase delays to both comb and annular array transducers and how such phase delays can be utilized to enhance or suppress selected guided wave modes.

The mathematical methods utilized here can be directly applied to any transducer array, which could include single-element or multi-element transducers or an array of single-element and/or multi-element transducers. Although a number of the cases here were simplified because of one or more levels of symmetry, the excitation spectrum for a nonsymmetric array is developed in the same manner and will yield results as a function of wave propagation direction, just as was done for the two-dimensional comb transducer and the phased comb array. Further complexities could also be taken into consideration, such as nonuniform loading across the area of the array (apodization), nonlinear phase delays, and nonuniform loading across each individual element.

19.5 Exercises

1. What are the advantages and disadvantages of an annular compared to a linear comb transducer?
2. Name at least two guided wave applications in which an annular array would be desirable over a linear comb array. Explain your reasoning for each example.
3. Explain the differences and similarities between the Fourier and Hankel transforms.
4. Is a given transducer array more likely to generate the A0 or S0 Lamb wave mode at low frequencies (well below the first cutoff) if its elements operate in the d_{33} (thickness) mode?
5. Design an annular array and a linear comb array to excite a guided wave mode with a phase velocity of 5500 m/s at 320 kHz with no phasing.
6. How can you design a linear or annular array actuator with a narrow wavelength bandwidth and small side lobes? What parameter is most critical?
7. Explain why the excitation spectrum of a phased annular array appears to be more complex than that of a phased linear array.

8. For a linear array, would applying positive time delays (i.e., the forward-most elements are excited later than the rearward-most elements) increase or decrease the primary wavelength generated by the array? Why?

9. Explain the significance of the Hankel function solutions of the first kind, $H_v^{(1)}$, and second kind, $H_v^{(2)}$, with regard to the annular array transducer solutions (or any circular-crested wave solution).

10. Describe the lines of constant excitation spectrum amplitude on a c_p vs. f dispersion curve for an array transducer (disregarding mode excitability).

11. For a linear comb array with element width 3 mm and element spacing 6 mm, what linear time delays would you apply to excite a guided wave mode with phase velocity $c_p = 3200$ m/s at 250 kHz?

12. Calculate and plot the excitation spectrum (in the wavenumber or wavelength domain) of a 5-element annular array with element width 2 mm, element spacing 4 mm, and inner radius 2 mm.

13. Calculate and plot the excitation spectrum (in the wavenumber or wavelength domain) of a 4-element linear comb array with element width 2 mm and element spacing 4 mm.

14. If the signals applied to each annular array transducer element can be independently controlled, an improved annular array design, based on the zero-order Bessel function of the first kind, $J_0(x)$, can be employed. Just as the linear comb array design is based on a sine wave, which describes plane wave propagation in one direction, the Bessel function annular array design is based on the Bessel function, which describes the propagation of waves propagating radially from a central point. However, for such an annular array to function properly, the amplitude of the signals applied to each element must be based on the amplitude of the respective peaks in the Bessel function. Derive the expression of the excitation spectrum for a Bessel function annular array and show that it is equal to

$$F^{ann}(k) = \frac{1}{k} \sum_{n=1}^{2N-1} (-1)^{n-1} A_m \frac{\alpha_n}{k_0} \cdot J_1 \left[\frac{\alpha_n}{k_0} k \right]$$

in which α_n is the nth root of the Bessel function, A_m is the amplitude of the mth peak of the Bessel function that will dictate the amplitude of the signal applied to the mth element of the Bessel function array, and k_0 is the primary wavenumber excited by the array. Show that this type of array provides very similar excitation in the wavenumber domain as a comparable comb array. Also show that without the application of the appropriate amplitude to each element, the excitation spectrum is less ideal. This concept for a phased Bessel annular array will be published soon, as the research is on-going.

15. What design parameter(s) of a finite linear comb array determine(s) the directionality of the guided waves generated by the array?

16. Using the same approach as that applied to the 2D comb array, develop an analytical expression in terms of k_x and k_y for the excitation spectrum of a regular 4×4 square grid of 5-mm square array elements separated from each other with 10-mm spacing in both directions.

19.6 REFERENCES

Rose, J.L. (1999). *Ultrasonic Waves in Solid Media*. New York: Cambridge University Press.

20 Introduction to Guided Wave Nonlinear Methods

20.1 Introduction

Up to this point we have described linear ultrasonics, that is, where the received signal is at the same frequency as the excitation. Now we consider nonlinear ultrasonics, where the received signal is not at the frequency of the excitation. The material is treated as weakly nonlinear elastic because the amplitude of the signal received at higher harmonics is very small relative to the excitation, which permits the use of a perturbation solution. The generation of higher harmonics in bulk solids has been studied for more than four decades, but the initial studies of higher harmonics in plates are much more recent. These studies are relevant because the amplitudes of higher harmonics have been shown to be sensitive to features of the microstructure of the material, whereas the primary harmonics are generally much less sensitive, or insensitive, to microstructural features such as dislocation density, precipitates, and cavities. This chapter introduces nonlinear methods for guided waves.

To maintain the best possible structural integrity of a component, it is highly desirable to detect damage at the smallest possible scale. Doing so with periodic nondestructive inspection or continuous structural health monitoring (SHM) enables tracking damage evolution over the service life of the structure, which can be used in conjunction with prognostics for condition-based maintenance and improved logistics. Nonlinear systems are known to be very good at indicating damage progression (e.g., Dace, Thompson, and Brashe 1991; Farrar et al. 2007; Worden et al. 2007). Generally speaking, linear ultrasonics with bulk waves can detect anomalies on the order of a wavelength. Ultrasonic guided waves can do significantly better in terms of wavelength, say $\lambda/40$ (e.g., Alleyne and Cawley 1992), but longer wavelengths are typically used to enable large penetration lengths. Nonlinear ultrasonics, where the received signal containing the information of interest is at a different frequency than the emitted signal, can provide sensitivity to microstructural changes. This anharmonic response is due to distortion of the waveform, which is attributed to changes in microstructural features like dislocation density, precipitates, and voids (i.e., features at the micron scale and below). It is impressive that Landau and Lifshitz (1986) have a qualitative description of anharmonic vibration in the 1954 edition of their elasticity book.

With contribution from Cliff Lissenden, Yang Liu, and Vamshi Chillara.

We will see that if the governing differential equations are not linearized, then harmonics can be generated at frequencies other than the excitation frequency (most notably at twice the excitation frequency), which is not possible for the linear equations. Harmonic generation is associated with nonlinear elasticity and finite amplitude waves having a sufficiently large displacement gradient. The generation of measurable harmonics is extremely useful because these harmonics are sensitive to the very microstructural features that cause nonlinear elasticity, which means that they can be related, not merely to macroscale defects, but potentially to precursors of them, such as dislocation density, nucleated voids, and precipitate coarsening. This chapter describes the generation of higher harmonics in isotropic solid media that are unbounded, but focuses on the more challenging, and potentially more rewarding, application of a waveguide. Specifically, higher harmonic generation in a plate, which has multiple dispersive modes, is discussed in a systematic manner. One of the features of higher harmonics that makes them intriguing is that the amplitude is cumulative with propagation distance. Combining a cumulative harmonic with the penetration power of guided waves has great potential for nondestructive evaluation (NDE) and eventually SHM.

20.2 Bulk Waves in Weakly Nonlinear Elastic Media

Consider first one-dimensional wave propagation in an unbounded isotropic elastic medium. As shown in Chapter 3, if the material is linearly elastic, then the field equations are simply:

$$\sigma_{,x} = \rho u_{,tt} \tag{20.1}$$

$$\sigma = E\varepsilon \tag{20.2}$$

$$\varepsilon = u_{,x} \tag{20.3}$$

where the variables σ, ε, and u represent stress, strain, and displacement, respectively, while ρ and E are the mass density and Young's modulus, respectively. The subscripts x and t following a comma denote partial differentiation with respect to the spatial and temporal variables, respectively.

Combination of the field equations leads to Navier's equation of motion in terms of displacements:

$$u_{,tt} = c^2 u_{,xx} \tag{20.4}$$

which admits harmonic solutions,

$$u = A_1 \sin(kx - \omega t) + B_1 \cos(kx - \omega t). \tag{20.5}$$

Here $c^2 = E/\rho$ is the longitudinal wave speed, $\omega = ck$ is the angular frequency, and k is the wavenumber. In this linear case, the only frequency generated is the one that is excited. However, in the case of nonlinear elasticity, higher harmonics are generated. Take, for instance, the stress-strain relation:

$$\sigma = E\varepsilon\left(1 + \frac{1}{2}\beta\varepsilon\right) \tag{20.6}$$

where β represents the nonlinearity parameter. The one-dimensional wave equation becomes nonlinear because of its association with the material nonlinearity,

$$u,_{tt} = c^2 \left(1 + \beta u,_x\right) u,_{xx} \qquad (20.7)$$

Note that geometric nonlinearities, or finite deformation gradients, are not included here for simplicity (but are included later). The fundamental harmonic solution of

$$u_1 = A_1 sin\left(kx - \omega t\right) \qquad (20.8)$$

is supplemented by a perturbation solution that leads to generation of higher order harmonics like

$$u_2 = \frac{\beta}{8}\left(A_1 k\right)^2 x cos2\left(kx - \omega t\right) \qquad (20.9)$$

and $u = u_1 + u_2$ as given by Truell, Elbaum, and Chick (1969). Clearly, the amplitude of this second harmonic, which is,

$$A_2 = \frac{\beta}{8}\left(A_1 k\right)^2 x \qquad (20.10)$$

is cumulative with propagation distance.

20.3 Measurement of the Second Harmonic

The nonlinearity parameter β is directly related to the amplitude ratio A_2/A_1^2 through Equation (20.10). Measurement of either the relative or absolute value of β requires proper instrumentation and due care to generate the purest possible waveform with minimal influences outside of the solid media itself. A relatively long sine wave pulse is preferred to concentrate the energy at a single fundamental frequency. A gated amplifier enables propagation of a finite amplitude stress wave. While on an absolute scale the amplitude is still quite small, the frequency and the particle velocity are high, resulting in a large displacement gradient. The microstructural features distort the wave as it propagates through and generate higher harmonics. While the distortion is not typically visible in the waveform, it is apparent in the frequency domain.

For example, the frequency spectrum for a borosilicate sample is shown in Figure 20.1. In Figure 20.1(a), the amplitude axis is truncated to emphasize the higher harmonics, while in Figure 20.1(b), the amplitudes are normalized with respect to the fundamental frequency and plotted on a dB scale. Notice that the odd harmonics are larger than the even harmonics. In this experiment the sample is a 38 mm-diameter cylinder 50 mm long and a forty-cycle toneburst excitation centered at 7 MHz is applied. Borosilicate is a glass with a very low nonlinearity parameter β. Another example of the frequency spectrum, for the nickel-based Alloy 617, is shown in Figure 20.2 for a 7 MHz excitation.

As indicated by Equation (20.10), the amplitude of the second harmonic increases linearly with the square of the fundamental wave amplitude, which is shown in Figure 20.3 for borosilicate. Equation (20.10) also indicates that the second

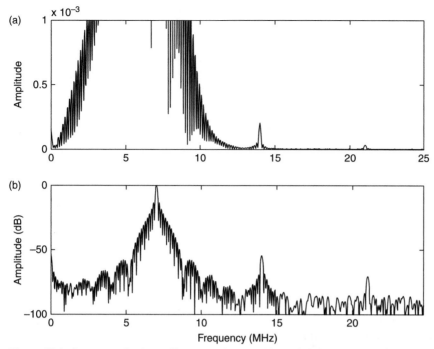

Figure 20.1. Spectrum for borosilicate specimen subjected to forty-cycle toneburst excitation centered at 7 MHz; (a)three harmonics are visible, (b) normalized spectrum.

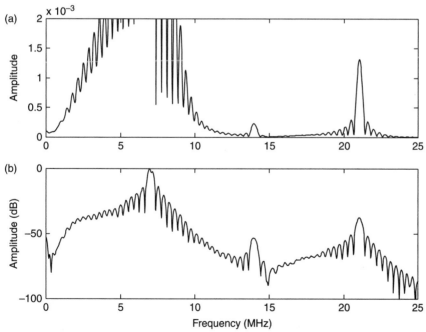

Figure 20.2. Spectrum for Alloy 617 specimen subjected to twenty-cycle toneburst excitation centered at 7 MHz; (a)three harmonics are visible, (b) normalized spectrum.

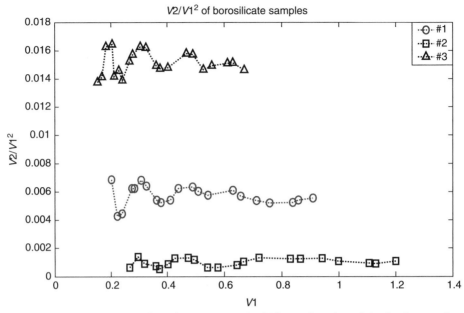

Figure 20.3. Qualitative nonlinearity parameter, V_2/V_1^2, as a function of the fundamental amplitude in terms of voltages for three different borosilicate samples. In each case, the value is small and its variation decreases as the fundamental amplitude increases.

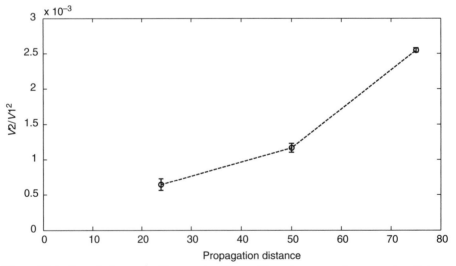

Figure 20.4. Cumulative second harmonic of borosilicate in terms of propagation distance.

harmonic is cumulative in that it increases linearly with propagation distance as shown in Figure 20.4 for borosilicate. Of course, the amplitude of the second harmonic is limited in that A_2 must remain significantly smaller than A_1 for the perturbation solution methodology to remain valid.

One of the challenges involved with measurement of harmonics is that conventional ultrasonic transducers apply and receive voltages based on the displacement or stress in the piezoelectric element. However, the couplant used to transfer energy between the transducer and sample is variable and its response is frequency dependent. Dace and colleagues (1991); Dace, Thompson, and Buck (1992); and Sun and colleagues (2006) provide experimental methods for calibration of couplant effects.

20.4 Second Harmonic Generation Related to Microstructure

Nonlinear ultrasonics techniques for NDE typically involve measuring the amplitude of a signal associated with a different frequency than the input signal. A selection of these techniques for assessing microscale damage is reviewed by Jhang (2009). According to Cantrell:

> In finite amplitude ultrasonics one generally measures the amplitude of harmonics generated by the nonlinear interactions of an initially pure sinusoidal sound wave with the material. The nonlinear interactions generally result from lattice anharmonicity (stemming from a nonquadratic interatomic potential) or from nonlinearities involving defects, microstructural features, or other disruptions in the lattice structure of the material. (2004: 364–5)

It's important to keep in mind that Hooke's law, which represents a linear relationship between stress and strain, is a first-order approximation to the attractive and repulsive forces between atoms. The constant of proportionality is Young's modulus, which represents the slope of the tangent of the net interatomic force relative to the atomic spacing. This is a gross simplification of the response of materials that contain millions of atoms with impurities and imperfections. Thus, it is always appropriate, but not always necessary, to use an interatomic potential with an order higher than quadratic.

As Equation (20.10) indicates, the amplitude of the second harmonic is proportional to the nonlinearity parameter β, which can in turn be related to microstructural evolution associated with thermal aging, creep, and fatigue, for example. Hikata, Chick, and Elbaum (1965) developed the first such relationship based on dislocation monopoles:

$$\beta_{mp} = \frac{24}{5} \frac{\Omega \Lambda L^4 R^3 E_1^2}{\mu^3 b^2} |\sigma|$$

where σ is stress, b is the Burgers vector, Λ is the dislocation density, Ω is the conversion factor from shear strain to longitudinal strain, R is the Schmid factor, L is the dislocation loop length, and E_1 and μ are the elastic and shear moduli. See Cantrell (2004, 2009) and Cantrell and Yost (2001) for further discussion on relating the nonlinearity parameter to dislocation mechanisms associated with initiation of fatigue cracks. There are many reports of experimental studies of fatigue initiation using the generation of higher harmonics (e.g., Na, Cantrell, and Yost 1996 and Kim et al. 2006). Figure 20.5 shows that simple thermal aging of Alloy 617 affects the generation of the second harmonic. The increase and then decrease in the amplitude of the second harmonic follows the microstructural evolution of precipitates observed by Chomette, Gentzbittel, and Viguier (2010).

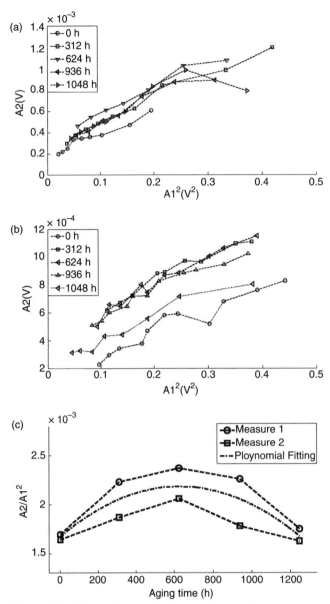

Figure 20.5. Effect of thermal aging on the amplitude of the second harmonic for Alloy 617; amplitude of second harmonic as a function of the square of the fundamental amplitude (a) test 1 and (b) test 2, (c) the nonlinearity parameter is proportional to A_2/A_1^2 and evolves with the microstructure.

20.5 Weakly Nonlinear Wave Equation

We now turn our attention to nonlinear methods for guided waves and present the problem in its full three-dimensional form using stress and strain variables defined in continuum mechanics. The goal is to set up and solve the boundary value problem of wave propagation in a plate. Key features of the solution are described and demonstrated experimentally in the subsequent Subsection 20.6.

Navier's equation of motion (Equation (20.4)) is applicable for infinitesimal displacements in linear elastic materials. In this section, the equation of motion is revisited, but without linearizing the strain-displacement relation and by employing a higher order strain energy function, which provides a nonlinear elastic material response. The Lagrangian formulation from continuum mechanics (e.g., Malvern 1969) is exploited using the Green-Lagrange strain tensor along with the first and second Piola-Kirchhoff stress tensors. The direct notation, where the order of terms is important, is used.

The Green-Lagrange strain tensor, E, is related to the displacement vector, u, by

$$E = \frac{1}{2}\left(H + H^T + H^T H\right) \tag{20.11}$$

where, for simplicity, the gradient of the displacement vector is denoted by

$$H = \nabla u. \tag{20.12}$$

The stress-strain relation for a hyperelastic material is obtained from the strain energy function, W, and taking care to use the proper definitions of stress and strain we write:

$$\tilde{T} = \frac{\partial W(E)}{\partial E}. \tag{20.13}$$

The reference configuration is natural for solid materials, so the second Piola-Kirchhoff stress tensor, $\tilde{T}$, is employed along with its conjugate strain E. The third-order strain energy function proposed by Landau and Lifshitz (1986) for a nonlinear hyperelastic solid is:

$$W = \frac{\lambda}{2}\left(tr[E]\right)^2 + \mu tr[E^2] + \frac{C}{3}\left(tr[E]\right)^3 + B tr[E]tr[E^2] + \frac{A}{3}tr[E^3] + \dots \tag{20.14}$$

where λ and μ are Lamé constants, A, B, and C represent the third-order elastic constants, and $tr[\ ]$ denotes the trace of the bracketed tensor. Norris (1998) and many recent authors use this strain energy function. Norris (1998) and Cantrell (2004) provide relations between these third-order elastic constants and those proposed by others. In this case, the stress-strain relation is:

$$\tilde{T} = \lambda tr[E]I + 2\mu E + C\left(tr[E]\right)^2 I + B tr[E^2]I + 2 B tr[E]E + A E^2 \tag{20.15}$$

where I is the second-order identity tensor.

The equation of motion is more naturally written in terms of the first Piola-Kirchhoff stress, T_o, than the second Piola-Kirchhoff stress (Malvern 1969). The first Piola-Kirchhoff stress gives the actual force on the deformed body, but it is relative to the reference configuration. It is a nonsymmetric tensor that is related to the second Piola-Kirchhoff stress through the deformation gradient,

$$F = I + H \tag{20.16}$$

$$T_o = F\tilde{T} \tag{20.17}$$

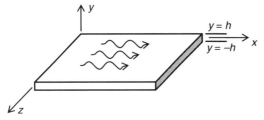

Figure 20.6. Schematic of plate with wave propagating in the *x*-direction.

The first Piola-Kirchhoff stress can be decomposed into linear and nonlinear components in terms of the displacement gradient:

$$\boldsymbol{T}_o = \boldsymbol{T}_o^L + \boldsymbol{T}_o^{NL} \tag{20.18}$$

$$\boldsymbol{T}_o^L = \lambda tr[\boldsymbol{H}]\boldsymbol{I} + \mu(\boldsymbol{H} + \boldsymbol{H}^T) \tag{20.19}$$

$$\boldsymbol{T}_o^{NL} = \left(\frac{\lambda}{2}tr[\boldsymbol{H}^T\boldsymbol{H}] + C(tr[\boldsymbol{H}])^2\right)\boldsymbol{I} + Btr[\boldsymbol{H}]\boldsymbol{H}^T + \frac{A}{4}\boldsymbol{H}^T\boldsymbol{H} + \frac{B}{2}tr[\boldsymbol{H}^2 + \boldsymbol{H}^T\boldsymbol{H}]\boldsymbol{I}$$

$$+(\lambda + B)tr[\boldsymbol{H}]\boldsymbol{H} + \left(\mu + \frac{A}{4}\right)(\boldsymbol{H}^2 + \boldsymbol{H}^T\boldsymbol{H} + \boldsymbol{H}\boldsymbol{H}^T) + O[\boldsymbol{H}^3] \tag{20.20}$$

where terms above second order have been neglected. Now the equation of motion in the reference configuration can be written as:

$$\nabla \cdot \boldsymbol{T}_o + \rho_o\boldsymbol{b} = \rho_o\ddot{\boldsymbol{u}} \tag{20.21}$$

where ρ_o is the mass density in the reference configuration, $\boldsymbol{b}$ is the body force, and the dots over the displacement variable represent time derivatives. Gol'dberg (1960) was the first to derive this nonlinear wave equation. Equation (20.21) can be rearranged in preparation for solution using a perturbation method:

$$(\lambda + \mu)\nabla(\nabla \cdot \boldsymbol{u}) + \mu\nabla^2\boldsymbol{u} + \nabla \cdot \boldsymbol{T}_o^{NL} = \rho_o\ddot{\boldsymbol{u}} \tag{20.22}$$

in the absence of a body force, $\boldsymbol{b}$. Additionally, the traction-free boundary conditions on the top and bottom surfaces of the plate are:

$$\boldsymbol{T}_o \cdot \boldsymbol{n} = 0 \text{ for } y = \pm h \tag{20.23}$$

where $\boldsymbol{n}$ is the outward unit normal vector to the surface (see Figure 20.6).

Considering weakly nonlinear wave motion, the nonlinear wave equation is solved using a perturbation method that decomposes the wave field into primary and secondary components:

$$\boldsymbol{u} = \boldsymbol{u}^{(1)} + \boldsymbol{u}^{(2)} \tag{20.24}$$

where

$$\boldsymbol{u}^{(1)} \gg \boldsymbol{u}^{(2)}, \tag{20.25}$$

which results in two separate linear boundary value problems (BVPs). The first-order approximation is:

$$(\lambda + \mu)\nabla(\nabla \cdot \boldsymbol{u}^{(1)}) + \mu\nabla^2\boldsymbol{u}^{(1)} - \rho_o\ddot{\boldsymbol{u}}^{(1)} = 0 \tag{20.26}$$

$$\boldsymbol{T}_o^{L(1)} \cdot \boldsymbol{n}_y = 0 \tag{20.27}$$

which is homogeneous and results in the Rayleigh–Lamb (RL) and shear horizontal (SH) wave solutions obtained in Chapters 6 and 14, respectively. This is the primary wave field. The second-order approximation is:

$$(\lambda + \mu)\nabla(\nabla \cdot \boldsymbol{u}^{(2)}) + \mu\nabla^2\boldsymbol{u}^{(2)} - \rho_o\ddot{\boldsymbol{u}}^{(2)} = -\boldsymbol{f}^{(1,1)} \tag{20.28}$$

$$\boldsymbol{T}_o^{L(2)} \cdot \boldsymbol{n}_y = -\boldsymbol{T}_o^{NL(1,1)} \cdot \boldsymbol{n}_y \tag{20.29}$$

which can be viewed as a forced waveguide, as both the differential equation and boundary conditions are nonhomogeneous. The forcing terms $\boldsymbol{f}^{(1,1)} = \nabla \cdot \boldsymbol{T}_o^{NL(1,1)}$ and $\boldsymbol{T}_o^{NL(1,1)}$ are obtained by substituting the primary wave field, which interacts with itself, into Equation (20.20). The solution to the second BVP is the secondary wave field. Auld's (1973) reciprocity relation can be used to prove orthogonality of the primary plate modes, which enables use of the normal mode expansion (NME) to solve the forced waveguide problem.

To solve the secondary problem, consider two propagating wave modes, (ω_a, k_a) and (ω_b, k_b), as in de Lima and Hamilton (2003). Self- and mutual interactions generate waves that propagate at $2\omega_a, 2\omega_b, \omega_a - \omega_b$, and $\omega_a + \omega_b$ with their associated wavenumbers. The generation of second harmonics is represented by simply not exciting (ω_b, k_b). The modal expansion (Auld 1973) is:

$$\boldsymbol{u}^{(2)}(x,y,t) = \frac{1}{2}\sum_{m=1}^{\infty} A_m(x)\boldsymbol{u}_m(y)e^{-i2\omega t} + c.c. \tag{20.30}$$

where c.c. represents the complex conjugate. Further, the modal displacement profile (wave structure) enables us to write:

$$\boldsymbol{v}_m(y) = \dot{\boldsymbol{u}}_m(y) \tag{20.31}$$

$$\boldsymbol{T}_m = \lambda tr\left[\nabla\boldsymbol{u}_m(y)\right]\boldsymbol{I} + \mu\left(\nabla\boldsymbol{u}_m(y) + \nabla^{\mathrm{T}}\boldsymbol{u}_m(y)\right) \tag{20.32}$$

The modal amplitudes, $A_m(x)$, can be determined for each mode m from:

$$4P_{mn}\left(\frac{d}{dx} - ik_n^*\right)A_m(x) = \left(f_n^{surf} + f_n^{vol}\right)e^{i2kx}, \quad m = 1, 2, \ldots \tag{20.33}$$

as in Auld (1973), where:

$$P_{mn} = -\frac{1}{4}\int_{-h}^{h}\left(\frac{\boldsymbol{v}_n^*}{2} \cdot \frac{\boldsymbol{T}_m}{2} + \frac{\boldsymbol{v}_m}{2} \cdot \frac{\boldsymbol{T}_n^*}{2}\right) \cdot \boldsymbol{n}_x dy \tag{20.34}$$

$$f_n^{surf} = -\frac{1}{2}\left(\boldsymbol{v}_n^* \cdot \boldsymbol{T}_o^{NL(1,1)}\right) \cdot \boldsymbol{n}_y \big|_{-h}^{h} \tag{20.35}$$

$$f_n^{vol} = \frac{1}{2}\int_{-h}^{h} \boldsymbol{v}_n^* \cdot \boldsymbol{f}^{(1,1)} \, dy \tag{20.36}$$

$$A_m = \frac{f_n^{surf} + f_n^{vol}}{4P_{mn}} \begin{cases} \dfrac{i}{k_n^* - 2k}\left(e^{i2kx} - e^{ik_n^* x}\right), & for\, k_n^* \neq 2k \\[2mm] xe^{i2kx}, & for\, k_n^* = 2k \end{cases} \tag{20.37}$$

The orthogonality of propagating and evanescent modes is imposed by:

$$P_{mn} = 0, \qquad for\, k_m \neq k_n^*. \tag{20.38}$$

20.6 Higher Harmonic Generation in Plates

The traction-free boundary conditions on the surface of a plate must be satisfied after consideration of the multiple reflections that occur from the top and bottom surfaces. RL waves and SH waves in plates have multiple modes that can, generally speaking, propagate over long distances. To investigate the ability of one mode to generate a cumulative second harmonic, we first examine where a mode exists at twice the excitation frequency in Section 20.6.1. In so doing, we further require synchronism – or phase matching – between the fundamental and higher harmonic. In Section 20.6.2, the power flux is analyzed to assess whether a mode can be excited as a higher harmonic. Both of these conditions:

- synchronism (phase matching)
- nonzero power flux between fundamental and the higher harmonic modes

are required to achieve internal resonance, in which the higher harmonic is cumulative. The synchronism requirement comes from Equation (20.37), where cumulative modal participation factors are only obtained if $k_n^* = 2k$ for twice the fundamental frequency. Thus, the phase velocities must be equal:

$$fundamental,\ c_p = \frac{\omega}{k}$$

$$second\,harmonic,\ c_p = \frac{2\omega}{2k}$$

Likewise, the power flux from the fundamental mode to the second harmonic is zero if the modal participation factors $A_m(x)$ are zero, which is the case if $f_n^{surf} + f_n^{vol} = 0$. Thus, internal resonance also requires nonzero power flux.

The group velocity matching condition is discussed in Section 20.6.3. Finally, sample laboratory results for RL and SH fundamental waves are presented in Section 20.6.4.

20.6.1 Synchronism

Consider a plate of thickness $2h$ and wave propagation in the x-direction as shown in Figure 20.6. Both RL and SH waves can propagate, with the RL dispersion relations given by:

RL waves

symmetric modes

$$\frac{\tan qh}{\tan ph} = \frac{-4k^2 pq}{\left(q^2 - k^2\right)^2} \qquad (20.39)$$

antisymmetric modes

$$\frac{\tan qh}{\tan ph} = \frac{\left(q^2 - k^2\right)^2}{-4k^2 pq} \qquad (20.40)$$

where

$$p^2 = \left(\frac{\omega}{c_L}\right)^2 - k^2, \quad q^2 = \left(\frac{\omega}{c_T}\right)^2 - k^2, \quad k = \frac{\omega}{c}. \qquad (20.41)$$

Presume that, for a fundamental mode (ω, k) to generate the mth harmonic, $(m\omega, mk)$ must also be a mode; that is, it satisfies the dispersion relation. Moreover, the harmonic is synchronized to the phase velocity of the fundamental mode. Consider, for example, the second harmonic, $(2\omega, 2k)$. Substituting 2ω for ω and $2k$ for k into Equation (20.41) for p and q gives:

$$p \xrightarrow{\text{yields}} 2p$$

$$q \xrightarrow{\text{yields}} 2q$$

and substituting these values into the right-hand side of Equation (20.39) for symmetric RL modes gives:

$$\frac{-4k^2 pq}{\left(q^2 - k^2\right)^2} = \frac{-4\left(2k\right)^2 \left(2p\right)\left(2q\right)}{\left(\left(2q\right)^2 - \left(2k\right)^2\right)^2} \qquad (20.42)$$

indicating that the RL dispersion relations are the same for (ω, k) and $(2\omega, 2k)$. The phase velocities are also the same. By analyzing the SH and RL dispersion relations, we can determine all possible situations where a second harmonic mode is synchronized with the fundamental mode. Table 20.1 provides the results. As an example, the case of a fundamental RL symmetric mode generating a second harmonic RL symmetric mode is detailed later. Chillara (2012) analyzed all of the cases.

The dispersion relations for symmetric RL modes are:

$$\frac{\tan qh}{\tan ph} = \frac{\tan 2qh}{\tan 2ph} = \frac{-4k^2 pq}{\left(q^2 - k^2\right)^2} \qquad (20.43)$$

which can be rewritten as:

$$\sin qh \cos ph \cos 2qh \sin 2ph = \cos qh \sin ph \sin 2qh \cos 2ph. \qquad (20.44)$$

Table 20.1. Phase matching

Fundamental Mode	Second Harmonic Mode	Frequency, ω	Phase Velocity	Comments
RL symmetric	RL symmetric	$\dfrac{n\pi c_L c_T}{h\sqrt{c_L^2 - c_T^2}}$	c_L	--
		--	c_T	--
		$\dfrac{\sqrt{2}n\pi c_L c_T}{h\sqrt{2c_T^2 - c_L^2}}$	$\sqrt{2}c_T$	Provided $\sqrt{2}c_T \geq c_L$
		--	--	Sym/Antisym intersections
	RL antisymmetric	$\dfrac{\sqrt{2}\,(2n+1)\pi c_T}{2h}$	$\sqrt{2}c_T$	--
		$\dfrac{(2n+1)\pi c_L c_T}{2h\sqrt{c_L^2 - c_T^2}}$	c_T	--
RL antisymmetric	RL antisymmetric	--	--	--
	RL symmetric	$\dfrac{\sqrt{2}\,(2n+1)\pi c_L c_T}{2h\sqrt{2c_T^2 - c_L^2}}$	$\sqrt{2}c_T$	Provided $\sqrt{2}c_T \geq c_L$
SH	RL symmetric	--	$> c_L$	Sym/Antisym intersections
		$\dfrac{n\pi c_L c_T}{2h\sqrt{c_L^2 - c_T^2}}$	c_L	--
		--	c_T	--
	RL antisymmetric	$\dfrac{n\pi\sqrt{2}c_T}{2h}$	$\sqrt{2}c_T$	--

By using the trig identities

$$\sin 2u = 2\sin u \cos u$$

$$2\cos^2 \frac{u}{2} = 1 + \cos u$$

Equation (20.44) can be rewritten as:

$$\sin qh \sin ph (\cos 2ph - \cos 2qh) = 0 \tag{20.45}$$

which has three possible solutions:

$$(i)\, \sin qh = 0 \overset{yields}{\rightarrow} qh = n\pi \tag{20.46}$$

$$(ii)\, \sin ph = 0 \overset{yields}{\rightarrow} ph = n\pi \tag{20.47}$$

$$(iii)\cos 2ph = \cos 2qh \overset{yields}{\rightarrow} qh \pm ph = n\pi \qquad (20.48)$$

where n is an integer. Considering case *(i)* where $qh = n\pi$ along with $ph \neq n\pi$, and from Equation (20.43) using that

$$\tan qh = 0 = -4k^2 pq \qquad (20.49)$$

there are three possibilities:

$$(i.a)\, k = 0 \overset{yields}{\rightarrow} \text{mode cutoffs} \qquad (20.50)$$

$$(i.b)\, p = 0 \overset{yields}{\rightarrow} \left(\frac{\omega}{c_L}\right)^2 - \left(\frac{\omega}{c}\right)^2 = 0 \overset{yields}{\rightarrow} c = c_L \qquad (20.51)$$

this phase velocity can be substituted into qh to determine the frequency:

$$qh = \sqrt{\left(\frac{\omega}{c_T}\right)^2 - \left(\frac{\omega}{c_L}\right)^2}\, h = n\pi \overset{yields}{\rightarrow} \omega = \frac{n\pi c_T c_L}{h\sqrt{c_L^2 - c_T^2}} \qquad (20.52)$$

$$(i.c)\, q = 0 \qquad (20.53)$$

can be shown by analysis of the RL dispersion relation (Equation (20.39)) in the limit as q→0 to not give a solution for case *(i)*. Solutions for cases *(ii)* and *(iii)* can also be obtained.

Table 20.1 indicates that second harmonics can only be generated:

- at mode crossings between RL symmetric modes and RL antisymmetric modes;
- when the phase velocity is equal to the longitudinal wave speed;
- when the phase velocity is equal to the transverse wave speed; and
- when the phase velocity is equal to $\sqrt{2}$ times the transverse wave speed.

While harmonics can also be generated at the cutoff frequencies, this is not useful for most applications.

20.6.2 Power Flux

The assessment of power flux from a fundamental mode to a mode activated by a higher harmonic relies on whether the symmetric or antisymmetric form of the fundamental mode is compatible with the functional form of the mode excited by a higher harmonic. Thus, a parity analysis is performed on the possible fundamental modes. Represent a generic symmetric function in terms of y by S and an antisymmetric function in terms of y by A. The displacement and displacement gradient fields are given in Table 20.2 for RL and SH modes (symmetric and antisymmetric modes are denoted with _S and _A, respectively). Evaluation of the f_n^{surf} (Equation (20.35)) and f_n^{vol} (Equation (20.36)) terms through $T_o^{NL(1,1)}$ (Equation (20.20)) indicates that fundamental RL and SH waves, whether they are symmetric or antisymmetric, can only generate cumulative second harmonics that are symmetric RL wave modes. This result is obtained because multiplying matrices or matrices and vectors having the same parity results in a symmetric matrix or vector, while if they have opposite

Table 20.2. Form of displacement vectors and displacement gradients by mode

Mode Type	Displacement, $\boldsymbol{u}$	Displacement Gradient, $\boldsymbol{H}$
RL_S	$\begin{Bmatrix} S \\ A \\ 0 \end{Bmatrix}$	$\begin{bmatrix} S & A & 0 \\ A & S & 0 \\ 0 & 0 & 0 \end{bmatrix}$
RL_A	$\begin{Bmatrix} A \\ S \\ 0 \end{Bmatrix}$	$\begin{bmatrix} A & S & 0 \\ S & A & 0 \\ 0 & 0 & 0 \end{bmatrix}$
SH_S	$\begin{Bmatrix} 0 \\ 0 \\ S \end{Bmatrix}$	$\begin{bmatrix} 0 & 0 & 0 \\ 0 & 0 & 0 \\ S & A & 0 \end{bmatrix}$
SH_A	$\begin{Bmatrix} 0 \\ 0 \\ A \end{Bmatrix}$	$\begin{bmatrix} 0 & 0 & 0 \\ 0 & 0 & 0 \\ A & S & 0 \end{bmatrix}$

Table 20.3. Power flux analysis of RL and SH modes

Fundamental Mode Type	$\boldsymbol{T}_o^{NL(1,1)}$	$\nabla \cdot \boldsymbol{T}_o^{NL(1,1)}$	f_n^{surf}	f_n^{vol}	Second Harmonic Mode
RL_S	$\begin{bmatrix} S & A & 0 \\ A & S & 0 \\ 0 & 0 & 0 \end{bmatrix}$	$\begin{Bmatrix} S \\ A \\ 0 \end{Bmatrix}$	$=0$ $=0$ $\neq0$ $=0$	$=0$ $=0$ $\neq0$ $=0$	SH_S SH_A RL_S RL_A
RL_A	$\begin{bmatrix} S & A & 0 \\ A & S & 0 \\ 0 & 0 & 0 \end{bmatrix}$	$\begin{Bmatrix} S \\ A \\ 0 \end{Bmatrix}$	$=0$ $=0$ $\neq0$ $=0$	$=0$ $=0$ $\neq0$ $=0$	SH_S SH_A RL_S RL_A
SH_S	$\begin{bmatrix} S & A & 0 \\ A & S & 0 \\ 0 & 0 & S \end{bmatrix}$	$\begin{Bmatrix} S \\ A \\ 0 \end{Bmatrix}$	$=0$ $=0$ $\neq0$ $=0$	$=0$ $=0$ $\neq0$ $=0$	SH_S SH_A RL_S RL_A
SH_A	$\begin{bmatrix} S & A & 0 \\ A & S & 0 \\ 0 & 0 & S \end{bmatrix}$	$\begin{Bmatrix} S \\ A \\ 0 \end{Bmatrix}$	$=0$ $=0$ $\neq0$ $=0$	$=0$ $=0$ $\neq0$ $=0$	SH_S SH_A RL_S RL_A

parity the result is antisymmetric. Differentiation and integration also affect the parity. The power flux parity analysis is greatly facilitated by writing the nonlinear component of the second Piola-Kirchhoff stress directly in terms of the displacement gradient (Equation (20.20)). It can be shown (as in the problems given at the end of this chapter) that $\boldsymbol{T}_o^{NL(1,1)}$ and $\nabla \cdot \boldsymbol{T}_o^{NL(1,1)}$ have the forms given in Table 20.3. These matrices and vectors enable assessment of whether f_n^{surf} and f_n^{vol} are nonzero, which is also shown in Table 20.3 for RL and SH modes. The conclusion is that RL and SH fundamental modes, both symmetric and antisymmetric, can generate cumulative second harmonics in RL symmetric modes. Cumulative second harmonics that are SH (symmetric or antisymmetric) or RL antisymmetric do not exist. The details of the parity analysis can be found in various forms by Chillara (2012), Müller and colleagues (2010), Srivastava and Lanza di Scalea (2009), de Lima and Hamilton (2003), and Deng (2003).

Without going through the mathematics, it may be surprising that power flux from fundamental SH waves to symmetric RL waves is nonzero. But this is indeed the case and is explored experimentally in Section 20.6.4. It can be shown (see problems given at the end of this chapter) that the power flux from a fundamental SH wave to a symmetric RL harmonic wave is due to the $H^T H$ and $H H^T$ terms in Equation (20.20). These quadratic terms in displacement gradient appear because of geometric nonlinearity (Equation (20.11)) and material nonlinearity (Equation (20.15)), illustrating the importance of including both types of nonlinearity.

20.6.3 Group Velocity Matching

As Equation (20.37) demonstrates, internal resonance and the resultant cumulative harmonic require phase matching and nonzero power flux from a theoretical basis. Müller and colleagues (2010) make a case from a practical standpoint for ensuring that the group velocities of the fundamental wave and the harmonic match as well. If the group velocities do not match, then the harmonic would separate from the fundamental wave over the propagation distance, thus complicating signal processing. Fortunately, the group velocity matching condition is satisfied in most of the cases where internal resonance occurs (Matsuda and Biwa 2011). In the case of a symmetric RL mode having $c_p = c_L$ that generates a second harmonic symmetric RL mode, the group velocity is:

$$c_g = \frac{c_L \left[\left(c_T / c_L \right)^2 + 8P \right]}{1 + 8P} \tag{20.54}$$

$$P = \frac{1 - \left(c_T / c_L \right)^2}{\left[2 - \left(c_T / c_L \right)^2 \right]^2}.$$

For an RL wave having $\sqrt{2} c_T < c_L$ that generates harmonic RL modes, the group velocity is simply $c_g = c_T / \sqrt{2}$. Matsuda and Biwa (2011) provide more details and examples.

20.6.4 Sample Laboratory Experiments

The generation of the second harmonic in 6061-T6 aluminum ($\rho = 2700$ kg/m³, $c_L = 6300$ m/s, $c_T = 3100$ m/s) plate is demonstrated with the aid of a high-power gated amplifier. In the first set of experiments an angle beam transducer is used to send S1 and A1 mode RL waves, while in the second set of experiments a magnetostrictive transducer is used to send an SH2 mode SH wave.

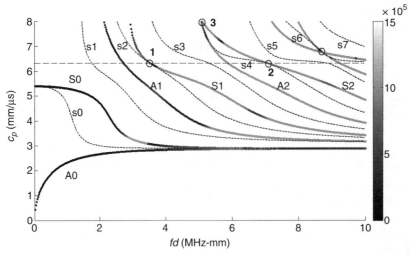

Figure 20.7. RL-S2 Internal resonance plot showing synchronous points as well as power flux intensity map to S2 mode superimposed on RL dispersion curves. Also shown by open symbols are the symmetric RL dispersion curves plotted at twice fd.

Fundamental RL Waves

The fd product of 3.56 MHz-mm is selected in order to excite the S1 wave mode at the synchronous phase velocity of the longitudinal wave speed ($c_p = c_L$, see Table 20.1). Figure 20.7 shows the power flux to the S2 mode from each mode shown as an intensity plot on the RL phase velocity dispersion curves. The power flux values are determined from Equations (20.35) and (20.36). Additionally, the symmetric RL modes at twice the frequency are depicted by hollow circles that indicate phase matching between the fundamental and the second harmonic where they intersect. The power flux is relatively large at $fd = 3.56$ MHz-mm, and the group velocities are also synchronous. Thus, the S1 mode at $fd = 3.56$ MHz-mm is in a state of internal resonance with the S2 mode, which should result in a second harmonic that is cumulative. We will call Figure 20.7 an RL-S2 internal resonance plot, where RL refers to the fundamental mode type and S2 defines the harmonic mode generated. Moreover, a similar plot of power flux (as in Figure 20.7) to any antisymmetric mode would result in values of identically zero. Notice that synchronous Lame modes where $c_p = \sqrt{2}c_T$ do not exist because for aluminum $\sqrt{2}c_T < c_L$ (see Table 20.1), which is the case for most conventional structural materials.

Regrettably, the S1 mode does not have good excitability from an angle beam transducer because, as shown in the wave structures of Figure 20.8, the value of the out-of-plane displacement at the surface of the plate is quite small. On the other hand, the in-plane displacement at the surface of the plate is substantial. Thus, a shear transducer is used as a receiver. Measurements were made on a 2 mm-thick plate at propagation distances of 50–300 mm, which are in the far field. Because of the inherent dispersion and multimode properties of guided waves, a short time Fourier transform (STFFT) is employed to create a time-frequency spectrogram of the wave field. Knowing the travel distance, time is converted to group velocity and the resulting spectrogram is shown in Figure 20.9(a). The group velocity dispersion curves are overlaid on the spectrogram to assist interpretation and to demonstrate

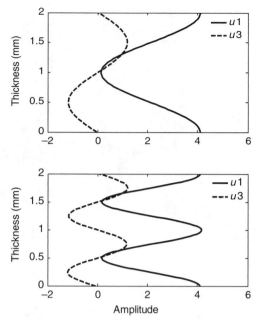

Figure 20.8. RL wave structures for (a) S1 mode at 1.78 MHz-mm and (b) S2 mode at 3.56 MHz-mm.

that the group velocities of the fundamental S1 and the secondary S2 modes are synchronized. Clearly, the fundamental S1 mode at the excitation frequency of 1.78 MHz dominates (unfortunately the A1 and A0 modes are also excited because of source influence and their greater excitability), but the second harmonic, which excites the S2 mode at 3.56 MHz, also represents a local peak – which is what the experiment was intended to demonstrate. Slices through the spectrogram of Figure 20.9(a) taken at the fundamental frequency of 1.78 MHz (intended S1 mode) and 3.56 MHz (second harmonic S2 mode) are shown in Figure 20.10. Notice that the amplitude of the second harmonic (A_2) has been multiplied by fifty to use a single-amplitude axis and that the intensities shown in the Figure 20.9 spectrograms are plotted based on a log scale. Figure 20.11 demonstrates that the second harmonic S2 from the fundamental S1 mode excitation increases linearly with propagation distance over the range of 50–300 mm, and is therefore cumulative as the theory predicts.

As a second RL wave example, consider activating the A1 mode at the same fd = 3.56 MHz-mm; this point is close to, but not exactly in-phase with, the second harmonic of the A2 mode. At these frequencies the A1 and A2 modes have significant out-of-plane displacement components on the surface. Thus the excitability with an angle beam transducer is good, and an angle beam transducer is also used as the receiver. The spectrogram in Figure 20.9(b) demonstrates that multiple higher harmonics are generated even without the internal resonance condition. The case of nearly matching phase velocities can lead to increasing amplitude of the second harmonic with propagation distance, albeit not a linear increase, as discussed by Müller and colleagues (2010). Furthermore, the analysis leading to the conclusion that second harmonic antisymmetric RL modes are not possible is based on a single excitation signal. In reality, the transducer excites a range of frequencies with a

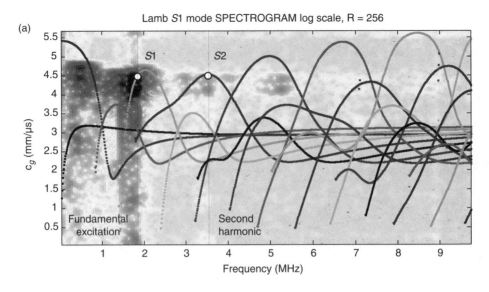

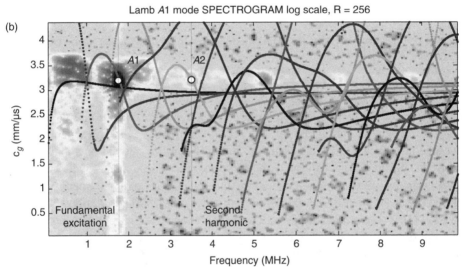

Figure 20.9. Spectrograms for fundamental RL modes (a) S1 and (b) A1 for a propagation distance of 100 mm.

range of phase velocities (see Chapter 13 on source influence); interactions between the components over the actual range of frequencies excited are not included in the analysis and could result in nonzero power flux to an antisymmetric second harmonic (Chillara 2012). Further analysis has shown that nonzero power flux to higher harmonic antisymmetric modes occurs because of interactions between all frequencies below that of the higher harmonic of interest (Srivastava and Lanza di Scalea 2009; Chillara 2012).

Fundamental SH Waves

An fd = 3.56 MHz-mm is once again imposed, but now the plate thickness is 6.35 mm so that the fundamental and second harmonics are 0.5606 and 1.1212 MHz, respectively. Magnetostrictive transducers are used to emit an SH2 mode (which is

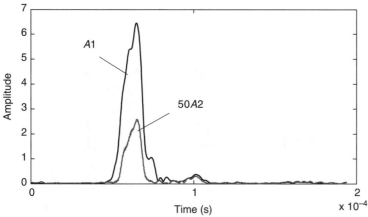

Figure 20.10. Wave amplitudes from spectrogram for fundamental RL mode S1 (Figure 20.9(a)) at the fundamental (A_1) and second harmonic frequencies (A_2).

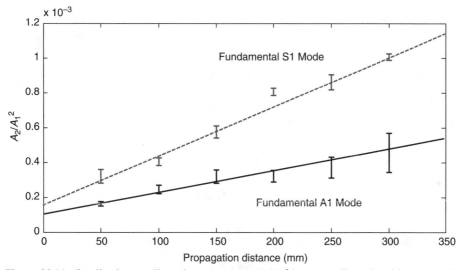

Figure 20.11. Qualitative nonlinearity parameter, A_2/A_1^2 increases linearly with propagation distance. The increase is significantly larger for the internal resonance condition (fundamental S1 mode to S2 harmonic).

symmetric) wave and receive both SH and RL waves. The transducer is constructed by bonding an iron-cobalt sheet to the aluminum plate with epoxy and then placing atop the iron-cobalt a meandering coil having a 10 mm-element spacing (which is also the wavelength generated) followed by a magnet. The polarization direction is dictated by the magnetic field bias relative to the current in the coil, thus both SH and RL waves can be received, but not at the same time. One advantage of the magnetostrictive transducer for these measurements is that the couplant, which complicates the relationship between the voltage received by a piezoelectric transducer and the actual displacement amplitude, is not used.

An SH-S2 internal resonance plot is shown in Figure 20.12. The power flux to the S2 RL mode is superimposed on the SH wave phase velocity dispersion curves in Figure 20.12. Phase matching occurs between a fundamental SH2 mode and the S2

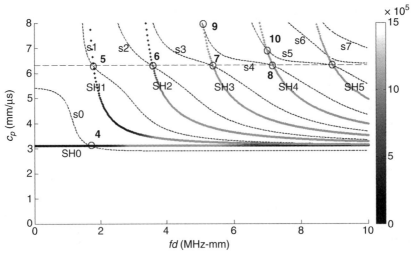

Figure 20.12. SH-S2 internal resonance plot showing synchronous points as well as power flux intensity map to S2 mode superimposed on SH dispersion curves. Also shown by open symbols are the symmetric RL dispersion curves plotted at twice *fd*.

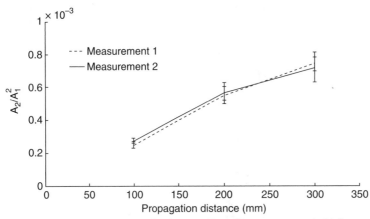

Figure 20.13. Increase in qualitative nonlinearity parameter, A_2/A_1^2, as a function of propagation distance.

RL second harmonic at $fd = 3.56$ MHz-mm as expected from Table 20.1. The power flux gets progressively higher as the order of the mode (and frequency) increases. The SH2 mode at $fd = 3.56$ MHz-mm is in internal resonance with the S2 RL mode and has a reasonable power flux. Thus, this mode is excited by a magnetostrictive transducer as described previously. Another magnetostrictive transducer is used to receive the SH mode, and then the bias of the magnetic field is rotated 90 degrees to receive the RL mode. The fundamental SH2 mode dominates at $fd = 3.56$ MHz-mm, but other velocities occur. It is clear that, as predicted because of the internal resonance condition, the S2 RL mode is generated at $fd = 7.12$ MHz-mm. Furthermore, second harmonic is cumulative for propagation distances between 100 mm and 300 mm as shown in Figure 20.13.

20.7 Applications of Higher Harmonic Generation by Guided Waves

Because of the complexity associated with the dispersive character of guided waves, and the complicated wave interactions that lead to higher harmonics, it is highly unlikely that the use of higher harmonic guided waves will be fruitful without understanding the underlying physics and modeling it mathematically. Through a series of advances starting around 1998, our understanding and modeling capabilities have arrived at a point where intelligent mode selection is possible (Liu, Chillara, and Lissenden 2013). The number of applications beyond laboratory demonstrations of principles, especially for materials characterization, NDE, and SHM, is expected to increase. Laboratory demonstrations of principles include: De Lima and Hamilton (2003); Deng, Wang, and Lv (2005); Bermes and colleagues (2007); Pruell and colleagues (2007, 2009); Lee, Choi, and Jhang (2008); Srivastava and Lanza di Scalea (2009); Matlack and colleagues (2011); and Liu and colleagues (2012). Otherwise, the solution of the nonlinear wave equation in waveguides has been exploited as a means to determine stresses in multi-strand cables (Nucera and Lanza di Scalea 2011a) and rail (Nucera and Lanza di Scalea 2011b) using acoustoelasticity.

20.8 Exercises

1. Based on Equation (20.20), describe why it is important to include both material and geometric nonlinearities in the formulation of the nonlinear wave equation.
2. Starting with the RL dispersion relation, determine the frequency at which the S2 secondary mode is synchronized with the S1 fundamental mode at the longitudinal wave speed. Also compare the group velocities.
3. Perform a parity analysis for symmetric SH wave fundamental modes and symmetric RL wave secondary modes to show that the power flux from the fundamental mode to the secondary mode is nonzero in this case. Start by determining the form of the displacement gradient, substituting this into Equation (20.20), and evaluating Equation (20.37), thereby confirming one case in Table 20.3.
4. Perform a parity analysis for symmetric SH wave fundamental modes and antisymmetric RL wave secondary modes to show that the power flux from the fundamental mode to the secondary mode is zero in this case. Follow a similar procedure as in problem 3.
5. What advantages do nonlinear guided waves have over linear guided waves for inspection? What disadvantages?
6. When performing measurements of the generation of a second harmonic in a material, very long toneburst excitations are typically used. Why is this the case?
7. Why is it necessary to match the phase velocity of the fundamental and harmonic?
8. List several applications where nonlinear guided waves may prove fruitful. What specific types of damage would the nonlinear guided waves be sensitive to in each of these applications?

20.9 REFERENCES

Alleyne, D. N., and Cawley, P. (1992). The interaction of Lamb waves with defects, *IEEE Trans. Ultra. Ferro. Freq. Control* 39: 381–97.

Auld, B. A. (1973). *Acoustic Fields and Waves in Solids*. London: Wiley.

Bermes, C., Kim, J. Y., Qu, J., and Jacobs, L. J. (2007). Experimental characterization of material nonlinearity using Lamb waves, *Appl. Phys. Letters* 90: 021901.

Cantrell, J. H. (2004). Fundamentals and applications of nonlinear ultrasonic nondestructive evaluation, in *Ultrasonic Nondestructive Evaluation: Engineering and Biological Material Characterization*, T. Kundu (Ed.). Boca Raton, FL: CRC Press, 363–433.

   (2009). Nondestructive evaluation of metal fatigue using nonlinear acoustics, in *Review of Progress in Quantitative Nondestructive Evaluation*, vol. 28, D. O. Thompson and D. E. Chimenti (Eds.). New York: Plenum Press, 19–32.

Cantrell, J. H., and Yost, W. T. (2001). Nonlinear ultrasonic characterization of fatigue microstructures, *Int. J. Fatigue* 23: S487–S490.

Chillara, V. (2012). Higher harmonic guided waves in isotropic weakly non-linear elastic plates, MS thesis in engineering mechanics, University Park: Pennsylvania State University.

Chomette, S., Gentzbittel, J. M., and Viguier, B. (2010). Creep behavior of as received, aged and cold worked inconel 617 at 850 °C and 950 °C, *J. Nuclear Materials* 399: 266–74.

Dace, G. E., Thompson, R. B., and Brashe, L. J. H. (1991). Nonlinear acoustics, a technique to determine microstructural changes in materials, in *Review of Progress in Quantitative Nondestructive Evaluation*, vol. 10B, D. O. Thompson and D. E. Chimenti (Eds.). New York: Plenum Press, 1685–92.

Dace, G. E., Thompson, R. B., and Buck, O. (1992). Measurement of the acoustic harmonic generation for materials characterization using contact transducers, in *Review of Progress in Quantitative Nondestructive Evaluation*, vol. 11B, D. O. Thompson and D. E. Chimenti (Eds.). New York: Plenum Press, 2069–76.

de Lima, W. J. N., and Hamilton, M. F. (2003). Finite-amplitude waves in isotropic elastic plates, *J. Sound Vibration* 265: 819–39.

Deng, M. (2000). Cumulative second-harmonic generation of generalized Lamb-wave propagation in a solid waveguide, *J. Phys. D: Appl. Phys.* 33: 207–15.

   (2003). Analysis of second-harmonic generation of Lamb modes using a modal analysis approach, *J. Appl. Phys.* 94: 4152–9.

Deng, M., Wang, P., and Lv, X. (2005). Experimental observation of cumulative second-harmonic generation of Lamb-wave propagation in an elastic plate, *J. Phys. D: Appl. Phys.* 38: 344–53.

Farrar, C. R., Worden, K., Todd, M. D., Park, G., Nichols, J., Adams, D. E., Bement, M. T., and Farinholt, K. (2007). *Nonlinear System Identification for Damage Detection*, LA-14353, Los Alamos National Laboratory.

Gol'dberg, Z. A. (1960). Interaction of plane longitudinal and transverse elastic waves, *Soviet Physics – Acoustics* 6: 306–10.

Hikata, A., Chick, B. B., and Elbaum, C. (1965). Dislocation contribution to the second harmonic generation of ultrasonic waves, *J. Appl. Phys.* 36(1): 229–36.

Jhang, K. Y. (2009). Nonlinear techniques for nondestructive assessment of micro damage in material: review, *Int. J. Precision Engng. Manuf.* 10(1): 123–35.

Kim, J. Y., Jacobs, L. J., Qu, J., and Littles, J. W. (2006). Experimental characterization of fatigue damage in a nickel-base superalloy using nonlinear ultrasonic waves, *J. Acoust. Soc. Am.* 120(3): 1266–73.

Landau, L. D., and Lifshitz, E. M. (1986). *Theory of Elasticity*. New York: Pergamon.

Lee, T. H., Choi, I. H., and Jhang, K. Y. (2008). The nonlinearity of guided wave in an elastic plate, *Mod. Phys. Letters B* 22(11): 1135–40.

Liu, Y., Chillara, V., and Lissenden, C. J. (2013). On selection of fundamental modes for generation of strong internally resonant second harmonics in plate, *J. Sound and Vibration* 332(19): 4517–28.

Malvern, L. E. (1969). *Introduction to the Mechanics of a Continuous Medium*. Upper Saddle River, NJ: Prentice-Hall.

Matlack, K. H., Kim, J. Y., Jacobs, L. J., and Qu, J. (2011). Experimental characterization of efficient second harmonic generation of Lamb waves in a nonlinear elastic isotropic plate, *J. Appl. Phys.* 109: 014905.

Matsuda, N., and Biwa, S. (2011). Phase and group velocity matching for cumulative harmonic generation in Lamb waves, *J. Appl. Phys.* 109: 094903.

Müller, M. F., Kim, J. Y., Qu, J., and Jacobs, L. J. (2010). Characteristics of second harmonic generation of Lamb waves in nonlinear elastic plates, *J. Acoust. Soc. Am.* 127(4): 2141–52.

Na, J. K., Cantrell, J. H., and Yost, W. T. (1996). Linear and nonlinear properties of fatigued 410Cb stainless steel, *Rev. Prog. Quant. Nondestr. Eval.* 15: 1479–88.

Norris, A. N. (1998). Finite-amplitude waves in solids, in *Nonlinear Acoustics*, M. F. Hamilton and D. T. Blackstock (Eds.). San Diego, CA: Academic Press, 263–77.

Nucera, C., and Lanza di Scalea, F. (2011a). Monitoring load levels in multi-wire strands by nonlinear ultrasonic waves, *Structural Health Monitoring* 10: 617–29.

(2011b). Nonlinear guided waves: theoretical considerations and applications to thermal stress measurement in continuous welded rails, in *Structural Health Monitoring 2011*, F. K. Chang (Ed.). Lancaster, PA: Destech Publications, 2521–8.

Pruell, C., Kim, J. Y., Qu, J., and Jacobs, L. J. (2007). Evaluation of plasticity driven material damage using Lamb waves, *Appl. Phys. Letters* 91: 231911.

(2009). A nonlinear-guided wave technique for evaluating plasticity-driven material damage in a metal plate, *NDT&E Int.* 42: 199–203.

Srivastava, A., and Lanza di Scalea, F. (2009). On the existence of antisymmetric or symmetric Lamb waves at nonlinear higher harmonics, *J. Sound Vibration* 323: 932–43.

Sun, L., Kulkarni, S. S., Achenbach, J. D., and Krishnaswamy, S. (2006). Technique to minimize couplant-effect in acoustic nonlinearity measurements, *J. Acoust. Soc. Am.* 120(5): 2500–5.

Truell, R., Elbaum, C., and Chick, B. B. (1969). *Ultrasonic Methods in Solid State Physics*. New York: Academic Press.

Worden, K., Farrar, C. R., Manson, G., and Park, G. (2007). The fundamental axioms of structural health monitoring, *Proc. R. Soc. A.* 463: 1639–64.

21 Guided Wave Imaging Methods

Guided wave imaging methods are extremely popular and useful for evaluating defect locations and sizes in guided wave inspection. Interpretation is easier and preferred when compared to studying complex RF waveforms. Images are generally reconstructed from the RF waveforms in a variety of ways. A few imaging methods are discussed here, including the following:

1. Guided Wave through Transmission Dual Probe Imaging
2. Defect Locus Map
3. Guided Wave Tomographic Imaging
4. Guided Wave Phased Array in Plates
5. Long-Range Ultrasonic Guided Wave Pipe Inspection Images

21.1 Introduction

Imaging methods in nondestructive evaluation (NHM) are preferred whenever possible, compared to results presented in RF waveform format. Imaging results are easier to read, to interpret, and to convince others of a result. RF waveforms can be complex and difficult to analyze, often calling for sophisticated signal processing and pattern recognition analysis. As a result, a discussion of several guided wave imaging methods are presented in this chapter.

21.2 Guided Wave through Transmission Dual Probe Imaging

A number of guided wave through transmission dual probe configurations have been used for defect imaging purposes. The two probes are fairly close together and scan from one to eight inches apart. The idea stems from an acousto-ultrasonic approach by Vary (1987). At that time, two normal beam probes were placed close to each other and an ultrasonic signature of the test piece was taken that often depicted a good or bad character of the component being tested. The method was also developed to assist in acoustic emission development programs where a sending transducer was used to imitate an acousto emission signal for subsequent capture and analysis by an acoustic emission sensor and analysis system.

A breakthrough paper by Rose, Ditri, and Pilarski (1994) clearly demonstrated that the acousto-ultrasonic technique was indeed a guided wave technique. Hence

(a) Guided wave air coupled scanning system

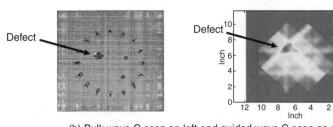

(b) Bulk wave C scan on left and guided wave C scan on right.

Figure 21.1. Air coupled inspection of impact delaminations in an orthotropic plate.

transducer size produced a source influence and modes were likely emitted at the highest end of the phase velocity scale of the phase velocity spectrum.

Since then, of course, angle beam and comb probes have often replaced the two normal beam probes for guided wave data acquisition for a line of sight or for actual scanning in order to get an improved phase and frequency spectrum to solve a particular problem. Some sample results are reported herein.

The first example of a guided wave scan with two transducers operating in a through transmission mode is illustrated in Figure 21.1 using air coupled sensors to inspect for delaminations in an orthotropic plate. An appropriate angle and frequency are selected to provide a wave structure with reasonable energy in the expected delamination area of the plate. In Figure 21.1(b), the guided wave scan of the delamination appears similar to that acquired with an ordinary bulk wave "C" scan.

The second example is reported by Yan and colleagues (2010). In this case, a twenty-three-layer structure is considered as illustrated in Figure 21.2. Nondestructive testing (NDT) for multilayered structures is challenging because of increased numbers of layers and plate thicknesses. In Yan and colleagues' paper, ultrasonic guided waves are applied to detect delamination defects inside a twenty-three-layer alcoa advanced hybrid structural plate. A semi-analytical finite element (SAFE) method generates dispersion curves and wave structures in order to select appropriate wave structures to detect certain defects. One guided wave mode and frequency is chosen to achieve large in-plane displacements and/or shear stress at regions of interest. The interactions of the selected mode with defects are simulated using finite element models. Experiments are conducted and compared with bulk wave measurements. It is shown that guided waves can detect deeply embedded damages inside thick multilayer fiber-metal laminates with suitable mode and frequency selection.

At the mode selected, an appropriate wedge angle and frequency are selected at a phase velocity of 5 km/s and 450 kHz at a fixed spacing illustrated in Figure 21.3.

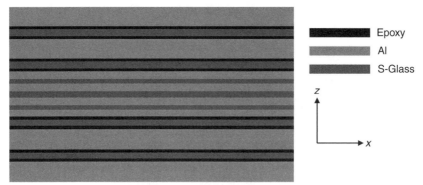

Figure 21.2. Lay-up sequence of the twenty-three-layer test specimen.

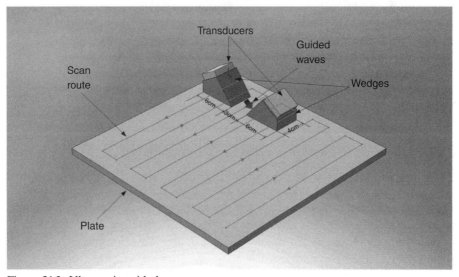

Figure 21.3. Ultrasonic guided wave scan.

Excellent results were obtained as shown in Figure 21.4. Almost identical results were obtained by scanning the plate from the bottom surface, because of the two defect locations and wave structure symmetry, near the fourth layer from the top and from the bottom, also in the spacial position shown.

The third example is associated with a titanium repair patch bonded onto an aluminum skin of an aircraft. Repairs are made by grinding out defects, filling with epoxy, and bonding a titanium plate over the area like a band-aid. The problem is complex because each repair is slightly different, so a knowledge base had to be established to successfully solve this problem. See Figure 21.5.

The handheld probe, shown in Figure 21.5, used in this task was a dual-element electromagnetic acoustic transducer with appropriate comb-type spacing configuration to produce a dominant in-plane displacement at the potentially weak interface to carry out the inspection. See Puthillath and Rose (2010) for details on the mode and frequency choice.

Using an encoder, a scan was produced as illustrated in Figure 21.6. Areas of concern are clearly pointed out, but additional work in signal feature extraction and

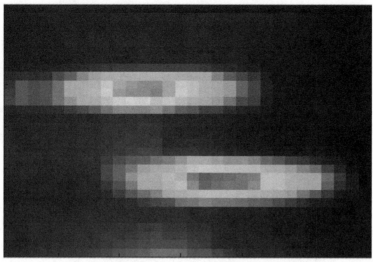

Figure 21.4. Image obtained by a guided wave SDC scan of the plate from the top surface.

Figure 21.5. Handheld guided wave dual EMAT through transmission probe scanner to determine bond integrity in repair patch inspection.

pattern recognition was required to decide if regions A, B, and C were acceptable or not because of the repair type. A number of physically based features of the guided wave waveform were studied, one idea of the study is depicted in Figure 21.7. Feature 3 value could find all of the poor bonds, but also had many false alarms, so a best-bet probability solution was required. Probability density functions could

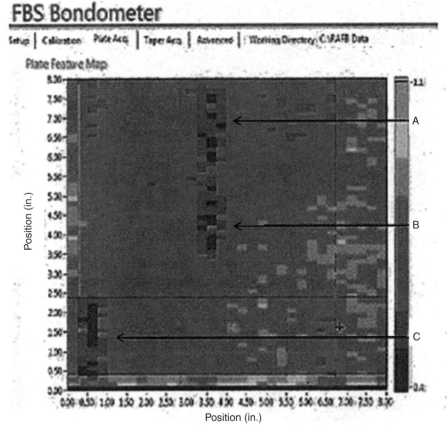

Figure 21.6. Guided wave scan to locate areas of interest in the bond integrity determination problem.

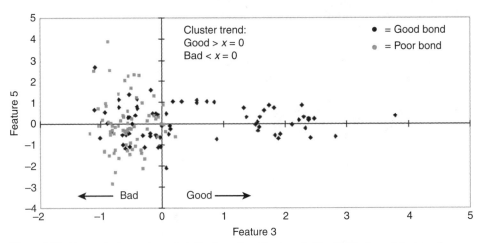

Figure 21.7. Example two-space plot. Feature 5 vs. Feature 3. Physically based feature for model analysis – pattern recognition. PDF of function (Duda, Hart, and Stork 2007).

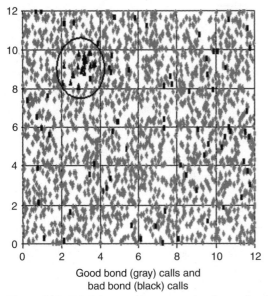

Figure 21.8. Decisions made on a region-by-region basis with the pattern recognition decision algorithm developed for this problem.

be used to explore the potential value of many benefits. The smaller the number used, the better, especially with a reduced number of data sets. For problem density function information and for topics on pattern recognition, see Duda, Hart, and Stork (2007).

Finally, an image like that is shown in Figure 21.8, which is a decision map. Using a committee vote pattern recognition concept, a final probabilistic decision can be made on bond integrity for the patch repair. A questionable area is shown in Figure 21.8, but the overall decision is good because of the huge good green calls. Obviously from a structural health monitoring (SHM) point of view, the circled region would be monitored more closely after a certain number of flights. More insight into this practical problem can be found in Appendix D, Section 2.11.

21.3 Defect Locus Map

It is possible to get an idea of the extent of ship hull penetration damage resulting from an impact or other situation by producing a defect locus plot with ultrasonic guided waves. This very simple concept can be quite useful. Mode and frequency choice would depend on plate thickness and knowledge of group velocity in that plate. A nondispersive area of the phase velocity region of the dispersion curve should be selected. A sample image can be produced as illustrated in Figure 21.9.

To make a defect locus map, guided waves are sent from a transducer in one direction. Then transducers are rotated about five degrees to generate waves in a different direction. If an annular or a semi-annular array is being used, no rotation is required. As the waves start to interact with the damage, scattering occurs so that the amplitude of the reflected waves are attenuated. The rotation continues until the

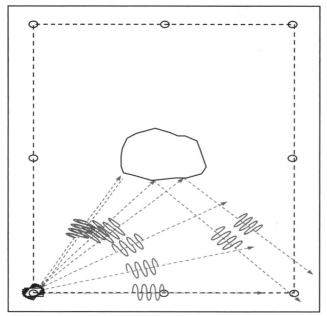

Figure 21.9. The process of generating defect locus map.

reflected signal with a maximum amplitude and minimum arrival time is captured or an annular array or a signal is immediately received.

A final result is illustrated in Figure 21.10. Because the group velocity is known along with the shortest arrival time of the reflected waves, an arc can be drawn with the radius equal to the distance from the test location to the boundary of a damage. This is repeated at each transducer in the grid. We can draw arcs at different test locations. The intersections of the arcs were calculated. To make a defect locus map closer to the actual damage, the middle points between intersections on the arcs are recalculated and a defect locus map can be obtained by connecting the middle points. See the pages by Song, Rose, and Whitesel (2003) for more detail.

21.4 Guided Wave Tomographic Imaging

Guided wave tomographic imaging has become quite popular in recent years. An indicator of damage via a reasonably good or at least a pseudo image is obtained from a SHM point of view. Early work is reported by Malyarenko and Hinders (2000); Jansen and colleagues (1994); Nagata and colleagues (1995); Prasad, Balasubramaniam, and Krishnamurthy (2004); Gao, Shi, and Rose (2005); and Bian and Rose (2005). Of many tomographic methods, emphasis here will be on the RAPID method (reconstruction algorithm for probabilistic inspection of damage). See Gao and colleagues (2005) and Hay and colleagues (2006). Material taken from those papers is presented here. First, the basic imaging principle is depicted in Figure 21.11. Note that the transducer placement could actually be over any shape or even be random. Of many features that could be used to produce a tomogram, time of flight, amplitude, and a signal difference coefficient are the most popular. Many other features could be used, for example, looking at the areas under an FFT in a ratio of a high-frequency content to

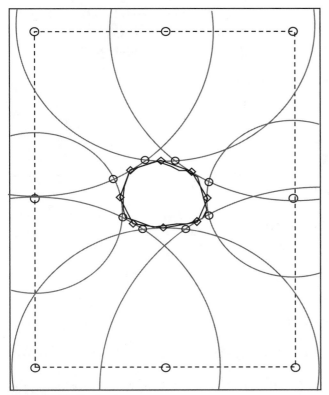

Figure 21.10. Defect locus map.

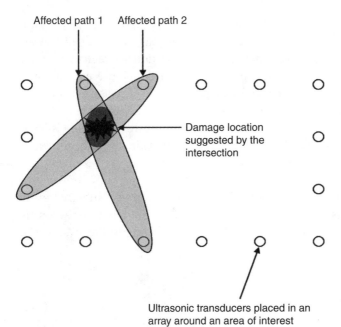

Figure 21.11. The tomographic imaging principle.

a low-frequency content using a suitable separation threshold value and starting and ending frequency value. Let's start with an example of the SDC.

The correlation coefficient, ρ, between two sets of data, s_j and s_k, is:

$$\rho = \frac{\text{Cov}(s_j, s_k)}{\sigma_{s_j} \sigma_{s_k}} \qquad (21.1)$$

where the covariance, Cov, is

$$\text{Cov}(s_j, s_k) = \frac{\sum_i^N (s_j(t_i) - \bar{s}_j)(s_k(t_i) - \bar{s}_k)}{N} \qquad (21.2)$$

and the standard deviations, σ_{s_j} and σ_{s_k}, are

$$\sigma_{s_j} = \sqrt{\sum_i^N (s_j(t_i) - \bar{s}_j)^2} \qquad \sigma_{s_k} = \sqrt{\sum_i^N (s_k(t_i) - \bar{s}_k)^2} \qquad (21.3)$$

then, the signal difference coefficient

$$\text{SDC} = 1 - \rho. \qquad (21.4)$$

A *data set* in this case refers to waveforms acquired from sensor pairs. The reference data set (s_j) is acquired immediately after initial installation. All future data (s_k) are compared to the initial data set (and/or any data set previously taken) to generate SDC tomograms.

Let's now consider the reconstruction algorithm for probabilistic inspection of damage (RAPID). Changes of the features extracted from an ultrasonic guided wave signal are related to a change in the original structural condition between two sensors. See Figure 21.12 for a sketch of the ray-affected area in the ellipse consideration process of RAPID. Therefore, the probability of defect occurrence at a certain point can be reconstructed from the severity of the signal change and its position relative to the sensor pair. The defect distribution estimation used in this algorithm is based on the following assumptions. (1) The defect distribution estimation for the entire reconstruction region can be expressed as a linear summation of all the effects from every possible transmitter-receiver pair. (2) The information from a transmitter-receiver pair contributes to the defect distribution estimation of a sub-area of the reconstruction region. This sub-area is the vicinity of the direct path between the transmitter and receiver. The specific formula used in the algorithm is shown in Equation (21.5):

$$P(x,y) = \sum_{k=1}^{N} p_k(x,y)$$

$$= \sum_{k=1}^{N} A_k \left(\frac{-1}{\beta - 1} R(x, y, x_{1k}, y_{1k}, x_{2k}, y_{2k}) + \frac{\beta}{\beta - 1} \right) \qquad (21.5)$$

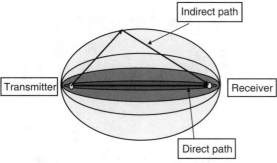

Figure 21.12. Concept of a ray affect area in RAPID reconstruction.

where

$$R\left(x, y, x_{1k}, y_{1k}, x_{2k}, y_{2k}\right) = \frac{\sqrt{\left(x - x_{1k}\right)^2 + \left(y - y_{1k}\right)^2} + \sqrt{\left(x - x_{2k}\right)^2 + \sqrt{\left(y - y_{2k}\right)^2}}}{\sqrt{\left(x_{1k} - x_{2k}\right)^2 + \left(y_{1k} - y_{2k}\right)^2}} \qquad (21.6)$$

Here, $P(x, y)$ is the estimation of the defect probability at position (x, y) within the reconstruction region. $p_k(x, y)$ is the estimation from the kth transmitter-receiver pair. N is the number of transmitter and receiver pairs in the sensor network. A_k is the input feature of the kth transmitter-receiver pair. R is the ratio of the total distance of the point to the transmitter and the receiver divided by the distance between the transmitter and the receiver. Here, (x_{1k}, y_{1k}) is the coordinate of the transmitter of the kth path. (x_{2k}, y_{2k}) is the coordinate of the receiver of the kth path. β is a scaling parameter, and it was set to 1.05. When $R = 1$, $p_k (x, y) = A_k$; when $R = \beta$, $p_k (x, y) = 0$. If a damage or material loss occurs in between sensors after the sensors were embedded and reference data were acquired, the set of signals will be affected. As a result, in the final defect distribution probability image, the defect point will have dominantly larger probability than the other points. Therefore, by applying a thresholding process, the estimated defect locations can be distinguished in the image.

A benefit of tomography can be seen in Figure 21.13. Note that where the circular eight sensor network was on the inside surface of a very complex structure. The corrosion patch was on the outside surface. Despite the complex reference signal, an excellent tomogram can be generated. NDT would have a very hard time finding a defect as no baseline would be available.

One additional tomogram is presented here. See Figure 21.14. Despite the complexity of the composite structure, with embedded sensors, an excellent tomogram is obtained showing impact damage.

Some interesting tomographic imaging results for pipe and elbow are covered in Van Velsor, Gao, and Rose (2007) and in Breon et al. (2007). Quite often, environmental changes can create confusion in interpretation when taking data to find defects or defect growth. As a result, compensation techniques are required. This work has a way to go, but Clarke and colleagues (2009), Croxford and colleagues (2010), and Yinghui and Michaels (2009) have reported excellent progress.

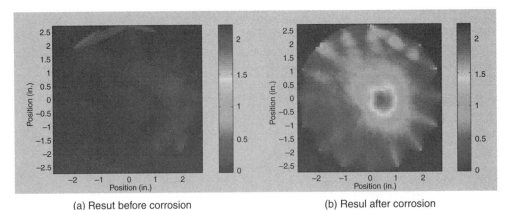

(a) Resut before corrosion   (b) Resul after corrosion

Figure 21.13. Sample ultrasonic guided wave tomographic results for corrosion on the wing of an aircraft. See plates section for color version.

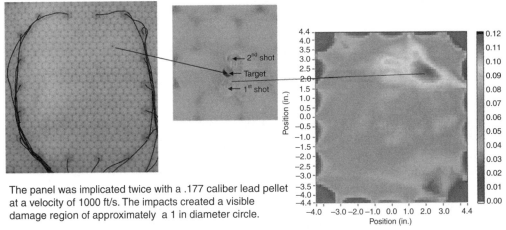

The panel was implicated twice with a .177 caliber lead pellet at a velocity of 1000 ft/s. The impacts created a visible damage region of approximately a 1 in diameter circle.

Figure 21.14. CT testing of ballistic damage to fabricated composite and the resulting guided wave tomogram. See plates section for color version.

21.5 Guided Wave Phased Array in Plates

A guided wave phased array system is presented for damage detection in plate and/or platelike structures. The system contains many transducers, which can be individually or simultaneously excited. The transducers may be placed closely together onto the structure to form a compact array and/or distributed on the structure for some distances away from each other in a random or orderly configuration. The system then contains a finite number of pulser and receiver channels. Furthermore, time delays can be used as input into each pulser and receiver channel. The time delays are used to steer the guided wave energy at a specific direction or to focus the energy at a specific location in the platelike structure. The system may combine the guided wave phased array technique with the guided wave computational tomography (CT) technique for damage imaging.

There are infinite numbers of possible guided wave modes in a platelike structure, as can be seen for a composite plate in Figure 21.15. These wave modes in a plate have

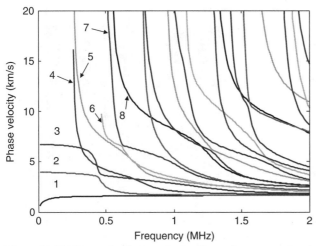

Figure 21.15. Example phase velocity dispersion curves for the 0° fiber direction in a sixteen-layer quasi-isotropic composite plate.

different phase and group velocities and energy distributions across the thickness, which may vary with frequency and/or excitation conditions. For guided wave beam steering or beam focusing, the aim is usually to excite guided wave modes with similar velocities. The guided wave modes with different velocities are considered unwanted wave modes and may result in significant wave energy traveling to directions other than the desired beam steering direction or create energy focal points other than at the desired focal point. Furthermore, the velocity differences may introduce coherent noise in guided wave damage detection applications. For instance, if the pulse-echo method is used to detect a single defect, the received signal may have multiple reflected wave packets because of the existence of wave modes with different wave velocities. The redundant wave packets coming from the unwanted wave modes may cause false alarms. To avoid the influence of unwanted wave modes, special transducers are needed that have the capability of dominantly exciting guided wave energy with the desired wave velocity while minimizing the energy of the unwanted wave modes. The design of such special transducers can be carried out based on theoretical calculations. Specially designed guided wave transducers include annular array transducers, time-delay annular array transducers, piezoelectric elements on angle wedges, electromagnetic acoustic transducers (EMATs), and magnetostrictive transducers.

With the energy of unwanted wave modes under control, time delays can be applied to the transducers to perform phased array beam steering or focusing. Each transducer in the array excites guided wave energy that can propagate in any direction. The direction of wave propagation can be controlled via a "phasing" approach. For example, Figure 21.16 shows a circular phased array probe geometry. By applying time delays to the phased array probe, the individual transducer elements can be "phased" to allow the guided wave energy to be steered in any direction. The steering direction can then be controlled to allow 360° scanning. This is essentially different from the guided wave array systems for plate structures that are presented in references (Wilcox 2003; Purekar et al. 2004; Giurgiutiu 2005; Fromme et al., 2006). In those systems, only one element of an array is pulsed at a time. As a result, there are no physically formed guided wave beams. The "beam

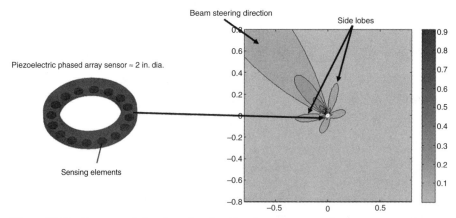

Figure 21.16. Conceptual drawing of a circular phased array probe geometry and a snapshot from a numerical simulation showing beam steering in one direction upon applying the appropriate time delays to a circular phased array sensor geometry. The appropriate time delays can be applied to sweep the beam steering direction around the entire structure for full coverage. This model was generated using a sixteen-element sensor array mounted in the center of an aluminum plate.

steering" or "focusing" of those arrays are conducted through post–data acquisition signal processing only. With real-time phased array technology, a physically formed beam of guided wave energy can be directed to different directions by varying the phase delays applied to the different elements of the phased array. Benefits of using the real-time phased array approach for guided wave inspection of plate structures include higher penetration power, better signal-to-noise ratio, and the capability of rapidly scanning selected directions and/or locations. Figure 21.16 also shows a snapshot from a numerical simulation illustrating how guided wave energy can be steered in one direction in a plate structure.

In the presented phased array system, hardware time delays are applied to physically form guided wave beams for different beam steering directions. For the syntheses of both the pulse-echo signals received by the elements of the phased array probe and the through-transmission signals received by the receiving array, instead of using the conventional delay-and-sum time domain approach, a back propagation wavenumber domain signal synthesis approach may be adopted. Using plate structures as an example and taking into account the guided wave dispersion and the wave divergence in the plate, the time signal at a point located in the far field of an array element can be approximately expressed as:

$$s'(t) = \frac{1}{\sqrt{x}} \int_{-\infty}^{\infty} S(\omega) e^{-ik(\omega)x} d\omega, \tag{21.7}$$

where $S(\omega)$ is the Fourier transform of the time domain guided wave input signal, x is the distance away from the array element, and k represents the wavenumber. The wavenumber k is a function of circular frequency ω for guided wave modes with dispersion. For the pulse-echo mode, the reflected guided wave signal introduced by a defect located in the far field of the array can then be approximately written as:

$$G_n(t) = \frac{\gamma \delta}{r_d} \int_{-\infty}^{\infty} S(\omega) e^{-ik(\omega)2r_d} e^{ik(\omega)d_n} d\omega, \tag{21.8}$$

where δ is the signal magnification coefficient introduced by the constructive interference of the signals generated by all of the phased elements, γ is the reflection coefficient, r_d is the distance from the defect to the center of the array, the subscript n represents that the reflection is received by the nth array element, and d denotes the propagation distance that needs to be compensated for beam steering to the angle where the defect locates. The wavenumber domain signal synthesis of the signals described by Equation (21.8) can be conducted using the following equation:

$$\sum_n B_n G_n(t) = \frac{\gamma \delta N}{r_d} \int_{-\infty}^{\infty} S(\omega) e^{-ik(\omega)2r_d} d\omega, \qquad (21.9)$$

where N is the number of array elements and B_n is the back-propagation term:

$$B_n = e^{-ik(\omega)d_n}. \qquad (21.10)$$

As shown in Equation (21.10), the dispersion relation of the guided wave modes is included in the back-propagation process so that the dispersion effects that could decrease defect detection resolution can be removed from the wavenumber domain signals. Note that Equation (21.9) can be implemented using FFT. The wavenumber domain signal synthesis is therefore also fast. An advanced deconvolution method can be combined with the wavenumber domain signal synthesis as well to suppress image artifacts caused by the side lobes of the phased array (Yan 2008).

A phased array scanning image obtained in an experiment on a 4- by 4-foot aluminum plate (1 mm thick) using the wavenumber domain signal synthesis is shown in Figure 21.17. A sixteen-element circular array was used in the experiment. The phased array was operated under a pulse-echo mode. The locations and shapes of the defects are indicated in the image for comparison. All defects were well detected and located.

Applications to Composite Materials

The presented guided wave phased array system can be used for anisotropic multilayer composite plates or platelike structures. Guided wave excitations become more complex when material anisotropy is involved. A Green's function–based theoretical method can be employed to study the guided wave excitations in composite platelike structures (Yan and Rose 2010). It has been shown that the amplitude and phase variations of the guided wave field excited by a point source applied normally to a composite plate are non-axisymmetric although the point source itself can be considered an axisymmetric loading. The angular dependences of the amplitude for mode 3 at 600 kHz and mode 1 at 160 kHz calculated using the Green's function–based method are given in Figure 21.18(a) and (b), respectively. The amplitude of mode 3 changes much more dramatically as compared with that of mode 1. The phased array beam steering directivity profile of a circular array for a composite plate can be calculated as:

$$p(\varphi) = \sum_n \alpha_g(\varphi) \exp\left\{-iR\left[\Phi_g(\varphi)\cos(\psi_n - \varphi) - \Phi_g(\varphi_0)\cos(\psi_n - \varphi_0)\right]\right\}, \quad (21.11)$$

16 element array mounted in
the center of the plate

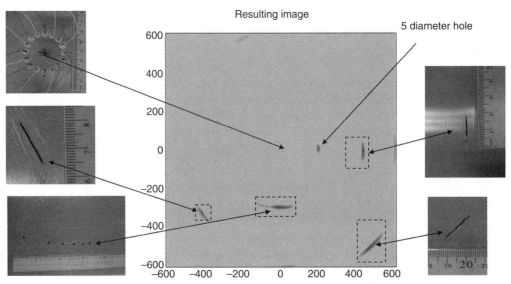

Figure 21.17. Experimental result of a phased array scanning image obtained by using the wavenumber domain back-propagation signal synthesis. A sixteen-element circular array mounted to the center of the plate was used. All defects were detected and located very well.

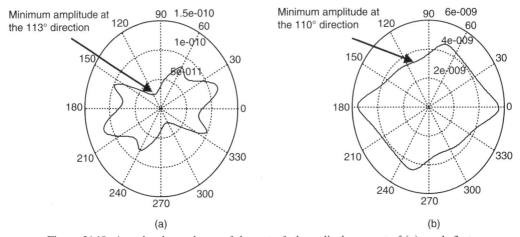

(a)   (b)

Figure 21.18. Angular dependence of the out-of-plane displacement of (a) mode 3 at frequency 600 kHz, (b) mode 1 at frequency 160 kHz, excited by a unit out-of-plane point source. It is shown that the amplitude variations of mode 3 are much more severe than mode 1.

where $\alpha_g(\varphi)$ represents the angular dependence of the guided wave amplitude; $\Phi_g(\varphi)$ is the corresponding angular dependence of phase variations, R denotes the radius of the array, ψ_n denotes the angular locations of the array elements, and φ_0 is the beam steering angle. Sample directivity profiles for mode 3 at 600 kHz and mode 1 at 160 kHz are given in Figure 21.19. From Figure 21.19(a), one can see that, although

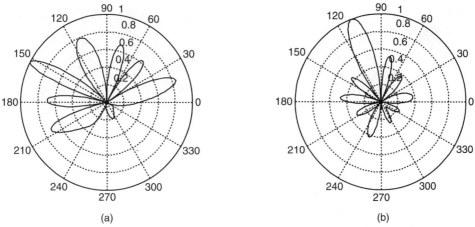

Figure 21.19. Phased array beam steering directivity for (a) mode 3 at 600 kHz, 113° beam steering direction; (b) mode 1 at 160 kHz, 110° beam steering direction. Note that the mode 3 beam steering fails, whereas for mode 1, the guided wave beam is well formed in the beam steering direction.

the beam steering direction is 113°, the strongest beam of the phased array output is close to the 150° direction. The beam steering fails. This is because at the 113° direction, the amplitude of mode 3 reaches its minimum, as shown in Figure 21.18(a). The large amplitudes of the excited wave in other directions form strong side lobes. By contrast, the mode 1 beam steering directivity profile for the 110° direction, which is the minimum amplitude direction for mode 1, demonstrates a good beam steering capability. This is due to the fact that the amplitude variations of mode 1 are much less severe than mode 3. It is therefore important to choose the wave mode with fewer amplitude changes in different directions to ensure good guided wave beam steering for all directions. It is also essential to know the directivity profiles when developing signal processing and defect imaging algorithms.

21.6 Long-Range Ultrasonic Guided Wave Pipe Inspection Images

For long-range guided waves, it becomes possible to produce a variety of different useful images. One interesting result is depicted in Figure 21.20. The lower image can be produced either by synthetic focusing showing a pipe rolled flat of channel or degrees around the circumference from 0° to 360 degrees versus axial distance along the pipe, in this case to fifty feet. Three defects are shown at the correct axial and circumferential locations. The indications between the first two defects is a weld. Defect 3 is actually three separate close by defects at the same circumferential location. The very last indication is from a back wall echo of the pipe and some mode conversion.

The image is presented at a particular frequency to achieve the best result. The upper figure is a fast frequency analysis (FFA) plot that shows a complete frequency scan from 20 to 85 kHz. This curve is extremely useful as certain defects are often found only by specific frequencies. In this case, you can see that 22 kHz should see all three defects. In fact, a simple video envelope of the RF waveform, shown in the middle figure, clearly shows the three defects using the single frequency of 22 kHz.

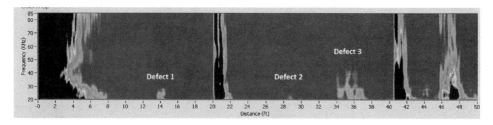

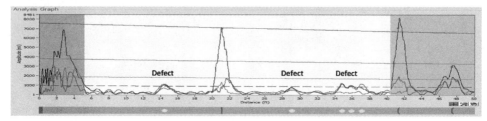

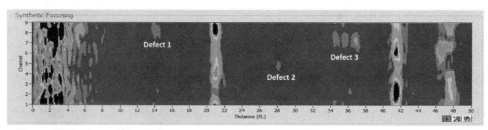

Figure 21.20. Sample fast frequency analysis plot, time domain waveform, and flattened pipe image showing defects and other features. See plates section for color version.

Excellent results were obtained and clear communication of the results can be easily made.

See Mu and Rose (2008) for more detail on the imaging process.

21.7 Exercises

1. How would you select a mode and frequency to do a defect locus map to estimate damage associated with an impact on a ship hull?
2. For a guided wave scan, how would you image weak zones in a lap splice joint of an aircraft?
3. Can you think of a few additional guided wave imaging possibilities for pipeline inspection?
4. When might you combine phased array and tomographic imaging?
5. Penetration power is always critical in inspection, and imaging with as few sensors as possible over a broad area is always desired. How would you increase penetration power on an aircraft wing? In a pipeline?
6. How would you arrange elements and delays for an annular array transducer element to achieve mode control and phased array scanning?
7. What guided wave design parameters would affect the results obtained in an acousto-ultrasonic signature experiment in a manufacturing scheme?
8. List some guided wave physically based features that could be used in tomographic imaging.

9. How would you handle skew angle effects in a phased array experiment? How about side lobe effects? Or phase velocity variation effects as a function of launch angle?

10. What are the benefits and disadvantages of a sparse array? How would you determine the minimum number of elements to use in a sparse array?

11. Tomographic imaging of multiple defects could cause problems. How could you improve the overall tomographic imaging process?

12. What are the benefits and disadvantages of using a signal difference coefficient (SDC)?

13. Could you design other probabilistic imaging techniques beyond the RAPID method?

14. What effect would coatings, soil, or insulation have on a pipe imaging technique?

15. Is it possible to change the actual imaging defect size shown in Figure 21.13 from less than 1 in^2 to say ¼ in^2 or 2 in^2?

16. See Figure 21.14. When might it be possible in guided wave tomography to image more than one defect in the tomographic zone being considered?

17. In Figure 21.20 excellent axial resolution is achieved as noticed by defect 3 that consists of 3 defects close together in a straight line. How is this achieved?

18. For the time domain signal shown in Figure 21.20, what frequency could have been selected to achieve a similar signal?

21.8 REFERENCES

Bian, H., and Rose, J. L. (2005). Sparse array ultrasonic guided wave tomography, *Materials Evaluation* 63(10): 1035–8.

Breon, L. J., Van Velsor, J. K., and Rose, J. L. (2007). Guided wave damage detection tomography for structural health monitoring in critical zones of pipelines, *Materials Evaluation* 65(12): 1215–19.

Clarke, T., Cawley, P., Wilcox, P. D., and Croxford, A. J. (2009). Evaluation of the damage detection capability of a sparse-array guided-wave SHM system applied to a complex structure under varying thermal conditions, *IEEE Transactions on Ultrasonics, Ferroelectrics and Frequency Control* 56(12): 2666–78.

Croxford, A. J., Moll, J., Wilcox, P. D., and Michaels, J. E. (2010). Efficient temperature compensation strategies for guided wave structural health monitoring, *Ultrasonics* 50(4–5): 517–28.

Duda, R. O., Hart, P. E., and Stork, D. G. (2007). *Pattern Classification.* 2nd ed. Wiley India Pvt. Limited.

Fromme, P., Wilcox, P. D., Lowe, M. J. S., and Cawley, P. (2006). On the development and testing of a guided ultrasonic wave array for structural integrity monitoring, *Trans. Ultrason., Ferroelect., Freq.* 53(4): 777–85.

Gao, H. (2007). *Ultrasonic Guided Wave Mechanics for Composite Material Structural Health Monitoring*, PhD thesis, Pennsylvania State University.

Gao, H., Shi, Y., and Rose, J. L. (2005). Guided wave tomography on an aircraft wing with leave in place sensors, review of quantitative nondestructive evaluation, New York, *AIP* 24: 1788–94.

Gao, H., Rose, J. L., Yan, F. et al., (2005). Ultrasonic guided wave tomography in structural health monitoring of an aging aircraft wing, *ASNT Fall Conference*, Columbus, Ohio, USA, October 17–21. Conference Proceedings, pp. 412–15.

Giurgiutiu, V. (2005). Tuned Lamb wave excitation and detection with piezoelectric wafer active sensors for structural health monitoring, *J. Intel. Mat. Sys. Struct.* 16: 291–305.

Hay, T. R., Royer, R. L., Gao, H., Zhao, X., and Rose, J. L. (2006). A comparison of embedded sensor Lamb wave ultrasonic tomography approaches for material loss detection, *Smart Materials and Structures* 15(4): 946–51.

Jansen, D. P., Hutchins, D. A., and Mottram, J. T. (1994). Lamb wave tomography of advanced composite laminates containing damage, *Ultrasonics* 32(2): 83–9.

Malyarenko, E. V., and Hinders, M. K. (2000). Fan beam and double crosshole Lamb wave tomography for mapping flaws in aging aircraft structures, *J. Acoust. Soc. Amer.* 108: 1631–9.

Mu, J., and Rose, J. L. (2008). Long range pipe imaging with a guided wave focal scan, *Materials Evaluation* 66(6): 663–6.

Nagata, Y., Huang, J., Achenbach, J. D., and Krishnaswamy, S. (1995). Lamb wave tomography using laser-based ultrasonics, *Rev. Prog. QNDE* 14: 561.

Prasad, S. M., Balasubramaniam, K., and Krishnamurthy, C. V. (2004). Structural health monitoring of composite structures using Lamb wave tomography, *Smart Mater. Struct.* 13: N73–N79.

Purekar, A. S., Pines, D. J., Sundararaman, S., and Adams, D. E. (2004). Directional piezoelectric phased array filters for detecting damage in isotropic plates, *Smart Mater. Struct.* 13: 838–50.

Puthillath, P., and Rose, J. L. (2010). Ultrasonic guided wave inspection of a titanium repair patch bonded to an aluminum aircraft skin, *International Journal of Adhesion & Adhesives* 30: 566–73.

Rose, J. L., Ditri, J., and Pilarski, A. (1994). Wave mechanics in acousto-ultrasonic nondestructive evaluation, *Journal of Acoustic Emission* 12(1/2): 23–6.

Song, W.-J., Rose, J. L., and Whitesel, H. (2003). An ultrasonic guided wave technique for damage testing in a ship hull, *Materials Evaluation* 60: 94–8.

Van Velsor, J. K., Gao, H., and Rose, J. L. (2007). Guided wave tomographic imaging of defects in pipe using a probabilistic reconstruction algorithm, *Insight* 49(9): 532–7.

Vary, A. (1987). The acousto-ultrasonic approach, NASA Technical Memorandum, Lewis Research Center, Cleveland, Ohio.

Wilcox, P. D. (2003). Omni-directional guided wave transducer arrays for the rapid inspection of large areas of plate structures, *IEEE Trans. Ultrason., Ferroelect., Freq.* 50(6): 699–709.

Yan, F. (2008). *Ultrasonic Guided Wave Phased Array for Isotropic and Anisotropic Plates*, PhD thesis, Pennsylvania State University.

Yan, F., and Rose, J. L. (2010). Defect detection using a new ultrasonic guided wave modal analysis technique (UMAT), in *Proceedings of the SPIE*, San Diego, CA, USA, March 7–11, 2010, vol. 7650, 76500R.

Yan, F., Xue Qi, K., Rose, J. L., and Weiland, H. (2010). Ultrasonic guided wave mode and frequency selection for multilayer hybrid laminates, *Materials Evaluation* 68(2): 169–75.

Yinghui, L., and Michaels, J. E. (2009). Feature extraction and sensor fusion for ultrasonic structural health monitoring under changing environmental conditions, *IEEE Sensors Journal* 9(11): 1462–71.

Ultrasonic Nondestructive Testing Principles, Analysis, and Display Technology

A.1 Some Physical Principles

It will be useful to review some widely used basic concepts in ultrasonic nondestructive evaluation (NDE) as a complement to the more detailed aspects of the mechanics and mathematics of wave propagation and ultrasonic NDE. Of first concern will be defining such fundamental ultrasonic field parameters as near field and angle of divergence. These will be followed by elements of instrumentation and display technology, along with aspects of axial and lateral resolution of an ultrasonic transducer. An excellent textbook on basic ultrasonics is Krautkramer 1990.

Wave velocity, one of the key parameters of wave propagation study, is the velocity at which a disturbance propagates in some specified material. Its value depends on material, structure, and form of excitation. Many different formulas for wave velocity are presented. The most widely used wave velocity value used in ultrasonic NDE is the bulk longitudinal wave velocity, generally thought of as directly proportional to the square root of the elastic modulus over density. Another common velocity is the bulk shear wave velocity, which is proportional to the square root of the shear modulus over density. These velocities are called bulk velocities. Bulk waves do not require a boundary for support. Guided waves, on the other hand, require a boundary for propagation. Many tables of wave velocity values for different materials are available in the literature.

The *particle velocity* in a material is also a useful wave propagation parameter. Note that, for a given material, the maximum wave velocity can be a thousand times greater than a maximum particle velocity. Particle velocity initiates motion once the wave reaches the specific particle being considered.

There are two basic modes of particle velocity motion with respect to the wave type generated within the material. For a longitudinal wave, the particle velocity direction is in the same direction as the wave vector or the wave velocity vector. For transverse or shear wave propagation, the particle velocity vector is at 90° to the direction of the wave vector. The transverse particle velocity vector can sometimes be called a vertical shear wave or a horizontal shear wave, depending on the coordinates used in the study. It should be pointed out that all kinds of wave motion (e.g., a guided wave consisting of a Lamb type wave, a surface wave, an interface wave) are actually made up of a superposition of longitudinal and shear wave particle velocity components.

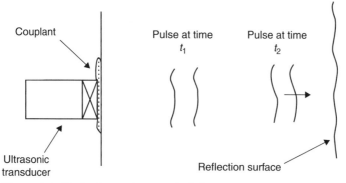

Figure A.1. Basic ultrasonic test principle.

We now outline some additional key parameters of wave propagation. *Attenuation* can come about from internal friction or energy absorption in a material, or it can also be geometric as in the case of spherical or cylindrical wave propagation. In general, attenuation produces a reduction in magnitude even if pulse duration is constant (nondispersive media). In special cases, however, where attenuation can be considered a function of frequency (dispersive media), pulse spreading can be observed, which obviously leads to magnitude reduction as well.

Scattering can produce both magnitude reductions and pulse spreading as a result of wave transmission from a transducer or from wave interaction with a small obstacle or flaw. Many approximations of scattering analysis are used in ultrasonic analysis. The simplest scattering theory utilizes Huygens's principle and spherical scattering integrated over the entire surface area of the obstacle. Attenuation of the $1/r$ type is considered at a point near the receiving transducer. More advanced theories include mode conversion and energy partitioning to shear, as well as phase and magnitude reduction considerations.

A variety of *interface waves* can be produced in certain materials. For example, imagine an angle beam impinging on an air–solid interface (surface wave), a fluid–solid interface (Scholte) wave, or a solid–solid interface (Stoneley) wave. Note that the depth of penetration of a surface wave equals approximately half to twice the wavelength. Surface waves can follow a curved surface, provided the radius of curvature is roughly greater than one half the wavelength. These topics are covered in Chapters 7 and 9.

Ultrasonic waves can be generated inside a material by placing a piezoelectric element in contact with the surface and then pulsing the element with an appropriate voltage-versus-time profile. The piezoelectric element converts electrical energy into mechanical energy by what is commonly known as the piezoelectric effect. Application of a suitable couplant material between the piezoelectric element and the material in question allows ultrasonic waves to propagate efficiently from the transducer element into the test material. The basic test principle is illustrated for a bulk wave in Figure A.1.

Good sound transmission is important in order to perform appropriate signal analysis. Amplification of the echoes recorded on a test instrument can occur either before the sound leaves the transducer or after it is received by the transducer. Ultrasonic wave frequencies are normally used in the range between 40 kHz and

Continuous wave (CW)

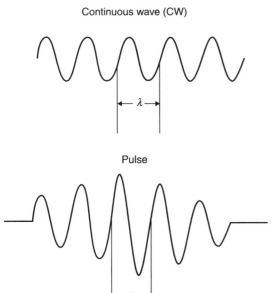

Pulse

Figure A.2. Basic ultrasonic waveforms.

500 MHz, although recent applications in concrete testing go even lower than 40 kHz and applications in acoustic microscopy often go beyond 500 MHz. Note that the audible region starts at close to 20 kHz.

Wavelength in a material can be calculated as a function of the input frequency f and wave velocity c, which is a characteristic of the material and structure in question:

$$c = f\lambda. \tag{A.1}$$

For example, here is the wavelength calculation for a 1.5-MHz transducer in water with $c = 1,500$ m/s:

$$\begin{aligned}
\lambda = \frac{c}{f} &= \frac{1,500 \text{ m/s}}{1.5 \times 10^6 \text{ cycles/s}} \\
&= 1,000 \times 10^{-6} \text{ m/cycle} \\
&= .001 \text{ m/cycle} \\
&= 1 \text{ mm/cycle.}
\end{aligned}$$

In steel, $c = 6,000$ m/s and so $\lambda = 4$ mm/cycle for the same 1.5-MHz transducer.

Pulse-echo ultrasonic techniques are often employed in NDE. In general, input waves may be either continuous wave (CW) or of pulse mode, as shown in Figure A.2. Note that λ in the pulse case is approximate because of the frequency bandwidth of the pulse. The wavelength λ agrees more closely with the formula $\lambda = c/f$ if the pulse is of narrowband frequency content and hence of many cycles (20 or more). For a highly damped pulse of (say) $1\frac{1}{2}$ to 3 cycles, the pulse is very broadbanded in frequency content and hence the λ value changes across the pulse, with the average

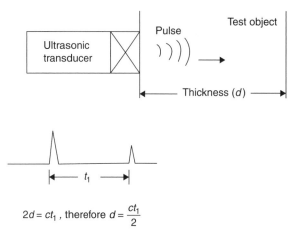

$2d = ct_1$, therefore $d = \dfrac{ct_1}{2}$

Figure A.3. Basic ultrasonic pulse-echo technique.

value being quite close to that obtained with the formula if the frequency is close to the center frequency of the pulse.

A simple example of a pulse-echo experiment is illustrated in Figure A.3. First, take a look at the pulse-echo contact ultrasonic measurement as shown in the figure. Results are displayed on a cathode ray tube (CRT), giving us a plot of amplitude versus time. Time can be measured from the oscilloscope display as t_1. The distance or thickness d of the test object can then be computed from the following formula: $d = ct_1/2$.

Several concepts related to these principles may be outlined as follows. Reflection factor analysis is essential in ultrasonic physics. A plane wave encountering an interface between two materials is divided into two components, one of which is reflected and one that is transmitted beyond the interface. Ultrasonic angle beams reflect and refract, or bend, at an interface. Critical angle analysis and analogies with optical systems are useful, along with a detailed explanation of mode conversion into longitudinal and shear waves.

An understanding of ultrasonic field analysis is critical in both nondestructive testing (NDT) and diagnostic ultrasound. Such topics – which include near-field and far-field analysis, ultrasonic beam angle of divergence, and pressure variations through the ultrasonic field – are useful in transducer design analysis, signal interpretation work, signal processing analysis, and also in understanding reflector and material classification. Although a detailed understanding of the analytical tools required in ultrasonic field analysis are not necessary for carrying out successful work in either NDT or diagnostic analysis, the basic philosophy on which these calculations are based is useful for understanding and appreciating the work tasks associated with advanced field analysis, transducer and instrument selection, signal display techniques, and problems associated with resolution, image improvement, and reflector classification. An understanding of ultrasonic field analysis also allows us to perform and modify tests on an interactive basis, thereby providing us with an effective feedback process for improving data acquisition, signal interpretation, and so forth.

Let's now review Huygens's principle and examine its role in ultrasonic field analysis and basic scattering theory. Scattering theory is associated with the calculation process of pressure or stress waves being reflected from a reflector.

Essentially, a plane wavefront could be approximated as an infinite number of spherical waves propagating from a source distribution along a straight line. Huygens's principle states that a finite scatterer in a material (considered in a fluid, initially) could be replaced by an infinite number of point sources placed over the surface area of that object. It is now possible to examine the wave propagation characteristics reflecting from the surface by mathematically adding together the contributions of all point sources at some particular desired point in the ultrasonic field. The summing process could take place by integration and consideration of a differential pressure function emanating from a differential element on the surface of the scatterer. An alternate approach exists where the scatterer could be replaced by a finite number of point sources and summing point by point to obtain the total solution. This approach is acceptable if the wavelength selected for the analysis is greater than the largest length of the finite element segment established on the surface of the scatterer. Aspects of Huygens's principle can be extended to include wave motion analysis in solid materials. The problem becomes difficult, however, because of the mode conversion and energy partitioning of both longitudinal and shear waves reflecting from the surface of an arbitrarily shaped scatterer. Much research has been focused on this subject; see, for example, Hueter and Bolt (1955), Kinsler et al. (1982), Pain (1993), and Pierce (1989).

A.2 Wave Interference

An understanding of wave interference is the key to understanding most aspects of ultrasonic wave propagation. Here we outline a brief explanation of the superposition process of waveforms that lead to constructive and destructive interference. First, consider the constructive interference problem illustrated in Figure A.4. If two point-source ultrasonic wave generators are stationed at positions 1 and 2, variation of ultrasonic pressure throughout the ultrasonic field will take place as a result of interference phenomena of the two wave fields as they propagate toward some point in question. In the constructive interference problem shown, the final solution at point 3 in space consists of the direct sum of the components coming from positions 1 and 2, as the two components arrive in phase at the point in question. Constructive interference can also take place at many other points in the ultrasonic field – namely, at any point where the distance between points 1 and 3 and points 2 and 3 differ by an integral number of wavelengths.

A destructive interference problem is shown in Figure A.5. In this case, the two components arrive at point 4 out of phase by half a wavelength. The net result in this case is total destructive interference. An intermediate interference problem for an arbitrary point 5 in space is illustrated in Figure A.6.

Throughout this textbook, the interference process (based on waveform superposition) is used repeatedly to explain various physical concepts of great value in ultrasonic analysis and in diagnostic ultrasound. Superposition phenomena are useful in understanding principles of frequency analysis and beam focusing as well as in multi-element transducer design.

A.3 Computational Model for a Single Point Source

A first-order approximation of the ultrasonic field in solid media is based on the computational model of a single point source in a fluid. Pressure variations resulting

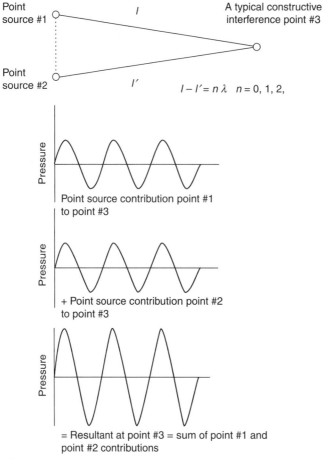

Figure A.4. A sample constructive interference problem.

from excitation of a piezoelectric element (or from ultrasonic wave interaction with a small reflector) can be calculated by considering the well-known point-source solution in a fluid in combination with Huygens's principle. A point-source pressure excitation in a fluid produces a spherical pressure field of the form

$$p(r,t) = \frac{A_0}{r} e^{i(kr - \omega t)}, \tag{A.2}$$

where $A_0 = \rho c U_0$, $\omega = 2\pi f$, U_0 is amplitude of the outgoing wave, c is the wave velocity, and ρ is density. Consider the pressure at the appropriate point B in Figure A.7, where r_1 and r_2 are much larger compared with d ($r_1 \gg d$, $r_2 \gg d$). This allows us to make the approximation $\delta_1 \approx \delta_2 = \delta$ and $(d \sin \theta)r_2 - r = r - r_1 = \delta$. The resulting pressure at the point B is the superposition from points S_1 and S_2:

$$
\begin{aligned}
p_B &= \frac{A_0}{r_1} e^{i(kr - k\delta - \omega t)} + \frac{A_0}{r_2} e^{i(kr - k\delta - \omega t)} \\
&\approx \frac{A_0}{r} e^{i(kr - \omega t)} \left[e^{ik\delta} + e^{-ik\delta} \right] \\
&= \frac{2A_0}{r} e^{i(kr - \omega t)} \cos[kd \sin \theta],
\end{aligned}
\tag{A.3}
$$

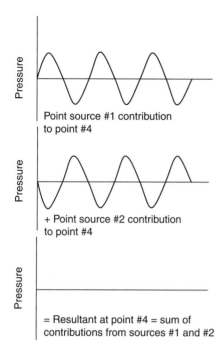

Pressure

Point source #1 contribution
to point #4

Pressure

+ Point source #2 contribution
to point #4

Pressure

= Resultant at point #4 = sum of
contributions from sources #1 and #2

$$l - l' = \left(\frac{2n-1}{2} \right) \lambda \quad n = 0, 1, 2, 3, \ldots$$

Figure A.5. A sample destructive interference problem.

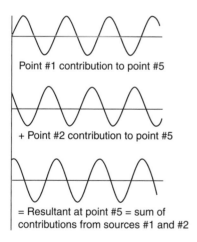

Point #1 contribution to point #5

+ Point #2 contribution to point #5

= Resultant at point #5 = sum of
contributions from sources #1 and #2

$$l - l' \neq n\lambda \quad n = 0, 1, 2,$$

$$l - l' \neq \left(\frac{2n-1}{2} \right) \lambda \quad n = 0, 1, 2, 3, \ldots$$

Figure A.6. A sample intermediate interference problem.

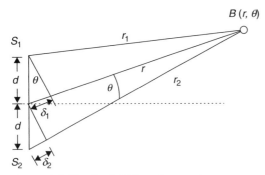

Figure A.7. The far-field interference from two equal sources.

where $r \gg d$ and

$$k = \frac{2\pi}{\lambda}. \tag{A.4}$$

The parameter δ governs the interference pattern. From (A.3), we can find nodal points where the pressure is almost zero – hence, an interference maxima. Nodal lines occur where

$$\cos(kd\sin\theta) = \cos\left(\frac{2\pi}{\lambda}(d\sin\theta)\right) = 0. \tag{A.5}$$

This leads to the relationship

$$\frac{2\pi}{\lambda}(d\sin\theta) = (2n+1)\frac{\pi}{2} \quad (n = 0, 1, 2, \ldots). \tag{A.6}$$

From (A.6) we obtain that

$$\sin\theta_n = \left(n+\frac{1}{2}\right)\frac{\lambda}{2d} \quad \text{(nodal lines)}. \tag{A.7}$$

For maximum value of interference, Equation (A.5) leads to

$$\cos\left(\frac{2\pi}{\lambda}(d\sin\theta)\right) = 1; \tag{A.8}$$

this gives directions for maxima interference $\sin\theta_n = n\lambda/d$ ($n = 0, 1, 2, \ldots$). From (A.3), we can calculate an amplitude for the arbitrary direction.

We will now visualize radiation from a circular piston. Suppose that a circular piston of radius a vibrates uniformly with some known or prescribed tone-burst displacement function. We can start with Huygens's principle and consider each element of the piston as a single source; we can then integrate over the surface because the amplitude at B is the result of superposition from all points. As in Figure A.8, we consider the resulting pressure at some point B obtained from all the elements of the piston surface.

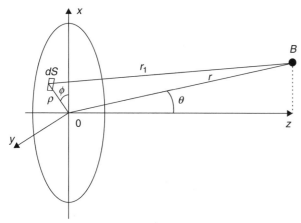

Figure A.8. Coordinate system used to derive characteristics of a flat cylindrical piston.

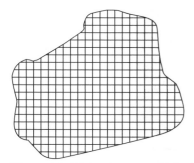

Figure A.9. Finite element grid for field computation.

The field due to a reflecting source can then be derived by integrating the point-source solution over the face of the source, which is actually a superposition process of all point-source contributions over the face of the ultrasonic transducer. The geometry of such a situation is shown in Figure A.8 for a cylindrical element. A numerical approximation procedure exists for obtaining a solution to the ultrasonic field variations from a piezoelectric source or reflecting element. A reflecting element could be divided into a finite number of elements of approximately equal area, as illustrated in Figure A.9.

The pressure variations generated from the arbitrarily shaped transducer element or reflecting element can then be calculated by summing the solutions (at a desired point in space) resulting from a finite number of individual solutions from each finite element on the reflecting surface. The magnitude variations can be calculated from the well-known point-source fluid solution. The phase variations, however, must be considered from a three-dimensional geometric point of view. Note that, as the size of the finite elements diminish, the resulting pressure function approaches the exact result that would be obtained by an exact mathematical integration process. Reasonable results and convergence to the true solution are found by selecting element sizes with dimensions less than the wavelength being considered in the field analysis problem. In fact, a general numerical approximation rule calls for wavelength λ to be greater than l, the largest linear dimension contained within the

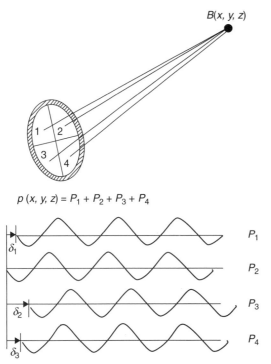

$$p(x, y, z) = P_1 + P_2 + P_3 + P_4$$

Figure A.10. Finite element numerical procedure.

finite element. For wavelengths larger than the maximum transducer diameter, the resulting ultrasonic field solution is simply the well-known point-source spherical wave solution in the fluid.

In order to demonstrate this numerical solution process, let us briefly consider the example presented in Figure A.10. If a solution is desired at point B in three-dimensional space from the piezoelectric element shown (containing just four elements), the calculation process would proceed as follows. The resulting solution would consist of a superposition process of four spherical wave solutions. Distances traveled from the piezoelectric element to point B must be calculated by trigonometry, resulting in values for $x_1, x_2, x_3,$ and x_4. Appropriate attenuation could then be calculated by letting the B values in the point-source solutions be equal to $x_1, x_2, x_3,$ and x_4. Note that the resulting solution at point B would consist of summing four solutions, each solution having slightly different arrival times and amplitudes: $p(x, y, z) = p_1 + p_2 + p_3 + p_4$. The resulting solution is obtained by adding the individual solutions together for each time position. Although the solution presented is for a CW excitation case, provisions for extensions into a pulse-type excitation can be made rather easily by way of Fourier series and transform analysis.

A.4 Directivity Function for a Cylindrical Element

A consideration of the coordinate system used to derive characteristics of a flat cylindrical piston (illustrated in Figure A.8) allows us to consider a more precise mathematical closed-form solution of the field emanating from a cylindrical element. The first approximation in this solution involves the representation of the actual distance between an arbitrary point on the surface of the disc source and an arbitrary point in the field.

Consider when point B is far from the piston. In this case, piston radius b is much smaller than r and r_1 ($r \gg b$, $r_1 \gg b$). The difference $(r - r_1)$ approximately equals the projection value of vector $\bar{\rho}$ on the vector $\overline{0B}$. As a result, we conclude that

$$r - r_1 = \rho \sin \theta \cos \varphi. \tag{A.9}$$

The pressure in the field point B (see Figure A.8) can be obtained by dividing the surface of the piston into an infinite number of elements dS. The pressure at the point B produced by element dS can be written as

$$dp = \frac{A_0 dS}{r_1} e^{i(kr_1 - \omega t)}, \tag{A.10}$$

where r and r_1 are large compared to the piston radius. By substituting (A.9) into (A.10), we obtain the following approximation (in polar coordinates):

$$dp = \frac{A_0}{r} e^{i(kr - \omega t)} e^{-ik\rho \sin \theta \cos \phi} \rho \, d\rho \, d\phi. \tag{A.11}$$

The total pressure is:

$$
\begin{aligned}
p &= \frac{A_0}{r} e^{i(kr - \omega t)} \int_0^b \int_0^{2\pi} e^{-ik\rho \sin \theta \cos \phi} \rho \, d\rho \, d\phi \\
&= 2\pi \frac{A_0}{r} e^{i(kr - g\omega t)} \int_0^b J_0(k\rho \sin \theta) \rho \, d\rho \\
&= \pi b^2 \frac{A_0}{r} e^{i(kr - \omega t)} \left[\frac{2 J_1(kb \sin \theta)}{kb \sin \theta} \right].
\end{aligned}
\tag{A.12}
$$

Some numerical results of the far-field pressure distribution are graphed in Figure A.11.

A detailed study of this approximation and resulting errors in field analysis is included in Rose (1975). Directivity functions for other multiple point or line source oscillators can be found in the literature; see, for example, Hueter and Bolt (1955).

The amplitude of the axial pressure distribution in the near field of the piston can be described (Kinsler et al. 1982; Pierce 1989) as:

$$A_{\text{ax}} = 2A_0 \left| \sin \left\{ \frac{1}{2} kz \left[\sqrt{1 + \left(\frac{6}{z} \right)^2} - 1 \right] \right\} \right|, \tag{A.13}$$

where A_{ax} denotes axial amplitude and z is the axial distance from the center of the piston. For $z/b \gg 1$,

$$\sqrt{1 + \left(\frac{b}{z} \right)^2} \approx 1 + \frac{1}{2} \left(\frac{b}{z} \right)^2. \tag{A.14}$$

In Equation (A.11), if $z/b \gg kb$ then we would be led to believe that the distance from the observation point is large compared with the piston radius and wavelength. In this case, the asymptotic axial amplitude of the pressure distribution is

$$A_{\text{ax}} = \frac{1}{4} A_0 \cdot \frac{b}{z} \cdot kb. \tag{A.15}$$

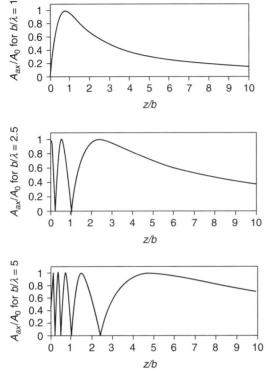

Figure A.11. Profile of the pressure distribution along the axial axis, where b is piston radius, λ is wavelength, and z is the axial distance from the piston.

As follows from (A.13), the near-field axial pressure distribution has strong interference phenomena. This effect occurs for

$$
\pi \frac{z}{\lambda}\left[\sqrt{1-\left(\frac{b}{z}\right)^{2}} -1\right]=\begin{cases}(\pi/2)(2n+1) & \text{for minima,}\\ \pi n & \text{for maxima,}\end{cases}\tag{A.16}
$$

where $n = 0, 1, 2, \ldots$. At the first maxima associated with $n = 0$ for z greater than this value, pressure decreases monotonically according to (A.15) as $1/r$; the first maxima is therefore a border between near and far field. Figure A.11 shows the amplitude of the near-field pressure distribution calculated from Equation (A.13).

A.5 Ultrasonic Field Presentations

Pressure variations in the ultrasonic field can be presented in a variety of ways. Although the distribution is three-dimensional, the practical presentation techniques are two-dimensional. Although many modified techniques are used and illustrated in the literature, two principle ones are most beneficial in the field analysis problem.

(1) *Axial pressure profile* – plots the maximum pressure of an ultrasonic waveform as a function of the axial coordinate emanating from the center line of a transducer element. The maximum pressure value is extracted as a peak-to-peak magnitude feature of the entire amplitude-versus-time profile as it passes the coordinate axis z. Sample curves are shown in Figure A.11.

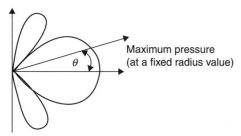

Maximum pressure
(at a fixed radius value)

θ

Figure A.12. Polar coordinate diffraction-type presentation.

(2) *Polar coordinate diffraction-type presentation* – plots maximum pressure against an angle θ. As we examine rays extending from the coordinate center point at this arbitrary angle θ, the maximum pressure value occurring at that angle can be measured along the radial coordinate, as shown in Figure A.12. This kind of presentation often produces side lobes of pressure energy that occur naturally because of constructive and destructive interference phenomena occurring in the superposition process of ultrasonic waveforms.

The characteristics of an ultrasonic transducer are caused by the interference of the point sources. For CW operation, there is a zone near the source that is characterized by large variations in field intensity (near field). Beyond this, the field intensity decreases smoothly (far field). Ultrasonic field data can be presented in a variety of ways. Both axial and transverse pressure profiles were presented here in an attempt to explain parameters for the near field and the angle of divergence field as well as aspects of computation based on the two profiles.

The polar coordinate presentation is an alternate and often useful format. This diffraction pattern provides us with useful information on beam angle of divergence. Even more importantly, the side lobe information and possible zero interaction pressures with scatterers at some off-angle position can easily be seen. In the plot of Figure A.12, the center line pressure (obtained at the coordinates $z = z_0, \theta = 0$) is arbitrarily set to unity. From this plot, angle-of-divergence values can easily be measured to (say) a zero pressure point, a 6-dB-down pressure point, and so forth. Note that, as the frequency and transducer radius b change, the diffraction pattern changes. As a rule of thumb, the following observation can be made: increasing f or b decreases the angle of divergence and increases the number of side lobes (see Figure A.13).

A.6 Near-Field Calculations

The near-field distance of a CW-type transducer is usually defined as that point on the axis of the transducer separating a region of intense oscillations from a region of a smooth intensity decay. The location of this point is the last of several local maxima. The farthest maximum for CW excitation occurs when

$$z = \frac{b^2}{\lambda} - \frac{\lambda}{4} = \frac{D^2 - \lambda^2}{4\lambda},$$

(A.17)

as illustrated earlier in this appendix. The last maximum is easily observed in Figure A.11.

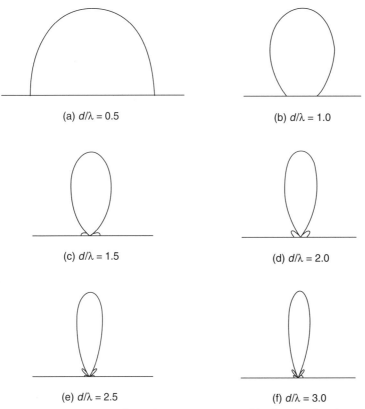

Figure A.13. Polar coordinate beam pressure profiles for circular pistons of various d/λ ratios, where d = diameter and λ = wavelength.

It is generally assumed that λ^2 can be neglected compared to D^2, in which case the near-field location is given by

$$N = \frac{D^2}{4\lambda}.$$

(A.18)

A.7 Angle-of-Divergence Calculations

One commonly used characteristic of the ultrasonic field is the beam angle of divergence. This characteristic is used as an indication of the relative intensity distribution of the beam. Basic textbooks show the ultrasonic field – confined to a cylinder having the same diameter as the transducer – extending from the transducer face to an axial distance of one near field. From this point, the field is generally considered (from a qualitative viewpoint) as being conical, with the apex at the center of the transducer face. From a quantitative viewpoint, the half-angle of divergence to a zero pressure point, based on the calculation of the first zero of the Bessel function for CW far-field work, yields

$$\sin \alpha = \frac{0.6\lambda}{b},$$

(A.19)

where b is the transducer radius.

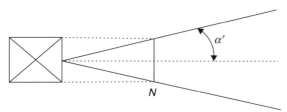

Figure A.14. Ultrasonic beam control.

A.8 Ultrasonic Beam Control

Ultrasonic field variations are controlled by varying transducer geometry, frequency, and size. Advanced topics on transducer design and beam control are presented in many texts. For now, basic elements of beam control can be evaluated by studying the near-field formula

$$N = \frac{D^2}{4\lambda} = \frac{D^2 f}{4c}$$

and the formula (A.19) for angle of divergence α: $\sin \alpha = 0.6\lambda/b$.

A small angle of divergence is usually desirable. Examining (A.19) shows that this can be obtained by using a small wavelength (which is the same as high frequency) and a transducer with large radius b. However, the problem with this approach is that the near field increases tremendously. In order to decrease the near field, it is desirable in most applications to avoid the confusion zone of constructive and destructive interference; hence it is necessary to *decrease* transducer diameter D (radius b) as well as frequency. Thus, beam control calls for a compromise in the selection of transducers for specific applications.

A more qualitative understanding of the relationship between near field and beam angle of divergence α can be obtained by studying the simplified diagram in Figure A.14. Clearly, as N increases, α' decreases (where α' is simply the angle in the box shown from the transducer diameter to the near-field point). Note that α' and α are different: α is used in a quantitative sense and α' in a qualitative sense.

A.9 A Note on Ultrasonic Field Solution Techniques

We shall now detail a comparison of two solution techniques as an exercise to illustrate the value and utility of a directivity function. The comparison – between approximate and exact (finite element) formulations – was conducted to determine type and magnitude of the errors introduced. A transverse field distribution was generated for each technique for CW b/λ ratios of 10 and 40; these distributions were computed at several axial distances. The results, shown in Figure A.15, indicate there are two types of errors that must be considered: signal amplitude variations along the center line z, and the spatial distribution profile at particular values of z.

The amplitude error does not become negligible until a distance of approximately "two near fields" is reached. (By the way, identical results were obtained for the case of $b/\lambda = 40$ and $b = 0.259$ inches, except that the relative pressure amplitude was four times less than the $b/\lambda = 10$ case.) If one of the signals being considered is within two near fields and the other is far from the transducer, a considerable error could

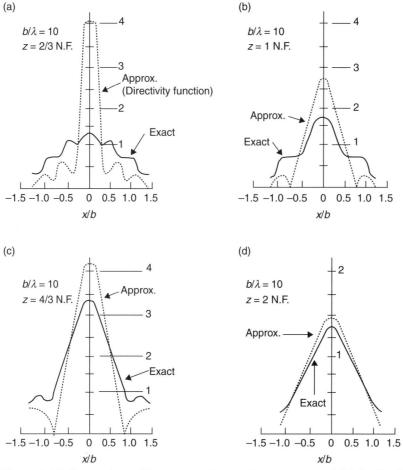

Figure A.15. Comparison of the exact and approximate pressure field distributions (N.F. denotes near-field distance).

be introduced by using the approximate equations. If the transverse distributions are considered then we will find that – although the error in magnitude of the axial intensity does not decrease to an acceptable level until a distance of two near fields is reached – the shapes of the distribution converge within a distance of one near field. The approximate solution is reasonable for axial distances greater than the near field of the transducer.

A.10 Time and Frequency Domain Analysis

In ultrasonic analysis it is essential to think in both time and frequency domain. See for example textbooks on Fourier Transforms for principles and concepts. See Table A.1 for basic information on this subject constantly useful in ultrasonic understanding.

A.11 Pulsed Ultrasonic Field Effects

The discussion has thus far been restricted to CW sources and the fields they generate. Most realistic applications of ultrasonics, however, deal with pulsed transducers. In many cases, the expected beam angle of divergence or near-field distance for a pulsed

Table A.1. Time and Frequency Domain essentials in ultrasonics

Time Domain	Frequency Domain
Continuous wave RF signal	Spike in frequency domain
5-20 RF cycle Gaussian envelope profile	Narrowband frequency content, a tone burst
1 ½ to 2 RF cycle Gaussian envelope profile	Broadband frequency content
A dirac delta spike	White noise or flat frequency response

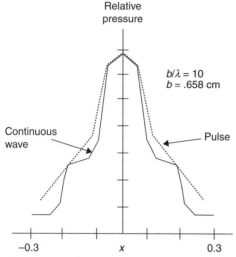

Figure A.16. Comparison of spatial distribution results for a pulse and continuous wave response function.

transducer is calculated using formulations based on a CW source. Yet this approach is flawed insofar as ultrasonic field distributions are a function of pulse shape.

Fourier analysis can enable a mathematical computation of the true beam angle of divergence. First, the Fourier components of the pulse are computed and then a traditional ultrasonic field is generated for each of the CW components. The intensity distribution of the pulse is the summation of the component distributions. It must be mentioned, however, that each of the components will have a slightly different pressure profile. Although all have the same magnitude when normalized on the transducer axis, the intensity of the components having higher D/λ values will increase more rapidly as the field point is moved away from the transducer axis. This implies that the pulse shape actually changes, since the spectral composition of the pulse is changing. The extent of this change can only be determined once the spectral content of the transmitted pulse is known.

Many other differences in ultrasonic field parameters exist for broadband transducers as compared with narrowband ones, and this should be carefully considered when doing ultrasonic field work. A qualitative comparison of spatial distribution results (transverse pressure profiles) for a continuous waveform versus a typical ultrasonic pulse form is shown in Figure A.16. A relatively smooth profile is obtained for the pulse waveform. Figure A.17 compares two axial pressure profiles, where the CW response function shows severe oscillations at values less than 3.0 inches. Once again, a smoother distribution occurs for the pulse waveform. In both cases, however, the near field is at 2.75 inches, which is close to the theoretical

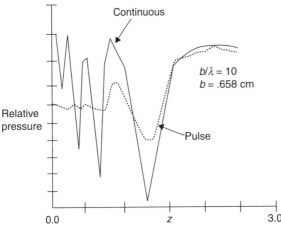

Figure A.17. Comparison of axial pressure results for a pulse and continuous wave response function.

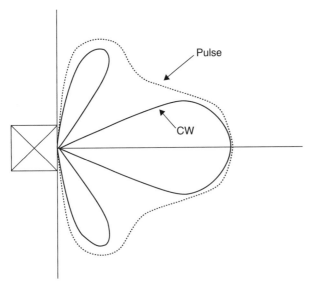

Figure A.18. Pulse effects on a diffraction pattern.

values using center-line frequency (from the frequency profile) as the nominal frequency in the calculations.

An ultrasonic field for an arbitrarily shaped ultrasonic waveform (a pulse) can be evaluated by considering the superposition of many different ultrasonic fields, one for each individual frequency component. As indicated earlier, irregularities associated with the axial and transverse profiles are generally reduced substantially by a smoothing effect. A diffraction pattern is likewise smoothed, as illustrated in Figure A.18.

We have calculated axial pressure profiles for six different ultrasonic response functions; the results are shown in Figures A.19A and A.19B. In each case, D/λ was selected close to 20 with a center-line nominal frequency of 2.25 MHz. As the ringing decreases, oscillations in the near field decrease; as the ringing increases, results approach those of the CW case. Additional details on the mathematical solution process associated with Fourier transform analysis are presented in many papers and textbooks.

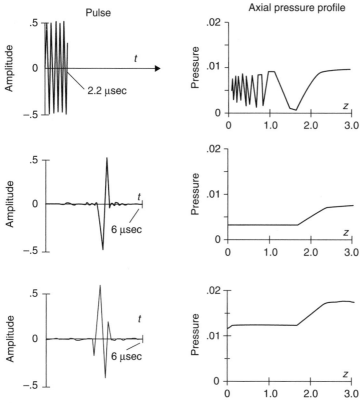

Figure A.19a. Pulse effects on axial pressure profile.

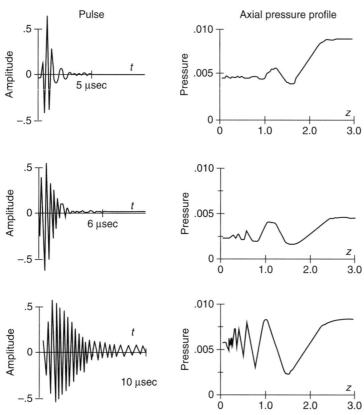

Figure A.19b. Pulse effects on axial pressure profile.

A.12 Introduction to Display Technology

Here we present a brief outline of display technology as an aid to understanding the basic elements of ultrasonic analysis. The four principal scanning types may be listed as follows.

(1) *A-scan*. This is an amplitude-versus-time display, samples of which are shown in Figure A.19.
(2) *B-scan*. This represents the brightness mode. An image of a cross section can be presented on a cathode ray tube (CRT) by keeping track of the amplitudes reflected from various portions of the cross section (see Figure A.20). Time gain compensation (TGC) is often used to present an improved image, whereby echoes appearing from greater distances are displayed as being greater in magnitude. Experience allows us to develop a nonlinear amplification function enabling improved image resolution. This technique is particularly useful in diagnostic ultrasound.
(3) *M-scan*. This is a time position scan that allows us to observe moving objects. A popular example involves motion within a body, as in scanning a moving cardiac wall by plotting echo position versus time.
(4) *C-scan*. This is a planar view at a particular depth range of the defects in that plane.

A sample C-scan result is presented in Figure A.21. Two Plexiglas plates were bonded together; the white areas indicate good bonding whereas black indicates no-bond areas. In this case, the intermediate echo is rather substantial in size. Note

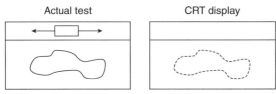

Figure A.20. B-scan concept.

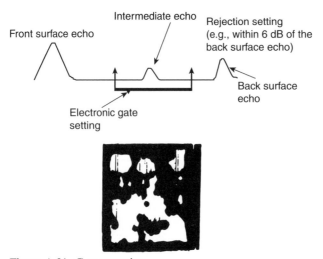

Figure A.21. C-scan testing.

that C-scan control maps with gray scaling can be obtained by rerunning a C-scan at different reject levels – say, at 4 dB, 6 dB, and 8 dB.

A.13 Amplitude Reduction of an Ultrasonic Waveform

We will now list the ways in which an ultrasonic waveform can be reduced in magnitude.

(1) The *reflection factor* at an interface between two materials can reduce waveform magnitude.
(2) Ultrasonic waveform magnitudes are reduced by *dispersion* due to geometric effects induced by cylindrical or plate-type structures or layered media. Such effects are actually a summation of reflection factor effects.
(3) Ultrasonic *beam spreading* causes a reduction in magnitude (see Section A.5 for additional details). Scattering from a reflector is also included in beam-spread amplitude reduction, since additional beam spreading occurs as an ultrasonic wave is reflected from a scatterer or from a foreign object or structural defect inside a material.
(4) *Attenuation* due to either spherical or cylindrical wave propagation can cause reduction in amplitude. This attenuation is nonabsorbing and nondispersive; it is due to geometric decay only.
(5) Attenuation due to *absorption* mechanisms includes internal friction and internal scattering, which also reduce wave magnitudes.
(6) Electronic *processing* (e.g., filtering and amplification) can also cause amplitude reduction.

A.14 Resolution and Penetration Principles

The resolution and penetrating power of an ultrasonic wave depends on the wavelength of excitation inside the material in question. Greater wavelengths or lower frequencies generally penetrate much further into a material and result in less absorption. Higher-frequency ultrasonic excitations with smaller wavelengths generally decay more rapidly inside a material, but resolution capability is improved. Now, we will address some basic definitions with respect to axial and lateral resolution of an ultrasonic transducer.

A.14.1 Axial Resolution

The smallest axial distance that can be resolved between two axially located reflections is called the *axial resolution* of the transducer; see Figure A.22. For example, axial

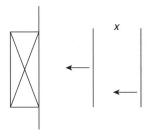

Figure A.22. Axial resolution definition.

(a) Low frequency, poorly damped

(b) High frequency, poorly damped

(c) High frequency, highly damped

Figure A.23. Axial resolution improvement.

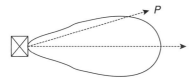

(a) Poor lateral resolution potential

(b) Good lateral resolution potential

Figure A.24. Lateral resolution improvement.

resolution is the smallest distance before superposition of the incident and reflected echoes makes an axial separation measurement impossible. A formula for axial resolution can be derived as a function of pulse duration. Axial resolution is defined as the capability of separating two point reflectors in an axial direction from the ultrasonic transducer. The best possible axial resolution can therefore come about by using as short a pulse as possible. This is illustrated in Figure A.23, where the best axial resolution is achieved for a high-frequency and highly damped ultrasonic transducer.

A.14.2 Lateral Resolution

We now move to lateral resolution improvement. *Lateral resolution* is defined as the ability to resolve two point reflectors in a lateral direction at some specific point inside the structure. It is clear that the narrowest possible ultrasonic beam should be used. Two polar coordinate profiles at a specified radius showing pressure versus angle are illustrated in Figure A.24. The narrow beam can be achieved by using a transducer of larger diameter or a higher frequency, or by using a focused transducer.

a.) Mechanical

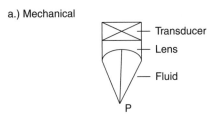

b.) Electronic

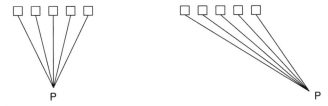

Figure A.25. Ultrasonic bulk wave focusing possibilities.

Lateral resolution will vary with depth because of ultrasonic beam divergence. The best lateral resolution will be achieved at an axial distance that is equal to either the near-field point or the design focal point of the ultrasonic transducer.

A.15 Phased Arrays and Beam Focusing

Phased array analysis in ultrasonic bulk wave inspection has become extremely popular this past decade because of the ease of use and rapid inspection speed. Almost everything in ultrasonic analysis can be linked to an understanding of wave interference phenomenon. So let's consider two examples of focusing, one mechanical and one electronic. See Figure A.25. For the mechanical method, the wave velocity in the fluid must be less than the wave velocity in the lens material or focusing will not occur. Now think of Huygen's principle and waves on the outer edge of the lens getting to the fluid media at a different time compared to the center of the lens, but now having the waves from all points on the transducer reaching point P at the same time.

For the electronic case, a comb type transducer configuration is used whereby each element of the comb can be activated electronically independently of the other elements. If you pulse the outer element of the comb transducer first, then move towards the center with later time delays, in such a manner that the waves arrive all at the same time at point P, the beam will then be focused at point P.

Suppose now that you want to focus or steer the beam to any region of the structure, it becomes possible to do so by applying electronic time delays across the elements in the comb transducer.

A.16 Exercises

1. Illustrate schematically the approximate qualitative change in a typical waveform as it travels from one position to another, showing (a) dispersion only, (b) attenuation as a function of frequency only, and (c) both dispersion and attenuation.

2. In a wave-scattering Huygens's principle exercise, how could you tell the difference between a flat reflector and a volumetric reflector? Also, how might you determine the size of each?
3. Calculate wavelength in steel and in water for a 1-inch transducer at 2.25 MHz.
4. Compute the near field and angle of divergence for a 1-inch-diameter transducer of 2.25 MHz in steel and in water.
5. How can you focus an ultrasonic transducer inside a solid material?
6. When using a finite element grid for field computation, how would you determine the mesh size?
7. Describe how to compute or find directivity functions in acoustics for transducers of arbitrary shape.
8. Explain the axial pressure profile smoothing process as a function of pulse frequency bandwidth.
9. Calculate axial resolution as a function of pulse duration.
10. Show how the zeros of the Bessel function could be used to estimate beam angle of divergence for a cylindrical piston source.
11. From the literature, list some directivity functions for different shaped transducers.
12. Make a sketch of a possible transverse pressure profile in a fluid for a continuous wave excitation and a narrow and broadband frequency bandwidth excitation.
13. Make sketches of the profiles referred to in Table A.1 for time and frequency domain profiles.
14. What is the ideal waveform to be used in ultrasonic bulk wave analysis? Why?
15. Make a sketch showing a phased array result to steer the beam to a 45° location from the normal to a loading surface.

A.17 REFERENCES

Hueter, T. F., and Bolt, R. H. (1955). *Sonics Techniques for the Use of Sound and Ultrasound in Engineering and Science*. New York: Wiley.

Krautkramer, J., and Krautkramer, H. (1990). *Ultrasonic Testing of Materials*, 4th ed. New York: Springer-Verlag.

Kinsler, L. E., Frey, A. R., Coppens, A. B., and Sanders, J. V. (1982). *Fundamentals of Acoustics*. New York: Wiley.

Pain, H. J. (1993). *The Physics of Vibrations and Waves*. New York: Wiley.

Pierce, A. D. (1989). *Acoustics: An Introduction to its Physical Principles and Applications*. Woodbury, NY: Acoustical Society of America.

Rose, J. L., and Meyer, P. A. (1975). Model for ultrasonic field analysis in solids, *J. Acoust. Soc. Am.* 57: 598–605.

Basic Formulas and Concepts in the Theory of Elasticity

B.1 Introduction

Elements of elasticity theory are fundamental to the studies of wave propagation in solid media. A brief review of the formulas and concepts are presented in this appendix for reference purposes. Many textbooks are available that cover the theory of elasticity, including such classic texts as Chou and Pagano (1992), Sokolnikoff (1956), and Timoshenko and Goodier (1987).

B.2 Nomenclature

Principal formulas and definitions may be listed as follows.

A = transformation matrix, E = Young's modulus,
F = body force, K = bulk modulus,
T = traction force, U = strain energy/volume,
u_i = displacement component, $\mu = G$ = shear modulus,
ε_{ij} = strain tensor, σ_{ij} or τ_{ij} = stress tensor,
ϕ = strain invariant (dilatation), θ = stress invariant,
ρ = density, τ = stress matrix,

v = Poisson's ratio, λ = Lamé constant $= \dfrac{2\mu v}{1-2v}$,

δ_{ij} = Kronecker delta $= \begin{cases} 1 & \text{if } i = j, \\ 0 & \text{if } i \neq j; \end{cases}$

u_i is the deformation in the x_i direction.
For the alternating tensor,

$$\epsilon_{ijk} \begin{cases} +1 & \text{all unequal, in cyclic order,} \\ -1 & \text{unequal, not in cyclic order,} \\ 0 & \text{any two indices equal.} \end{cases}$$

The strain tensor is

$$\varepsilon_{ij} = \frac{1}{2}\left(\frac{\partial u_i}{\partial x_j} + \frac{\partial u_j}{\partial x_i}\right) = \varepsilon_{ji},$$

and the rotation tensor is

$$-\omega_k = \omega_{ij} = -\omega_{ji} = \frac{1}{2}\left(\frac{\partial u_i}{\partial x_j} + \frac{\partial u_j}{\partial x_i}\right);$$

the dilatation is

$$\phi = \varepsilon_{ii} = \frac{\partial u_i}{\partial x_i} = \varepsilon_{11} + \varepsilon_{22} + \varepsilon_{33}.$$

Cartesian coordinate vector operators are

$$\nabla = \frac{\partial}{\partial x}\hat{i} + \frac{\partial}{\partial y}\hat{j} + \frac{\partial}{\partial z}\hat{k} \quad \text{(gradient)},$$

$$\nabla^2 = \frac{\partial^2}{\partial^2 x} + \frac{\partial^2}{\partial^2 y} + \frac{\partial^2}{\partial^2 z} \quad \text{(Laplacian)};$$

cylindrical coordinate vector operators are

$$\nabla = \frac{\partial}{\partial r}\hat{e}_r + \frac{1}{r}\frac{\partial}{\partial \theta}\hat{e}_\theta + \frac{\partial}{\partial z}\hat{e}_z \quad \text{(gradient)},$$

$$\nabla^2 = \frac{\partial^2}{\partial^2 r} + \frac{1}{r}\frac{\partial}{\partial r} + \frac{1}{r^2}\frac{\partial^2}{\partial \theta^2} + \frac{\partial^2}{\partial^2 z} \quad \text{(Laplacian)}.$$

In cylindrical coordinates, $u_r, u_\theta,$ and u_z are deformations in the r, θ, and z directions (respectively), and strain tensor components are

$$\varepsilon_{rr} = \frac{\partial u_r}{\partial r},$$

$$\varepsilon_{\theta\theta} = \frac{1}{r}\frac{\partial u_\theta}{\partial \theta} + \frac{u_r}{r},$$

$$\varepsilon_{zz} = \frac{\partial u_z}{\partial z},$$

$$\varepsilon_{r\theta} = \frac{1}{2}\left(\frac{1}{r}\frac{\partial u_r}{\partial \theta} + \frac{\partial u_\theta}{\partial r} - \frac{u_\theta}{r}\right),$$

$$\varepsilon_{rz} = \frac{1}{2}\left(\frac{\partial u_r}{\partial r} + \frac{\partial u_r}{\partial z}\right),$$

$$\varepsilon_{\theta z} = \frac{1}{2}\left(\frac{\partial u_\theta}{\partial z} + \frac{1}{r}\frac{\partial u_z}{\partial \theta}\right).$$

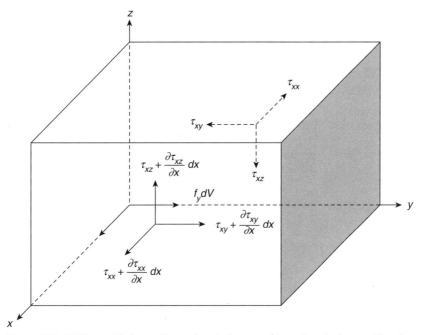

Figure B.1. Differential three-dimensional element. (A surface is denoted by the axis to which it is perpendicular. The stresses shown are on the positive surfaces. On the opposite or negative surfaces, the stresses are in the opposite directions.)

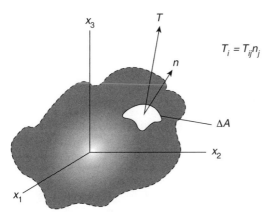

$$T_i = T_{ij}n_j$$

Figure B.2. Differential surface area element (**n** is the normal unit vector to the area ΔA; T is not in the direction of **n**).

The stress tensor τ_{ij} is a symmetric tensor that can be written as

$$\tau_{ij} = \tau_{ji} = \begin{bmatrix} \tau_{11} & \tau_{12} & \tau_{13} \\ \tau_{21} & \tau_{22} & \tau_{23} \\ \tau_{31} & \tau_{32} & \tau_{33} \end{bmatrix}$$

(see Figure B.1). Note also that, in Cartesian coordinates, $\tau_{11} = \tau_{xx}$ and so forth (i.e., indices $1, 2, 3$ map to axes x, y, z, resp.). The strain tensor ε_{ij} is also symmetric and can be written as

$$\varepsilon_{ij} = \begin{bmatrix} \varepsilon_{11} & \varepsilon_{12} & \varepsilon_{13} \\ \varepsilon_{21} & \varepsilon_{22} & \varepsilon_{23} \\ \varepsilon_{31} & \varepsilon_{32} & \varepsilon_{33} \end{bmatrix}.$$

Finally, the dot product $\bar{a} \bullet \bar{b} = a_i b_i$; the cross product $\bar{a} \times \bar{b} = \bar{c}(C_i = \epsilon_{ijk} \, a_j b_k)$.

B.3 Stress, Strain, and Constitutive Equations

Consider the application of stress on a differential element, as in Figure B.1. The relationship between traction and stress is $T_i = \tau_{ij} n_j$ (see Figure B.2). A general stress–strain relationship for anisotropic media is

$$\tau_{ij} = \lambda \delta_{ij} \phi + 2\mu\varepsilon_{ij} = \frac{Ev\phi}{(1+v)(1-2v)} \delta_{ij} + \frac{E}{1+v} \varepsilon_{ij},$$

where $\tau_{ij} = C_{ijkl}\varepsilon_{kl}$ (Hooke's law) and dilatation $\phi = \varepsilon_{11} + \varepsilon_{22} + \varepsilon_{33}$. In terms of strain we have

$$\varepsilon_{ij} = \frac{\lambda}{2\mu(3\lambda+2\mu)} \delta_{ij}\theta + \frac{1}{2\mu} \tau_{ij} = \frac{1+v}{E} \tau_{ij} - \frac{v}{E} \theta\delta_{ij},$$

where $\theta = \tau_{11} + \tau_{22} + \tau_{33}$ is the stress invariant and $\tau_{ik} = 2G\varepsilon_{ik} + \lambda\theta\delta_{ik}$. The equilibrium equations of elasticity can be written as

$$\frac{\partial \tau_{ij}}{\partial x_j} + F_i = 0.$$

B.4 Elastic Constant Relationships

Also note the following elastic constant relationships for an isotropic material:

$$E = \frac{\mu(3\lambda+2\mu)}{\lambda+\mu} = \frac{\lambda(1+v)(1-2v)}{v}$$

$$\mu = \frac{E}{2(1+v)}, \quad \lambda = \frac{vE}{(1+v)(1-2v)}, \quad v = \frac{\lambda}{2(\lambda+\mu)}.$$

The strain energy is

$$U = \frac{1}{2}(\lambda+2\mu)[\varepsilon_{xx}^2 + \varepsilon_{yy}^2 + \varepsilon_{zz}^2]$$
$$+ 2\mu[\varepsilon_{yx}^2 + \varepsilon_{x2}^2 + \varepsilon_{yz}^2] + \lambda[\varepsilon_{xx}\varepsilon_{yy} + \varepsilon_{zz}\varepsilon_{xx} + \varepsilon_{zz}\varepsilon_{yy}].$$

In terms of stress,

$$U = \frac{1}{2E}(\tau_{xx}^2 + \tau_{yy}^2 + \tau_{zz}^2) - \frac{v}{E}(\tau_{xx}\tau_{yy} + \tau_{yy}\tau_{zz} + \tau_{xx} + \tau_{zz}) + \frac{1}{2\mu}(\tau_{xy}^2 + \tau_{xz}^2 + \tau_{yx}^2).$$

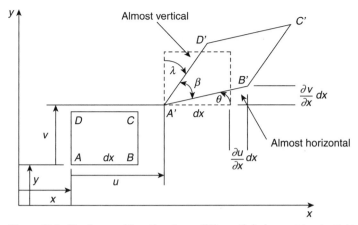

Figure B.3. Strain consideration for a differential element in elasticity.

B.5 Vector and Tensor Transformation

We now review briefly the particulars of vector and tensor transformation for rotating coordinate systems. Consider A as an orthogonal matrix of direction cosines between two different coordinate systems, primed and unprimed.

For a first-order tensor, $x'_i = a_{ij}x_j$, where $a_{ij} = \cos(x'_i, x_j)$. Note that i is a free index and that the js appearing twice on the right-hand side of the equation indicate summation according to the Einstein summation rule. In vector notation, $\bar{x}' = A\bar{x}$, where

$$A = \begin{array}{c} x' \\ y' \\ z' \end{array} \begin{array}{ccc} x & y & z \\ \left[\begin{array}{ccc} a_{11} & a_{12} & a_{13} \\ a_{21} & a_{22} & a_{23} \\ a_{31} & a_{32} & a_{33} \end{array} \right] \end{array}$$

and also (if needed) $x_i = a_{ji}x'_j$.

For a second-order tensor, $\tau'_{ik} = a_{ij}a_{kl}\tau_{jl}$ or $\tau' = A\tau A^{-1}$ for matrices in a similarity transformation.

B.6 Principal Stresses and Strains

We now consider the eigenvalue problem for principal stress and principal strain computation. The principal stresses are given by

$$|\tau_{ij} - \tau\delta_{ij}| = 0$$

and the vectors A_j by

$$|\tau_{ij} - \tau\delta_{ij}|[A_j] = 0.$$

The principal strains are given by the determinant equation

$$|\varepsilon_{ij} - \varepsilon\delta_{ij}| = 0,$$

and the eigenvectors A_j are given by

$$|\varepsilon_{ij} - \varepsilon\delta_{ij}|[A_j] = 0.$$

B.7 The Strain Displacement Equations

In this section we examine the derivation of the strain displacement equations

$$\varepsilon = [\underbrace{\varepsilon_x, \varepsilon_y, \varepsilon_z}_{\substack{\text{normal} \\ \text{strains}}} \underbrace{\gamma_{yz}, \gamma_{xz}, \gamma_{xy},}_{\substack{\text{engineering} \\ \text{shear strains}}}]^T.$$

For infinitesimally small deformations, consider translation and deformation of a two-dimensional element, as shown in Figure B.3. Note that

$$
\begin{aligned}
\varepsilon_x &= \frac{A'B' - AB}{AB} \frac{A'B' - dx}{dx}, \\
\varepsilon_y &= \frac{A'D' - AB}{AD} \frac{A'D' - dy}{dy},
\end{aligned}
\quad (*)
$$

$$\gamma_{xy} = \frac{\pi}{2} - \beta = \theta - \lambda \ \text{(since counterclockwise is positive)}.$$

Point B is displaced by

$$u + \frac{\partial u}{\partial x} dx, \quad v + \frac{\partial v}{\partial x} dx;$$

point D is displaced by

$$u + \frac{\partial u}{\partial y} dy, \quad v + \frac{\partial v}{\partial y} dy;$$

Therefore, in trying to solve for ε_x, we need $A'B'$. From $(*)$ we have

$$(A'B')^2 = [dx(1 + \varepsilon_x)]^2.$$

Separately,

$$(A'B')^2 = \left(dx + \frac{\partial u}{dx} dx \right) + \left(\frac{\partial u}{dx} dx \right)^2.$$

Therefore, $(A'B')^2 = (A'B')^2$ implies that

$$\varepsilon_x^2 + 2\varepsilon_x + 1 = 1 + 2\frac{\partial u}{\partial x} + \left(\frac{\partial u}{\partial x} \right)^2 \left(\frac{\partial v}{\partial x} \right)^2$$

For small strain and displacement, note that

$$\left(\frac{\partial u}{\partial x} \right)^2 \to 0 \quad \text{and} \quad \left(\frac{\partial v}{\partial x} \right)^2 \to 0$$

and that second-order effects are zero. Hence

$$\varepsilon_x = \frac{\partial u}{\partial x}$$

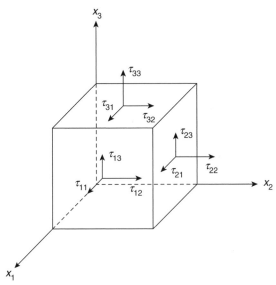

Figure B.4. Differential element used to derive the equilibrium equation and the governing three-dimensional wave equation.

and, similarly,

$$\varepsilon_y = \frac{\partial v}{\partial y}, \quad \varepsilon_z = \frac{\partial \omega}{\partial z}.$$

Now consider engineering shear strain:

$$\tan\theta \approx \theta = \frac{\dfrac{\partial v}{\partial x}\,dx}{dx + \left(\dfrac{\partial u}{\partial x}\right)dx}$$

$$= \frac{\dfrac{\partial v}{\partial x}\,dx}{dx\left(1 + \dfrac{\partial u}{\partial x}\right)},$$

where $\partial u/\partial x$ is small when compared with unity. Therefore,

$$\theta = \frac{\partial v}{\partial x}.$$

Similarly,

$$\lambda = \frac{-\partial u}{\partial y} = -\tan\lambda = \frac{\dfrac{\partial u}{\partial y}\,dy}{dy\left(1 + \dfrac{\partial v}{\partial y}\right)}$$

and so

$$\gamma_{xy} = \theta - \lambda = \frac{\partial u}{\partial y} + \frac{\partial v}{\partial x}.$$

These are the well-known strain displacement equations from elasticity. Note the linear approximations and small strain assumptions.

B.8 Derivation of the Governing Wave Equation

In light of the aspects of elasticity discussed so far, we may now consider how to derive Navier's governing wave equation. Stresses are shown on each face in Figure B.4, which depicts a cubic solid (six sides formed by three sets of parallel planes). Use also Figure B.1. The sum of the stresses is zero for the static case in (x, y, z)-coordinates, or equal to ma (from Newton's second law). Looking at the x direction, we see that $\tau_{xy} = \tau_{yx}$, $\tau_{yz} = \tau_{zy}$, and so forth. Note that $\sigma_x = \tau_{xx}$. For the static case we have

$$\sigma_x + \frac{\partial \sigma_x}{\partial_x} dx - \sigma_x - \tau_{yx} + \frac{\partial \sigma_{yx}}{\partial y} dy - \tau_{zx} + \tau_{zx} + \frac{\partial \sigma_{zx}}{\partial z} dz + f_x dV = 0.$$

Therefore,

$$\frac{\partial \sigma}{\partial x} + \frac{\partial \tau_{xy}}{\partial y} + \frac{\partial \tau_{xz}}{\partial z} + f_x = 0 [x, y, z]$$

(here, $\{x, y, z\}$ signifies the cyclic permutations required to obtain three equations). These are the equilibrium equations of elasticity.

Assuming a free index i and summing over j, we may write

$$\sigma_{ij,j} = 0, \quad \sigma_{11,1} + \sigma_{12,2} + \sigma_{13,3} = 0$$

for $i = 1, 2, 3$. We can also let $1, 2, 3$ become x, y, z as follows:

$$\partial \frac{\tau_{xx}}{\partial x} + \partial \frac{\tau_{xy}}{\partial y} + \partial \frac{\tau_{xz}}{\partial z} = 0.$$

Given $F = ma$ for motion, it follows that $\sigma_{ij,j} = \rho \ddot{u}_i$ for $i, j = 1, 2, 3$.

B.9 Anisotropic Elastic Constants

Consider the following general linearly elastic constitutive equation for anisotropic elastic constants: $\sigma_{ij} = C_{ijkl}\varepsilon_{kl}$. The term C_{ijkl} is a rank-4 tensor with $3^4 = 81$ components; however, because of symmetry, only 21 coefficients are needed to describe a general anisotropic body. Stress and strain are also symmetric. The four subscripts of C_{ijkl} can be reduced to two as follows.

For the tensor notation: 11 22 33 23 32 31 13 12 21
Use the abbreviated matrix notation: 1   2   3   4   5   6

This allows us to consider the simple matrix multiplication $\tau = C\varepsilon$, where τ is a 6×1 matrix, C is 6×6, and ε is 6×1:

$$
\begin{bmatrix} \tau_{11} \\ \tau_{22} \\ \tau_{33} \\ \tau_{12} \\ \tau_{23} \\ \tau_{13} \end{bmatrix} = \begin{bmatrix} C_{11} & C_{12} & \cdot & \cdot & \cdot & C_{16} \\ C_{21} & C_{22} & & & & \\ \cdot & & \cdot & & & \\ \cdot & & & \cdot & & \\ \cdot & & & & \cdot & \\ C_{61} & & & & & C_{66} \end{bmatrix} \begin{bmatrix} \varepsilon_{11} \\ \varepsilon_{22} \\ \varepsilon_{33} \\ \varepsilon_{12} \\ \varepsilon_{23} \\ \varepsilon_{13} \end{bmatrix}
$$

The number of constants for various types of anisotropic materials may be listed as follows:

- 21 constants for triclinic materials;
- 13 constants for monoclinic materials;
- 9 constants for orthorhombic or orthotropic materials;
- 7 or 6 constants for tetragonal materials;
- 7 constants for trigonal materials;
- 5 constants for hexagonal or transversely isotropic materials;
- 3 constants for cubic materials;
- 2 constants for isotropic materials.

We now summarize the elastic constant stiffness coefficient matrices for a few selected anisotropic materials. For more detail on anisotropy, see Auld (1990) or Pollard (1977).

Triclinic: 21 constants

C_{11}	C_{12}	C_{13}	C_{14}	C_{15}	C_{16}
	C_{22}	C_{23}	C_{24}	C_{25}	C_{26}
		C_{33}	C_{34}	C_{35}	C_{36}
			C_{44}	C_{45}	C_{46}
				C_{55}	C_{56}
					C_{66}

Monoclinic: 13 constants (standard orientation)

C_{11}	C_{12}	C_{13}	0	C_{15}	0
	C_{22}	C_{23}	0	C_{25}	0
		C_{33}	0	C_{35}	0
			C_{44}	0	C_{46}
				C_{55}	0
					C_{66}

Orthorhombic: 9 constants

C_{11}	C_{12}	C_{13}	0	0	0
	C_{22}	C_{23}	0	0	0
		C_{33}	0	0	0
			C_{44}	0	0
				C_{55}	0
					C_{66}

Tetragonal: 7 constants

$$
\begin{bmatrix}
C_{11} & C_{12} & C_{13} & 0 & 0 & C_{16} \\
 & C_{11} & C_{13} & 0 & 0 & -C_{16} \\
 & & C_{33} & 0 & 0 & 0 \\
 & & & C_{44} & 0 & 0 \\
 & & & & C_{44} & 0 \\
 & & & & & C_{66}
\end{bmatrix}
$$

6 constants

$$
\begin{bmatrix}
C_{11} & C_{12} & C_{13} & 0 & 0 & 0 \\
 & C_{11} & C_{13} & 0 & 0 & 0 \\
 & & C_{33} & 0 & 0 & 0 \\
 & & & C_{44} & 0 & 0 \\
 & & & & C_{44} & 0 \\
 & & & & & C_{66}
\end{bmatrix}
$$

Trigonal: 7 constants

$$
\begin{bmatrix}
C_{11} & C_{12} & C_{13} & C_{14} & -C_{25} & 0 \\
 & C_{11} & C_{13} & -C_{14} & C_{25} & 0 \\
 & & C_{33} & 0 & 0 & 0 \\
 & & & C_{44} & 0 & C_{25} \\
 & & & & C_{44} & C_{14} \\
 & & & & & \frac{1}{2}(C_{11} - C_{12})
\end{bmatrix}
$$

Hexagonal: 5 constants

$$
\begin{bmatrix}
C_{11} & C_{12} & C_{13} & 0 & 0 & 0 \\
 & C_{11} & C_{13} & 0 & 0 & 0 \\
 & & C_{33} & 0 & 0 & 0 \\
 & & & C_{44} & 0 & 0 \\
 & & & & C_{44} & 0 \\
 & & & & & \frac{1}{2}(C_{11} - C_{12})
\end{bmatrix}
$$

Cubic: 3 constants

$$
\begin{bmatrix}
C_{11} & C_{12} & C_{12} & 0 & 0 & 0 \\
 & C_{11} & C_{12} & 0 & 0 & 0 \\
 & & C_{11} & 0 & 0 & 0 \\
 & & & C_{44} & 0 & 0 \\
 & & & & C_{44} & 0 \\
 & & & & & C_{44}
\end{bmatrix}
$$

Isotropic: 2 constants

C_{11}	C_{12}	C_{12}	0	0	0
	C_{11}	C_{12}	0	0	0
		C_{11}	0	0	0
			$\frac{1}{2}(C_{11}-C_{12})$	0	0
				$\frac{1}{2}(C_{11}-C_{12})$	0
					$\frac{1}{2}(C_{11}-C_{12})$

The anisotropic factor A is commonly used to evaluate the level of anisotropy:

$$A = \frac{2C_{44}}{C_{11}-C_{12}};$$

if the material is isotropic then $A = 1$.

A number of effective modulus theories are available to reduce an inhomogeneous multilayered composite material to a single homogeneous anisotropic layer for wave propagation and strength considerations. See Chapter 17 for more detail.

B.10 Exercises

1. Expand the equation for motion in Cartesian coordinates:

$$\sigma_{ij,j} = \rho \ddot{u}_i$$

B.11 REFERENCES

Auld, B. A. (1990). *Acoustic Fields and Waves in Solids*, 2nd ed., vols. 1 and 2. Malabar, FL: Kreiger.

Chou, P. C., and Pagano, N. J. (1992). *Elasticity: Tensor, Dyadic, and Engineering Approaches*. New York: Dover.

Pollard, H. F. (1977). *Sound Waves in Solids*. London: Pion Ltd.

Sokolnikoff, I. S. (1956). *Mathematical Theory of Elasticity*. New York: McGraw-Hill.

Timoshenko, S. P., and Goodier, J. N. (1987). *Theory of Elasticity*. 3rd ed. New York: McGraw-Hill.

Physically Based Signal Processing Concepts for Guided Waves

C.1 General Concepts

Interesting processing concepts are discussed quite often in guided wave studies. In guided wave analysis, it is often useful to think on a frequency basis rather than to employ more common time domain thinking. As a result, Fourier transform analysis is commonplace. An analytic envelope for a Hilbert transform is also useful. The short time Fourier transform (STFFT) can also provide significant insight into the studies of guided wave response functions encountered in different situations. Physical insight into a resulting image of a spectrogram and its relationship to a group velocity dispersion curve along with the wavelet transform and its ability to see when certain frequency packets arrives as a function of time, as well as its relationship to a group velocity dispersion curve. On the other hand, the 2-D Fourier transform (2DFFT) relates to portions of a phase velocity dispersion curve. Quite often, the partial images generated by these transform techniques can provide us with an indication of damage in a structure by shifting and other indicators.

This appendix gives an overview of some of these transform techniques that can also be used as a basis for extracting features that provide insight into important characteristics of guided waves. In particular, they provide data for constructing portions of the relevant dispersion curves, identifying the modes that are actually propagating, and providing a physical explanation of certain aspects of ultrasonic guided wave propagation.

The transform methods are:

Fourier Transform	(FFT)
Short Time Fourier Transform	(STFFT)
2-D Fourier Transform	(2DFFT)
Wavelet Transform	(WT)

The FFT is used to extract frequency and energy content. It is also used as the basis for the STFFT and the 2DFFT. The FFT, STFFT, and WT are applied to a single waveform and therefore do not provide information on the phase characteristics of the propagation. The 2DFFT requires a collection of spatially segmented waveforms that are periodically spaced and thus it retains phase information.

With significant contribution from Michael J. Avioli.

The FFT identifies propagating frequencies but does not indicate where particular frequencies occur temporally. The WT not only identifies the frequency content but also the location in time where particular frequencies occur.

The STFFT and the WT enable partial construction of the related group velocity dispersion curves whereas the 2DFFT enables partial construction of the related phase velocity dispersion curves.

C.2 The Fast Fourier Transform (FFT)

The Fast Fourier Transform (FFT) is a special algorithmic implementation of the Discrete Fourier Transform (DFT) that decreases the computational time of the DFT.

The DFT is $F(\Omega_k) = \sum_{i=1}^{N} f(t_i) e^{-j\frac{2\pi}{N}\Omega_{k-1} t_{(i-1)}}$ where t is time and Ω is frequency

with $k = 1, 2, ..., N;$ $\quad i = 1, 2, ..., N$

The FFT is typically implemented using a subroutine with the following input variables:

Freal(i) – Real part of the input function (e.g., waveform)
Fimag(i) – Imaginary part of input function (e.g., zero for real functions)
N number of points – Typically a power of 2; $N = 2^n$ where n = integer

$$i = 1, 2, ..., N$$

For example,
FFT Freal( ), Fimag( ), N

If the input function has only M points, where M is not a power of two, then zeroes can be added to the function, points $M + 1$ through N, where N is a power of $2 > M$.

$f(1), f(2), ..., f(M), f(M + 1), f(M + 2), ..., f(N) = f(1), f(2), ..., f(M), 0, 0, ... 0$
$f(M + 1), f(M + 2), ..., f(N) = $ "Zero padding"

The subroutine returns the real part of FFT as the array Freal ( ) and the imaginary part of the FFT as array Fimag ( ).

Figure C.1 shows a representative waveform and Figure C.2 shows the FFT of it.

The FFT returns positive frequency values in array elements 1 through $(N/2) + 1$.
The negative frequency values are array elements $(N/2) + 2$ through N.
The frequency domain sampling rate is the time domain frequency sampling rate, f_{samp}, divided by N;

$$\Delta f = \frac{f_{samp}}{N} \ or \ \Delta f = \frac{1}{\Delta t \, N} \ \text{where } \Delta t \text{ is the time sampling rate.}$$

The Fourier magnitude spectrum, $S(j)$, is obtained by

$$S(j) = \sqrt{Freal(j)^2 + Fimag(j)^2} \quad j = 1, 2, ..., N$$

For positive frequencies only, $j = 1, 2, ..., N/2+1$

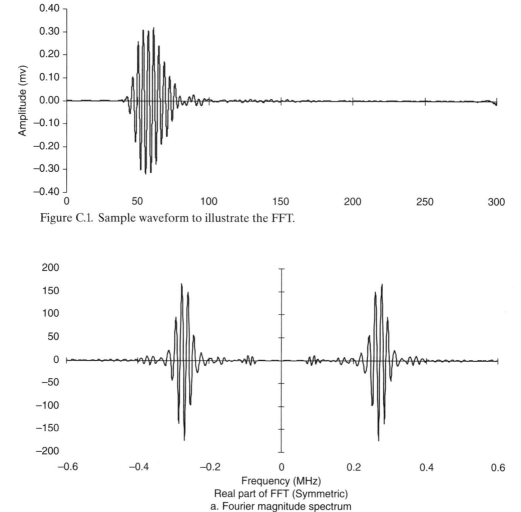

Figure C.1. Sample waveform to illustrate the FFT.

Real part of FFT (Symmetric)
a. Fourier magnitude spectrum

Imaginary part of FFT (Antisymmetric)
b. Fourier phase angle [unwrapped]

Figure C.2. The real and imaginary parts of the waveform shown in Figure C.1. Note the symmetries of the real and imaginary parts.

The Fourier phase angle is obtained by

$$\theta(j) = \tan^{-1}\left(\frac{\text{Fimag}(j)}{\text{Freal}(j)}\right) \quad j = 1, 2, \ldots, N$$

For positive frequencies only, $j = 1, 2, \ldots, N/2+1$

For $j = 1$, corresponding to frequency zero, the value is called the "DC component." There is no imaginary part.

The resultant phase is "wrapped," meaning that it varies between $-\dfrac{\pi}{2}$ and $\dfrac{\pi}{2}$.

This is the limited range that a computer can resolve for the inverse tangent function.

The phase is "unwrapped" by adding or subtracting integer multiples of 2π at points of phase discontinuity.

Note that the magnitude spectrum is a symmetric function for real functions and the Fourier phase is an odd function for real functions.

Figure C.3 shows the Fourier magnitude spectrum and the Fourier phase angle, wrapped and unwrapped.

It is important to note that a high sampling rate lowers the frequency domain sampling rate and a low sampling rate increases the frequency domain sampling rate. If one, for instance is calculating features from the frequency domain magnitude spectrum, one must ensure the magnitude spectrum is adequately represented. See Figure C.4 for an example.

A good rule of thumb is to sample at ten times as high as the expected peak frequency. Sampling higher than this results in oversampling, as is the case in Figure C.4(a).

The inverse FFT is implemented in the following manner:

$$\text{Freal}(j) \rightarrow \left(\frac{1}{N}\right)\text{Freal}(j)$$
$$\text{Fimag}(j) \rightarrow \left(-\frac{1}{N}\right)\text{Fimag}(j); \quad j = 1, 2, \ldots, N$$

FFT Freal( ), Fimag( ), N

That is, the FFT is conjugated and the real and imaginary parts are each divided by N. The (forward) FFT subroutine is then called. Note that the same FFT routine is used for both forward FFTs and inverse FFTs.

A good example of the MatLab[†] implementation can be found at http://www.mathworks.com/help/techdoc/ref/fft.html.

For LabView[‡] users, the site http://zone.ni.com/devzone/cda/tut/p/id/4541 provides an excellent tutorial.

[†] MatLab is a trademark of The MathWorks, Inc., Natick, Mass. U.S.A.
[‡] LabView is a trademark of National Instruments, Inc., Austin, Texas, U.S.A.

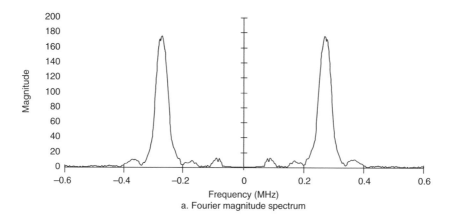

a. Fourier magnitude spectrum

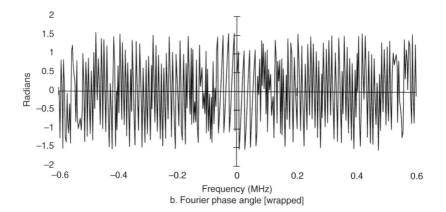

b. Fourier phase angle [wrapped]

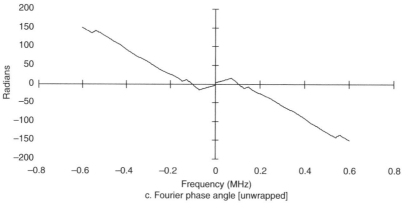

c. Fourier phase angle [unwrapped]

Figure C.3. a. Fourier magnitude spectrum. b. Fourier phase [wrapped]. c. Fourier phase [unwrapped]

C.2.1 Example FFT Use: Analytic Envelope

The analytic signal is defined as

$$a(t) = u(t) + j\,h(t)$$

where $u(t)$ is the real part and the imaginary part; $h(t)$ is the Hilbert transform of $u(t)$. $u(t)$, for example, could be an ultrasonic waveform.

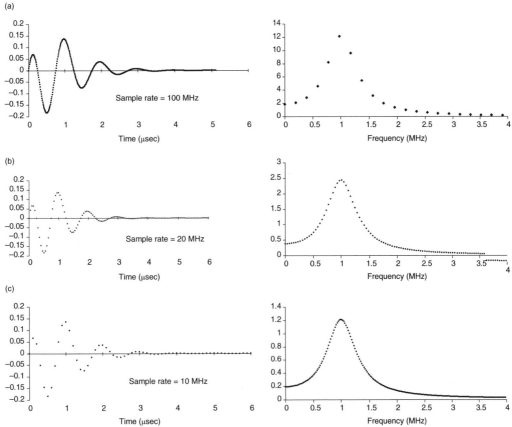

Figure C.4. 1 MHz (peak frequency) waveform (a) sampled at 100 MHz (b) sampled at 20 MHz (c) sampled at 10 MHz.

The analytic envelope, $e(t)$, is defined as

$$e(t) = \sqrt{u(t)^2 + h(t)^2}$$

The following steps show how to arrive at the analytic envelope of a waveform $u(t)$.

FFT $ureal(\ ), uimag(\ ), N$ $\qquad\qquad$ $ureal(\ ) = u(t), uimag(\ ) =$ array of zeroes

Calculate $H(k)$ as

$H(1)_R = ureal(1); H(1)_I = 0$ $\qquad\qquad$ DC component
$H(k)_R + H(k)_I = 2\ ureal(k) + j\ 2\ uimag(k)$ $\qquad$ $k = 2, 3, ..., (N/2) + 1$
$H(k)_R + H(k)_I = 0 + j\ 0$ $\qquad\qquad$ $k = (N/2) + 2$ through N

Set up for inverse FFT,

$H(\mathrm{k})_R \rightarrow ureal(k)/N$ $\qquad\qquad$ $k = 1, 2, 3, ..., N$
$H(\mathrm{k})_I \rightarrow -\ uimag(k)/N$ $\qquad\qquad$ $k = 1, 2, 3, ..., N$

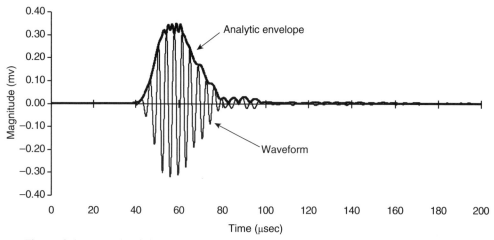

Figure C.5. Example of the analytic envelope of a waveform.

Get the inverse FFT

FFT $H(\ )_R, H(\ )_I, N$

Then

$u(\mathrm{t}) = H_R(t)$, the original waveform
$h(t) = H_I(t)$, the Hilbert transform of $u(t)$

The analytic envelope is then

$$e(t) = \sqrt{H_R(t)^2 + H_I(t)^2}$$

Basically, the DC component of the input waveform remains unchanged. The waveform energy from the negative frequency range is transferred to the positive frequency range. Because of the spectral symmetry, multiplication by two performs this transfer. The analytic signal, therefore, has a one-sided spectrum. Figure C.5 shows an example of an analytic envelope.

C.2.2 Example FFT Use: Feature Source for Pattern Recognition

The Fourier magnitude spectrum can be used as source for features that could be used in a pattern recognition setting. Figure C.6 shows an example of possible features that can be defined on the magnitude spectrum.

C.2.3 Discrete Fourier Transform Properties

Linearity $\qquad a f_1(i) + b f_2(i) \Leftrightarrow a F_1(k) + b F_2(k); \quad F(k) = \mathrm{FFT}(f(i))$

Time reversal $\qquad f(-i) \Leftrightarrow F^*(k) \quad ;^* = \mathrm{Conjugate}$

Time shift $\qquad f(i - i_0) \Leftrightarrow e^{-j \frac{2\pi}{N} k i_0} F(k)$

Convolution $\qquad \displaystyle\sum_{i=0}^{N-1} f(i) g(n-i) \quad ; n = 0, 1, \ \dots, N \quad \Leftrightarrow \quad F(k) G(k)$

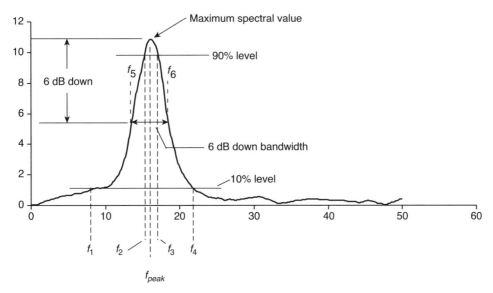

Figure C.6. A Fourier spectrum is shown with features that can be derived from it.

$$\text{Function multiplication} \quad x(i)\,y(i) \quad \Leftrightarrow \quad \frac{1}{N}\sum_{m=0}^{N-1} X(m)\,Y(k-m) \quad ; \quad k = 0,1,\dots,N$$

$$X(m) = \text{FFT}\big(x(i)\big), \; Y(k) = \text{FFT}\big(y(i)\big)$$

C.3 The Short Time Fourier Transform (STFFT)

The STFFT is used to extract time frequency information from a signal. In music, in particular, it is important to know what note (frequency) occurs at what point in time. In guided wave technology, the frequency content of a signal as a function of time is important. Such information can be used to identify propagating modes.

The STFFT is implemented by sequentially windowing contiguous sections of a signal. Figure C.7 illustrates this concept. The FFT is calculated for each windowed section. Thus the frequency content for windowed section is made visible.

To make the sampling of the signal quasi-continuous, the windows are overlapped by an amount called the *skip distance*. Figure C.8 defines this term and others.

The window function is important for two reasons. First, if no window is used, with section-by-section partitioning, additional frequency content called *leakage* is introduced into the frequency content. Leakage frequencies do not exist in the original signal but rather are an artifact of a square window truncation. Figure C.9 illustrates this for a tone burst at a fixed frequency of 2.5 MHz.

Second, the window length determines both the time and frequency resolution. A long window will produce poor time resolution but good frequency resolution while a short window will provide good time resolution but poor frequency resolution.

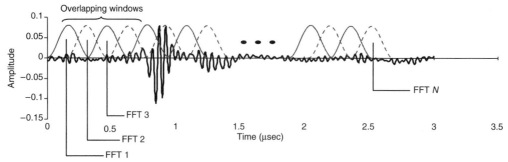

Figure C.7. Waveform partitioning using overlapping window functions. The FFT of each windowed section of the waveform is taken. The resulting collection of FFTs is assembled to form a spectrogram.

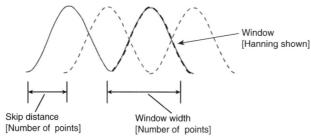

Figure C.8. Definition of terms used to describe STFFT.

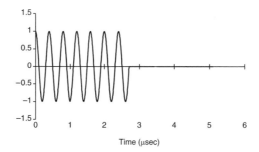

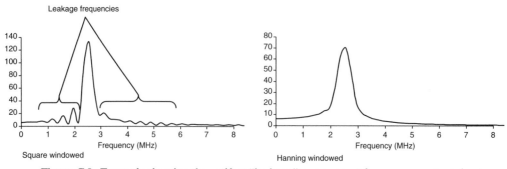

Figure C.9. Example showing the artifact "leakage" or apparent frequency content due to windowing rather than actual content.

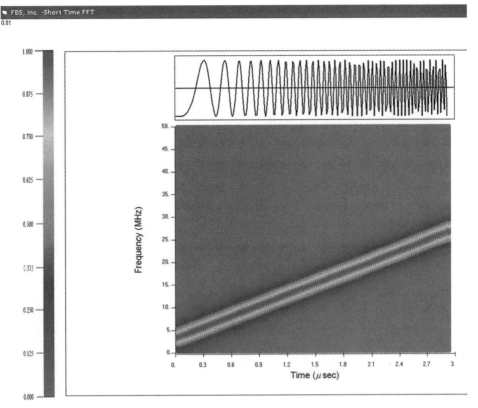

Figure C.10. STFFT spectrogram of a chirp signal showing the spectrogram tracking the frequency change in the signal as a function of time. See plates section for color version.

The result of STFFT analysis is a graphical display called a *spectrogram*. Figures C.10 and C.11 show the spectrograms of a chirp signal and a stepped frequency (5, 10, 15, and 20 MHz) signal. These examples are provided to help the reader with the interpretation of spectrograms.

Figure C.12, to be compared with the signal in Figure C.11, shows good time resolution but poor frequency while the signal in Figure C.12 exhibits good frequency resolution with poor time resolution. This result shows the effect of window length. A "short" window (e.g., 32 points) provides better time resolution but poor frequency resolution. A "long" window (e.g., 128 points) provides better frequency resolution but poor time resolution.

Figure C.13 shows the STFFT of a typical signal.

Because the STFFT and WT both extract the time/frequency representations of a signal, each produces a similar decomposition. Figure C.14 shows a comparison of STFFT and WT of a signal.

The STFFT (and WT) can be used to verify theoretically derived dispersion curves and to identify propagating modes. An example is provided in Section C.5.

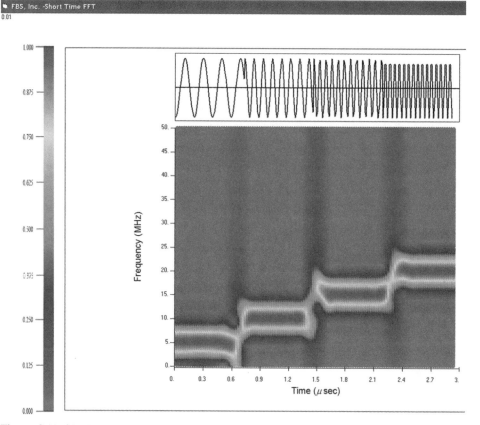

Figure C.11. STFFT spectrogram of a stepped frequency signal showing the spectrogram tracking the frequency change in the signal as a function of time. Window length = 32; Poor frequency resolution, good time resolution. See plates section for color version.

C.3.1 Example: STFFT to Dispersion Curves

Because only a limited range of frequencies can be produced by a transducer excited at a particular frequency, only certain modes, corresponding to group velocity dispersion curve segments, will propagate. In addition, other physical constraints may limit the excitability of certain modes. Figure C.15 shows the concept.

Therefore, only segments of the governing group velocity dispersion curves will be available via the STFFT or WT. A number of these segments can be obtained by frequency sweeping. The individual STFFTs (for each excitation frequency) can be compiled into one image to provide an enhanced comparison with theoretical dispersion curves.

Figure C.16 shows such a compilation for a ¾″-thick aluminum plate. The theoretical curves have been overlaid on the image to compare experimentally obtained dispersion curve hot spots with theoretical dispersion curves.

Thus the STFFT (and WT) can provide insight into the validity of theoretical dispersion curves and also identify the modes that actually propagate.

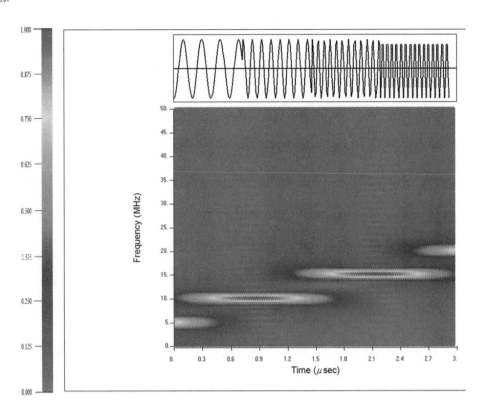

Figure C.12. STFFT spectrogram of a stepped frequency signal showing the spectrogram tracking the frequency change in the signal as a function of time. Window length = 128; Good frequency resolution, poor time resolution. See plates section for color version.

C.4 The 2-D Fourier Transform (2DFFT)

The composition of guided waves varies with distance and time or with wave velocity and frequency. This is obvious from the phase term, $\phi = kx - \omega t$, where k = wavenumber, x = distance, ω = frequency, and t = time.

Setting the phase derivative $\dfrac{d\phi}{dt}$ to zero results in the relationship $\dfrac{dx}{dt} = \dfrac{\omega}{k}$ which is the phase velocity, c_p.

Representation: if we use wavenumber instead of velocity, we will have reciprocal variables

$$\text{because } c_p = \frac{\omega}{k}, \ k = \frac{\omega}{c_p} = \frac{2\pi}{\lambda} \Rightarrow \frac{1}{\text{distance}}, \ x \Rightarrow \text{distance}$$

$$\text{and } \omega = 2\pi f \Rightarrow \frac{1}{\text{time}}, \ t \Rightarrow \text{time}.$$

Also, remember that dispersion curves are only representations of a complicated propagation process. The curves are only the expected values of velocity for a given

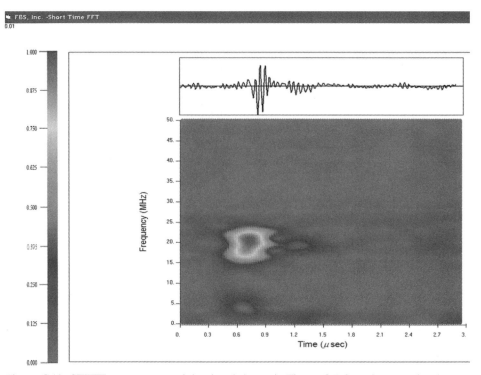

Figure C.13. STFFT spectrogram of the signal shown in Figure C.7. See plates section for color version.

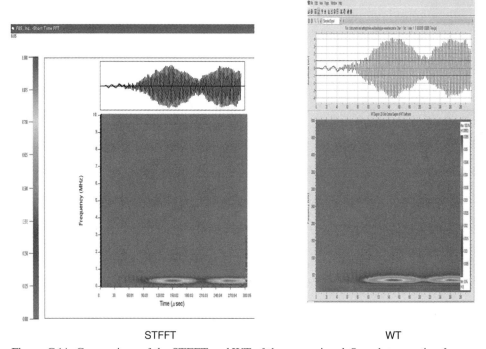

STFFT   WT

Figure C.14. Comparison of the STFFT and WT of the same signal. See plates section for color version.

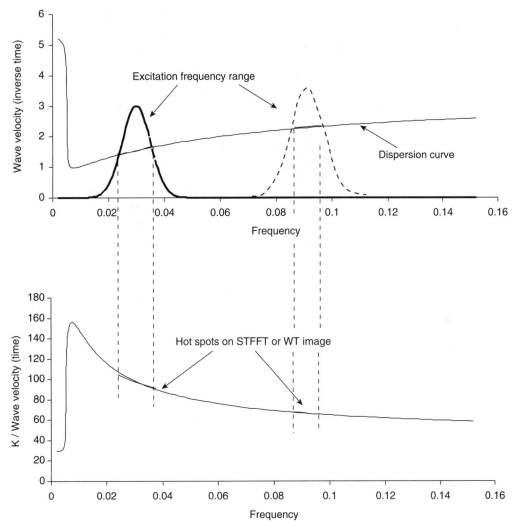

Figure C.15. Illustration of the concept of limited mode excitation due to transducer bandwidth and physical constraints to excitability.

frequency. The terms used to describe this are *phase velocity spectrum* (distribution) and *frequency spectrum*.

In the time-distance domain, propagation is represented as $f(x, t)$. Dispersion curves represent propagation in the reciprocal domain as $g(k, \omega)$. We know we can measure $f(x, t)$. How do we get to $g(k, \omega)$? [Experimental dispersion curves].

Two possibilities are the 2DFFT and the WT. The WT is discussed later in this appendix.

The 2DFFT of a function $f(x, t)$ can be defined as

$$g(k, \omega) = \int f(x, t) \, e^{-j(kx + \omega t)} \, dt \, dx$$

which can be rewritten as

$$g(k, \omega) = \int h(x, \omega) \, e^{-j(kx)} \, dx$$

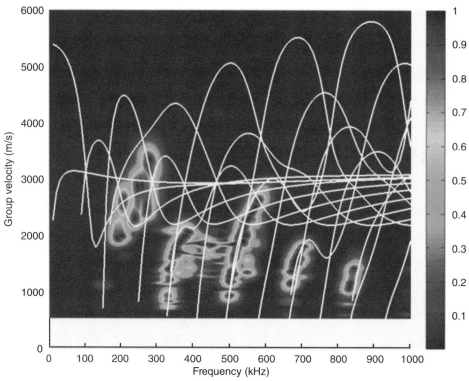

Figure C.16. A compilation of STFFTs created by frequency sweeping a ¾″-thick specimen of aluminum plate. See plates section for color version.

with

$$h(x,\omega) = \ f(x,t)\, e^{-j\omega t}\, dt$$

For the discrete case,

$$G(k_m,\omega_n) = \sum_{p=0}^{N-1}\sum_{q=0}^{M-1} F(x_p,t_q)\, e^{-2\pi j\,(k_m x_p + f_n t_q)}$$

or

$$G(k_m,\omega_n) = \sum_{p=0}^{N-1} H(x_p,\omega_n)\, e^{-2\pi j\,k_m x_p}$$

with

$$H(x_p,\omega_n) = \sum_{q=0}^{M-1} F(x_p,t_q)\, e^{-j\,\omega_n t_q}$$

$m = 0, 1, 3, \ldots, M$ [Number of time points]
$n = 0, 1, 3, \ldots, N$ [Number of spacial points]

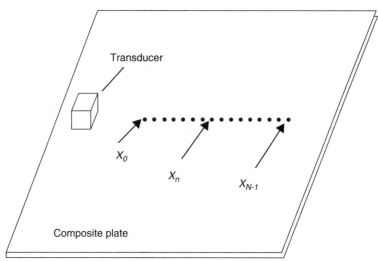

Figure C.17. Data acquisition protocol for 2DFFT data. A composite plate is shown as an example.

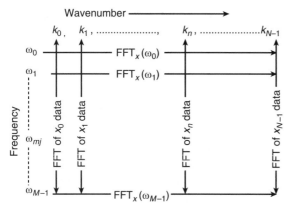

Figure C.18. Schematic representation of the 2DFFT process. The FFTs of the waveform data from each spatial location are calculated first. The spatial FFTs are performed on those FFTs.

That is, the FFT of each time-based waveform is taken at each spacial location. The FFT is then taken of the resulting ensemble of FFTs. Figure C.17 shows an example of spacial data locations for a composite plate.

The second step for obtaining experimental phase velocity dispersion curves is the calculation of the FFTs of the spacial data for each frequency. See Figure C.18.

The third step for obtaining experimental dispersion curves is the conversion of frequency/wavenumber data into frequency/phase velocity data via the transformation shown below.

$$(\omega_i,\, k_i) \longrightarrow \frac{\omega}{k} = \frac{2\pi f}{\dfrac{2\pi}{\lambda}} = \lambda\, f = c \longrightarrow (f_i,\, c_i)$$

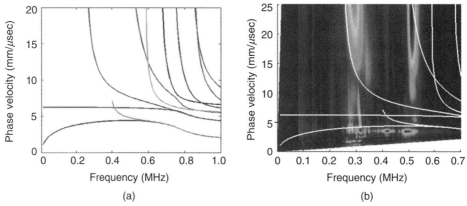

Figure C.19. Comparison of (a) phase velocity dispersion curves with (b) 2DFFT results generated from a composite plate. See plates section for color version.
Software for 2DFFT can be found in Brigham (see references).
MatLab[†] and LabView[‡] also have software that supports the 2DFFT.

Figure C.19(b) shows a color-coded 2DFFT magnitude plot with phase velocity dispersion curves superimposed on it. Note that the STFFT and the WT provide insight into group velocity dispersion curves whereas the 2DFFT provides phase velocity dispersion curve information. The STFFT and the WT use only one waveform from one location; no phase information is available.

C.5 The Wavelet Transform (WT)

Thus far, we have presented the 2DFFT and the STFFT. Now we present the wavelet transform (WT). Recall that the motivation for studying these transforms was to produce a physically based image of a time domain signal that provided information about propagating modes and their location on the appropriate dispersion curves. The WT maps the time-amplitude domain into a translation (τ) – scale (s) domain. $(t, a) \rightarrow (\tau, s)$. This will be made clear later in this appendix. The scale, s, as will be seen, varies inversely as frequency. The higher the frequency, the lower the scale. Figure C.20 shows the concept.
The definition of the WT shows how the above representations come about.
    The WT is defined as

$$\gamma(s,\tau) = \int f(t)\, \psi_{s,\tau}^{*}(t)\, dt$$

WT
s = scale
τ = translation

$$s \sim \frac{1}{f}$$

$$f(t) = \int \int \gamma(s,\tau)\, \psi_{s,\tau}(t)\, d\tau\, ds$$

[†] MatLab is a trademark of The MathWorks, Inc., Natick, Mass. U.S.A.
[‡] LabView is a trademark of National Instruments, Inc., Austin, Texas, U.S.A.

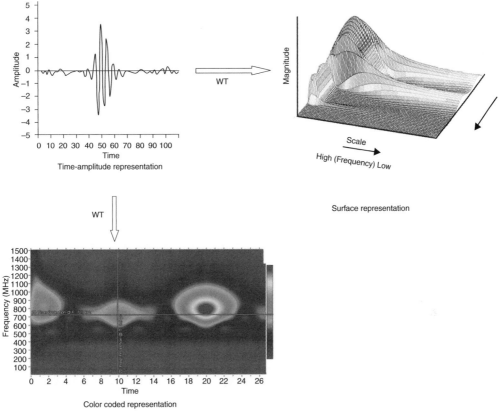

Figure C.20. Illustration of the results of using the WT. The WT takes a function (or waveform) from the time-amplitude domain to the translation-scale domain or to the time-frequency domain. Further, for a fixed distance (the location where the waveform was collected relative to the transmitter), a dispersion curve representation can be obtained. See plates section for color version.

Inverse WT

$\gamma(s, \tau)$ is the WT of $f(t)$.

$\psi_{s,\tau}(t)$ is the kernel of the transformation and is called the *mother wavelet*. For complex wavelets, "*" means conjugation.

$$\psi_{s,\tau}(t) = \frac{1}{\sqrt{s}} \psi \frac{t - \tau}{s} \qquad \text{where } \psi(t) \text{ is a wavelet.}$$

The variable τ moves the wavelet through the function or waveform $f(t)$ and the variable s stretches or shrinks the wavelet, essentially making it lower or higher in frequency.

Examples of wavelets are given in Figure C.21.

As an example, the Gaussian WT would be

$$\gamma(s, \tau) = \frac{1}{\sqrt{2\pi s}\, \sigma} \int f(t)\, e^{-\frac{1}{2}\left(\frac{t - \tau}{s\sigma}\right)^2} dt$$

$$\psi\left(t\right)=\frac{1}{\sqrt{2\pi}\sigma}\,e^{\frac{-t^2}{2\sigma^2}}$$

Gauss function

$$\psi\left(t\right)=\frac{1}{\sqrt{2\pi}\sigma^3}\left(e^{\frac{-t^2}{2\sigma^2}}\left(\frac{t^2}{\sigma^2}-1\right)\right)$$

"Mexican hat" based on gauss

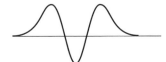

$$\psi\left(t\right)=e^{iat}\cdot e^{\frac{-t^2}{2\sigma}}$$

Morlet wavelet   Real part

Imaginary part

Figure C.21. Sample wavelets. To implement a WT, t must be replaced by $\frac{t-\tau}{s}$ and the resulting function multiplied by $\frac{1}{\sqrt{s}}$.

Figure C.22 shows a "Mexican Hat" wavelet for various translations, τ, and for various scales, s.

Note that in each of the subfigures of Figure C.22, the wavelet is multiplied with the waveform $f(t)$ and then integrated. Where strong matches occur, the $\gamma\left(s,\tau\right)$ will have a large amplitude, and where there is a weak or no match, $\gamma\left(s,\tau\right)$ will have a very low value.

Essentially, the waveform $f(t)$ is dissected into time and frequency (scale) components. This dissection is important for guided wave analysis in the sense that guided wave modes (wave packets with a particular frequency [scale] content) travel at different speeds. Identification of those modes that propagate well in the specimen at hand can be facilitated by the use of the WT.

Many of the investigators in the field of guided wave technology use a time frequency representation rather than the translation-scale representation. Correlation with dispersion curves is made much simpler using the time frequency representation.

Figure C.23 shows representation with an example pair of time and frequency coordinates, 19.9 μsec and 791 kHz. If the distance from the transmitter is known, then a dispersion curve point can be calculated. For example, if the wave packet peak was known to arrive at 127 mm (or 5") from the transmitter, c_g = 127 mm/19.9 μsec = 6.38 mm/μsec. The dispersion curve coordinates would be (791 kHz, 6.38 mm/μsec).

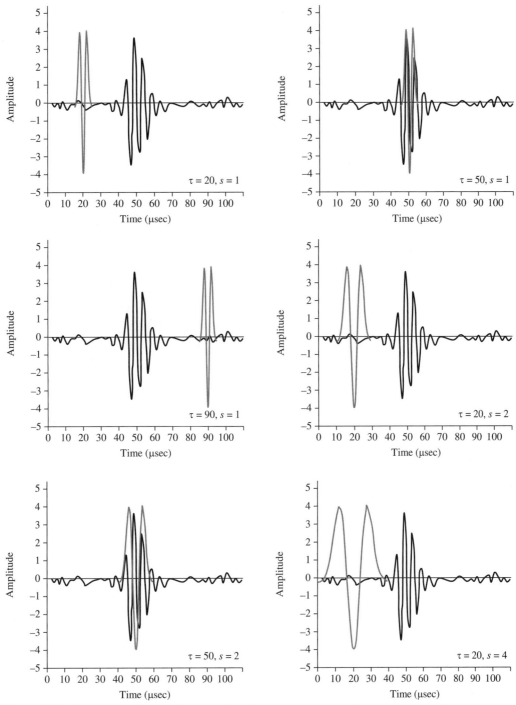

Figure C.22. Illustration of translation, τ, and scaling, s, when using a Mexican hat wavelet.

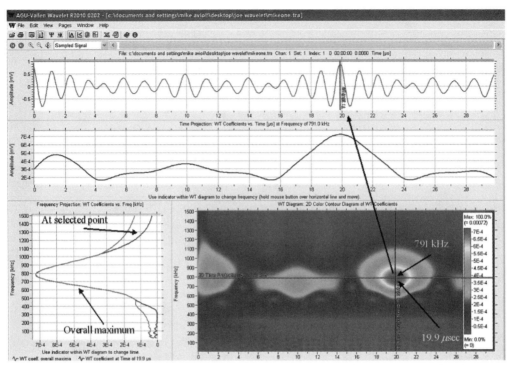

Figure C.23. Example WT with a pair of time-frequency coordinates shown. Using the arrival distance of the selected wave packet, the group velocity, c_g, can be calculated to obtain the dispersion curve coordinates (f, c_g). [AGU-Vallen Wavelet free software.] See plates section for color version.

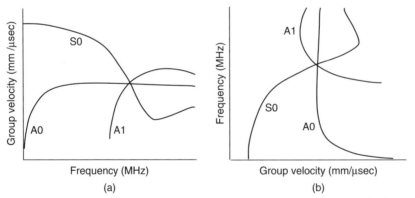

Figure C.24. a. Standard dispersion curve representation for group velocity b. Representation for comparison with the WT. See Figure C.25.

The (free) software used for the example of Figure C.23 is available at http://www.vallen.de/downloads. Both MatLab[†] and LabView[‡] offer WT capabilities.

Figure C.24 shows how dispersion curves be can related to a color-coded WT. The normal representation dispersion curves are rotated 90° counterclockwise and overlaid on the WT. Figure C.25 shows a sample result.

[†] MatLab is a trademark of The MathWorks, Inc., Natick, Mass. U.S.A.
[‡] LabView is a trademark of National Instruments, Inc., Austin, Texas, U.S.A.

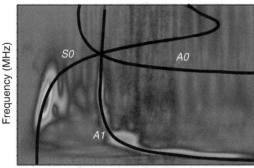

Figure C.25. Illustration of rotated group velocity dispersion curves superimposed on a WT. The red hot spot shows that the **S0** mode is the strongest propagating mode with **A1** as a weaker propagating mode. See plates section for color version.

C.6 Exercises

1. Describe the data acquisition process to achieve (a) Figure C.16, (b) Figure C.19, and (c) Figure C.25.
2. Discuss how the results shown in Figure C.16, C.19, and C.25 could be used in a material defect detection situation in a plate.

C.7 REFERENCES

Alleyne, D. N., and Cawley, P., (1990). *A 2-dimensional Fourier transform method for quantitative measurement of Lamb modes.* IEEE Ultrasonics Symposium – 1143.

Brigham, E. O. (1988). *The Fast Fourier Transform and Its Applications.* Englewood Cliffs, NJ: Prentice Hall.

Hamstad, M. V., Gallagher, A. O., and Gary, J., *A Wavelet Transform Applied to Acoustic Emission Signals: Part 1: Source Identification.* National Institute of Standards and Technology, Materials Reliability Division (853), Boulder, CO; http://www.engr.du.edu/ profile/pdf/20–039.pdf.

Hayes, M. H. (1999). *Digital Signal Processing.* New York: McGraw-Hill.

Rabiner, L. R., and Schafer, R. W. (1978). *Digital Processing of Speech Signals.* Englewood Cliffs, NJ: Prentice Hall.

Rose, J. L. (1999). *Ultrasonic Waves in Solid Media.* New York: Cambridge University Press.

The Wavelet Tutorial Part I by Robi Polikar at http://users.rowan.edu/~polikar/wavelets/ wtpart1.html.

The Wavelet Tutorial Part II by Robi Polikar at http://users.rowan.edu/~polikar/WAVELETS/ WTpart2.html.

Vallen Systeme (Free wavelet software) http://www.vallen.de/downloads.

APPENDIX D

Guided Wave Mode and Frequency Selection Tips

D.1 Introduction

Computation plays a critical role in the development of any ultrasonic guided wave inspection system. Beyond hardware and software development, modeling analysis is essential for evaluating various designs and ultimately relates to making a good choice from the hundreds of test points available on the dispersion curves. The process is illustrated in Figure D.1.

First, phase and group velocity dispersion curves must be calculated for a particular waveguide structure being considered. Wave structure computation is also critical. This early part of the process is the analytical section. This part leads to a mode and frequency choice based on experience, further modeling of wave interactions with certain defects in specific structures, or an intuitive selection and evaluation.

The actuator design could be, for example, a normal or angle beam device, or a comb-type array. A piezoelectric, electromagnetic acoustic, magnetostrictive, laser, or controlled mechanical impact could be considered for closeness to other mode points in an attempt at mode isolation, or phase velocity spectrum influences from a source influence, or of course a center frequency and frequency spectrum. See Figures D.2 and D.3.

Figures D.2 and D.3 depict phase velocity dispersion curves with conceptual phase and frequency spectrums for a simple platelike structure, illustrating the difficulty of generating only a single mode in a structure. Quite often, guided wave signals appear noisy but the information is not noise, it is coherent wave propagation from different modes. Nevertheless, proper transducer design can narrow the phase velocity and frequency spectrums and also reduce side lobes in the spectrums if they are present.

The most important issues are mode and frequency choice and sensitivity to a special inspection situation.

The actuator design can then be tested with suitable FEM models and computation. Success here could lead to key experiments, or failure could prompt additional wave structure analysis and new actuator designs in a feedback process to optimize a design. If key experiments are successful, the process leads to a system design. If experiments are not good enough, additional feedback is required as shown in Figure D.1 in an attempt to get to a final system design.

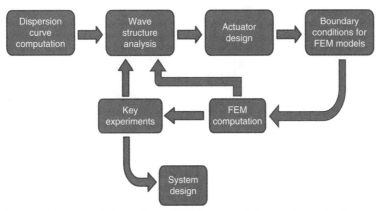

Figure D.1. The hybrid analytical FEM approach for solving guided wave problems.

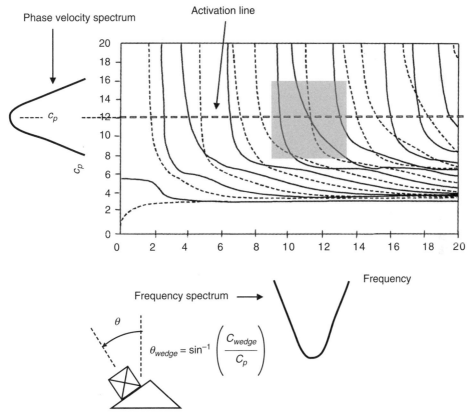

Figure D.2. Source influence for a typical angle beam excitation or an ability to generate a specific mode and frequency.

As a result of the significance of mode and frequency choice for solving new critical guided wave problems, a number of sample problems are discussed in this appendix:

1) A surface-breaking defect
2) Mild corrosion and wall thinning
3) Transverse crack detection in a rail head
4) A weak interface in a bonded repair patch
5) Examining water-loaded structures

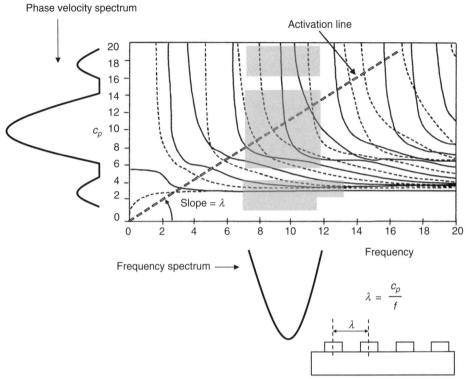

Figure D.3. Source influence for a typical comb transducer excitation or an ability to generate a specific mode and frequency.

6) Frequency tuning
7) Ice detection
8) Deicing
9) Focusing in pipe inspection
10) Aircraft problems of lap splice, tear strap, and skin-to-core delamination
11) Coating delamination and axial crack detection
12) Multilayer structures

Note that the solutions presented are not necessarily unique. Multiple solutions are possible because of the many mode and frequency choices available. Optimization is generally not possible because of the required numerical solutions as opposed to analytical solutions. Even the choice of sensitivity variable could be different, including, for example, in-plane displacement at a particular region or out-of-plane displacement, a stress variable, an energy variable, a wave velocity parameter, or energy absorptions.

These discussions are presented to enhance creativity and logic for finding solutions to inspection problems that might benefit from a guided wave approach. The thought considerations behind certain inspection concepts are also outlined in some of the problems.

D.2 Mode and Frequency Selection Considerations

A proper mode and frequency choice from the phase velocity dispersion curves are essential to guided wave system development in order to move forward to an appropriate actuator design.

Of hundreds of points on a phase velocity dispersion curve, which should be selected and used for achieving the best sensitivity for finding a particular defect? Several sample problems are discussed next to provide the reader with ideas for tackling these problems where a guided wave solution could be valuable.

D.2.1 A Surface-Breaking Defect

Let's start by considering a fairly simple problem of finding and sizing a surface-breaking crack. Early work on mode and frequency selection based on certain features and wave structures was reported by Ditri, Rose, and Chen (1991). At any frequency, a finite but possibly large number of Lamb wave modes could propagate in a layer. Each mode, when excited, will produce a different deformation field inside the layer; that is, the particle displacements and velocities as well as the stress and strain fields will all vary with depth inside the layer in a different manner for each mode. In fact, even the same mode at a different frequency (times thickness) will cause different field distributions.

Ditri, Rose, and Chen (1991) investigated the possibility of using, as a selection criteria, a mode's energy distribution across the thickness of the layer. This is only one of many criteria that may be tested, but it was chosen because it had a somewhat appealing intuitive basis behind it.

Studies of energy distribution across the layer can be utilized, therefore, for mode and frequency selection to carry out crack detection and depth studies. To find a small surface-breaking crack, sufficient energy must be at the surface of the plate. Also, to find an almost-through crack approaching 100 percent in depth, the wave structure must have sufficient energy on the opposite side of the plate.

When a Lamb wave mode propagates in a layer, it becomes a means by which energy is transported along the layer. The rate at which a given mode transports energy per unit cross-section area of the layer can be written in terms of the stress and velocity fields as:

$$< P_z > = -\frac{1}{2}\left\{V^*{}_y \tau_{yz} + V^*{}_z \sigma_{zz}\right\} \tag{D.1}$$

where the asterisk denotes complex conjugation, $<\cdot>$ denote averages over one period, y (thickness direction) and z (wavevector direction) are the in and out of plane coordinates, respectively, and all field variables are independent of x (because of a plane-strain problem). $<P_z>$ is actually the time-averaged projection of the Poynting vector onto the z coordinate direction, and is a function of the depth coordinate, y. The velocity and stress fields that appear in Equation (D.1) can be found analytically, and subsequently used in the equation.

Integrating Equation (D.1) across the thickness of the layer yields the time-average power flow in the z-direction per unit width in the x-direction, denoted by $<P_{nn}>$; see Auld (1990).

$$< P_{nn} > = \int_{-\frac{b}{2}}^{\frac{b}{2}} < P_z > dy \tag{D.2}$$

Percent energy versus percent depth for an A1 mode at 6.35 MHz-mm and an S2 mode at 4.78 MHz-mm showed the A1 mode was more sensitive to smaller depth cracks, both modes with similar sensitivity to a 50 percent crack, and S2 with a greater sensitivity to very large cracks.

The problem just covered could be more complicated for sizing if mode conversion occurs at the defect, which is possible depending on crack angle and surface characteristics. Difficulties also occur for 3-D situations compared to the 2-D plane strain just considered. A consideration of plane or volumetric defects is also important. Any mode at the impinging frequency could reflect from the defect. On the other hand, if any mode and frequency is selected that has a constant energy value across the thickness, defect depth would be proportional to amplitude in the 2-D plane-strain case. This could occur for the shear horizontal (SH) lowest order mode, below the first cutoff frequency of the dispersion curve for the structure.

Advanced methods for 3-D defects would first consider characterization, planar versus Volumetric, for example, then defect width, perhaps by a focusing method, and then finally depth based on a cross-sectional area consideration. Proper calibration would also be necessary as a first step in the sizing process.

D.2.2 Mild Corrosion and Wall Thinning

Let's consider a guided wave health monitoring scheme as illustrated in Figure D.4 for mild corrosion depicted in small thickness changes over some length L.

We draw from a paper by Gao and Rose (2010) for some discussion.

Let's think about guided wave sensitivity to general corrosion. Different from a vertical crack defect, the general corrosion defect will not produce a significant amount of reflection. However, the transmission time will be affected by the damage because the thickness reduction changes the group velocity of wave propagation, hence time of flight variation and the severity of corrosion or erosion metal loss will be considered. This information can be extracted from the group velocity dispersion curves as a function of frequency times thickness. A search of the group velocity dispersion curve space would show that a sloped region should work best, as it shows

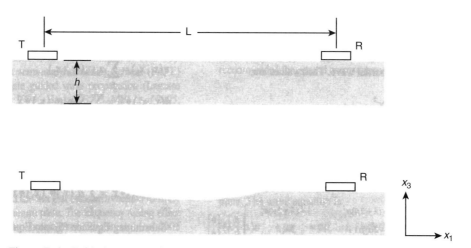

Figure D.4. Guided wave monitoring scheme for general corrosion.

a sharp contrasted region versus thickness for a particular frequency. Gao and Rose (2010) demonstrate that the time of flight shift is proportional to the cross-sectional area of the corrosion. Gao and Rose (2007) also considered the excitability of the points in the group velocity dispersion curve space. An excitability dispersion curve is derived from normal mode expansion (NME) theory.

The NME technique is based on a reciprocity relation in elasticity. The basic idea of NME is to express the actual wave field as a superposition of orthogonal guided wave mode solutions. The general theory of NME in an elastic and piezoelectric plate is described in a textbook by Auld (1990). Ditri and Rose use the NME technique for guided wave excitation in isotropic plates and pipes (1992, 1994).

It is shown that, for surface-mounted transducers, the excitability can be defined as the normalized particle velocity at the excitation surface as shown in Figure D.4, in direction x_3. Excitability in the x_1-direction was eliminated because it was smaller than in the x_3-direction for many points. Excitability in the x_2-direction was considered zero. The results show that to get a strong and sensitive signal, the transducer ought to have a dominantly normal loading in the x_3-direction.

Gao and Rose (2010) introduced the concept of a *goodness curve*, which is defined as a product of the group velocity, the excitability, and the sensitivity $s = -\dfrac{1}{v^2}\left(\dfrac{dv}{dh}\right)$.

The result indicates that the guided wave modes with a large group velocity slope and a small group velocity value are the most sensitive modes. Guided wave mode candidates are recommended from the goodness dispersion curve after evaluating the group velocity, excitability, and sensitivity dispersion curves. These include the S0 mode around 1 MHz, A1 mode around 2 MHz, S0 mode around 2.8 MHz, and A2 mode around 3.8 MHz for the 2 mm-thick aluminum plate.

Note that this approach could be followed in other examples of mode and frequency selection, but the hypothesis of exploring the intuitive group velocity change is quite useful. Smart experiments will always be needed with appropriate parameters while trying to get to a good result.

Note that an alternate approach to this problem would be to simply evaluate a group velocity measurement as an indicator of thickness reductions in some smart, theoretically driven experiment in an SHM mode. If group velocity is used, keep in mind that the starting thickness must be known and a follow through in an SHM sense will be required as thickness measurement via an inverse problem from group velocity does not provide us with a unique solution. See Luo and Rose (2003a,b) for help in approaching the problem in this manner.

By examining tone burst mode crossings when sweeping frequency at a particular phase velocity or angle via the Cremer hypothesis, in through transmission, or by a Fourier transform of a pulse in through transmission, it becomes possible to estimate the thickness of a platelike structure (via an angle or normal beam approach at a suitable phase velocity value or a suitable coil or magnet spacing of a wavelength for an electromagnetic acoustic transducers (EMAT) or tooth spacing for a comb-type transducer).

The idea of mode crossing for thickness measurement was presented by Rose (1999). See Figure D.5.

For additional information on modeling for flaw sizing, see Rose, Pelts, and Cho (2000).

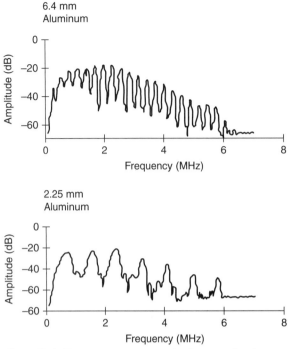

Figure D.5. Frequency sweep results for two aluminum plates of different thicknesses.

An interesting approach to thickness measurement, primarily as an academic exercise for now, would be to consider a cutoff frequency measurement to come up with a thickness value. See Zhu and colleagues (1998) for details along with other measurement concepts.

Characteristics of a phase velocity dispersion curve of phase velocity versus a frequency thickness (fd) product, including information on mode cutoff frequencies could be used to measure the thickness of a test object through transmission. Select a mode, say A2, beyond the A0 and S0 mode that do not have cutoff frequencies. Select a point on A2 just to the right of its cutoff frequency. Consider now the following at the point selected.

1. Incident angle is $\theta = \sin^{-1}\left(c_{wedge} / c_p\right)$
2. fd at point selected is 5.5 MHz-mm
3. fd at nearby cutoff frequency is 5.0 MHz-mm
4. Assume that initial plate thickness is 1.62 mm

Wall thinning now takes place and measurements are taken to determine thickness. The following steps are taken.

5. From an initial narrow band tone burst signal of f = 5.5/1.62 = 3.4 MHz sweep frequency slowly till the A2 mode disappears, which is an indication of moving just below the cutoff frequency of the A2 mode.
6. Suppose we measure a value of 3.6 MHz.
7. Because at A2 cutoff value is 5.0, the new thickness can be calculated. fd = 5.0, therefore d = 5.0/3.6 = 1.39 mm, which corresponds to a 14 percent wall loss.

D.2.3 Transverse Crack Detection in the Head of a Rail

The goal of this study was to develop a guided wave method that could detect transverse cracks in the head of a rail, even with random shelling, typical of spalling in a Hertzian contact stress loading situation. Ordinary bulk wave measurements are often not effective for detecting such transverse cracking. Following the strategy in Figure D.6, the first step was to obtain dispersion curves in a rail. Hayashi, Song, and Rose (2003) accomplished this task with the SAFE technique, but found so many close-by dispersion curves that it proved difficult to analyze. Subsequently, Lee, Rose, and Cho (2008) tackled the problem by examining wave structures at various regions of the dispersion curve. See Figure D.7.

In Figure D.6, the areas of concern are listed as 30–1, 30–2, 30–3, 30–4, and so on, as noted in the figure. The wave structures at specific points in the dispersion curve space were examined to find special points that would have maximum energy, for example, on the surface, in the head, in the web, or in the base of a rail. It was found that at position 30–1, as it moves to higher frequencies, this lower order mode energy will propagate in the head only, especially from 30 to 200 kHz. This would work well for finding shelling or transverse cracks in the head of a rail. We could call this lower order mode a pseudo-Rayleigh surface wave mode.

It was shown that the impingement onto the surface of a rail head could be induced through an angle beam or comb-type EMAT sensor probe on the head surface with the wave structure result as illustrated in Figure D.7. Time delays are incorporated in the FEM model to illustrate the approach times of the waves at the interface between the angle beam probe and the rail head itself. Computations have been carried out to show that this model can be used to simulate ultrasonic wave transmission from an angle beam probe or EMAT into a rail structure. A sample wave structure along the pseudo-Rayleigh surface wave mode (30–1, 60–1, to 200–1) is illustrated in Figure D.7.

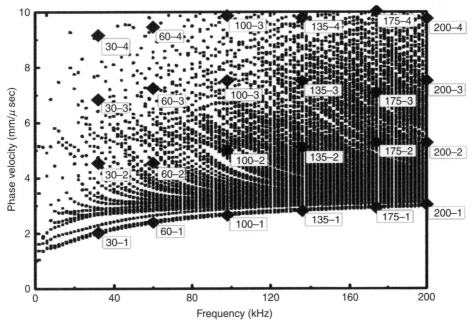

Figure D.6. Phase velocity dispersion curve for rail.

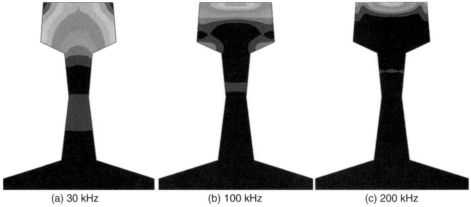

(a) 30 kHz   (b) 100 kHz   (c) 200 kHz

Figure D.7. Wave structures at different frequencies showing that the energy of higher frequency (200 kHz) is concentrated on the top surface of a rail head and the energy of lower frequency (30 kHz) is distributed over the entire area of a rail head.

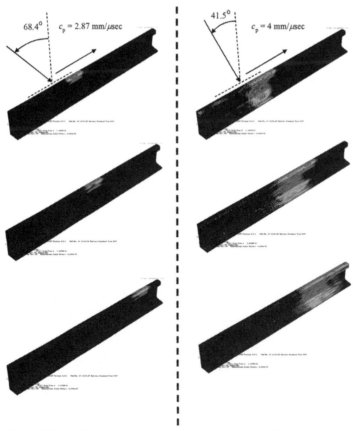

Figure D.8. Rail coverage as a function of mode selection.

A study of the wave structure produced excellent results as shown in Figure D.8, where the wave propagates over many time increments.

Once the proper mode and frequency are selected, one can return to Figure D.1 to proceed with an effective design for finding defects of interest in a rail.

D.2.4 Repair Patch Bonded to an Aluminum Layer

Adhesive bonding inspection problems have been addressed for years. Cohesive situations have adequate solutions via ultrasonic "C" scan testing or vibration mode bond testers. Adhesive or interface situations, including the elusive "kissing bond" issue, are more complex. Experience has shown that getting shear waves into the interface, or possible in-plane displacement waves onto the interface, can often solve the problem.

Puthillath and Rose (2010) studied the problem of the ultrasonic guided wave inspection of a titanium repair patch bonded to an aluminum aircraft skin. They reported the phase velocity dispersion curves for a titanium – epoxy – aluminum test specimen. Superimposed onto the dispersion curve in Figure D.9 is the in-plane interfacial displacement at the Al-epoxy interface. There are test regions with strong in-plane displacement. The regions at 2.5 MHz and high phase velocity were selected for the test system because of mode separation and reasonable source influence profile for a comb-type EMAT sensor. See Figure D.10 for a source influence result. The system was designed and excellent test results were obtained. Tests were conducted in a through transmission text mode with sensors just a few inches apart. Comb spacing was a wavelength dictated by the activation line slope on the phase velocity dispersion curve from the origin to the point in question at 2.5 MHz with a phase velocity of 18 km/s, for an EMAT coil spacing of 7.2 mm.

D.2.5 Water-Loaded Structures

A major concern in many inspection problems is an inability to inspect structures covered with water splashes or perhaps even structures that are completely

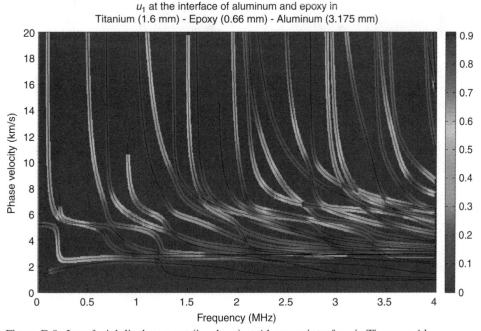

Figure D.9. Interfacial displacement (in-plane) at Al-epoxy interface in Ti-epoxy-Al specimen. See plates section for color version.

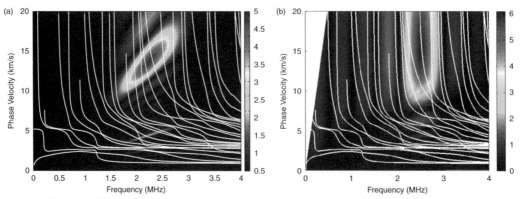

Figure D.10. The white lines are the guided Lamb-type wave phase velocity dispersion curves for the repair patch. (a) Source influence of λ = 6.36 mm (0.25") comb loading using four elements each 1.58 mm wide and supplied with 2.5 MHz tone burst voltage for 3 cycles on the range of phase velocities and frequencies excited. (b) Source influence of a 6 mm diameter transducer mounted on a 10° acrylic angle beam wedge and supplied with a 2.5 MHz tone burst input voltage for 5 cycles on the range of phase velocities and frequencies excited. See plates section for color version.

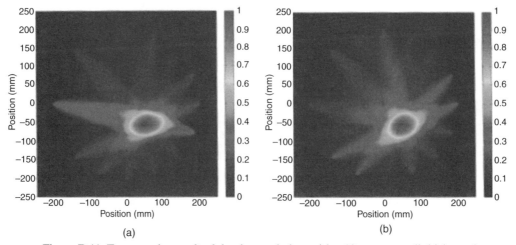

Figure D.11. Tomography result of the dry steel plate with a 10 percent wall thickness loss corrosion defect (a) A1 mode at 2.5 MHz (b) S1 mode at 2.8 MHz. See plates section for color version.

submerged. Energy absorption and imaging artifacts can occur as the guided wave ultrasound can leak into the water.

Presented in Chapter 6 on guided waves in plates was a section on finding dominant in-plane displacement at the surface of a plate so that energy leakage into water would not occur because of shear loading of the fluid. See Pilarski, Ditri, and Rose (1993). It was found that total in-plane displacement on the surface of a plate was possible wherever the dilatational velocity value along the phase velocity ordinate intersected the symmetric modes in the plate at S1, S2, S3, and so forth. Hence, the frequency and phase velocity were known and actuators, either angle beam or comb type, could be designed to generate this mode. A sample tomographic image of a small corrosion defect on a dry plate shows successful imaging with two different modes in Figure D.11. On the other hand, an incorrect image appears in Figure D.12(a) for the A1 mode. A water drop appears as a defect because the out-of-plane component leaks into the water drop. An excellent image, however, when

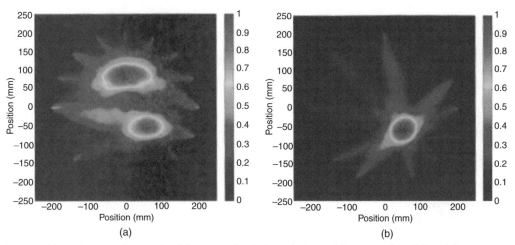

Figure D.12. Tomography result of the water-loaded steel plate with a corrosion defect (a) A1 mode at 2.5 MHz (b) S1 mode at 2.8 MHz. See plates section for color version.

using the proper mode is illustrated in Figure D.12(b). Only the corrosion defect is seen. The water drop is not seen.

Note that successful imaging could also come about by using a lower-frequency S0 mode, which also has dominant in-plane displacement on the plate surface. A SH guided wave would also produce excellent results.

Let's consider another instance of an in-plane and out-of-plane example, this time on the inside surface of a pipe. Suppose one wants to find the percent of gas entrapment inside a pipeline in a power plant. (Pumps cannot be turned on with too much air in the line as serious damage could occur.) It was found that a transducer ring mounted at two points of a pipeline could be used to determine the air content between the two points. The pipeline length being considered could be up to, say, 100 feet and could also contain elbows along the pipeline section. Note that some in-plane displacement is needed that will not leak into the fluid in order to reach a transducer receiver position. An out-of-plane component is also needed to leak into the fluid. Neither component leaks into the gas-contained sections. A sample result is illustrated in Figure D.13 for an energy feature. Note that calibration would be necessary using both a fluid-filled and an empty pipe. A suitable mode and frequency choice would also be important following both theoretical and experimental guidelines.

D.2.6 Frequency and Other Tuning Possibilities

Utilization of dispersion curves and wave structures places us into a limited parameter space in searching for a guided wave solution; theoretically driven experimentation is often necessary to close the door on obtaining a solution. Even a follow through to FEM analysis is not perfect because the structural and defect geometric properties are not precisely known for the model. One of the most powerful ideas for mode and frequency choice considerations is to allow perturbations or tuning in phase velocity and frequency in the phase velocity dispersion curve space. One of the most useful experimental procedures is to carry out frequency tuning in a search for defects. Frequency tuning can even be used to generate phase velocity dispersion curves for

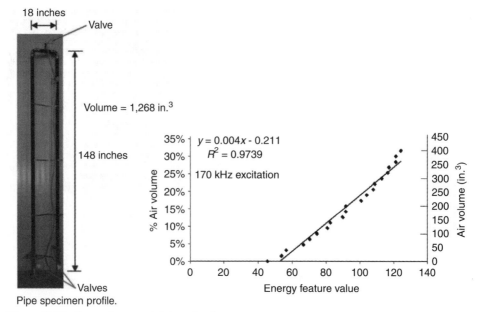

Pipe specimen profile.

Figure D.13. Gas entrapment determination.

a particular waveguide. Whether moving across a horizontal line at a fixed incident angle/phase velocity, or across a line of slope λ from the origin of the phase velocity dispersion curve, a pulse amplitude of narrow frequency bandwidth will oscillate when tuning, such that maximum values will be associated with mode crossings. By changing the incident angle or slope λ, repeating this process can lead to the development of a complete phase velocity dispersion curve provided the frequency bandwidth of the actuator is adequate.

Response from a defect is often like searching for a wave resonance with the scattering components from the defect summing to an excellent constructive interference wave pattern. Some defects are seen over a very large range of frequencies and others have very sharp wave resonance. The situation is analogous to the thinking associated with vibrational resonance. Hence, a small frequency increment is required in the frequency tuning process to have the best chance at finding a defect. Interpolation can create problems as unusual defect types could be missed.

Besides frequency tuning, many other guided wave and actuator parameters could be used to produce improved constructive interference results. For example, at a fixed frequency, the variables of incident angle or the use of time delays to effectively vary λ for a comb-type transducer could be finely tuned to seek out a better constructive interference test result.

In addition, imagine the use of a partial loading situation that leads to natural focusing in a pipe. Tuning could come from loading length variations, circumferential position changes, frequency changes for a fixed set of parameters for partial loading, or even distance variation tuning to peak any interference that might occur.

Many other tuning ideas could be considered. For a few examples, see Rose and Soley (2000); Shin and Rose (1998); Barshinger, Rose, and Avioli (2002); Zhang, Gavigan, and Rose (2006); and Rose and colleagues (2005).

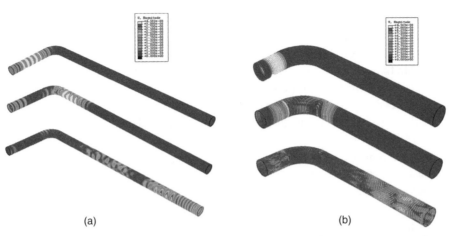

75 kHz L(m, 2) wave group propagating in a 2" schedule 40 elbowed steel pipe.	75 kHz L(m, 2) wave group propagating in a 16" schedule 30 elbowed steel pipe.

Figure D.14. Guided wave propagation in elbowed pipes with different sizes. See plates section for color version.

Of many interesting results from mode, frequency, and test parameter tuning, excellent outcomes can be obtained in the case of trying to find defects beyond an elbow. Study the result in Figure D.14 and see if you can think of a tuning solution.

In general, an axisymmetric wave in long-range pipe guided wave inspection cannot detect small defects reliably beyond an elbow. Mode conversion occurs at the elbow, hence generating many non-axisymmetric or flexural modes to travel beyond the elbow. Notice a sample result in Figure D.14. In (a), axisymmetric waves are again generated beyond the elbow. On the other hand, in (b) only flexural modes are generated, leaving lots of blind spots beyond the elbow. Natural focusing takes place where defect detection would be strong, but at certain places, a defect would not be seen. This can be overcome by frequency tuning as the natural focal points and blind spots constantly shift around and along the pipe. It can be shown, however, that for a reasonable frequency range for tuning, chances of finding all of the defects at different positions becomes quite high.

D.2.7 Ice Detection with Ultrasonic Guided Waves

This problem developed as a result of wanting to perform ice detection over a large area of a fixed-wing aircraft or a helicopter rotor blade with ultrasonic guided waves. Another goal was to avoid confusing ice detection with glycol fluid used in deicing grounded aircraft or rotorcraft or with rain water or other contaminants. The approach was to utilize guided wave propagation in the skin structure so that mode and frequency selection would use a wave structure such that in-plane displacement at the surface would leak into ice, leak very little into somewhat viscous glycol, and not leak at all into water. On the other hand, the wave structure of out-of-plane displacement on the outer surface would leak into ice, glycol, and water. Hence utilizing both modes and comparing leakage rates would enable us to detect ice. Both modeling and experiment would be used to accomplish this. As a result, a search of the phase velocity dispersion curve space for dominant in-plane displacements on the outer surface would take place as well as for the dominant out-of-plane displacement on the outer surface. A single mode point with both components could also possibly be used. This concept is reported in a patent by Rose and colleagues (1997). Note that dominant in-plane displacement on the outer plate surface occurs for a plate at the intersection point of the dilatational velocity value on the phase velocity dispersion curve vertical axis with the symmetric modes S1, S2, S3, and so forth at those particular frequency values. This point is demonstrated by Pilarski, Ditri, and Rose (1993).

Note that points on the phase velocity dispersion curve for a plate along the S0 mode for very low frequency below the first cutoff frequency, and that are practically nondispersive in character, are also of dominant in-plane displacement at the plate surface.

Gao and Rose (2009) reported an alternative approach to ice detection and classification utilizing ultrasonic SH guided waves. Ice type and thickness is modeled and specific features, including phase velocity dispersion curve shifts, are used in an inverse sense to solve a variety of different ice detection problems. See a sample problem in Rose (1999).

In reality, care must be taken to analyze patch-type ice, uniform thickness ice, nonuniform thickness ice, and possible sliding ice. Dispersion curve shifts and multi-mode impingement could be useful in an overall surface contaminant evaluation.

D.2.8 Deicing

Deicing a structure with ultrasonic guided waves also becomes possible with a proper mode and frequency choice. The idea is to get maximum shear stress at the interface between the substrate (or airfoil) and ice patch or layer. Cracks or delamination at the interface can lead to complete deicing in a short period of time. Oscillating shear stress reversals can also create fatigue quickly at the interface. Because ultrasonic signals could now lead to a vibration situation with multiple reflections occurring because of thousands of cycles, maximizing shear stress values could vary with position and frequency. Ideas of multiple actuators to achieve focusing when combined with frequency tuning can eventually provide enough energy for complete deicing of a structure. See a sample deicing in Figure D.15.

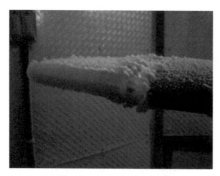
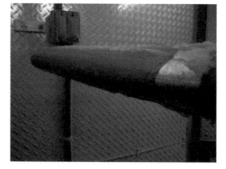

(a) No ultrasonic ice protection   (b) With ultrasonic ice protection

Figure D.15. Photograph showing airfoil leading edges. Without ultrasonic ice protection a 0.431" ice layer accreted to the leading edge.

Photographs detail rotating impinging icing tests recently completed in Pennsylvania State University's Adverse Environment Rotating Test Stand (AERTS) facility. The room containing the rotating stand is a freezer and icing nozzles from NASA were installed in the ceiling of the room to allow specific control of icing conditions. The first photograph shows the leading edge that had no ultrasonic actuation applied during the tests. Note the glaze ice formation. The second photograph shows the leading edge that was under ultrasonic actuation. Note this leading edge is clean of ice formation.

D.2.9 Real-Time Phased Array Focusing in Pipe

Associated with wave structure in Figure D.1 could be angular profiles in hollow cylindrical structures or pipelines. When studying dispersion curves for a pipe, the presence of non-axisymmetric (or flexural) modes are quite obvious. Families of flexural modes are associated with and are close to axisymmetric modes on the phase velocity dispersion curves. Flexural modes are created by partial loading around the circumference of a pipe. Mathematical details were covered in two earlier chapters (Chapters 10 and 16), but thinking and analysis here is associated with system designs. Let's consider a sample problem studying angular profiles of an 80 kHz T (M, 1) mode group in an eight-inch schedule 40 steel pipe over a range of fifteen feet to twenty-nine feet. See Figure D.16. The waves were excited by a 45° wide transducer located at $\theta = 0°$ and $z = 0$ feet. What can we learn from this figure? The angular profiles change significantly as the flexural modes propagate in the pipe. Imagine the use of eight 45° wide transducers around the circumference of a pipe. If we could pulse them independently with time delays between the eight segments, it might be possible to focus the energy at some specific point along the pipe. In fact, this was done and is reported in earlier chapters. Let's think further. Focusing would always occur, but it may not be effective if zero content is coming from some of the segments. Hence certain profiles would be desired. From the angular profiles shown in Figure D.16, which distances would achieve the best focusing results? Can you see any natural focusing points at certain distances along the pipe? While on the subject of natural focusing, do you think that you could inspect a pipe with partial loading only? The answer is perhaps yes, by changing frequency as the angular profiles are a strong function of frequency. The natural focal points would move around and possibly cover the whole pipe, especially if done individually with all eight segments. How could we improve the possibility of getting complete coverage? Change circumferential loading lengths as well. Each again has a different angular profile. Now, back to real-time phased array focusing. How can you guarantee a good focal point at anywhere you want in the pipe? The answer is to use a variety of circumferential segment sizes and frequencies, to achieve focusing, and to use redundancy in the focusing algorithm by multiple focus parameters to accomplish focusing and its inherent benefits of improved penetration power, defect detection sensitivity, and accurate circumferential sizing.

So clearly define your problem and then proceed to Figure D.1 in an attempt at system design. For additional information, see Rose and colleagues (2003); Li and Rose (2001a,b; 2002; 2006); and Sun, Zhang, and Rose (2005).

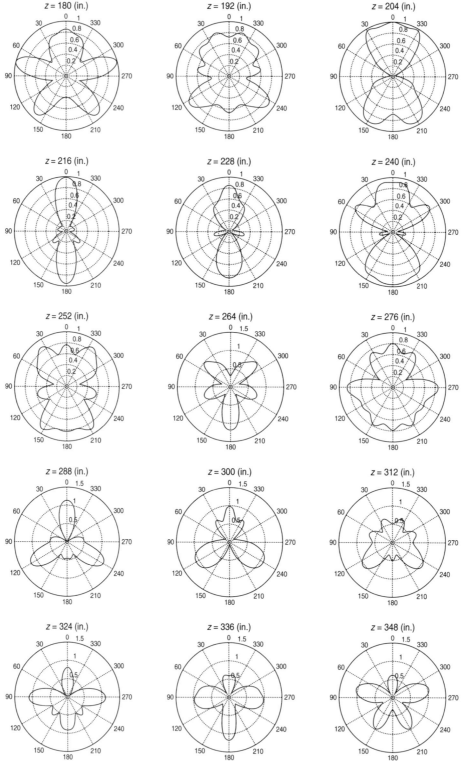

Figure D.16. Angular profiles of 80 kHz T(m, 1) mode group in an eight-inch schedule 40 steel pipe over distance range: 15ft ~ 29ft. The waves were excited by a 45° wide transducer located at θ = 0° and z = 0ft. (Note that with closer axial distances you could imagine a continuous angular profile variation as the waves moved down the pipe; only small changes would occur for one position to the next small position.)

D.2.10 Aircraft, Lap Splice, Tear Strap, and Skin-to-Core Delamination Inspection Potential

A few interesting ideas on mode and frequency selection for inspection problems associated with aircraft components are discussed here. Let's consider four sample problems.

a) Fuselage wall thinning
b) Lap splice joints
c) Tear straps
d) Honeycomb structures

a) Fuselage Wall Thinning

Traditional bulk wave ultrasonics can be used to examine wall thickness, but guided waves also offer an approach. Section D. 2 lists some ideas. The idea presented earlier of examining mode crossings via sweeping frequency with a tone burst pulse

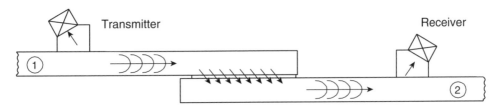

(a) Ultrasonic through-transmission approach for lap splice joint inspection

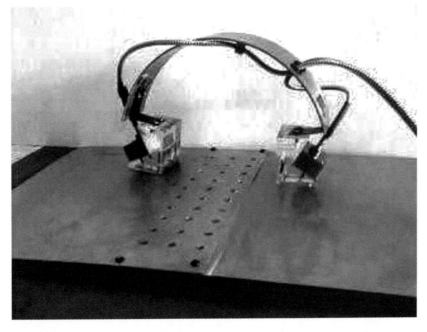

(b) Double spring "hopping probe" used for the inspection of a lap splice joint

Figure D.17. A lap splice inspection sample problem.

or applying a Fourier transform to a pulse can provide data that allow you to extract thickness information from an examination of the phase velocity dispersion curve at a particular wedge incident angle for a particular phase velocity valve. See Rose and Soley (2000) for more detail.

b) Lap Splice Joints

See Figure D.17 for a lap splice joint inspection concept. The overall idea is illustrated in Figure D.17(a). The logic is quite simple. The amount of energy traveling from transmitter to receiver provides an indication of the quality of the lap splice joint. Because the system is being used in a through-transmission mode, the rivets' influence is minimal. Though simple, one must select an appropriate mode and frequency that allows sufficient energy to leak into the lower skin structure. An incorrect mode and frequency choice can lead to a false alarm as little or no energy would reach the receiver. Examine mode structures to solve this problem. See Rose and Soley (2000) for more detail.

c) Tear Straps

The idea here is similar to that presented for lap splice inspection to allow sufficient energy to leak into the tear strap from the fuselage, but this time a pulse echo mode is recommended whereby energy from the end of the tear strap is reflected and leaks back into the fuselage to the receiver. See Figure D.18. The transmitter can

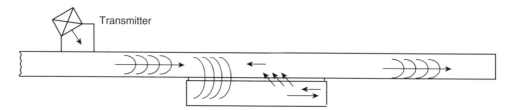

(a) Ultrasonic pulse-echo approach for tear strap inspection

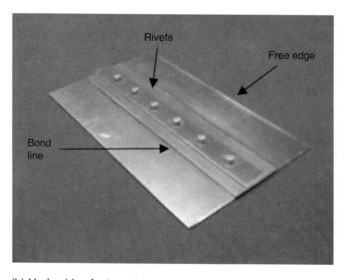

(b) Underside of a tear strap

Figure D.18. Tear strap inspection sample problem.

search for tear straps without having precise location information. See Rose and Soley (2000).

d) Honeycomb

The inspection process here is illustrated in Figure D.19. Guided wave energy travels from the transmitter to the receiver over some fixed distance. Energy leakage into the honeycomb reduces the amount of energy that gets to the receiver. As a consequence, the signal is strong for complete delamination. Again, though a simple concept, a suitable mode and frequency must be selected to allow leakage into the honeycomb to occur. Mode and frequency selections could come from the phase velocity dispersion curve associated with a plate model, a layer on a half-space model, or a clever theoretically driven experiment examining aspects of mode

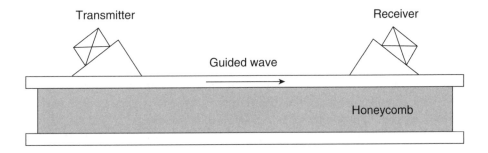

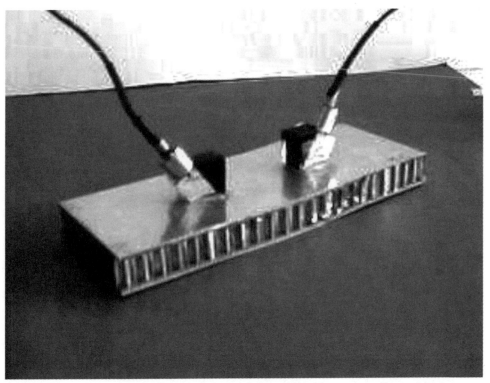

Figure D.19. Ultrasonic guided wave inspection concept for skin-to-core adhesive bond evaluation of a honeycomb structure (almost no signal for good bond, large signal for poor bond).

and frequency tuning to obtain the best possible result for a particular honeycomb structural configuration. See Rose and Soley (2000) for more detail.

D.2.11 Coating Delamination and Axial Crack Detection

These two problems are addressed from a circumferential guided wave point of view. See Chapter 11 on circumferential guided waves for additional information.

a) Disbond Detection

Let's consider the coating disbond detection problem first. A schematic to the debond problem and feature solutions is shown in Figures D.20, D.21, D.22, and D.23. The approach is to use EMAT sensors to send circumferential guided waves around the circumference of the pipe from the inside surface. See Figure D.24. Special modes and frequencies that are sensitive to the presence of coatings are used to determine coating integrity.

Three features were considered, the first being a wave velocity around the circumference of the pipe. It turns out that with coating bonded correctly, the group wave velocity is reduced compared to the value for an uncoated pipe. Hence, this change in time of flight (TOF) provides an indication of debond extent. Consequently, why not find the mode and frequency that give the greatest difference in TOF between these two extreme situations? This would make it easy to estimate debonding length around the circumference. Sample results of TOF are illustrated in Figure D.23. Note that the TOF decreases as debond length increases. Beyond this primary feature of time of flight, two additional features worked well. Note that redundancy in solving any problem is always good. The use of multiple features improves reliability and overall probability of detection. One feature was signal attenuation, which obviously is impacted as more energy is absorbed by the

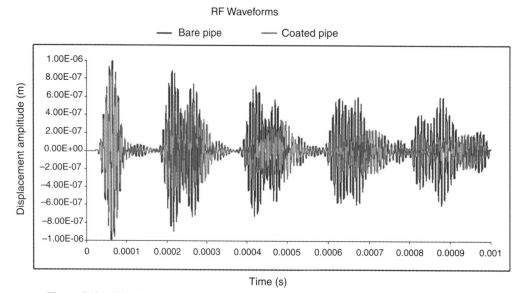

Figure D.20. Waveforms obtained from numerical modeling – SH circumferential wave propagation in an eight-inch diameter ¼″-thick bare pipe and the same pipe with a 3 mm coal tar enamel coating. See plates section for color version.

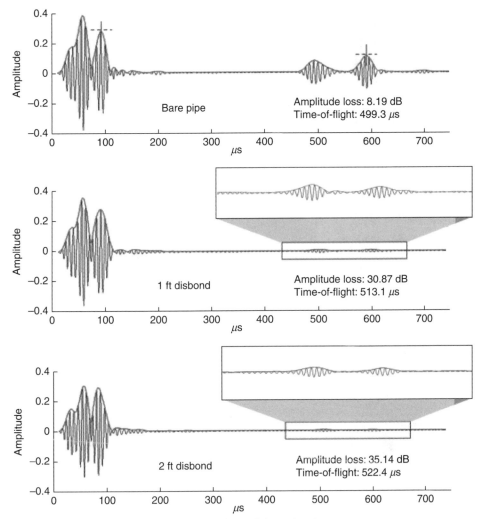

Figure D.21. Time- and amplitude-based disbond detection features.

coating. Another feature was a lost high-frequency component percentage. Higher frequencies are more readily absorbed by the coating material, and of course are not absorbed in the coating disbond situation.

b) Axial Crack Detection

Circumferential guided waves were also used to tackle this problem. In this case, the axial crack can be considered with normal incidence to the crack surface to provide a maximum reflection factor value. Zhao and Rose (2003) present details in a plate example. Reflection and transmission factor calculations were carried out for the crack model with results to be applied to a circumferential guided wave in a pipe. Keep in mind that mode conversion occurs as a guided wave impinges onto a defect. To satisfy boundary conditions at the defect, mode conversion takes place in all of the modes in the phase velocity dispersion curve that are present at the impinging frequency. The energy partition into the different modes can be considered from a

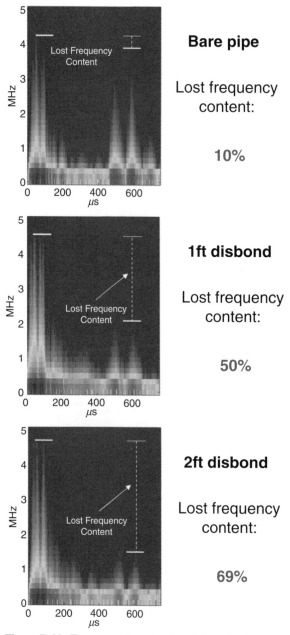

Figure D.22. Frequency-based disbond detection feature. See plates section for color version.

NME point of view. A vertical line at the impinging frequency crosses all of the modes to be considered.

Both longitudinal and SH type loading were considered. The reflection factor calculation for a longitudinal mode showed that beyond the first cutoff frequency value too much mode conversion was present and the results were a mess. Lower-frequency values did show some promise for solving the problem, but lower frequency did not provide acceptable sensitivity for finding small cracks. On the other hand, calculations for the SH mode produced excellent reflection factor curves over a very

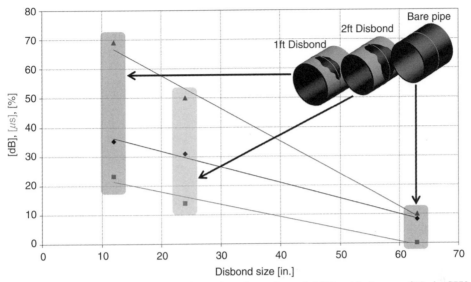

Figure D.23. Coating disbond detection using circumferential SH guided waves (Mode: SH0, Frequency: 130 kHz, Source: EMAT, Pipe: 20in S10 with coal tar coating). See plates section for color version.

Figure D.24. EMAT configuration to inspect from the inside of a pipe.

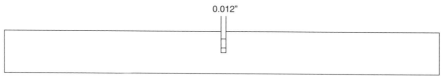

Figure D.25. Modeling statement of a guided wave striking elliptical crack-like defects of 10 percent, 20 percent … 90 percent through plate thickness.

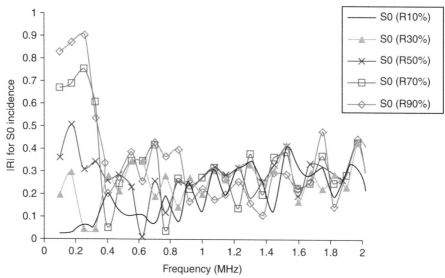

Figure D.26. Calculated reflection coefficients for S0 mode under S0 incident for 0.012" (0.3 mm) elliptical crack-like defect and 10 percent, 30 percent, 50 percent, 70 percent, and 90 percent through plate thickness depth.

large frequency range as a result of much less mode conversion occurring beyond the first cutoff frequency. As a result, SH circumferential guided waves were selected for studying the axial crack detection problem in a pipe. See sample results in Figures D.25, D.26, and D.27. Note that for SH circumferential guided waves, almost any frequency value could be used for the inspection as long as penetration power was sufficient.

D.2.12 Multilayer Structures

The most important item in the inspection of multilayer structures with ultrasonic guided waves is to ensure wave structures have sufficient energy at each interface to carry out a reliable inspection. Multimodes and frequency might be required to do a complete inspection of all interfaces. The feature could be energy in-plane displacement, shear stress, or something else. Many problems have been studied with this approach. See, for example, Rose, Zhu, and Zaidi (1998) and Yan and colleagues (2010).

D.2.13 Concluding Remarks

Several approaches have been presented on the subject of guided wave mode and frequency selection for improved detection sensitivity and/or penetration power. Quite often, precise material properties are unknown and hence accurate dispersion

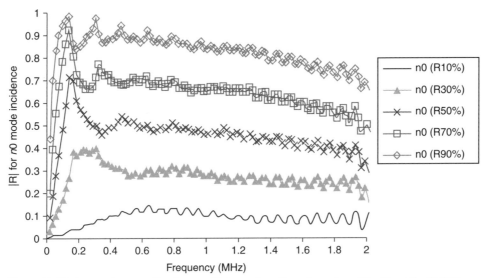

Figure D.27. Approximate reflection coefficients for $n0$ mode under $n0$ incident for 0.012"
(0.3 mm) elliptical defect length and 10 percent, 30 percent … 90 percent through plate
thickness depth.

curves are not available. An interesting technique has just been proposed by Philtron
and Rose (2014) that makes use of an ultrasonic phased array comb transducer
to sweeping phase velocity and frequency space. This perturbation technique in
frequency and phase velocity can significantly improve defect detection sensitivity
and/or penetration power. Newer techniques will surely evolve in the future to
improve even further the highly important topic of mode and frequency choice to
solve specific problems. See Figure D.28.

D.3 Exercises

1. Beyond the problems discussed in this appendix, can you think of another
   problem and come up with ideas on mode and frequency choice? Consider,
   for example, a subsurface defect in the inner or outer race of a ball bearing. A
   base defect in a rail. A cohesive void in an adhesive bond situation. A guided
   wave approach to weld inspection. Bond integrity of plasma spray on an aircraft
   carrier. Or concrete surface cracking on an aircraft runway or some other useful
   inspection problem that you may have observed.
2. What features beyond sensitivity to a certain defect should be considered in an
   inspection system design?
3. Why does frequency tuning provide better results in an inspection?
4. What other parameters could be tuned in an inspection to improve overall
   probability of detection?
5. Why is redundancy in guided wave feature choice a benefit in inspection?
6. Which guided wave features could be utilized in composite material fabrications?
   Or by for-service evaluation?
7. How would you tackle a problem in prognostics, a step beyond diagnostics,
   that is, how would you determine the effects of certain defects on structural
   performance?

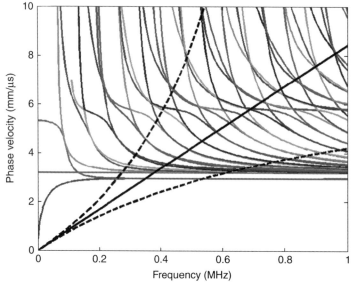

Figure D.28. Activation lines for a linear, multi-element comb transducer with in phase (thick solid line) and 1 μs linear time delay (dashed curved lines) element excitation. Phase velocity dispersion curves for a 25 mm steel plate are also shown.

8. How would you choose piezoelectrics, electromagneto-acoustics, magneto-strictive, laser loading, or controlled mechanical impact to solve certain problems with emphasis on mode and frequency choice? Or other criteria?

9. In system design, because active system choice is required, where do you think energy-harvesting methods might be useful compared to battery use? List a few energy-harvesting methods.

10. In system design, where do you think wireless methods for sensor activation and data acquisition might be required?

11. Two common guided wave mode excitation methods are the angle beam and linear comb transducers. Both of these excitation methods have unique excitation patterns, and their own strengths and weaknesses. Consider the excitation regions by these two methods in phase velocity–frequency (c_p–f) space. Explain fully using a sketch and appropriate formulas.

12. For a comb transducer, if the elements are excited out of phase, and more specifically, with a linear time delay to successive elements $(0, 1\tau, 2\tau, 3\tau, \ldots)$, the activation line curves are shown in Figure D.28. Using this property, we can effectively sweep through a region of cp–f space with a single transducer by applying different time delays (and frequencies), without changes in the physical location, set up, or couplant condition of the transducer. An example of straight (no delay) and curved (linear delay) activation lines are shown in Figure D.1 for $s = 8.43$ mm and $\tau = 1\mu$s. The activation lines are plotted with the phase velocity dispersion curves for a 25 mm (1 inch) steel plate. Derive the equations needed to adjust time delays for a mode sweep. Explain fully the results shown in Figure D.28.

13. From the particular expanded red regions in Figure D.10, why did this occur?

14. See Figures D.11 and D.12. Clearly explain why the water spot image could be eliminated.

15. In Figure D.14 is shown the possibility of an axisymmetric wave to reform as it passes a 90° elbow, but in some cases, it does not reform. Explain why and how inspection beyond an elbow could still be carried out.

16. In Figure D.20, what is happening differently in the bare pipe versus the coated pipe? Could any ideas here be used to detect coating delamination and how could improved sensitivity be achieved?

17. In Figure D.22, what would the highest frequency be to propagate around the circumference of the pipe being studied for (a) 6" disbond length and (b) 1.5 ft disbond length?

18. What is the advantage of using all 3 features considered in Figure D.23?

D.4 REFERENCES

Auld, B. A. (1990). *Acoustic Fields and Waves in Solids*, 2nd ed., vols. 1 and 2. Malabar, FL: Kreiger.

Barshinger, J., Rose, J. L., and Avioli, Jr., M. J. (2002). Guided wave resonance tuning for pipe inspection, *Journal of Pressure Vessel Technology* 124: 303–10.

Ditri, J. J., and Rose, J. L. (1992). Excitation of guided elastic wave modes in hollow cylinders by applied surface tractions, *J Applied Physics* 72(7): 2589–97.

(1994). Excitation of guided waves in generally anisotropic layers using finite source, *ASME J Applied Mechanics* 61(2): 330–8.

Ditri, J. J., Rose, J. L., and Chen, G. (1991). Mode selection guidelines for defect detection optimization using Lamb waves, in *Proceedings of the 18th Annual Review of Progress in Quantitative NDE*, vol. 11, pp. 2109–15. New York: Plenum.

Gao, H., and Rose, J. L. (2007). Multifeature optimization of guided wave modes for structural health monitoring of composites, *Materials Evaluation* 65(10): 1035–41.

(2009). Ice detection and classification on an aircraft wing with ultrasonic shear horizontal guided waves, *IEEE Transactions on Ultrasonics, Ferroelectrics, and Frequency Control* 56(2): 334–44.

(2010). Goodness dispersion curves for ultrasonic guided wave based SHM: a sample problem in corrosion monitoring, *The Aeronautical Journal of the Royal Aeronautical Society* 114(1151): 49–56.

Hay, T. R., Wei, L., Rose, J. L., and Hayashi, T. (2003). Rapid inspection of composite skin-honeycomb core structures with ultrasonic guided waves, *Journal of Composite Materials* 37(10): 929–36.

Hayashi, T., Song, W. J., and Rose, J. L. (2003). Guided wave dispersion curves for a bar with an arbitrary cross-section, a rod and rail example, *Ultrasonics* 41: 175–83.

Lee, C. M., Rose, J. L., and Cho, Y. (2008). A guided wave approach to defect detection under shelling in rail, *NDT&E International*, Elsevier Ltd.

Li, J., and Rose, J. L. (2001a). Excitation and propagation of non-axisymmetric guided waves in a hollow cylinder, *J. Acoust. Soc. Am* 109(2): 457–64.

(2001b). Implementing guided wave mode control by use of a phased transducer array, *IEEE Transactions on Ultrasonics, Ferroelectrics and Frequency Control* 48(3): 761–8.

(2002). Angular-profile tuning of guided waves in hollow cylinders using a circumferential phased array, *IEEE Transactions on Ultrasonics, Ferroelectrics and Frequency Control* 49(12): 1720–9.

(2006). Natural beam focusing of non-axisymmetric guided waves in large-diameter pipes, *Ultrasonics* 44: 35–45.

Luo, W., and Rose, J. L. (2003a). Lamb wave thickness measurement potential with angle beam and normal beam excitation, *Materials Evaluation* 62(8): 860–6.

(2003b). Guided wave thickness measurement with EMATS, *Insight* 45: 735–9.

Philtron, J. H., Rose, J. L. (2014). Mode Perturbation Method for Optimal Guided Wave Mode and Frequency Selection, to be published in J. Ultrasonics.

Pilarski, A., Ditri, J. J., and Rose, J. L. (1993). Remarks on symmetric Lamb waves with dominant longitudinal displacements, *J. Acoust. Soc. Am.* 93(4) (part 1): 2228–30.

Puthillath, P., and Rose, J. L. (2010). Ultrasonic guided wave inspection of a titanium repair patch bonded to an aluminum aircraft skin, *International Journal of Adhesion & Adhesives* 30: 566–73.

Rose, J. L., (1999). *Ultrasonic Waves in Solid Media*. Cambridge, UK: Cambridge University Press.

Rose, J. L., Pelts, S., and Cho, Y. (2000). Modeling for flaw sizing potential with guided waves, *Journal of Nondestructive Evaluation* 19(2): 55–66.

Rose, J. L., Pilarski, A. B., Hammer, J. M., Peterson, M. T., and Readio, P. O. (1997). "Contaminant detection system," U.S. Patent 5629485, May 13.

Rose, J. L., and Soley, L. E. (2000). Ultrasonic guided waves for anomaly detection in aircraft components, *Materials Evaluation* 50(9): 1080–6.

Rose, J. L., Sun, Z., Mudge, P. J., and Avioli, M. J. (2003). Guided wave flexural mode tuning and focusing for pipe testing, *Materials Evaluation* 61: 162–7.

Rose, J. L., Zhang, L., Avioli, Jr., M. J., and Mudge, P. J. (2005). A natural focusing low frequency guided wave experiment for the detection of defects beyond elbows, *Journal of Pressure Vessel Technology* 127: 310–16.

Rose, J. L., Zhu, W., and Zaidi, M. (1998). Ultrasonic NDE of titanium diffusion bonding with guided waves, *Mat. Eval.* 56(4): 535–9.

Shin, H. J., and Rose, J. L. (1998). Guided wave tuning principles for defect detection in tubing, *Journal of Nondestructive Evaluation* 17(1): 27–36.

Sun, Z., Zhang, L., and Rose, J. L. (2005). Flexural torsional guided wave mechanics and focusing in pipe, *Journal of Pressure Vessel Technology* 127: 471–8.

Wilcox, P. D., and Lowe, M. J. S. et al. (2001). Mode and transducer selection for long range Lamb wave inspection, *J Intelligent Material Systems and Structures* 12: 553–65.

Yan, F., Xue Qi, K., Rose, J. L., and Weiland, H. (2010) Ultrasonic guided wave mode and frequency selection for multilayer hybrid laminates, *Materials Evaluation* 68(2): 169–75.

Zhang, L., Gavigan, B. J., and Rose, J. L. (2006). High frequency guided wave natural focusing pipe inspection with frequency and angle tuning, *Journal of Pressure Vessel Technology* 128: 433–8.

Zhao, X., and Rose, J. L. (2003). Boundary element modeling for defect characterization potential in a wave guide, *International Journal of Solids and Structures* 40: 2645–58.

Zhu, W., Rose, J. L., Barshinger, J. N., and Agarwala, V. S. (1998). Ultrasonic guided wave NDT for hidden corrosion detection, *Res. Nondestr. Eval.* 10(4): 205–25.

Index

UNIVERSITY
OF
GLASGOW
LIBRARY